Henry Trimen, William T. Thiselton-Dyer

Flora of Middlesex

a topographical and historical account of the plants found in the county - with

sketches of its physical geography and climate

I0047043

Henry Trimen, William T. Thiselton-Dyer

Flora of Middlesex
a topographical and historical account of the plants found in the county - with sketches of its physical geography and climate

ISBN/EAN: 9783337272807

Printed in Europe, USA, Canada, Australia, Japan

Cover: Foto ©berggeist007 / pixelio.de

More available books at **www.hansebooks.com**

FLORA OF MIDDLESEX:

A TOPOGRAPHICAL AND HISTORICAL ACCOUNT

OF

THE PLANTS FOUND IN THE COUNTY;

WITH SKETCHES

OF ITS

Physical Geography and Climate,

AND OF THE PROGRESS OF

MIDDLESEX BOTANY DURING THE LAST THREE CENTURIES.

BY

HENRY TRIMEN, M.B. (Lond.), F.L.S.

Botanical Department, British Museum, and Lecturer on Botany, St. Mary's Hospital;

AND

WILLIAM T. THISELTON DYER, B.A.

Late Junior Student, Christ Church, Oxford;
Professor of Natural History, Royal Agricultural College, Cirencester.

WITH A MAP.

LONDON:

ROBERT HARDWICKE, 192 PICCADILLY.

1869.

PREFACE.

THE object of this book is to give a complete and accurate catalogue of the plants which have at any time been recorded to grow in Middlesex, either as natives or in a more or less completely naturalised state, to indicate the special localities where they have been found, and to trace the history of their discovery. The existence of many of these is attested only by records in scarce or little-known books, or by the original specimens preserved in old collections. As the amount of material relating to the subject is very large, much of our work has consisted in collecting and reducing into order the widely scattered records; and in this, though we cannot hope to have avoided all error, we hope that no blunders of magnitude have been committed or important facts overlooked.

The Cryptogamia—with the exception of the Ferns and their allies—have, from want of material, been much less fully treated ; lists of some of the Orders are given in the Appendix, but the Algæ and lower Fungi have been entirely omitted. It is hoped that those who have made a special study of these plants will be induced to publish something more complete.

The history of the earliest labourers in any scientific field has an exceptional interest. We who succeed to their labours cannot fail to appreciate the intelligence of men who, at a time with no past example, and little encouragement, laid the foundations of so much of our present knowledge. The Historical Sketch on pp. 363—400, drawn up partly from unpublished sources, contains biographies of several botanists of whom little has hitherto been known but the names, and may in part be regarded as supplementary to Dr. Pulteney's well-known *Sketches of Botany*.

To the numerous correspondents who have helped us in the compilation of our material we offer our thanks. Those botanists to whom we are chiefly indebted are enumerated on p. 11; but the Rev. W. W. Newbould requires special mention here, for his care in helping to correct the proof-sheets. To the Hon. J. Leicester Warren for his help in the Rubi, to Mr. Worthington G. Smith for his list of Hymenomycetous Fungi, and to the Rev. James M. Crombie for his note on the Lichens, our acknowledgments are also due.

The size of this volume may appear to be excessive in comparison with the importance of the subject treated in its pages; it has indeed considerably exceeded the bulk originally contemplated, but, with the exception of some localities which the peculiarities of the county seemed to render it desirable to put on record, there is, we believe, very little irrelevant matter; indeed, some care has been expended on concentrating the material.

From those who have ever been occupied in endeavouring to estimate, each at its true value, the mass of alleged facts which go to make up a local Flora, we need scarcely ask for indulgence. We have, in all cases, endeavoured to be just to others as well as to ourselves.

August, 1869.

CONTENTS.

———◇———

INTRODUCTION.

I. TOPOGRAPHY :—   PAGE

    1. POSITION   ix

    2. SIZE   ix

    3. SHAPE   ix

    4. BOUNDARIES   .   .   .   x

    5. ELEVATION OF SURFACE .   .   .   x

    6. GEOLOGY   .   .   xi

    7. DRAINAGE   .   .   .   xvii

    8. SOIL AND AGRICULTURE .   .   .   xxv

    9. WOODS   xxvii

    10. HEATHS, COMMONS, AND WASTE LANDS .   xxviii

II. CLIMATE :—

    1. TEMPERATURE   xxx

    2. RAINFALL   .   .   .   xxxiv

III. DIVISION INTO BOTANICAL DISTRICTS   .   .   xxxvii

THE FLORA.

    PLAN OF THE FLORA   .   .   .   1

    LIST OF BOOKS, MSS., AND HERBARIA CONSULTED .   .   3

            CORRESPONDENTS   11

PHANEROGAMIA :—

    DICOTYLEDONES   12

    MONOCOTYLEDONES   266

    CRYPTOGAMIA   335

SUMMARY :—   PAGE

    NUMBER OF SPECIES   345

    COMPARATIVE FREQUENCY AND EXTINCT SPECIES   .   345

    COMPARISON WITH THE 'CYBELE BRITANNICA'   .   346

                 ADJACENT COUNTIES   .   .   348

                  THURMANN'S 'ESSAI'   .   357

SKETCH OF THE PROGRESS OF BOTANICAL INVESTIGATION IN

    MIDDLESEX ; WITH BIOGRAPHICAL NOTICES   .   .   363

APPENDIX :—

    I. LIST OF MUSCI   .   .   .   401

    II.   HEPATICÆ   .   .   .   404

    III.   CHARACEÆ   .   .   .   405

    IV.   LICHENES   .   .   .   405

    V.   HYMENOMYCETOUS FUNGI   .   .   .   407

ADDITIONS AND CORRECTIONS   .   .   .   421

INDEX TO GENERA IN THE FLORA   .   .   .   426

INTRODUCTION.

I.—TOPOGRAPHY.

1. POSITION.—Middlesex, the metropolitan county of England, is situated in its south-east corner, and though containing the port of the capital, and having a tidal river for its southern boundary, is wholly inland. It is included between the parallels of 51° 22′ and 51° 42′ N. lat., and 0° 2′ E. and 0° 31′ W. long.

2. SIZE.—With the exception of Rutland (141 square miles), Middlesex is the smallest of the English counties. According to the *Agricultural Returns* (1868, p. 16), its total area is 180,136 statute acres, or 281·46 square miles. Compared with the whole area of England and Wales, it stands in the proportion of 1 : 207.

About one-sixth of the county is occupied by the metropolis. From a botanical point of view, this is equivalent to an absolute diminution of area to that extent. In 1868, 108,956 statute acres were under cultivation, including bare fallow and permanent pasture (*Agricultural Returns*, p. 16) ; adding to this 31,524 acres occupied by the metropolis (*Population Returns*, 1861), and subtracting the sum from the whole area, the remainder, 3,656 acres, one-fiftieth of it, gives the area outside the metropolitan districts which is not under cultivation. This must consequently include all portions of the surface which have hitherto remained comparatively undisturbed, and have consequently more or less preserved their natural vegetation.

3. SHAPE.—The shape of Middlesex is a very irregular parallelogram. In the north-east it is penetrated by a spur of Hertfordshire, while in the south-west a large portion projects into Surrey. The greatest length is from Chertsey Bridge, in the south-west, to the north-east corner of the county near Waltham Abbey, Essex, 28 miles ; the greatest breadth from near Rickmansworth, in the north-west, to the Isle of Dogs, in the south-east, 17 miles. The county corresponds roughly, therefore, to a rectangle 20 miles long by 14 broad. The

perimeter of such a figure would of course fall short of the actual irregular outline, which is about 115 miles * (*Camden*).

4. BOUNDARIES.—On three sides Middlesex is separated from adjacent counties by rivers. To the south, the Thames divides it from Surrey and Kent. The last county is only adjacent by its north-west corner for about 3 miles, from a point opposite Bow Creek to the west side of Deptford, corresponding in Middlesex to the eastern and southern sides of the Isle of Dogs. The whole length of the southern boundary is about 45 miles.† On the east, Middlesex is divided from Essex by the Lea. On the west it is bounded by another tributary of the Thames, the Colne. From Staines to Harefield this separates it from Buckinghamshire, and from Harefield to Mill End, about 3 miles further north, from Hertfordshire. Both eastern and western boundaries are about 18 miles in length.‡ At Mill End the northern boundary begins, and, dividing the county in its whole length from Herts, is very irregular in its outline, although following no marked physical or topographical features in the country. The eastern half of the northern boundary lies about 5 miles farther north than the western half. A line drawn from Mill End to Southgate would cut the western half no less than six times, although the boundary would be nowhere distant more than about a mile on either side of such a line. From Southgate the boundary takes a north-west direction, returning upon itself through Chipping Barnet, so as to almost insulate the spur of Hertfordshire containing Totteridge and East Barnet. A mile to the west of Chipping Barnet it turns northward for about 4 miles, and then proceeds in an almost straight line to Waltham Cross.

5. ELEVATION OF SURFACE.—The general surface of the county is a plain, sloping gradually southwards to the Thames. The northern border is consequently the highest ground, reaching from 400 to 500 feet. The surface itself is usually undulating, except near the rivers. The whole country, however, in the south-west of the county, lying south of the road from Brentford to Colnbrook, is remarkably uniform. It is nowhere more than 20, and usually not more than 10, feet above the Thames level at Staines Bridge; nor is it more than from 3 to 5 feet above the surface of the water in the intersecting rivulets.

Near the centre of the northern boundary are the eminences of Deacon's Hill, Brockley and Elstree Hills. From these the high ground extends east to Highwood and Mill Hill (and to Totteridge and Barnet, in Herts), and west to Stanmore and Harrow Weald

---

* Fisher makes it only 100.   † 43, Middleton's *Survey*.   ‡ *Encyc. Brit.*

Heaths and Pinner Hill. Stanmore Heath and Highwood, the highest parts of this range of hills, are also the points of greatest altitude in the county.

There is another shorter and somewhat lower range of hills about 4 miles from the Thames, to the north-east of London. These are the Highgate and Hampstead Hills. Their highest point is that part of Hampstead Heath near where Jack Straw's Castle stands. Besides these there is no other considerable eminence in Middlesex, except the isolated hill on which the old church of Harrow stands.

The following heights above the mean sea level have been supplied from the Records of the Ordnance Survey, through the kindness of its director, Colonel Sir Henry James :—

	Feet
County boundary stone at Hart's Bourn (on Bushey and Watford Road)	504·104
Stanmore Heath, guide-post at north corner of Bentley Priory .	492·920
Rising Sun public-house, Highwood Hill	443·491
Hampstead Heath, mark on fourth milestone from St. Giles's pound	441·337
Hampstead Heath, mark on brick wall 7 chains east of Castle Hotel	440·583
Hampstead Heath, at junction of road to Child's Hill, near Castle Hotel	436·491
West entrance to Pinner Hill House, Watford Road . . .	436·425
Original toll-house, Highgate	424·065
Mr. Foyne's (grocer) house, east side of road, Highgate . .	417·014
St. Michael's church-tower, Highgate (base)	415·880
Harrow-on-the-Hill church-tower (base)	405·518
Gate at entrance to Mill Hill Church	402·048
Entrance to Mr. Johnson's house, Mill Hill	401·528
South entrance to Pinner Hill House	381·810
Guide-post at junction of roads to Mount Pleasant, Highwood Hill	297·581
New church, Harrow-on-the-Hill	292·611
Mark on post of wooden seat on top of Primrose Hill . . .	219·154
South-east angle of wall of reservoir on Barrow Hill . . .	161·083
Junction of Barnsbury Park and Thornhill Road, Islington .	140·959
West entrance of Trinity Church, Cloudesley Square, Islington .	136·202

The ground in Middlesex covered by London has generally a downward slope from north-west to south-east, but it is much undulated, though the levels have been a good deal altered in making streets and buildings.

6. GEOLOGY.—Only a very brief summary of the geological features of the county can be given here. For fuller information Mr. Whitaker's *Memoir*, describing sheet 7 of the Geological Survey, and the papers quoted by him, should be referred to. The following account has been mainly drawn up from these sources :—

With the exception of two small portions of chalk at its edge, the whole of Middlesex is included in the great tertiary district which has been sometimes termed the London basin.

The *CHALK* forms the western slope of the high ground about Harefield, extending in a narrow slip from the northern boundary of the county to Bungers Hill. There is also a piece about a mile long forming the bottom of a small valley between Warren Gate and South Mims. Near Pinner the chalk is reached by a shaft about forty feet deep, by means of which it is worked for lime-burning.

LOWER EOCENE.—The *Thanet Sands*, the lowest of the tertiary series, is nowhere exposed at the surface, and is only found in making deep wells. It is important as a water-bearing bed. Under London it is from 13 to 44 feet thick, but thins out westward.

The *Woolwich and Reading beds* of Prestwich (Plastic Clays of other writers) consist of mottled clays and sands with rolled flints from the chalk, and are remarkable for presenting, unlike other Eocene deposits, identical characters in the tertiary basins of London, Hampshire, and Paris.

'On the eastern side of the Colne, the Reading beds rise from beneath the alluvium of the river at a point nearly two miles north of Uxbridge, being followed a short way further north by the chalk.' (*Whitaker*, p. 16.) Near Harefield both mottled plastic clay and sand have been found. Sands belonging to these beds are also found on both sides of the small valley between Mims Wood and South Mims.

There are two inliers of these beds in the county surrounded by London clay, the denudation of which has laid them bare in consequence of elevation produced by some local disturbance. The most westerly of these is on the north-east and east of Ruislip Wood, and, with an area of about half a square mile, forms 'in great part a rough common-land, quite different in character from the surrounding clay-hills. It seems to be bounded on the south by the large reservoir (p. 28).* Just south of the Green Lane are some old sand pits. Along the northern edge of the wood, plastic clay has been got, and other parts yielded sand. In some places the beds are hidden by *drifted loam*. The Pinner inlier is of a more irregular shape, and rather larger. It follows the course of the brook from Wood Hall southwards to Pinner Grove. At the lime-kiln south of the first place, the chalk is reached by the shaft before mentioned, through the Woolwich beds. Close to the shaft is a sand pit (p. 29). A very small inlier exists

---

* The pages refer to Whitaker's *Memoir.*

also along the course of a small brook about a quarter of a mile west of Pinner station.

The *London Clay* in the tertiary district westward of London consists, according to Prestwich, of tenacious brown and bluish-grey clays with layers of septaria. He also remarks that it everywhere maintains throughout its mass a nearly uniform mineral character. 'Its boundary line rises from below the alluvium at a spot about a mile and a half north of Uxbridge' (p. 49). East of the Colne the London clay forms the greater part of the well-marked tertiary escarpment extending north-east from Rickmansworth to Hatfield. Some way south-east it rises into a second and higher ridge, often capped with gravel (p. 96).

Immediately at the base of the London clay, a bed of sand, never more than a few feet in thickness, is commonly found ('basement bed,' Prestwich). It has been observed near the Ruislip Wood inlier of the Woolwich beds. The upper beds are also more sandy than the general mass, but are seldom found, owing to denudation, at any great distance from the overlying Bagshot beds. The Archway Road at Highgate originated in an attempt to drive a tunnel through the hill; but this was abandoned on account of the sandy nature of the upper beds, rendering them incapable of supporting an arch of the dimensions required. (Conybeare and Phillips' *Geology of England and Wales*, p. 25.)

MIDDLE EOCENE.—The *Lower Bagshot Sands* are represented in Middlesex by three outliers, forming the tops of Harrow, Highgate, and Hampstead Hills. Mr. Whitaker gives the following account of these:—

'The Harrow is the smallest, and reaches from the north of the church, southwards to Mount Pleasant, and is very narrow at its northern end. The largest and best-marked forms all the highest ground of Hampstead, and gives rise to the well-known gorse-covered heath. The irregularly-winding boundary of the sand is well marked both by the sharp rise of the ground, and by the water thrown out from these permeable beds by the underlying water-tight London clay; * but the northern end of the outlier, in Bishop's Wood, is rather hidden. The sand spreads from the western end of Lord Mansfield's kitchen garden, on the road to Highgate, westward nearly to the high road at Child's Hill, and southward to Rosslyn House. There are many pits on the heath, and they show light-coloured sand with thin layers of clay here and there: so much indeed has

* One spring at Hampstead is chalybeate.

the sand been dug away that but little of the original surface of the
ground has been left. The outlier at Highgate spreads from a little
north of "The Wrestlers" southwards to the higher part of the
cemetery, and then eastwards a little beyond the Archway. Besides
these outliers there may be some other small patches of the Bagshot
sand, as on the top of Parliament Hill, between Hampstead Heath and
Kentish Town' (pp. 61, 62). Coarse iron sandstone occurs in the sand
at Harrow and Highgate.

POST-PLIOCENE.—*Southern Boulder Clay.*—According to Mr. Prestwich, Boulder clay occurs on the Finchley Hills (*R. Geol. Soc.* vol. xii.
p. 133). It has also been met with on Muswell Hill.

High-Level Gravel.—At Stanmore Hill a drift of pebbles (rounded
flints) in sand and loam spreads over the whole of the hill-top. At
Potter's Bar there is also gravel. At Hendon there is pebble gravel
on the hill-top, and some of a fine nature is found in places on the
Bagshot outlier at Hampstead (p. 69).

At Hadley and Barnet, and thence along the ridge westward, the
London clay is capped with gravel containing angular flints (p. 70).

Between Muswell Hill and Finchley Common a gravel deposit
exists. It is about three-quarters of a mile wide, about 150 yards
broad, and from 15 to 20 feet deep. Immediately beneath the
vegetable soil is a bed about 14 feet thick, containing water-worn
fragments of granite, porphyry, micaceous sandstone, mountain
limestone, coal, lias, oolite, and chalk, with many of the characteristic
fossils of these formations. The most abundant are lias and chalk,
the latter sufficiently so as to give the whole accumulation a chalky
character. Beneath it is a bed, about 6 feet thick, of red pebble
gravel resting on London clay (pp. 70, 71). The uppermost of these
gravels is composed of the boulders of the southern boulder clay, the
cementing materials having been separated from the heavier portions
by running water, to form the accumulations of brick earth which are
occasionally found on the higher ground. 'A pebbly brick earth is
found on the London clay in many places: thus, near Bentley, west
of Stanmore, there is brick earth twelve feet thick, not so sandy as
that of the Thames valley, and which burns to a red brick.' (*Whitaker,*
p. 68.)

The Rev. J. C. Clutterbuck remarks, that the porous character of the
Lower Bagshot beds, and Higher Level Gravels which cap, more or less,
all the higher elevations, permit the percolation of water, which, being
upheld by the subjacent impervious clay, is thrown out in springs;
amongst which are the small perennial sources of the River Brent,

Yedding, and other brooks. The presence of water in these gravels has attracted a population which is remarkably deficient in many parts of the county—as, for example, the line of the Yedding brook between Harrow and Uxbridge, where in many square miles scarcely a habitation is to be found. (*Journal R. Agricultural Society*, second series, vol. v. p. 4.) In dry seasons these watercourses are nearly dry.

Valley Drifts.—Gravel, brick earth, and alluvium are one or more present in the valley of the Thames and its tributaries. They are of a later age than the gravel and brick earth of the high grounds, and are found not only at the bottoms of the valleys (low-level gravels) but often reaching some way up the flanks, and even capping the tops, of the low hills bounding them (terrace gravels). Each terrace forms a more or less flat tract sometimes worn through by side valleys. As many as three admit of being traced, including the low-level gravel. The gravel and brick earth of the Thames Valley is a very important feature in the geology of Middlesex, occupying as it probably does a third of the whole county. It covers all the southern part bounded by the Colne, Thames, and Lea, and to the north, approximately, by a line drawn from Hillingdon to Hanwell through Acton and Paddington to Stoke Newington. The first part of this line runs nearly parallel with the turnpike road from Uxbridge to London, and about half a mile to the north of it. The whole breadth on both sides of the river of this tract, in the western part of the county, is 4 miles or more. According to Mr. Prestwich, the great bulk of the gravel is composed of subangular chalk flints derived directly from the chalk, with a considerable number of rolled flint pebbles from the tertiary beds, and pebbles of quartz, slate, and other older rocks, from the conglomerates of the new red sandstone of Worcestershire and Warwickshire. The whole deposit is more or less arenaceous, and often contains beds of sand; it is rarely more than 20 feet thick. Gravel spreads over Uxbridge Common, and forms the top of the hill. It is also found between Southall and Hanwell, and at Staines and Hounslow Heath. There is a gravel flat, about a quarter of a mile broad, between Greenford and Perivale. 'At the top of Notting Hill there is a high terrace of gravel, separated from the lower terrace of Kensington by a narrow strip of London Clay. This terrace is the cause of the springs and moist ground in the southern part of Kensington Gardens' (p. 86). The gravel has been denuded through for a short way down the valley of the Serpentine, the water of which is held up by the London Clay. Along the Fleet Valley the gravel has also been denuded, and the springs which have been thrown out at its junction with the underlying clay, 'have given their names to the thickly-housed districts that were the means of their destruction—Clerkenwell, Bagnigge wells, and Coldbath fields' (p. 90). The boundary of the gravel of the Lea Valley is 'in great part doubtful,' but it passes from Stoke Newington, west of Tottenham, Edmonton, and Enfield, to Waltham Cross, and 'from Enfield to Clay Hill it follows' the line 'of the new river on its west' (p. 92). South-east of Southgate there is gravel which may be either high-level or a higher terrace of the Lea gravel (p. 71). The shallow wells of London get their water from the gravel, and from it the springs, besides those already mentioned, formerly numerous in London, especially at Islington and its neighbourhood, have been thrown out along the valleys. They now mostly supply pumps or pass into sewers. Such were Holy Well, Skinner's Well, Dame Annis de Cleare, Perillous (or Peerless) Pond. Some of them, as Shadwell (S. Chad's well), had medicinal qualities. Bagnigge Wells and Islington Spa, or new Tonbridge Wells, were cathartic and chalybeate.*

The brick earth overlies the gravel, but is not always present; it is a brown sandy clay or loam, very valuable not only for brick making, but also for the fine soil it affords the market gardener. Mr. Prestwich regards the brick earth as a deposit from flood waters. For a detailed account of the relations of the brick earth to the gravel, and the exact limits of these deposits, reference must be made to Mr. Whitaker's *Memoir*. In many places in the south-west of the county, the appearance of the surface has been much altered by the extensive working of the brick earth. At West Drayton and thence to Southall there is a great sheet of it, giving rise to many and extensive brick-yards (p. 86); near Drayton and elsewhere it has been quite worked out, and the gravel laid bare. Brick earth, forming the middle terrace, extends between East Acton and Notting Hill, from Wormwood Scrubs to the Thames, and is less sandy and not unlike London clay. Near Highbury the brick earth is very thick, nearly 30 feet of it having been found. At Newington it has been worked away, and the houses built of bricks made from it rest on the gravel (p. 89).

'In the British Museum there is a flint weapon of the spear-headed form, which we are told was found with an elephant's tooth at Black Mary's, near Gray's Inn Lane. In a letter dated 1715, printed in Hearne's edition of Leland's *Collectanea*, vol. i. p. 73, it is stated to have been found, in the presence of Mr. Conyers, with the skeleton

---

* Other springs were on the London clay, *e.g.*, Highwood Hill and Kilburn (cathartic), St. Pancras and Acton.

of an elephant (Evans, *Archæologia*, 1860). So many bones of the elephant, rhinoceros, and hippopotamus have been found in the gravel on which London stands, that there is no reason to doubt the statement as handed down to us. Fossil remains of all these three genera have been dug up on the site of Waterloo Place, St. James's Square, Charing Cross, the London Docks, Limehouse, Bethnal Green, and other places, within the memory of persons now living. In the gravel and sand of Shacklewell, in the north suburbs of London, I have myself collected specimens of the *Cyrena fluminalis* in great numbers, with the bones of deer and other mammalia.' (Lyell, *Antiquity of Man*, 1863, pp. 160, 161.)

Alluvium.—Along the Thames there is comparatively little of this modern deposit. Westminster is built upon alluvium, and was once an island (Thorney Island). Alluvial tracts also occur in the Isle of Dogs, and for about a mile above the mouth of the Lea. East of the county there are broad alluvial flats.

7. DRAINAGE.—Middlesex is wholly included in the Thames basin, the entire area of which is about 5,162 square miles, or 3,303,680 statute acres.* The greatest length of the Thames basin is, according to the Rivers Commission Report, from Trewsbury Mead in Gloucestershire to the estuary; its greatest breadth from Priors Marston, in Warwickshire, to Fernhurst, in Sussex. Besides Middlesex, which forms almost exactly one-nineteenth part of it, it includes the whole or nearly the whole, of Bucks, Herts, Surrey, Berks, Oxon, about one-fourth of Essex, one-sixth of Kent and Wiltshire, one-third of Gloucestershire, and small portions of Warwickshire, Northampton, Bedford, Sussex, and Hampshire.

The Thames is usually stated to take its rise† in the oolitic limestone rocks of the Cotswold Hills, the escarpment of which is the western boundary of its basin. It is not derived from any definable spring at its head giving out all, or nearly all, its waters, but is fed by many small rivers and brooks.‡

The following account is taken from the *Penny Cyclopædia*: 'The spring which has commonly been regarded as the head of the Thames is about three miles S.W. of Cirencester, near a bridge over the Thames

---

* *Report of Rivers Commission*, p. 8.

† It is customary to regard a river as originating at the place where what is looked upon as its main stream commences. A river is, however, after all, as ordinarily conceived, a rather arbitrary notion. In nature the main stream is only part of a great system of minutely ramifying streams. The edge of the basin in which these originate is the real source rather than any one point upon the edge. A river system is no better represented by its main stream than a leaf by its mid-rib.

‡ Bravender in *Proceedings of Cotswold Club*, 1867.

and Severn Canal, which is called "Thames-head bridge;" but that
which is to be regarded as the true head of the Thames is about three
or four miles south of Cheltenham.   Two streams rise, one from
fourteen springs at what is popularly called the Seven Wells, and
the other from four springs near Ullen Farm, the westernmost of
which springs is the real (i.e. the remotest) head of the river : both
streams rise on the south-eastern slope of the Cotswolds, and
form by their junction, about a mile from their respective sources, the
river Churn, a name the element of which is embodied in the ancient
name Corinium, and modern name Cirencester, of the town by which
it flows, and in the name of two villages, North and South Cerney,
which are near it.   At Cricklade, nineteen to twenty miles from its
source, the Churn joins the commonly reputed Isis or Thames, the
length of which above the junction is only about ten to eleven miles.'

Mr. Bravender remarks (*Proceedings of Cotswold Club*, 1867) that,
according to the Ordnance map, the Thames commences in Wiltshire,
at a place called Waterhay Bridge, about a mile below Ashton
Keynes, and about 3 miles west of Cricklade.   Above this point
it is called Swillbrook, and is joined by the brook running from the
spring in Trewsbury Mead near Cirencester, mentioned above.

The whole length of the Thames is 201 miles, and the portion of
it bounding Middlesex is about 45 miles long, following a serpentine
course from Staines to Blackwall.

The tide ascends to Teddington about 18 miles above London
Bridge, and the following species, partial to brackish water, occur in
Middlesex : some, however, are now no longer to be found so high : —

Cochlearia anglica.   Samolus Valerandi.
Althæa officinalis.   Scirpus triqueter.
Apium graveolens.   carinatus.
Petroselinum segetum.   Tabernæmontani.
Aster Tripolium.   Carex divisa.
Sonchus palustris.   Sclerochloa distans.

The following distances along the river from London Bridge are
taken from the *Rowing Almanack*, 1868 :—

		m.	f.				m.	f.
Westminster	. .	1	7	Kingston	.	.	20	2
Battersea	. . .	4	7	Hampton Court	.	.	23	1
Putney	. . .	7	2	Sunbury	.	.	26	2
Hammersmith	. .	9	0	Halliford	.	.	28	6
Kew	. . . .	13	0	Shepperton	.	.	29	2
Richmond	. . .	15	7	Chertsey	.	.	32	1
Twickenham	. .	17	2	Staines	.	.	35	7

The Thames is liable to heavy floods, which spread over the valley on either side. 'The flood waters are sometimes out in winter for weeks together. In summer, when extreme floods occasionally occur, they are most injurious, spoiling the crops, and sometimes even sweeping them bodily away.'* In extreme floods at Kingston the water rises as much as 11 feet 10 inches above summer level, and it very frequently rises from 4 to 6 feet above summer level.† After heavy rains the flood water at Kingston is first felt from the Mole, which drains the eastern half of Surrey. The up-country flood is not felt till a little time afterwards.‡

To the winter floods may perhaps be attributed the introduction of plants elsewhere characteristic of a calcareous soil into the parts of Middlesex adjoining the Thames from the chalk districts to the west; for example:—

Cerastium arvense.	Campanula glomerata.
Spiræa Filipendula.	Salvia Verbenaca.
Scabiosa Columbaria.	Calamintha Acinos.
Picris hieracioides.	Nepeta Cataria.
Carduus acaulis.	Avena pubescens.

Scilla autumnalis and *Allium oleraceum* are also confined to the Thames bank, to which their bulbs may possibly have been carried by the stream.

The depth of water in the river at summer level is very variable in different places from the irregularity of the bed. According to one of the witnesses examined before the Rivers Commission, 'it is all in hills and holes.'§ The depth in the Kingston reach varies from eighteen inches to four or five feet; or, where ballast has been dredged, to eight or nine feet. As the water becomes shallow, weeds become more abundant. Below Teddington the level of low water has been much depressed by the removal of Old London Bridge, and the deepening of the channel below Richmond. This allows the water to flow off with greater rapidity, leaving the upper part of the course comparatively dry. The level of low water at Richmond Bridge has been shown to have sunk three feet since its construction.‖

The principal tributaries of the Thames between Staines and Blackwall, or that part of its course which bounds Middlesex on the right or Surrey side, are the Wey, Mole, Hog's Mill River, and Wandle.

The Wey drains the west of Surrey, and joins the main river at Weybridge ¶; while the Mole, which joins it at East Molesey, drains

---

* *Report of Rivers Commission*, p. 28.   † *Evidence*, p. 178.   ‡ Ibid. p. 184.
§ *Evidence*, p. 185.   ‖ *Report of Royal Commission*, p. 29.
¶ The naturalised *Impatiens fulva* has been carried from mid Surrey by the Wey into the Thames, and thence into its Middlesex tributaries (see p. 71). It thus affords a

the eastern half of the county. The Hog's Mill River, flowing into the Thames at Kingston, and the Wandle at Wandsworth, only drain the north-east corner.

The streams of Middlesex require more minute description, as the division of the county into districts for botanical purposes is founded upon them. The *Colne* and its tributaries drain the south-west of Hertfordshire. It rises on the north side of the hills separating Middlesex from Herts, by several sources. A small stream belonging to the Colne system commences at a point midway between Elstree and Barnet, and enters Middlesex at Derham Park; it then takes a north course, and draining the country lying to the west of the Barnet and Hatfield Road, passes east of South Mims, and re-enters Herts between that village and North Mims; it has, however, no outlet; 'but the water is lost in swallow-holes in the chalk,* that thus receives the drainage of some twenty square miles of country' (*Whitaker*, p. 20). Other small streams originate in Middlesex from the north side of Brockley Hill and on the east of Stanmore Heath, and flow into the reservoir at Elstree,† a small portion of which is in Middlesex. The water of this reservoir flows north as a small stream (and is by some considered the true Colne), and joins the Colne at Colney Street in Herts.

After uniting with several large streams (the Ver and the Gade from Herts, and the Chess from Buckinghamshire), and having become a navigable river, the Colne reaches Middlesex at Rickmansworth, from which point it forms the county boundary for the remainder of its course. It enters the Thames at a point between Egham and Staines, by three branches. Just above Staines it sends off another branch, which passing Littleton flows into the Thames between Sunbury and Walton Bridge.

The chief tributary to the Colne during its course along the west border of Middlesex is a stream draining Harrow Weald Common, Pinner, and Ruislip, and joining it below Cowley. Ruislip Reservoir, which receives a few rivulets from the north, is connected with this stream. It occupies part of what was Ruislip Common. The water is carried from it in an artificial stream along the course of the Cran into the Paddington Canal near Southall.

The Colne divides, during its course through the low ground called

good example of the interdependence of the parts of a river system with reference to their floras.

* In Potterell's Park. See *Fl. Herts*, p. xxxii.
† Also called Aldenington Reservoir.

'moors' lying along the western border of Middlesex, into several irregularly anastomosing streams, which intersect and feed the Grand Junction Canal between Rickmansworth and West Drayton. Near the latter place a branch, called the 'Old River,' leaves the Colne to flow south-east to Baber Bridge, where it joins the Cran. The 'New River,' which also leaves the Colne near West Drayton, is an artificial stream (or partly so), made in the reign of William III. to supply Bushey Park with water.

The *Cran* or Crane is a small stream, called at its upper part the Yedding or Yeading Brook, and originates in several rivulets near Headstone Farm and the low country between Pinner and Harrow. These unite to form a small stream, which flows south by Yedding Green to Cranford, draining the adjacent country. From this point its course is south across Hounslow Heath to Baber Bridge on the Staines Road (where it is joined by the 'Old River' from the Colne). Gradually curving east to Twickenham, it flows north-east into the Thames a little above Isleworth. The Cran is much increased in size after its union with the 'Old River.' The conjoint streams flow together to within a mile of Twickenham, where the main body of water flows along an artificial canal, locally called the 'Duke's River,' to Sion Park, whilst only a limited quantity is allowed to proceed along the old bed of the stream.

The *Brent* commences by several small rivulets, which, flowing down from the hill in Herts on which Totteridge stands, and from the north-east side of Highwood Hill in Middlesex, unite to form a small stream which, flowing between Totteridge and Whetstone, forms the boundary of Herts and Middlesex. It enters the latter county about a mile north of Finchley, and passing west of that place receives a small brook from Mill Hill on the west, and several from the north slopes of Hampstead and Highgate on the east. Leaving Hendon on the west, it flows south-west to Kingsbury, before reaching which it is joined by a tributary from the north, bringing the surface drainage of the south slope of the Brockley, Elstree, and Highwood Hills, and the country between Edgware and Hendon. At the junction of this stream a large reservoir has been formed. After passing Kingsbury, the Brent receives a large branch at Wembly Park, formed by streams originating at Harrow Weald, the east side of Harrow Hill, &c., and also streams from the country about Wilsdon, Holsden, and adjacent parts. From this point the Brent flows west as far as Greenford, where it turns abruptly to the south, and, passing Hanwell, flows into the Thames at Brentford.

A small stream beginning between West Acton and Gunnersbury, and flowing into the Thames at Hammersmith, is marked in all the old maps.

Eel-brook, Hell-brook, or Hill-brook, flows south from between Hammersmith and Kensington, and passing between Walham Green and Little Chelsea, flows into the Thames opposite Battersea.

The *West Bourne* begins near the hamlet of West End, Hampstead, passes by Kilburn to Westbourne Park, Bayswater, through Kensington Gardens and Hyde Park, across Knightsbridge, along the west side of Sloane Street into the Thames a little below Chelsea Hospital. In Norden's map (1610) it is represented flowing through the centre of Hyde Park. It afterwards contributed to form the piece of water called the Serpentine, which was made about 1730. According to Middleton, the Serpentine was principally supplied with water from a conduit in a close of meadow-land near Bayswater belonging to the Bishop of London, but partly by springs. More recently it has been supplied from a well at the north end, the West Bourne being carried across the Park as a separate sewer.

The *Fleet* takes its rise on the south side of the Hampstead and Highgate Hills. It flows south, passing Kentish Town, Old St. Pancras, King's Cross (where Battle Bridge was built over it), the House of Correction, Clerkenwell, Field Lane, the foot of Holborn Hill, and Farringdon Street, to flow into the Thames at Blackfriars Bridge. From Kentish Town its course its now entirely subterranean. It was originally called the River of Wells,* the Turnmill or Tremill Brook; more recently the Fleet Dike or Ditch, and now the Fleet sewer.

The Fleet was formerly navigable up to Holborn Bridge at the foot of Holborn Hill. According to Stow (*Survey*, Thom's ed. p. 7), Oldborne or Hilborne was a small rivulet which broke out 'about the place where the bars do now stand, and it ran down the whole street till Oldborne Bridge and into the River of Wells or Turnemill Brook.' But in Domesday Book (i. 127 *a*), 'Holeburne' is applied to the Fleet above Holborn Bridge at Clerkenwell, &c., possibly from its running in a hollow.

In the London district there were several other small streams, now converted into sewers.

The *Lang Bourne*, so called, according to Stow, 'from the length thereof,' but of which in his time 'no sign . . . remaineth more than the name,' rose from a spring near Magpie Alley, adjoining St. Cathe-

* From the number of springs thrown out along its course, in consequence of the denudation of the gravel. See p. xvi.

rine Coleman. It ran west through Lombard Street as far as St. Mary Woolnoth, then turned south and united with the Wall Brook. From the stream overflowing and rendering the road swampy, 'the way was called Fenchurch Street near the church' (Noorthouk's *History of London*, 1782).

Walbrook entered the city between Bishopsgate and Moorgate through the wall, whence its name (*Stow*, Thom's ed. p. 45). After many turnings it flowed through the street now called Walbrook to the river at Dowgate. Formerly, says Stow, ' a fair brook of sweet water,' but in 1598 it had been some time arched over, and ' the trace thereof hardly known to the common people.'

These old watercourses of the metropolitan district have long since ceased to possess any place among the physical features of the county, and have been merged into the general sewer system. It is, however, desirable to trace out their former history, as their names occur in the writings of the older botanists, many of whose collecting grounds, as well as the streams that drained them, have been absorbed by the growth of London.

The *Lea* rises in the Chiltern Hills in Bedfordshire, flows south-east through Herts, the eastern two-thirds of which is drained by it and its tributaries; then flows south, separating Essex from Herts and Middlesex (which it first touches at its north-east corner a little below Waltham Cross), to fall into the Thames at Bow Creek, a little east of Blackwall.

The Lea receives several brooks from the east part of the county, the largest being Salmon's Brook, rising by surface drainage from the country formerly called Enfield Chase, east of the road from Barnet to Hatfield. Another small tributary drains what was formerly Finchley Common, and the country about Southgate and Colney Hatch (and East Barnet in Herts). The main stream of the Lea itself has been rendered navigable for about eight miles from the Thames. Beyond this a canal runs parallel to it through Tottenham, Edmonton, and Enfield Marshes to Waltham. The Lea is connected with the Paddington Canal by a cut called *Sir G. Duckett's Reservoir*, and with the Thames at Limehouse by another called the *Limehouse Cut*.

In addition to the natural rivers and streams of Middlesex there are many artificial watercourses which, especially in so small a county, deserve description amongst the other physical features. As the habitats and means of distribution of water-plants, their names are often mentioned, and in this respect they are probably of equal importance with the natural watercourses themselves. Indeed, these last have in many cases been so much modified for the sake of the mills and navigation, that the present condition of those which were at first natural, and those which have always been artificial, is not now in many cases very different.

The *New River* has especially, from the length of time since it was made, almost acquired the appearance of a natural stream. The water was admitted September 29, 1613, by Sir Hugh Middelton, who commenced the work February 20, 1608, when the scheme had been given up as impracticable by others. It nominally commences about half-a-mile south-west of Ware in Herts, in a spring called Chadwell, but the large proportion of its water (as much, it is said, as nineteen-twentieths) is let in from the Lea a little below the spring. The whole length of the New River is thirty-six miles, of which twenty-four miles are in Middlesex. It enters the county near Waltham Cross, and follows a very serpentine course south by Enfield, Winchmore Hill, Wood Green, Hornsey, and Highbury, to Islington and the New River Head at the corner of Amwell Street and Penton-ville Road. The latter part of this course from Canonbury is now bricked over, but a very few years back it was an open canal in front of Duncan Terrace and Colebrook Row, Islington.

'The boarded river,' often mentioned in the writings of the older metropolitan botanists, was part of the course of the New River close to Highbury Vale, near Stoke Newington, where it leaves the parish of Hornsey and enters that of Islington. The water was here carried in an aqueduct made of wood lined with lead, and supported by strong timbers standing on piers of brick. The trough was 462 feet long and 17 feet high. This was found to need such constant repairs that it was removed in 1776 by raising a clay bank under-neath it. A bridle road, which passed under the 'boarded river,' was carried over it afterwards by a bridge. A little stream also passed under it, and, receiving waste water from it, ran into the Lea at Old Ford under the name of *Hackney Brook.* A similar trough for the New River existed at Bush Hill near Enfield, which was removed about 1784. (See *Gentleman's Magazine,* 1784, and Nelson's *History of Islington.*) The New River was also carried in a trough over a hollow, made by digging gravel, in the City Road, and filled up in 1803.

The New River is one of the water supplies of London; other canals are used for purposes of navigation alone.

The *Grand Junction Canal,* from the north (begun 1794), is con-nected with the Colne from Rickmansworth to West Drayton, keeping generally within the Middlesex boundary. From West Drayton it is carried east to Norwood (crossing the Cran north of Cranford), and joins the Brent below Hanwell. The level of the canal falls 114 feet from Rickmansworth to the Thames. The *Paddington Canal* (opened 1801) connects the Grand Junction with the metropolis. It leaves it at Bull Bridge, near where it crosses the Cran, and passes Northolt, Apperton, Twyford, and Kensal Green. It is on the same level throughout. The *Regent's Canal* (opened 1820) continues the system from Paddington entirely round the north of London to the Thames at Limehouse. It passes through two tunnels, one under Maida Hill, the other under Islington and the New River. It is nine miles long, with twelve locks and a fall of eighty-four feet.

The *Kensington Canal* (opened 1828), three-quarters of a mile long, runs from Chelsea meadows to Kensington.

8. SOIL AND AGRICULTURE.—In few counties, says Mr. Clutterbuck, is the meadow and arable land so nearly divided, or the extent so clearly defined ; and, though not without exceptions, the surface occupied by the London clay and the valley drifts respectively, determines the extent under grass and under the plough * (*Jour. Roy. Agr. Soc.*, second series, vol. v. p. 9). The part of the county in which the London clay is at or near the surface, consists of gently rising hills with small valleys gradually worn away by the surface drainage. This is sufficiently rapid to prevent serious stagnation from the absence of subsoil drainage. Mr. Clutterbuck states that in the farms on the clay all operations are made subservient to hay-making for the London market. ' Very many of the fields are laid out in small and not very convenient enclosures, sometimes overgrown with timber in the hedgerows, which are generally of white or black thorn. The elm timber is often unduly shredded, and the oaks, especially on the sheer London clay, are of stunted growth. As a rule, the higher ground, or that on the outcrop of the plastic clay, is best clothed with timber.' (*Ibid.* p. 10.)

Mr. Caird speaks of the land in the neighbourhood of Wilsdon, about five miles south-east of Harrow, as exceedingly stiff and undrained. 'The tenants do not seem anxious to have their land thoroughly underdrained, and yet it is so wet that they cannot put stock on it after October. During an open winter the fields have a pleasant green appearance, looking richer than they really are, their summer produce being only one and a half to two loads of hay per

* According to the Agricultural Returns for 1868, of 108,956 acres, the total acreage under cultivation, 75,006 acres, or more than two-thirds, are under grass.

acre.'* Middleton says, and the practice is not likely to have changed,
that 'the cattle are removed on the tenth day of October or sooner, if
rain falls in quantity sufficient to wet the ground, as it is well known
that wherever a bullock makes a hole with his foot in this kind of
soil, it holds water and totally destroys every vestige of herbage,
which is not quite replaced till several years after the hole has
disappeared. The rest of the aftermath is eaten off by sheep, which
are continued till the second day of February.' According to Mr.
Clutterbuck, the practice if not the prejudices of the county are
against the removal of the water by under-drainage, lest the land
should become too dry. He says that the hay of Middlesex must
be allowed to be often of good quality. Hendon produces the hay
which has the best name in the market, and the Hendon Bent (*Cynosurus cristatus*), the Crested Dog's-tail, is well known to the dealers,
and is seldom found except on good meadow-ground (p. 25).

The valley drifts form 'wide flats slightly sloping upwards from
the river, often broken by low terraces and cut through by the side
valleys' (*Whitaker*, p. 97). There is a considerable breadth of low
meadow adjoining the western and south-western boundary, part of
it flooded at certain seasons by the waters of the Thames and Colne.
To the east there is the same condition on the banks of the Lea, such
as Hackney Marsh ; and meadow land interspersed with garden cultivation, which is found in that district, is again varied by some part
being under ordinary arable cultivation (*Clutterbuck*, p. 14). With
these exceptions almost the whole of the south-western portion of the
county south of the Uxbridge and London Road is covered by a fine
sandy loam (brick earth), beneath which is the low-level gravel
resting on the London clay. Professor Wrightson describes the
district south of the Brentford and Colnbrook Road as fine, deep, and
dry land. North of it the soil is a light sandy loam from eighteen
inches to five feet deep, and possessing that happy medium of texture
which fits it alike for the production of every kind of corn, pulse, and
roots. The tradition that Queen Elizabeth would have none other
than bread made from wheat grown at Heston, is still preserved in the
neighbourhood.† To this day the finest qualities of wheat, chiefly
Chidham, are grown on the brick earth of Middlesex, and it is
for the wheat grown on this district, which extends far into Buckinghamshire, that Uxbridge market is famed (*Clutterbuck*, p. 7).

---

* Caird's *English Agriculture*, 1850–51.

† Norden says that it was reported that Queen Elizabeth had 'the manchets for her
Highness' own diet from Heston' (*Spec. Brit.* 25). See also Camden (Gough's ed.), ii. 2.

The removal of brick earth in the manufacture of bricks in the neighbourhood of Heston and elsewhere is constantly changing the condition of the surface. The brickmaker pays a royalty, varying from 1s. to 2s. 6d., for every thousand of bricks, and also agrees to replace the tilth and level the surface. In this district it is often difficult to find an outfall for land drainage, especially when the brick earth is removed in brick-making. The brick-maker has, however, to provide for the under drainage, though sometimes osiers are planted where water is found on reaching the gravel (*Clutterbuck*, pp. 25–28).

The market gardens and orchards which formerly occupied the suburban districts to the west and north-east of London along the Lea, are giving way to the advancing wave of buildings more rapidly than they are replaced. Middleton speaks of them as lining the road for seven miles between Kensington and Twickenham; and the fruit trees in the gardens of houses along this road show, by still standing in line, that they originally belonged to market gardens. Market-garden cultivation is, however, gradually occupying the arable farms. The land is sometimes planted as an orchard with trees at wide intervals, or is deep ploughed with a heavy dressing of manure, planted with potatoes, cabbages, or other coarser vegetables; all kinds of garden produce are gradually introduced, and the arable farm becomes a market-garden. Manure is brought from London by the carts which take the produce to market, and the application of this, combined with deep trenching, so as to bring to the surface, after the removal of a crop, the soil which a few months before had been turned down with a large quantity of dung, is the great secret of maintaining the continued fertility of the ground. Garden cultivation will more and more deprive the south-western part of Middlesex of its agricultural character as far as arable culture is concerned (*Clutterbuck*, p. 18).

9. Woods.—A considerable extent of Middlesex was originally covered with wood, especially in the centre and north-west of the county, where the subsoil is clay, with sometimes a thin coating of upper level gravel. Watling Street was probably the first inroad made into it. In 1170–1182 it was an immense forest, full of wild animals (FitzStephen's *Survey*). It was not disafforested till the reign of Henry III. (1218).

The woods have since been constantly decreasing in extent, and there are but few pieces of original wood land remaining. The chief are Highgate Wood, Bishop's and Turner's Woods on the north slope

of the Hampstead Hills, Scratch Wood beyond Edgware, and numerous woods to the north of Ruislip (Duck's Hill, Copse Park, Mad Bess,
and North Riding Woods). These are all on a soil of stiff yellow clay.
The woodlands near Colney Hatch (including Hollick Wood and
others) were grubbed up when the Lunatic Asylum was built, which
now occupies their site. A piece of wood, partly plantation, lies
between Winchmore Hill and Southgate, and is sometimes named
from the one, and sometimes from the other place. Enfield Chase formerly contained much wood. The hills about Hornsey were originally
wooded, but were cleared at the end of the last century, and laid down
to grass (*Middleton*).

Mr. Clutterbuck thinks that the existing woods are not likely to be
brought into cultivation, from the nature of the soil. He describes
the growth as consisting of hazel, hornbeam, and blackthorn, from
which the best and straightest are cut into stakes and headers for
making hedges; the practice has of late decreased. At present, with
the exception of the larger poles, very much of the growth is made
into faggots for domestic use, which meets with a ready sale (*Journ.
Royal Agr. Soc.* 1869). The trees, chiefly oak and hornbeam, do not
grow to any size. Ken Wood, which stands higher and on sandy
soil, is also said to be of original native growth; if so, it presents
considerable differences from the adjacent woods, containing many
large beeches. At Harefield some fine trees, chiefly beeches, grow
naturally in the old woodlands called Old Park Woods, on the
chalk.

The wooded land in the north of London now exists only in name.
Besides Great St. John's Wood, to the north-west there was also a
Little St. John's Wood, which abutted on the New River a little
north of Highbury Wood ('woods near the Boarded River'—*R.
Syn.* iii. 411). One half of Highbury Wood was grubbed up in
1650 (Tomlins' *Islington*). Cream Hall in Highbury Vale is the
farm-house (*temp.* Charles II.) which was built on the site of Highbury Wood when it was finally cleared.

10. HEATHS, COMMONS, AND WASTE LANDS.—While the clay ground
of the north of the county and the slopes of the hills were wooded,
the sandy tops of the hills and the flat loamy and gravelly districts of
the south-west were heaths. The area of heath was probably larger
originally than that of wood, and extensive portions retained the
character of open heath up to quite recent times. In 1807 the commons of Middlesex were roughly estimated at about 8,700 acres,
which included also the water meadows of the Colne and Lea (Middle-

ton's *View*, 2nd edition, p. 113). The publication of Middleton's book
in 1797 greatly stimulated the inclosure of waste lands. He considered their neglect a direct encouragement to vagrancy and crime,
and a reprehensible waste of the resources of the country. The
result of his expostulations was that in the five years preceding 1807,
20,653 acres of waste lands, of which 7,895 acres belonged to commons properly so called, were inclosed. Since then nearly all the
remaining spaces have also been brought under cultivation. At
the present time not more than one-eighth of the area given above
still exists as common, and even this is not likely to remain
undiminished.

Beginning on the western side of the county, Harefield Common
(200 acres in 1807) is only represented by a small piece of heathy
land by the roadside on the opposite side to the entrance to Harefield
Grove, and by the village green which formed part of it. Ruislip
Common is entirely inclosed, as are also Uxbridge Common (350
acres in 1807) and Hillingdon Heath (150 acres in 1807).

Hounslow Heath is now only represented by the Drilling Ground
(south of the Staines Road beyond Hounslow), some ground south-
west of it towards Hanworth, a rough piece of land adjoining the
Cran below Baber Bridge, another small piece by the same river in
Dockwell Lane, Heston parish, and a few wide roadsides and gravel
pits. Hounslow Heath contained in 1754, according to Roque's map,
6,658 acres. In 1789, 350 acres in Stanwell parish were inclosed,
and in 1802, 1200 or 1400 more in the parishes of Teddington, Hanworth, and Feltham. In Faden's map, 1802, it is represented as
extending from Harlington and Stanwell south-east to Twickenham,
Teddington, and Bushey Park (more than 6 miles) and from Hounslow
to Feltham (about 3 miles). Sunbury Common is inclosed, except
the parts of it known as Ashford and Littleton Commons (each,
according to Middleton, 250 acres in 1807).

In the north of the county Harrow Weald Common, of which
695 acres were inclosed in 1805, is now only represented by a small
piece of its north and highest part. It is on a strong gravel (high-
level). Formerly connected with this common were Bushey Heath
(Herts) and Stanmore Heath. The former has been inclosed. The
latter is the highest ground in the county, a little over 500 feet above
sea level. It is a fine open piece of heath, sloping to the north-east.
Finchley Common, inclosed between 1819 and 1829, occupied 1,243
acres (*Roque*, quoted by Middleton). Two-thirds of it, on the west
side, was wet clay; the remainder dry, and covered with furze.

Hampstead Heath, less than 100 acres in 1807, remains almost untouched, except a few small inclosures.

In the west of the county Enfield Chase was mostly inclosed before 1787. In 1803, 1,500 acres were inclosed, as well as 2,746 acres of common fields. The names of various lodges and gates are the only local evidences of its former extent. South Mims inclosure (1,000 acres) formed part of it. Edmonton Common (1,231 acres) was inclosed before Enfield Chase. The waste places in the immediate neighbourhood of London have been gradually absorbed by the extension of buildings. Tothill Fields were beyond Westminster, extending to the river; Vincent Square occupies part of the ground, which also included the site of Millbank Penitentiary. Wormwood Scrubs (Wormholt Common, Wermer Wood Common) is still open ground, south of Kensal Green, though 60 acres have been enclosed; it was formerly a wood of 200 acres. Old Oak Common is a part of it. Eel-brook, or Hell-brook Common, a small piece of ground between Little Chelsea and Walham Green, is gradually becoming surrounded by buildings, and most of the less common plants which it produced have disappeared through draining. Chelsea Common occupied the ground between the King's Road and Brompton, on which Marlborough Road, etc. now stands.

II.—CLIMATE.

1. TEMPERATURE.—The *mean* or average temperature of a place is the first thing to be determined in discussing its climate. This might be done theoretically, by first determining the mean temperature of each day for a considerable period, say three years, by observing the temperature every alternate hour of the twenty-four, and then dividing the sum by twelve. The mean temperature of each month is obtained by dividing the sum of the mean daily temperatures by the number of days it contains; and of the year, by dividing the sum of the mean monthly temperatures by twelve. Monthly means can, however, be determined from a single daily observation, as the monthly mean of observations taken at a particular hour differs from the monthly mean temperature by a quantity which is constant for each particular month, and can be applied as a correction. It was formerly usual to determine the mean temperature of a month by taking the mean of the maxima and minima, but this gives a result which is too great for every month except December. The mean annual temperature at the sea level becomes less by $0.9°$, or about one degree for

an increase of one degree of latitude, and an elevation of 100 yards also causes a decrease of one degree of temperature. Hence, if T be the mean temperature of Greenwich (alt. 155 ft., lat. 51·5°) at the sea level, T', the mean temperature of any place not very remote from Greenwich, in England for example, may be obtained approximately from the following formula :—

$$T'=T+(51\cdot5° - l)\times0\cdot9-h\times\cdot00345,$$

where l is the latitude of the place and h its height above the sea level.

The mean temperature of Middlesex will be only subject to a variation of about 0·3° from range of latitude; and as this takes place about equally on either side of the parallel of Greenwich, the mean temperature of that place will represent with great accuracy the mean temperature of the county. The mean temperature of Greenwich has been determined from twenty-eight years' observations, ending 1868, to be 49·4°. At Chiswick, from observations made in the garden of the Royal Horticultural Society for seventeen years ending 1842, the mean temperature was found to be 49·88° (Daniell's *Meteorology*, vol. ii. p. 364). From observations made at Tottenham and Stratford from 1817 to 1830, it was found by Howard to be 49·651° (*Climate of London*, vol. i. p. 40).

The mean temperature of London itself was found by Howard to be somewhat higher than that of the adjoining country from, as he supposed, 'the effect of the population and fires' (*Ibid.* p. 236). The difference is, however, now known to be due to the proximity of the Thames. In discussing the observations made at the rooms of the Royal Society at Somerset House (*Phil. Trans.* 1850), Mr. Glaisher found that ' upon the whole year the excess of temperature at Somerset House was 1·2°.' He considers that 'one part of these differences is owing to the difference of elevation, and will probably amount to about 0·3°; the greater part of the remaining difference is most probably owing to the vicinity of the water of the Thames, whose temperature during the night hours at all seasons of the year is several degrees higher than that of the air. Those parts of London situated near the River Thames are somewhat warmer upon the whole year than the country; but those parts of London which are situated at some distance from the river do not enjoy higher temperatures than those due to their latitudes ' (p. 580). For example, 'no certain difference exists between the temperatures of the air at Greenwich and at St. John's Wood ' (p. 578). The temperature of Fleet Street is 0·7° higher than Greenwich (p. 578 ; see also Drew's

Meteorology, pp. 97–100). The mean temperature of the northern parts of the county, where the ground rises to 400 feet or more, would be on that account rather more than a degree colder than that of the county generally.

It is well known that the character of vegetation alters both with altitude and latitude; and, since the mean temperature of a place is influenced by both of these, some connection may be inferred between its vegetation and its mean temperature. Mr. Watson, in the *Cybele Britannica*, has divided the surface of the island into six zones, each covering a range of about 3° of mean annual temperature, and having a flora marked by certain peculiarities. All south-east England, including Middlesex, belongs to the inferagrarian zone, of which *Clematis Vitalba* is characteristic. The annual range of temperature, however, exerts a more direct influence on the flora of a place. The summer heat may be sufficient to satisfy the needs of plants which are either annuals or send up annual stems, where the climate is extreme and the mean temperature low; while their existence may not be possible in a place with a higher mean temperature but a more equable climate. Thus, Mr. Baker remarks that ' on the Andes, where the mean temperature is nearly the same all the year round, they cannot grow grain much above 7,000 feet, where the annual mean is 55°. In Britain we have to stop at about 44°, and in Switzerland they stop at 40°; but in Norway wheat goes up to the 64th, oats to the 65th, rye to the 67th, and barley to the 70th parallel of latitude, where the annual mean temperature is 32° or a little less. They can grow grain in places at least 12° lower in mean temperature than we can in England, and to get in Europe a mean of 55° we should have to go to Madrid or Milan' (*Flora of Northumberland and Durham*, p. 56).

The great characteristic of British climate, as compared with other European countries in the same latitude, is the comparatively small range of temperature between winter and summer. This absence of extremes either of heat or cold is due to the influence of the sea, and distinguishes an *insular* from a *continental* climate. Mr. Glaisher has deduced from registers kept from 1771 to 1849 the following mean temperatures for the seasons at Greenwich (*Phil. Trans.*, 1850, p. 594):—

Spring .	.	.	. March to May	.	.	. 46·4
Summer	.	.	. June „ Aug.	.	.	. 60·0
Autumn	.	.	. Sept. „ Nov.	.	.	. 49·3
Winter .	.	.	. Dec. „ Feb.	.	.	. 37·6
		Mean	.	.	.	. 48·3

The difference between winter and summer is only 22·4°. Compare this with the range at Munich, 34·5°; at Madrid, 33·5°; and at

Milan, 37°. Mr. Baker (*loc. cit.* p. 51) remarks that the winter is not materially colder at Newcastle than at London; but all the other seasons are. The summer is warmer in the interior of the south-east of England than anywhere else in Britain, rising in some places to a little over 60°. The autumn, both in the north and south, comes very near to the average of the whole year, being generally a little above it. The influence of water in cutting off extremes —even in comparatively so inconsiderable a body as the Thames— is also well seen in comparing the mean daily range of temperature at Greenwich Observatory and Somerset House (*Glaisher*, p. 606) :—

	Greenwich	Somerset House
January	8·5 . . .	6·8
February	9·9 . . .	7·9
March	13·4 . . .	10·2
April	16·7 . . .	12·3
May	18·9 . . .	14·8
June	19·4 . . .	15·5
July	18·4 . . .	14·9
August	17·6 . . .	14·1
September	17·0 . . .	12·5
October	13·3 . . .	9·9
November	10·5 . . .	8·0
December	9·0 . . .	6·4

Extremes of Cold and Heat.—'A night on which Fahrenheit's thermometer remains for some hours below zero is, in this climate, a rare occurrence: probably not above five of them fall within a century' (Howard, *Climate of London*, vol. ii. p. 289). The lowest temperature hitherto recorded, −6·5 occurred in the night of December 24, 1796, at Edmonton (vol. i. p. 26). In the winter of 1794–95, the thermometer descended to −2. On February 9, 1816, the thermometer recorded −5 at Tottenham. Watson quotes the following temperatures for the night of January 19, 1838: Chiswick −4·5; Hampstead Heath, −4; Kensington, −4; London Royal Society −11 (*Cyb. Br.* vol. iv. p. 172). This was the lowest temperature recorded at Chiswick between 1826 and 1842 (Daniel's *Meteorology*, vol. ii. p. 364). The highest temperature which has been recorded is 97°; this was north of London, July 18, 1825 (*Howard*, vol. iii. p. 194). At Greenwich 96·6° was reached July 22, 1868, and at Chiswick, 94·4° in 1836. 'Continued frost,' remarks Howard (vol. i. p. 239), 'in winter, is always an exception to the general rule of the climate. The winter even passes, occasionally, almost without a frost: in return for which we have, at uncertain intervals, a rigorous season of many weeks' duration, attended with the deep snows and clear atmosphere common to more northern latitudes.' The mean temperature of winter does not fall below freezing point.

Plants which would not bear complete exposure to frost will often survive, with slight shelter, frosts of short duration; and near the western coasts, where the influence of the sea has greater effect, especially in mitigating the winter, comparatively tender plants flourish in the open air throughout the year. A warm winter is an essential condition for the existence of tender plants with *perennial* stems. In the neighbourhood of London, on the other hand, the semi-spontaneous exotic plants which belong to the vegetation of climates with a higher mean temperature are necessarily *annuals.* Many of these are more abundant in some years than in others, a warm spring being essential to allow them to reach maturity before the first frosts.* Provided that the summer heat is sufficient to allow them to ripen their seeds, annuals are capable of a more extended northern duration than perennials. With regard to perennials the following remarks may be quoted from Mr. Baker :— ' In general terms, the polar limit of species liable to be killed by frost, runs across Europe, from N.W. to S.E., diagonally with the parallels of latitude; and to sum up, in a single comprehensive phrase, the relations of the British to the Continental flora, we may say, that the north limits of the plants, from the nature of the case, as regulated by temperature, radiate from our island, like the spokes of a wheel from the axis.' (*Flora of Northumberland and Durham,* p. 57.)

2. RAINFALL.—Stations on the west coasts seldom have less than 40 inches rain, but, passing from west to east, the fall decreases gradually if the country is level. At about three-fourths of the distance across the country, the minimum of 20 is reached, and then the fall increases again until we arrive at the east coast, where the fall is 25 inches. (Symons, *Rain,* p. 55.) Thus Barnstaple, Chapel-en-le-Frith, Plymouth, Shetland, Cork, Waterford, have an approximate mean annual rainfall of 40 inches. Ackworth, Epping, Hereford, Horncastle, Oxford, Sunderland, Elgin, 25 inches. Bushey, Cobham, London, Norwich, Thirsk, Hawarden, Edinburgh, 24 inches. Lincoln, Southwell, Stamford, 20 inches (p. 52).

The mean annual rainfall at Chiswick, deduced from observations made from 1826 to 1842, was 24·16 inches (Daniell's *Meteorology,* vol. ii. p. 316). The greatest rainfall during that time was 30·97 in 1841. The least 18·87 in 1840. At Tottenham, from Howard's

* See ' Exotic Plants about London in 1865 ' (*Seem. Journ. of Bot.* iv. 147).

Greenwich Meteorological Averages, deduced from 28 Years' Observations, ending 1868, made at the Royal Observatory.

Months	Barometer	Mean Temperature					Weight of		Amount of Rain on the Ground: 53 Years
		Highest by Day	Lowest by Night	Daily Range	Of Month	Dew Point	Vapour in a Cubic Foot of Air	Cubic Foot of Air	
	inches	°	°	°	°	°	grains	grains	inches
January . .	29·746	43·1	33·4	9·7	38·1	34·9	2·4	554	1·8
February . .	29·799	45·3	33·9	11·4	39·1	34·9	2·4	553	1·6
March . .	29·746	49·8	35·2	14·6	41·6	36·4	2·5	550	1·6
April . .	29·764	57·5	39·1	18·4	46·9	40·5	2·9	543	1·7
May . .	29·777	64·7	44·3	20·4	53·1	45·6	3·5	542	2·2
June . .	29·807	71·2	50·2	21·0	59·2	50·8	4·2	531	2·0
July . .	29·802	73·9	52·9	21·0	61·8	53·6	4·6	528	2·6
August . .	29·785	72·8	53·1	19·6	61·3	53·8	4·6	529	2·4
September . .	29·817	67·7	49·2	18·5	57·3	51·2	4·2	534	2·4
October . .	29·701	58·6	43·9	14·7	50·4	46·3	3·7	539	2·8
November . .	29·766	49·1	37·4	11·7	43·9	39·8	2·8	548	2·4
December . .	29·815	45·4	35·8	9·6	40·6	37·3	2·6	552	1·9
Means . .	29·777	58·3	42·4	15·9	49·4	43·8	3·4	542	Sum 25·4

Table showing the Mean Temperature of each Year from 1771 to 1868.

Year	Mean Temperature	Year	Mean Temperature	Year	Mean Temperature
1771	45·4	1804	49·5	1837	47·3
1772	47·1	1805	47·7	1838	46·4
1773	46·6	1806	50·5	1839	47·7
1774	47·7	1807	48·3	1840	47·8
1775	50·0	1808	48·1	1841	48·7
1776	48·3	1809	48·0	1842	49·6
1777	48·2	1810	48·7	1843	49·4
1778	49·2	1811	49·6	1844	48·7
1779	51·2	1812	46·5	1845	47·6
1780	48·8	1813	47·2	1846	51·3
1781	49 8	1814	45·8	1847	49·6
1782	45·5	1815	49·0	1848	50·2
1783	48·0	1816	46·4	1849	49·9
1784	45·1	1817	47·7	1850	49·3
1785	46·5	1818	50·8	1851	49·2
1786	45·8	1819	49·3	1852	50·6
1787	48·1	1820	47·4	1853	47·7
1788	47·9	1821	49·3	1854	48·9
1789	46·7	1822	51·0	1855	47·1
1790	48·1	1823	47·3	1856	49·0
1791	48·1	1824	48·3	1857	51·0
1792	48·0	1825	49·6	1858	49·2
1793	47·9	1826	49·9	1859	50·7
1794	48·9	1827	48·5	1860	47·0
1795	47·2	1828	50·1	1861	49·4
1796	47·8	1829	46·6	1862	49·6
1797	47·2	1830	47·8	1863	50·3
1798	48·6	1831	50·4	1864	48·5
1799	45·7	1832	49·1	1865	50·3
1800	48·3	1833	49·0	1866	49·8
1801	49·0	1834	51·0	1867	48·6
1802	48·0	1835	49·2	1868	51·6
1803	48·2	1836	48·1		

observations during the years 1813–1819, the mean rainfall was
25·46 inches. The greatest fall was 32·37 inches in 1816. (Howard,
Climate of London, vol. i. pp. 100–101.) At Greenwich the rainfall,
deduced from 28 years' observations, was 25·4 inches.

'One-third, or 33 per cent. above or below the average, is usually the
extreme difference. All years may fairly be set down as wet or
dry in which the fall is 20 per cent. either in excess or in defect
(*Symons*, p. 55).

The two preceding tables were kindly supplied by Mr. Glaisher.
They give the most complete information obtainable up to the
present time for the climate of London. The table of annual mean
temperatures, extending over all but a century, is the result of the
elaborate labours of Mr. Glaisher in combining the registers kept
at the Royal Society's apartments in Somerset House, at Epping, and
at Lyndon in Rutlandshire, with those kept at the Royal Observatory.

III.—DIVISIONS OF MIDDLESEX FOR BOTANICAL PURPOSES.

Although the county is not extensive it is convenient to divide it.
The various localities are more easily seen when grouped under
different divisions, and thus the flora of any particular part more
readily ascertained. But what is of more consequence, such a
method is the only way to ensure a systematic examination of the
whole surface, and to express with any accuracy the rarity or commonness of each species.

It is evident that a plant occurring in all the districts into which
a given tract of country may be divided is *more widely diffused* in that
tract than another found in any number of such districts less than the
whole; so far such a plant may be said to be *more common* than the
other. But it may happen that the former species occurs in but one
or two spots in every district, whilst the latter is 'found in very
numerous places in but one district; so far, then, as that district is
concerned, the latter species is the more common. The difference
between general commonness and local abundance can be readily
expressed by the use of district-divisions, and it will be perceived
that the more numerous the divisions, the more accurately (supposing
the examination to be complete) can we express comparative frequency; so that if a tract of country could be completely examined
throughout its whole area, and as many districts made as there are
square yards in it, a statement of the districts in which any species
occurred would almost perfectly express its true distribution. In this

case, too, we should find that the apparent distinction between general and local abundance almost entirely disappeared, the two groups of facts on which it depends gradually becoming merged into one.

The value of botanical districts is well exemplified in Mr. Watson's *Cybele Britannica*, in which the whole of Great Britain is so treated. To this comprehensive work, and especially to the fourth volume, we refer our readers for full explanations of the philosophy of plant-distribution.

We have divided the county of Middlesex into seven districts, and the principle followed has been to adhere strictly to the natural drainage, i. e. to make the land drained by each stream (when not too small) a separate division, which, when necessary, is divided again. It results from this, that the districts are irregular in shape and not very equal in size, and it may be asked what advantages are gained by such divisions when the county might be easily cut up into nearly equal spaces with regular and easily-defined boundaries. To this there are several answers :—(1.) It has been frequently shown by geographical botanists, that hills or watersheds separate floras, which are formed naturally by river basins. (2.) The system was followed out for the primary divisions in the *Cybele Britannica* with the best results. (3.) Inasmuch as a county is a tract of country with often purely artificial boundaries, it is very desirable that its flora should be examined and recorded with reference to some fixed natural con-formation applicable to all such tracts, so that, neighbouring counties being similarly treated, their adjacent districts may fit together and form complete river-basins. (4.) In particular, the *Flora Hertford-iensis* is drawn up on the same plan, and with this, as the boundary between the two counties is entirely artificial, it is important that the present Flora should correspond. Of course, in so small a county as Middlesex it is scarcely to be expected that results of any import-ance can be obtained; still it seems right to carry out what is believed to be a sound principle to its furthest application.

An account of the rivers and streams of the county has been already given (see pp. xvii.—xxv.) and the districts into which, by their means, it is divided are these :—

I. Upper Colne.	IV. Upper Brent.
II. Lower Colne.	V. Lower Brent.
III. Cran.	VI. Lea.

VII. Metropolitan (drained by the Fleet, West Bourne, Wall Brook, &c.).

The area of each district averages about forty square miles; their boundaries are shown on the accompanying map, and may be accurately traced on the Ordnance survey.

I. *Upper Colne.*—*N.* The north boundary of the county from its western extremity to Deacon's Hill near Elstree; *W.* The west boundary of the county from its northern extremity to Iver Court (in Bucks); *S.* A road from the Colne passing through Skrill and Colham Green into the Uxbridge Road; *E.* very irregular, from the point last mentioned along the Uxbridge Road in a north-west direction for a short distance, then north along the road to Ickenham, thence north-east passing along the high ground south of Ruislip and Pinner to Pinner Park Farm, Hatch End, and Harrow Weald: from this point it is continued up Brook's Hill, crosses the grounds of Bentley Priory, south of the mansion, and following the high ground across Brockley Hill, joins the north boundary at Deacon's Hill. There is also an outlying part of this district, the most northern part of Middlesex, which, as it were, locks into Herts; the county boundary limits this on all sides, except the east, where it is bounded by the high road from Barnet through Hadley and Potter's Bar.

This district is more diversified in soil and situation than any other: chalk is found in it alone, and it contains the highest ground in the county.

II. *Lower Colne.*—*N.* the south boundary of district I.; *W.* the west boundary of the county from Iver Court to its south extremity; *S.* the south boundary of the county from its western extremity to Twickenham; *E.* from the point last mentioned along the road westward through Fulwell towards Hanworth, thence north following the road which crosses the Windsor (South Western) Railway, and passes through Hatton and Harlington to Hayes, and so into the Uxbridge Road, which it follows north-west to join the north boundary.

This district is remarkably flat and uniform, with a sandy or gravel soil; a great many situations favourable for the growth of the aquatics are found in it.

III. *Cran.*—*W.* the east of district II., and east of district I., as far north as Harrow Weald; *S.* that of the county from Twickenham Ait to the west side of Sion House grounds; *E.* from that point along the road westward through Spring Grove and Lampton, and northward through Heston, Norwood, Southall, Northolt, and Wood End to Harrow, thence it passes near Headstone Farm to Harrow Weald, where it joins the north extremity of the western boundary; the district is thus pointed at its northern extremity.

Most of this district is flat and lies low, the north part is thinly inhabited and laid out in grass meadows, the south part was formerly an extensive heath, but is now mostly cultivated.

IV. *Upper Brent.—N.* that of the county from Deacon's Hill on the west to the Barnet Road north of Whetstone; *W.* the east of I. as far south as Harrow Weald, and the east of III., as far south as Harrow; *S.* the road from Harrow through Sudbury and Apperton, thence along the North-Western Railway, and Hampstead Junction line to Kensal Green station; *E.* the Hampstead Junction line from Kensal Green station to a point near the Edgware Road station, thence north-east following the high-ground at Shootup Hill through Fortune Green by Burgess Hill, to the western corner of the north Heath, Hampstead, thence along the south-western side of the heath, and north along the made road across the heath to the 'Spaniards,' and so by Hampstead Lane between Bishop's Wood and Ken Wood grounds to Highgate, from this point north along the road to Muswell Hill, thence west to Finchley, and follows the high road through Whetstone to the northern boundary. (A small portion of Herts comes into the basin of the Brent, though included in *Fl. Herts.* in that of the Lea; it contains the eminence on which Totteridge stands, and the valley between it and Barnet.)

This is the central district of the county, in contact with all the others except II. It is a well-cultivated, fertile valley on the clay, at one time forest land, and still containing many small woods and copses. The hills forming its boundaries are capped with sand and gravel, and are heaths.

V. *Lower Brent.—N.* the south of IV.; *W.* the east of III. as far north as Harrow: *S.* that of the county from the western side of Sion House Grounds to Hammersmith Bridge; *E.* from this point along Bridge Road, Brook Green Lane, Shepherd's Bush Lane and Wood Lane, along the eastern side of Wormwood Scrubs to the Kensal Green Railway station.

The south-western part of this district is part of the metropolis, and building is rapidly extending into it.

VI. *Lea.—N.* that of the county from the high-road north of Potter's Bar to its eastern extremity; *W.* the east of the outlying part of I., the very irregular county boundary from Barnet to near Whetstone and the east of IV., as far south as Highgate; *S.* from Highgate along Hornsey Lane through Crouch End to Hornsey Church, thence across the Green Lanes to West Green, and through Tottenham High Cross, to the county boundary; *E.* that of the county from its northern ex-

tremity to Tottenham High Cross. (A small piece of Herts, containing the village of East Barnet, really forms part of this drainage
district.)

The southern part of this district is quite suburban in character, the
northern portion was formerly open forest land, but is now all enclosed
and brought into cultivation.

VII. *Metropolitan.*—*N.* the east of IV. from Kensal Green Station
to Highgate, and the south of VI., from Highgate to the county
boundary ; *W.* the east of V.; *S.* that of the county from Hammersmith Bridge eastward ; *E.* that of the county from Tottenham High
Cross southward.

This district is in no respect in a natural condition. The elevation,
drainage, and other physical conditions have been quite altered and
diverted, with the exception of a few places in the extreme north and
west of the district. It contains the ancient cities of London and
Westminster, and the greater part of the districts and parishes which
make up modern London.

THE FLORA.

In the following pages, after some necessary lists of books and corre-spondents, the Flowering Plants and Ferns of Middlesex are systematically enumerated. In the arrangement of natural families, genera, and species, the last (6th) edition of Babington's *Manual* is, with a very few exceptions, followed; but many plants considered in that work as species are here reckoned as sub-species or varieties.

Each species is treated on the following plan :—

I. In the first line, and printed in thick type, stands the botanical (Latin) name of the plant as given either in Babington's *Manual*, Syme's *English Botany*, or *The London Catalogue of British Plants* (ed. 6); and whenever the names in these three standard works differ, those not here adopted are given as synonyms. When the plant possesses an English name, it follows the Latin one; but artificial names, or mere translations of the scientific titles, have not been perpetuated. It is only the commonest, the most conspicuous, and the useful species that have any real vernacular name.

II. The second paragraph contains the names of the species (including the names employed by the ante-Linnæan observers, from Turner to Black-stone) used in the books, MSS., &c. in which it is mentioned as found in Middlesex, when such names differ from the one adopted in this *Flora*. These synonyms are usually arranged in chronological order.

III. In the third paragraph will be found :—(*a.*) A reference to Mr. H. C. Watson's *Cybele Britannica*, and to the *Compendium* of that book so far as printed, in which nearly all that is known of the distribution of the species will be found. (*b.*) A reference to a figure of the plant. As far as at present published, Syme's *English Botany* is chiefly quoted; but the plates of Curtis's *Flora Londinensis* are often used, and the original edition of *English Botany*, with its *Supplement*, J. Curtis's *British Entomology*, and a few older works, are occasionally referred to. In the *Cyperaceæ*, Reichenbach's excellent figures in the *Icones Fl. Germ.*, vol. viii. are quoted; in the *Gramineæ* (of which there are no very good figures published), Lowe's *British Grasses*; and in the Ferns, Moore's *Octavo Nature-printed British Ferns*, and Newman's *History*, ed. ii. Usually the most characteristic

B

figure has been selected; but a portrait of a specimen actually gathered in the county, or at all events near London,* is, when possible, quoted.

IV. The fourth paragraph contains:—(a.) An indication of the usual sort of situation to which the species is partial, and a general statement of its *comparative frequency* in the county *as a whole.* For the expression of the latter condition six graduated terms are employed :—

Very common.—Generally distributed; in numerous localities in all the districts.

Common.—Distributed th.oughout all the districts, but less equally and in fewer localities in some or all.

Rather common.—Widely distributed, but in scattered localities, and not necessarily in all the districts.

Rather rare.—In some only of the districts, and showing decided local tendencies.

Rare.—In a few (about eight to four) scattered localities, or confined to a single district.

Very rare.—In a very few (three to one) localities.

It must, however, be understood that the signification of such terms cannot be strictly defined or very accurately applied.† (b.) The duration of the plant: annual, biennial, perennial, shrub, tree. (c.) The months in which it is usually found in flower. All these particulars refer exclusively to the species as an inhabitant of Middlesex.

V. The localities in which the species has been observed to grow then follow. These are grouped under the seven districts elsewhere explained and defined, each of which, distinguished by its number in Roman figures, forms a distinct paragraph. Except when the authors are responsible for a statement, the authority on which the locality depends is always quoted, separated from the locality by a semicolon, and printed in italics; when several statements are separated from one another by semicolons, the authority quoted after the last refers to all.

Special localities, however, are not given for the 'very common' species, except in the metropolitan district (VII.), where, from their peculiar character and fugitive existence, they possess a special interest.

VI. In the last paragraph will be found the date of the *first record* of the species as an inhabitant of Middlesex, with the name of the discoverer: these statements are, of course, liable to be superseded by the discovery of other sources of information.

Following this, and concluding the account, are placed any points of interest connected with the history of the plant as a native of the county, degree of naturalization, &c. ; and, in the case of the extinct species, the date

---

* It is to be regretted that the practice of stating whence the individual plant, from which the drawing was made, was obtained, is not more generally followed. Mr. Syme commenced his work with such useful notes to the new plates, but abandoned them after the first volume.

† 'Abundant,' 'plentiful,' &c., are *local* terms, and do not refer to *distribution* ; a 'very rare' species may be ' abundant' in one or two places.

when they were last noticed. A star, *, prefixed to the Latin name indicates that the species has probably been introduced into Middlesex at no very remote period, but is now more or less completely naturalized.

The native and naturalized (*) plants are treated in accordance with the above plan, and are besides numbered consecutively throughout; but there are also many species noticed which are undoubtedly of exotic origin, and, though introduced into our county, not naturalized in it. These have no number prefixed, and do not enter into the Flora proper. Their names are printed in ordinary type, they are treated less fully, and the localities run on, each district not having a separate paragraph.

Following the Systematic Catalogue are some lists and summaries which explain themselves, and a comparison of the Middlesex Flora with that of the whole of Great Britain, and of each of the surrounding counties.

SIGNS AND ABBREVIATIONS.

[ ]. Localities enclosed in square brackets are those in which the plant is now *certainly or very probably extinct.* When the whole account of a species is so enclosed it probably no longer inhabits the county.

( ). Statements enclosed in round brackets are such as we have reason to believe *erroneous, or admitting of great doubt.* It is thought better to notice all such printed statements, taking care to distinguish them, than to pass over them altogether, and so incur the charge of neglecting to become acquainted with what has been already done in our field of observation.

* Introduced, and more or less completely naturalized. (See above.)

! After a locality means that the authors have either seen the plant growing there, or received *fresh-gathered* specimens from the spot.

v. s. means that a *dried specimen* has been seen by the authors, collected in the locality indicated.

A. Annual.

B. Biennial.

P. Perennial.

LIST OF BOOKS, MSS., AND HERBARIA CONSULTED AND QUOTED IN THE 'FLORA.'

[When there are two or more editions of a book, those not quoted are within round brackets ( ).]

Ann. and Mag. Nat. Hist.—The Annals and Magazine of Natural History, 1st series, 1841–47; 2nd series, 1848–57; 3rd series, 1858–67; and 4th series, 1868, and in progress. [This is an amalgamation and continuation of the new series of *Mag. Nat. Hist.* (q. v.) and *Ann. Nat. Hist.*, formerly *Mag. of Zool. and Bot.*]

B. G.—The Botanist's Guide through England and Wales. By Dawson Turner, F.R.S., and Lewis Weston Dillwyn, F.R.S. Lond. 1805.

Bab. Man.—Manual of British Botany, &c. By Charles Cardale Babington, M.A., F.R.S., &c. Lond. Ed. i. 1843. Ed. ii. 1847. Ed. iii. 1851. Ed. iv. 1856. Ed. v. 1862. Ed. vi. 1867.

Bauhin.—Caspari Bauhini Pinax Theatri Botanici. Basiliæ. 1623. (Reprinted, 1671.)

Baxter.—British Phænogamous Botany. By William Baxter. 6 vols. Oxford. 1834–43.

Benth.—Handbook of the British Flora. By George Bentham, F.R.S. Lond. (Ed. i. 1858.) Ed. ii. 1866.

Blackst. Fasc.—Fasciculus Plantarum circa Harefield sponte nascentium. Lond. 1737. [By John Blackstone.]

Blackst. MSS.—Notes by John Blackstone in a copy of *R. Syn.* ii. and in a copy of *Johnson's Mercurius Botanicus.* About 1734–40.

Blackst. Spec.—Specimen Botanicum quo Plantarum plurium rariorum Angliæ indigenarum loci natales illustrantur. Auth. J. Blackstone. Lond. 1746.

Bot. Chron.—The Botanist's Chronicle. Nos. 1–17 [all published]. Lond. Nov. 1863—March 1865. [Edited by Alex. Irvine.]

Bot. Gaz.—The Botanical Gazette. Edited by A. Henfrey, F.L.S. 3 vols. Lond. 1849–51.

Brit. Phys.—Botanologia; the British Physician, or the Nature and Vertues of English Plants. By Robert Turner, Botanolog. Stud. Lond. 1664.

Brit. Ent.—British Entomology. By John Curtis, F.L.S. 16 vols. Lond. 1824–39. [Quoted for figures of plants.]

Budd. Herb.—v. *Herb. Budd.*

Budd. MSS.—'*Methodus nova,* &c.' By Adam Buddle. (Sloane MSS., 2970–2980.) With references to his Herbarium. About 1700–1708.

Burnett.—Medical Botany, &c. By John Stephenson, M.D., F.R.S., and James M. Churchill, F.L.S. New ed., edited by Gilbert T. Burnett, F.L.S. 3 vols. Lond. 1834–36. (Original ed. 1828–31.)

C. B. Pin.—v. *Bauhin.*

Camd. Brit.—v. *Forst. Midd.* and *Pet. Midd.*

Cat. Lond.—A Catalogue of scarce Plants found in the neighbourhood of London. 1813. [By Joseph Cockfield.]

Clusius.—Rariorum Plantarum Historia. Antv. 1601.

Coles.—Adam in Eden; or, the Paradise of Plants. By William Coles. Lond. 1657.

Comp.—Compendium of the Cybele Britannica. By Hewett Cottrell Watson. Part I. (pp. 200). Thames Ditton. 1868. [Not yet published.]

Cooper.—Flora Metropolitana; or, Botanical Rambles within thirty miles of London. By Daniel Cooper. Lond. 1836.

Coop. Supp.—Supplement to the Fl. Met. Lond. 1837.

Cullum.—Floræ Anglicæ Specimen, imperfectum et ineditum, anno 1774 inchoatum. [Auctore Thoma Gery Cullum, Baroneto.]

Curt. Br. Grasses.—Practical Observations on British Grasses. By William Curtis. Ed. ii. Lond. 1790.

Curt. Cat.—A Catalogue of the Plants growing wild in the Environs of London. Lond. 1774. [By William Curtis.]

Curt. F. L.—Flora Londinensis ; or, Plates and Descriptions of such Plants as grow wild in the Environs of London. By William Curtis. 6 fasciculi. Lond. 1777–1798. [490 plates.] v. *Hook. Curt.*

Cyb. Br.—Cybele Britannica ; or, British Plants and their Geographical Relations. By Hewett Cottrell Watson. 4 vols. Lond. 1847–59.

Cyb. Br. Supp.—First Part of a Supplement to the ' Cyb. Br.' Lond. 1860. [Not published.]

Dicks. H. S.—Hortus Siccus Britannicus ; being a collection of dried British Plants, &c. By James Dickson, F.L.S. 19 fasciculi. Lond. 1793–1802.

Dill. = Dillenius.—v. *R. Syn.* iii.

Don.—General History of Dichlamydeous Plants. By George Don. 4 vols. Lond. 1831–38.

Doody MSS.—Notes by Samuel Doody in a copy of *R. Syn.* ii. About 1695–1705.

E. B.—English Botany ; or, Coloured Figures of British Plants. By James Edward Smith, M.D. ; the figures by James Sowerby. 36 vols. Lond. 1790–1814.

E. B. Supp.—Supplement to E. B. Lond. Vol. i. 1831. Vol. ii. 1834. Vol. iii. 1843. Vol. iv. 1849. Vol. v. (6 Nos.), 1865.

E. Fl.—v. *Smith E. Fl.*

Faulkner.—Historical and Topographical Account of Fulham. By T. Faulkner. Lond. 1813.

Fl. Brit.—v. *Smith Fl. Brit.*

Fl. Bucks.—Flora of Buckinghamshire. By James Britten. Wycombe. 1867. [Not published.]

Fl. Camb.—Flora of Cambridgeshire. By C. C. Babington, M.A., F.R.S., &c. Lond. 1860.

Fl. Essex.—The Flora of Essex. By Geo. Stacey Gibson, F.L.S. Lond. 1862.

Fl. Harrow.—v. *Melvill.*

Fl. Herts.—Flora Hertfordiensis. By the Rev. R. H. Webb, M.A., and the Rev. W. H. Coleman, M.A. Lond. 1849. A *Supplement* in 1851. A *2nd Supplement* in 1859.

Fl. Lond.—v. *Curt. F. L.*

Fl. Metrop.—v. *Cooper.*

Fl. Rust.—v. *Mart. Fl. Rust.*

Fl. Surrey.—Flora of Surrey. Compiled by James Alex. Brewer. Lond. 1863.

Forst. Midd.—Rare plants found in Middlesex. In vol. ii. of Richard Gough's edition of Camden's *Britannia*. Lond. 1789. (And subsequent editions.) [Compiled by Edward Forster.]

Francis.—An Analysis of the British Ferns and their Allies. By G. W.

Francis. Lond. Ed. i. 1837. (Ed. ii. 1842. Ed. iii. 1847. Ed. iv. 1850 ?) Ed. v. revised and enlarged by A. Henfrey. 1855.

Galpine.—Synoptical Compendium of British Botany. By John Galpine, A.L.S. Salisbury. 1806. (Ed. ii. Liverpool, 1819. Ed. iii. Lond., 1829.)

Ger.—The Herball, or General Historie of Plantes. Gathered by John Gerarde. Lond. 1597.

Ger. em.—v. *Johns. Ger.*

Gibs. Camd.—v. *Pet. Midd.*

Gough's Camd.—v. *Forst. Midd.*

Gray.—A Natural Arrangement of British Plants according to their relations to each other. By Samuel Frederick Gray. 2 vols. Lond. 1821.

Graves.—Monograph of the British Grasses. By George Graves. 5 Nos. all published. Lond. 1822-23.

Herb. Budd.—Herbarium collected by Adam Buddle at the end of the Seventeenth and beginning of the Eighteenth Centuries. Contained in 14 vols. (54 & 114–126) of Sir H. Sloane's Herbarium in the British Museum.

Herb. G. & R.—Collected by W. F. Goodger and Richard Rozea in 1815-25. In possession of Mr. Varenne, of Kelvedon, Essex.

Herb. Hardw.—Collected between the years 1840 and 1855. In possession of Mr. Robert Hardwicke.

Herb. Harr.—Collected in 1863 by Mr. J. C. Melvill. Now in the library of Harrow School.

Herb. Linn. Soc.—The Smithian Herbarium and Herbarium of British Plants in the possession of the Linnæan Society. Contains many of Joseph Wood's, A. B. Lambert's, and other botanists' specimens.

Herb. Mus. Brit.—The Herbarium of British Plants in the British Museum. Founded on Edward Forster's and J. Sowerby's collections; to which have been added Mrs. Robinson's and Miss Atkins' Herbaria, and donations from numerous botanists.

Herb. Pet.—Collected by James Petiver, at about the same period as Herb. Budd., but further into the eighteenth century. Contained in 3 vols. (150-152) of the Sloane Herbarium in the British Museum. The labels to the specimens are mostly printed, and cut out from *R. Syn.* ii., *Botanicum Anglicum, Hortus Siccus Pharmaceuticus, Botanicum Londinense,* &c. Many lie loose, and some are wrongly affixed. When written they are mostly copies of *Budd. MSS.*

Herb. Rudge.—Collected by Samuel Rudge in about 1800–20. Chiefly about Elstree; apparently contains many garden plants. Preserved in the British Museum.

Herb. Young.—Collected by J. Forbes Young, M.D., and other botanists, from 1833 to 1844. Preserved at Kew.

Hill.—Flora Britannica sive Synopsis Methodica Stirp. Brit. By John Hill, M.D. Lond. 1760.

Hook. B. Fl.—The British Flora. By Sir W. Jackson Hooker. Lond. (Ed. i. 1830. Ed. ii. 1831.) Ed. iii. 1835. (Ed. iv. 1838. Ed. v. 1842.) And by G. A. Walker-Arnott (Ed. vi. 1850. Ed. vii. 1855). Ed. viii. 1860.

Hook. Curt.—Flora Londinensis. Enlarged by W. J. Hooker, M.D., vols. iv. and v. Lond. 1821–28. [212 additional plates.]

How.—Phytologia Britannica. Lond. 1650. [By William How, M.D.]

Huds.—Gulielmi Hudsoni, F.R.S., Flora Anglica, exhibens Plantas per regnum Angliæ sponte crescentes distributas secundum systema sexuale. Lond. 1762. Ed. ii. 2 vols., 1778 (and in one vol., 1798).

Irv. H. B. P.—Handbook of British Plants. By Alex. Irvine. Lond. 1858.

Irv. Lond. Fl.—The London Flora ; containing a concise description of the Phænogamous British Plants which grow spontaneously in the vicinity of the Metropolis. By Alex. Irvine. Lond. 1838.

Irv. MSS.—' A List of Plants growing within a two miles' radius of Hampstead Heath.' 1825–1834. MS. ined.

Jenkinson.—A Generic and Specific Description of British Plants, &c. By James Jenkinson. Kendal. 1775.

Johns. Eric.—Ericetum Hamstedianum, sive Plantarum ibi crescentium observatio habita anno eodem [1629] 1 Augusto. Descripta studio et opera Thomæ Johnsoni. Lond. 1629.

Johns. Enum.—Enumeratio Plantarum in Ericeto Hampstediano locisque vicinis crescentium. Lond. 1632.

Johns. Ger.—The Herball, &c., by John Gerarde ; very much enlarged and amended by Thomas Johnson. Lond. 1633. (Again in 1636.)

Journ. Bot.—v. *Seem. J. of B.*

Knapp.—Gramina Britannica. By J. L. Knapp, F.L.S. Lond. Ed. i. 1804. (Ed. ii. 1842, little more than a transcript of ed. i.)

Lightf. MSS.—Notes by John Lightfoot in a copy of Hudson's *Flora Anglica*. Written about 1780.

Lindley.—Synopsis of the British Flora. By John Lindley, Ph.D., &c. Lond. (Ed. i. 1829.) Ed. ii. 1835. (Ed. iii. 1840.)

Linn. Soc. Journ.—Journal of the Linnæan Society. Lond. 1857. And in progress.

Linn. Soc. Trans.—Transactions of the Linnæan Society. Lond. 1791. And in progress.

Linn. Sp. Plant.—Caroli Linnæi Species Plantarum exhibentes Plantas rite cognitas, &c. Holmiæ. (Ed. i. 1753.) Ed. ii. 1762.

Lob. Adv. Nov.—Stirpium Adversaria Nova . . . . authoribus Petro Pena et Matthio de Lobel. Lond. 1570.

Lob. Ill.—Stirpium Illustrationes [by Lobel] accurante Gul. How, M.D. Lond. 1655.

Lob. Obs.—Matthiæ de Lobel Stirpium Observationes. Antv. 1576.

L. Cat.—The London Catalogue of British Plants. Lond. (Ed. i. 1844. Ed. ii. 1848. Ed. iii. 1850. Ed. iv. 1853. Ed. v. 1857.) Ed. vi. 1867.

Loud. Arb. et Frut.—Arboretum et Fruticetum Britannicum. By J. C.
   Loudon. 4 vols. Lond. 1838.
Lowe.—A Natural History of British Grasses. By Edward J. Lowe.
   Lond. 1858.
M. & G.—Indigenous Botany, &c. By Colin Milne, LL.D., and Alexander
   Gordon. Vol. i. all published. Lond. 1793.
Macreight.—Manual of British Botany. By D. C. Macreight. Lond.
   1837.
Mag. Nat. Hist.—The Magazine of Natural History. Conducted by J. C.
   Loudon. 9 vols. 1829–36. New series, conducted by E. Charles-
   worth. 4 vols. 1837–40 [continued as *Ann. and Mag. Nat.
   Hist.*, q. v.].
Mart. App. P. C.—Plantæ Cantabrigienses . . . . to which are added Lists
   of the more rare plants growing in many parts of England and
   Wales. By Thomas Martyn, M.A. Lond. 1763.
Mart. Fl. Rust.—Flora Rustica. By Thomas Martyn. 4 vols. Lond.
   1792–94.
Mart. Mill. Dict.—The Gardener's and Botanist's Dictionary . . . . By the
   late Philip Miller, F.R.S. . . . . to which are now first added, A
   complete Enumeration and Description of all Plants hitherto Known.
   . . . . By Thomas Martyn, B.D., F.R.S. 2 vols. Lond. 1807.
   [Is a 9th edition of Miller's *Gard. Dict.*, first published in 1724.]
Mart. Tourn.—Tournefort's History of Plants growing about Paris . . . .
   translated into English and accommodated to the plants growing in
   Great Britain by John Martyn, F.R.S. 2 vols. Lond. 1732.
Melv. or *Melvill.*—The Flora of Harrow. By J. C. Melvill. Lond. 1864.
Merrett.—Pinax Rerum Naturalium Britannicum. Auth. Christophero
   Merrett. Lond. 1666. (Reprinted 1667.)
Mill. Bot. Off.—Botanicum Officinale ; or, a Compendious Herbal. By
   Joseph Miller. Lond. 1722.
Mill. Dict.—v. *Mart. Mill. Dict.*
Moore.—Octavo Nature-printed British Ferns. By Thomas Moore. 2 vols.
   Lond. 1859.
Moris. Umb.—Plantarum Umbelliferarum distributio nova. Auth. Roberto
   Morison. Oxonii. 1670.
Moris. Hist. Ox.—Plantarum Historia Universalis Oxoniensis. Pars ii.
   1680. Pars iii. 1699. Ed. Jacob Bobart. (Pars i. never published,
   though exists in MS. at Oxford.)
New B. G.—The New Botanist's Guide to the Localities of the Rarer Plants
   of Britain. By H. C. Watson. Vol. i. 1835. Vol. ii. 1837.
Newman.—A History of British Ferns and allied Plants. By Edward
   Newman, F.L.S. Lond. (Ed. i. 1840.) Ed. ii. 1844. (Ed. iii.
   1854. Ed. iv. 1865.)
Newt. MSS.—'Mr. [James] Newton's MS. notes as set down in his "Catalogus Plant. Angliæ."' Copied by an unknown hand in a copy of
   R. Cat. ii. Refer to about 1680.

Park Hampst.—Topography and Natural History of Hampstead. By J. J. Park. Lond. 1818.

Park. Parad.—Paradisi in Sole. Paradisus Terrestris; or, Garden of all sorts of Pleasant Flowers, &c. [By John Parkinson.] Lond. 1629. (Ed. ii. 1656.)

Park. Theat.—Theatrum Botanicum; the Theater of Plants; or, an Herball of large extent. Collected by John Parkinson. Lond. 1640.

Parnell.—The Grasses of Britain. By Richard Parnell, M.D., &c. Illustrated by figures drawn and engraved by the Author. Edinburgh and London. 1845.

Pet. Bot. Lond.—Botanicum Londinense; or, the London Herbal, &c., in vol. iii. of the 'Monthly Miscellany; or, Memoirs for the Curious.' By James Petiver. Lond. 1709.

Pet. Gram. Conc.—Graminum, Muscorum, Fungorum, Submarinorum, &c., Britannicorum Concordia, pp. 12 all published? By James Petiver. Lond. 1716.

Pet. Herb.—v. *Herb. Pet.*

Pet. H. B. Cat.—Herbarii Britannici clariss. D. Raii Catalogus cum Iconibus ad vivum delineatis. By James Petiver. Lond. Tab. 1–50, 1713. Tab. 51–72, 1715.

Pet. Midd.—More rare Plants growing wild in Middlesex, communicated by Mr. James Petiver. In Edmund Gibson's translation of Camden's 'Britannia.' Lond. 1695. (And later editions.)

Pet. Mus.—Museum Petiverianum a Jacobo Petiver, F.R.S. Centuriæ X. Lond. 1695–1703.

Phyt.—The Phytologist; a Popular Botanical Miscellany. Conducted by George Luxford, A.L.S. 5 vols. June, 1841—July, 1854.

Phyt. N. S.—The Phytologist; a Botanical Journal. Edited by Alexander Irvine. [A new series.] 6 vols. May, 1855—July, 1863.

Pluk. Alm.—Almagestum Botanicum, sive Phytographiæ Pluc'netianæ Onomasticon. [By Leonard Plukenet.] Lond. 1696.

Pluk. Amalth.—Leonardi Plukenetii Amaltheum Botanicum, i.e., Stirpium indicarum alterum Copiæcornu. Lond. 1705.

Pluk. Mant.—Almagesti Botanici Mantissa. [By Leonard Plukenet.] Lond. 1700.

Pluk. Phyt.—Leonardi Plukenetii Phytographia; sive Stirpium illustriorum et minus cognitarum Icones. Partes i. et ii. 1691. Pars iii. 1692.

R. Cat.—Catalogus Plantarum Angliæ. Opera Joannis Raii, M.A., F.R.S. Lond. Ed. i. 1670. Ed. ii. 1677.

R. Fasc.—Fasciculus Stirpium Brit. post editum *Plant. Ang. Cat.* observat. a Joanne Raio et ab amicis; cum synonymis et locis natalibus. Lond. 1688.

R. Hist.—Historia Plantarum. By John Ray. Lond. Vol. i. 1686. Vol. ii. 1688. Vol. iii. [supplementary], 1704.

R. Syn.—Synopsis Methodica Stirpium Britann. auct. Joanne Raio. Lond. Ed. i. 1690. Ed. ii. 1696.

R. Syn. iii.—Joannis Raii Synopsis Meth. Stirp. Brit. Editio tertia.
 [Edited by J. J. Dillenius, M.D.] Lond. 1724.
Robson.—The British Flora. By Stephen Robson. York. 1776.
Rudge.—v. *Herb. Rudge.*
Seem. J. of Bot.—The Journal of Botany, British and Foreign. Edited by
 Berthold Seemann, Ph.D., F.L.S. Lond. 1863. And in progress.
Smith E. Fl.—The English Flora. By Sir James Edward Smith, M.D.,
 F.R.S. 4 vols. Lond. 1824–28.
Smith Fl. Brit.—Flora Britannica. Auctore J. E. Smith, M.D. 3 vols.
 Lond. 1800–4.
Smith MSS.—Notes by Sir J. E. Smith on the original drawings for *E. B.*
 preserved in the British Museum. 1790–1814.
Sole.—Menthæ Britannicæ. By William Sole. [Twenty-four plates of
 Mints, with descriptions.] Bath. 1798.
Syme E. B.—English Botany. Edited by John T. Boswell Syme. Ed. iii.
 with descriptions of all the species by the Editor. Lond. 1863.
 And in progress.
Turn.—William Turner's Herball. Part 1. Lond., 1551. (Part 2. Cologne,
 1562. Part 3. Cologne, 1566.) The whole together, ed. ii. of part 1,
 Cologne, 1568.
Turn. Names.—The Names of Herbes in Greke, Latin, Englishe, Duche,
 and Frenche, wyth the commune names that Herbaries and Apote-
 caries use. Gathered by William Turner. Lond. 1548.
Turner (Robert).—v. *Brit. Phys.*
Waring.—A Letter from Richard Waring, F.R.S. . . on some plants found
 in several parts of England. In Philos. Trans. vol. lxi. 1770.
Watson.—v. *Comp. ; Cyb. Br. ; Cyb. Br. Supp. ; New B. G.*
Wilson.—Synopsis of British Plants. By John Wilson. Newcastle.
 1744.
Winch. MSS.—Notes, by Nathaniel J. Winch, in a copy of *Smith Fl. Brit.*
 in the Linnæan Society's Library. Written 1800–30.
With.—A Botanical Arrangement of British Plants, &c. By William
 Withering, M.D., F.R.S. Birmingham. (Ed. i. 2 vols.. 1776.)
 Ed. ii., including a new set of references to figures, by Jonathan
 Stokes, M.D., 3 vols., 1787–92. Ed. iii. 4 vols., 1796. (Ed. iv. By
 William Withering, junr., D.D., F.L.S. 1801. Ed. v. 1812. Ed. vi.
 1818.) Ed. vii. 1830. (Ed. viii. By William Macgillivray. 1840.)

LIST OF THE PRINCIPAL PERSONS WHO HAVE CON-
TRIBUTED INFORMATION BEARING ON THE FLORA.

Baker, John Gilbert, F.L.S.; Richmond, Surrey. (Localities near the
Thames.)

Bennett, J. J., F.R.S., V P.L.S.; Lewisham, Kent. (Dried specimens of
Middlesex plants, collected 1815–25.)

Bloxam, Rev. Andrew, M.A.; Twycross, Leicestershire. (Names of some
Rubi; localities near Hampton Court, 1867.)

Britten, James; High Wycombe, Bucks. (Lists of plants about London.)

Cherry, John; Stratford, Essex. (Lists of Brentford, Hackney, and
Edmonton plants, 1866–68.)

Church, A. H., M.A.; Cirencester. (Localities about Southgate.)

Cole, Arthur B.; London. (Localities, from various parts.)

Davies, William; London. (Localities, from Hampstead.)

Farrar, Rev. F. W., M.A., F.R.S.; Harrow. (Additions to *Harrow Flora*.)

Fox, Rev. H. E., M.A.; Oxford. (Localities about London.)

Griffiths, A. W.; London. (Localities from Ruislip and Harefield.)

Grugeon, Alfred; London. (Localities from Hackney, Dalston, &c.)

Hemsley, W. B.; Sussex. (Localities in Districts III. and V., 1861–63;
Extracts from Borrer's and Bromfield's Herbaria at Kew.)

Hind, Rev. W. M., M. A.; Pinner. (Additions to the *Harrow Flora*,
localities about Pinner, &c.)

Irvine, Alexander; Chelsea. (Loan of a MS. Flora of Hampstead; localities about Chelsea, &c.)

Lawson, Marmaduke A., M.A., F.L.S.; Prof. Bot. Oxford. (Various localities.)

Lees, Edwin, F.L.S.; Worcester. (Localities at Perivale.)

Masters, M. T., M.D., F. L. S.; Isleworth. (Localities in District V.)

Melvill, J. C.; London. (Several localities and notes.)

Morris, James, M.D.; London. (His Herbarium, collected in 1845–50.)

Newbould, Rev. W. W., M.A., F.L.S.; London. (Very numerous localities,
loan of books, critical suggestions, and much help in many ways.)

Pamplin, William, A.L.S.; Llandderfel, Merionethshire. (Valuable notes
on London plants, loan of books, &c.)

Savage, Charles; London. (Localities near London.)

Smith, Rev. G. E., M.A.; Malvern. (A few localities near London.)

Smith, W. G., F.L.S.; London. (Localities at Hampstead, &c.)

Syme, J. T. Boswell, F.L.S.; Balmuto, Fife. (Localities near Hampstead, &c.)

Tucker, Robert, M A.; London. (Localities near Acton and Enfield.)

Varenne, E. G.; Kelvedon, Essex. (Localities from his Herbarium, 1827–30,
and from collections of Messrs. Goodger and Rozea.)

Warren, Hon. J. B. Leicester, M.A., F.L.S.; London. (Localities in
London and at Hanwell; notes on Rubi, &c.)

Watson, H. C.; Thames Ditton, Surrey. (Localities about Sunbury.)

Wollaston, G. B.; Chiselhurst, Kent. (A few localities.)

PHANEROGAMIA.

DICOTYLEDONES.

RANUNCULACEÆ.

CLEMATIS, *Linn.*

1. C. Vitalba, *L.*   *Traveller's Joy.*
Viorna, Ger. em. (Blackst.).
Cyb. Br. i. 70; iii. 372; Comp. 79. Curt. F. L. f. 4.
Climbing over hedges and bushes; rather rare. Shrub. July, August.
    I. Harefield, frequent!; *Blackst. Fasc.* 111. Near Uxbridge, plentifully;
        Jenkinson, 118, and *Phyt. N. S.* i. 65. Woodready. Waxwell,
        Pinner.
    II. Bet. W. Drayton and Yewsley!; *Newb.* By Kingston Bridge.
    III. At the back of Kneller Park.
    V. Horsington Hill.
    VI. Bet. Finchley and Whetstone; bet. Tottenham and Finchley; *Herb.*
        Hardw. On both sides of Bury Street, Edmonton, bet. the railway
        crossing and the footpath to the church.
    VII. [In the way from Chelsea to Fulham; *Forst. Midd.*] [Edgware Road,
        1830; *Varenne.*]
    First record: *Blackstone*, 1737. Perhaps not indigenous to some loca-
lities, but certainly native in I. Prefers a calcareous soil, but is by no
means restricted to it.

THALICTRUM, *Linn.*

2. T. flavum, *L.*   *Meadow Rue.*
T. sive T. majus, Ger. em. (Blackst.).
Cyb. Br. i. 73; iii. 373; Comp. 80. Syme E. B. i. t. 8.
Sides of streams and ditches; rather rare. P. June, July.
    I. Harefield!, frequent; *Blackst. Fasc.* 98. By the Colne, bet. Harefield
        and Rickmansworth, Rev. E. Hodgson; *Fl. Herts.*
    II. Common by the Thames; Hampton, Sunbury, bet. Strawberry Hill
        and Teddington, &c.

III. By the river, Twickenham; *Rev. T. Firminger, MSS.* Isleworth; *Herb. G. & R.*

IV. Hampstead; *Irv. MSS.*

V. Greenford; *Cooper,* 108. Near Sawyer's boat-yard, Hammersmith; *Cole.*

VII. Kilburn; *Varenne.* Ditch in market-garden bet. Chelsea and Fulham; *Britten.* Hackney marshes, 1865; *Grugeon (v. s.).*

First record: *Blackstone,* 1737. Two forms occur, apparently vars. a and β of Syme E. B.

ANEMONE, *Linn.*

3. A. nemorosa, *L.* *Wood Anemone. Wind Flower.** *A. nemorum fl. alb. et purpurascente, Ger. em.* (Blackst.). Cyb. Br. i. 74; Comp. 80. Syme, E. B. i. t. 11.

Woods, hedgebanks, and bushy places; common. P. March—May.

I. Harefield!; *Blackst. Fasc.* 6. Eastcott; *Melvill,* 1. Pinner.

III. Roxeth; *Melvill,* 1.

IV. Hampstead Heath!; *Johns. Eric.* Kingsbury; *Cole.* Near Finchley!; *Newb.* Bentley Priory. Bishop's and Turner's woods.

V. Bet. Brentford and Hanwell; *Hemsley.* Apperton; *Cole.* Wormholt Scrubs; *Britten.*

VI. Hadley!; *Newb.* Bush Hill, Edmonton. Winchmore Hill Wood.

VII. Marylebone fields, 1817; *Herb. G. & R.* West End Lane, Ken Wood.

First record: Johnson, 1629. A state *'with crimson petals'* (IV.). Bentley Priory, W. M. H.; *Melvill,* 2.

[*A.* apennina, *L. Ranunculus nemorosus fl. purp.-cæruleo, Park.* (Ray). Cyb. Br. i. 75. Syme E. B. i. t. 10. Near Harrow-on-the-Hill, Mr. Du Bois; *R. Syn.* iii. 259. Probably from a garden. The station has never been verified. (See *Melvill,* 2.)]

[Adonis autumnalis, *L. Red Maithes. Pheasant's Eye. Flos Adonis fl. rubro, Ger. em.* (Blackst.). *Adonis annua* (Mart.). Cyb. Br. i. 76; iii. 373; Comp. 80. Curt. F. L. f. 2. Among the corn about Acton, Mr. Watson; *Blackst. Spec.* 23, quoted in *Mart. App. P. C.* 64, and *Huds.* ii. 239. A casual introduction, not since noticed. At one time a favourite garden annual, and 'annually cried about our streets under the name of *Red Morocco.*' (*Curt. F. L.*)]

MYOSURUS, *Linn.*

4. M. minimus, *L.* *Mouse-tail.* *Cauda muris* (Ger.). Cyb. Br. i. 76; Comp. 81. Curt. F. L. f. 4.

---

* *Ranunculus alpinus albus,* in a wood called Hampstead Wood; *Ger.* 805. May be this, but the figure represents the foreign *R. aconitifolius,* L.

Wet fields and roadsides ; rare. A. May—June.
IV. Pinner Drive, a single plant, F. W. Farrar ; *Melvill*, 2.
V. Near an orchard close to Perivale Farm ; *Lees*.
VI. Field bet. Edmonton and a house thereby, called Pims ; *Ger.* 346.
Edmonton, J. Woods ; *B. G.* 403. Bet. Edmonton and Enfield, 1819;
E. T. Bennett (v. s.).
VII. [Way from London to Hampstead, on a barren ditch bank ; *Ger.* 346.
In Chelsea meadow ; *Merrett*, 24. In a sloughy lane near the Devil's
house going to Hornsey ; *Pet. Midd.* About Hornsey ; *Blackst.*
MSS. Near Marylebone Park and meadows behind KentishTown
Chapel ; Lane from Copenhagen house to Kentish Town, Dr. Wilmer ;
Blackst. Spec. 57. Kentish Town ; *Dicks. Hort. Sic.* About Islington,
Paddington, and Pancras ; *Curt. F. L.* Wet fields by the Harrow
Road, near where the G. W. R. Terminus now stands, 1825 abund-
antly ; *Pamplin.* Harrow Road, 1833 ; *Herb. G.* and *R.* (Ken Wood,
Hunter ; *Park Hampst.* 29.)]

First record : *Gerarde*, 1597 ; also first as a British plant. Not met
with recently about London, though formerly common; the plant at
Pinner Drive was probably brought from elsewhere, with soil used
in making the road.

RANUNCULUS, *Linn.*

5. ? R. Drouetii, *Schultz.*
Cyb. Br. Supp. 49, 77 ; Comp. 81. Syme E. B. i. t. 21.

Ponds and ditches ; rather common ? P. April—June.
I. Pinner, W. M. H. ; *Melvill*, 2. Harefield ! ; *Newb.* Ruislip Moor.
V. Greenford (and ' *R. trichophyllus* '), W. M. H. ; *Melvill*, 2. Near
Shepherd's Bush Station ! ; *Newb.*
VI. Near Hadley ! ; *Warren.* Whetstone.
VII. Green Lanes, Newington ! ; *Newb.* New West End, Hampstead.

First record : *Hind*, 1860. It is difficult to decide whether the Middlesex
plant should be referred to this or *R. trichophyllus*, Chaix, if, indeed,
it be referable to either. There seems, however, to be but one form
in the county.

6. R. peltatus, *Fries (Syme).*   *Water Crowfoot.*
Ran. aquaticus hepaticæ facie, Lob. (Johns.). *R. aquatilis, Ger. em.*
(Blackst.). *R. aquatilis, L. (in part.).*
Cyb. Br. i. 77 ; Comp. 81.

a. R. floribundus, Bab. E. B. S. 2969, reproduced in Syme E. B. i. t. 18.
Ponds and still parts of streams ; very common. P. April—June.'
This is the plant to which, till lately, the name *R. aquatilis* was generally
applied by London botanists ; it is the commonest aquatic Ranunculus
in Middlesex. Professor Babington notices (*E. B. S. loc. cit.*) it as
' common on the north side of London.'

Occurs in all the districts commonly (except II.), and is especially frequent in I. IV. and VII. When growing on mud, its 'floating' leaves are often absent, and the flowers smaller. The form of the leaves in luxuriant specimens not unfrequently closely approaches that of *R. heterophyllus*, *Fries*; which plant, however, we have not observed in Middlesex.

A large plant, found at Harefield by Rev. W. W. Newbould, appears to be *R. pseudo-fluitans* of Syme. (See Syme *E. B.* i. 34.) The *R. fluitans* of *Melvill*, 2, seems the same. (See *Herb. Harr.*)

β, *R. peltatus*, E. B. S. 2965, reproduced in Syme E. B. i. t. 17 (drawn from a Hampstead specimen).

I. In many places: Stanmore Heath; Eastcott; Elstree reservoir; Woodready; South Mims.

III. Hatch End (a small-flowered form). Roxeth; *Hind.*

IV. Near Finchley!; *Newb.*

V. River Brent at Hanwell!; *Warren.*

VII. Hampstead Well Pond, 1862; Salter and Newbould; *E. B. S. loc. cit.* Ponds near Hornsey Wood.

This seems less common in the county than *R. floribundus*, with which it is often found, and from which we can see no marks sufficient to distinguish it. We have therefore united them, following Syme in the nomenclature.

First record: *Johnson*, 1629.

7. R. circinatus, *Sibth.*
Ran. aquat. alb., circinat. tenuiss. divis. fol., fl. ex alis long. pedic. innexis, Pluk. Phyt. (Blackst.).
Cyb. Br. i. 79; iii. 373 and 521; Comp. 81. Syme E. B. i. t. 15.

Ponds and slow streams; rare. P. June—August.

I. In a bog on Uxbridge Moor, plentifully; *Blackst. Fasc.* 83.

II. Staines Common. Bushey Park!; *Newb.* Queen's River, Hampton, abundant.

III. (Brook on road from Pinner to Harrow; *Melvill*, 2.*)

VI. Pond in garden of Warren Lodge, Edmonton.

VII. [Ditches in Tothill Fields; *Blackst. MSS.*] Reservoir, side of London canal, nearly opposite the gas-works; *Cherry (v.s.).*

First record: *Blackstone*, 1737.

8. R. fluitans, *Lam.*
Ran. aquat. alb. affine millefolium, maratriphyllon fluitans, Chabr. (Blackst.). *R. peucedanifolius, All.*
Cyb. Br. i. 78; iii. 374; Comp. 82. Syme E. B. i. t. 16.

* The specimen labelled '*R. circinatus*' in *Herb. Harr.*, being *R. Drouetii*, this locality is rendered doubtful.

Running water, rarely ponds; rather common. P. 'June—July.'
I. In Harefield River, abundantly; *Blackst. Fasc.* 85. Grand Junction
   Canal, Harefield. Colne, near Yewsley !; *Newb.*
II. Streams, Stanwell Moor.
III. Duke's River, Isleworth, and Twickenham, abundant, but does not
    flower. Cran on Hounslow Heath.
V. Pond near Kew Bridge Railway Station !; *Newb.*
VI. In the Lea, E. F.; *Herb. Mus. Brit.* Brook near Warren Lodge,
    Edmonton (this is, perhaps, *R. Bachii*, Wirtg.).
VII. River Lea, Hackney.

First record: *Blackstone.* 1737. Abounds in all the tributaries to the
Thames in this county, growing luxuriantly, but rarely flowering. On
Hounslow Heath we noticed a state growing on mud, in which the
segments of the leaves were short and curled, but which passed gradually into the deep-water form.

9. R. hederaceus, *L.*   *Ivy-leaved Water-Crowfoot.*
R. hed. aquaticus, Lugd. (Johns.).
Cyb. Br. i. 80 ; iii. 374 ; Comp. 82. E. B. 2003, reproduced in Syme
E. B. i. t. 26 (drawn from a Middlesex specimen).

Muddy places, marshes and ponds, common. P. May—August.
I. Harefield ; *Blackst. Fasc.* 83. Harrow Weald Common ; *Melvill,* 3.
   Stanmore Heath. Eastcott.
III. Hatch End. Hatton. Heston. Hounslow Heath. Whitton.
IV. Hampstead Heath !; *Johns. Enum.* Harrow Weald ; *Herb. Harr.*
V. Chiswick ; *Lawson.* Acton Green !; *Newb.*
VI. Hadley !; *Warren.*
VII. [Hyde Park and Tothill fields; *E. B.* Hyde Park, 1817 ; *Herb.*
    G. & R.] Walham Green ; *Morris* (*v. s.*) New West End and
    South Heath, Hampstead.

First record: *Johnson,* 1632. The large deep-water form is *R. cœnosus*,
Guss. (non Bab.). A figure is added in Syme E. B. i. t. 26.

10. R. sceleratus, *L.*   *Round-leaved Water-Crowfoot.*
R. rotundifol. forte apium risus (Johns.). *R. palustris rotundifol., Ger.*
em. (Blackst.).
Cyb. Br. i. 90 ; iii. 574 ; Comp. 85. Curt. F. L. f. 2.

Ditches, ponds, and wet places; very common. P. May—September.
In all the districts.
VII. Parson's Green. Near Chelsea Hospital. Ditch bet. Kensington
    Gardens and Hyde Park. Highbury. Victoria Park. Isle of
    Dogs. Kentish Town, &c.

First record: *Johnson,* 1629. Accidentally omitted in Melvill's *Harrow*
Flora.

11. R. Flammula, *L.*   *Lesser Spearwort.*

R. flammeus minor, Ger. em. (Blackst.).

Cyb. Br. i. 84; Comp. 83. Curt. F. L. f. 6.

Wet places and ponds, especially on heaths; common. P. June—August.

   I. Ditches on Harefield Common, plentifully!; *Blackst. Fasc.* 84. Harrow Weald Common; *Melvill,* 3. Eastcott.

  II. Bushey Park, by the ponds.

 III. Whitton Park. Hounslow Heath.

 IV. Hampstead Heath!; *Johns. Eric.* Finchley; *Newb.*

  V. Hanwell!; *Newb.* Brentford; *Cherry.*

 VI. Hadley!; *Warren.* Southgate, F. Y. Brocas; *Herb. Mus. Brit.* Forty Hill, Enfield.

VII. South Heath, Hampstead.

First record: *Johnson,* 1629.

(R. Lingua, *L. Great Spearwort.* Cyb. Br. i. 85; Comp. 83. Syme E. B. i. t. 31. IV. Hampstead Heath; *Cooper,* 101. It has been recently reported to us from the same place; but on inquiry *R. Flammula* was found to be the plant intended.)

Recorded from Iver Heath, Bucks; *Blackst. Fasc.,* 84, and Totteridge Green, Herts; *Phyt. N. S.* i. 391 and ii. 156; both spots close to our boundary, but in the latter introduced by Mr. Mackay.

12. R. Ficaria, *L.*   *Pilewort. Lesser Celandine.*

Chelidonium minus, Ger. em. (Johns. Blackst.).

Cyb. Br. i. 83; Comp. 83. Curt. F. L. f. 2.

Damp, shady places, common. P. March—May.

   I. Harefield!; *Blackst. Fasc.,* 18. Ruislip. Pinner.

  II. Staines.

 III. Hounslow Heath. Harrow Grove; *Herb. Harr.*

 IV. Hampstead Heath; *Johns. Eric.* Willesden; *Warren.* Near Harrow. Bishop's Wood. Stanmore.

  V. Chiswick, &c.!; *Newb.* Hanwell, abundant; *Warren.* Acton; Harlesden Green; *Cole.*

 VI. Hadley!; *Warren.* Whetstone. Edmonton.

VII. [Marylebone fields, 1817; *Herb. G. & R.*] Hackney; *Cherry.* Kensington Gardens, 1868; *Warren* (*v. s.*). Hornsey Wood; Kentish Town, &c.

First record: *Johnson,* 1629. The *var. β incumbens* of *Syme E. B.* occurred at (IV.) Hampstead. At (I.) Harefield we noticed a state with narrowly lanceolate petals and numerous bulbels in the axils of the leaves; and Blackstone (*Fasc.* 18) records it with double flowers.

13. R. auricomus, *L.*   *Goldilocks.*

R. nemorosus vel sylvaticus, fol. rotundo, C. B. Pin. (Blackst.).

Cyb. Br. i. 86; iii. 574; Comp. 83. Curt. F. L. f. 2.

Woods and hedges; rather common. P. April, May.

I. Harefield; *Blackst. Fasc.* 84. Pinner.

III. Hatch End.

IV. Hampstead; *Pet. H. B. Cat.* Bishop's Wood, abundant. Bentley Priory grounds. Hedges near Harrow. Mill Hill, Hendon; *Morris (v. s.).* Willesden; Finchley; *Newb.*

V. Near Ealing; *Hemsley.* Hanwell, near the Brent!; *Warren.*

VI. Whetstone. Colney Hatch. Enfield. Hadley!; *Warren.*

VII. [Marylebone fields, 1817; *Herb. G. and R.*] [On the muddy bank of the Thames, Isle of Dogs; *Herb. Brit. Mus.*] Hedge by Belsize Lane, 1860. West End Lane, 1868.

First record: *Petiver,* 1713. Rare, if not absent, in the S.W. parts of the county.

14. R. acris, *L.*   *Upright Buttercup.*
R. surrectis cauliculis, (Ger.). *R. pratensis erectus acris, C. B. Pin.* (Blackst.).
Cyb. Br. i. 87; Comp. 84. Curt. F. L. f. 1.

Meadows and roadsides; very common. P. June—August.
Distributed commonly through the county.

VII. The meadows round London contain quantities of it, and it may be gathered in Kensington Gardens, Victoria Park, on Primrose Hill, and similar places.

First record: *Gerarde,* 1597. [A *double-flowered variety* was noticed 'in the field next the Theater * by London;' *Ger.* 804.]

15. R. repens, *L.*   *Creeping Buttercup.*
R. pratensis repens, Park. (Blackst.).
Cyb. Br. i. 87; Comp. 84. Curt. F. L. f. 4.

Wet fields, sides of streams and ditches, and roadsides; very common. P. May—August.
In all the districts.

In VII. it is frequently found in the parks, squares, and other places, e.g., St. Paul's Churchyard. [A small form grew on the river wall of the Temple gardens previously to the embankment of the river.]

First record: *Blackstone,* 1737. *With double flowers.* IV. Northwick Walk, 1861; Football Field, near Harrow; *Melvill,* 3. Buddle distinguished a very large plant, nearly glabrous, as *Ran. pratensis repens major et glaber.* By the Thames side abundantly; *Herb. Budd.* 121. 30. It is figured in *Pet. H. B. Cat.* 38, *fig.* 8. [*R. Caleyanus* was found 'near London, in dry places, particularly in the late Mr. Caley's

---

* The first public theatre in or near London; built about 1570 in Shoreditch near Holywell. (See Stow's *Survey,* ed. i. (1598), p. 349; and Collier's *Annals of the Stage,* i. 227, and iii. 265.)

garden at Bayswater;' *Don Dichlam. Plants*, i. 37. This plant is said to have had a reflexed calyx. Mr. Syme considers it ' probably only a form of *repens*' (*E. B.* i. 41). We have seen no specimens.]

16. R. bulbosus, *L.* *Bulbous Buttercup.*
R. pratensis rad. verticilli modo rotunda, C. B. Pin. (Blackst.).
Cyb. Br. i. 88; iii. 374; Comp. 84. Curt. F. L. f. 1.
Rich meadows and pastures; very common. P. May—June.
Though less common than the two last, this species is also generally distributed.
VII. The meadows round London abound with it. Hyde Park. Isle of Dogs. Highbury. Kensington Gardens, 1868; *Warren* (*v. s.*).
First record: *Johnson*, 1632. *With white flowers*; VII. Brompton Cemetery, by the catacombs, 1859-63, J. Britten; *Phyt. N. S.* vi. 592.

17. R. hirsutus, *Curt.*
R. rectus, fol. pallidioribus hirsutis, J. B. (Merrett). *Ran. annuus trilobatus lucidus* (Doody).
Cyb. Br. i. 89; iii. 374; Comp. 84. Curt. F. L. f. 2.*
Waste land in damp places; very rare. A. June—September.
V. Acton Green !, 1867; *Newb.*
VII. Below Hampstead, in the meadows betwixt the town and the heath; *Merrett*, 102. [About Paddington; *Doody, MSS.*] Field bet. Hampstead and Kentish Town, E. H. Button; *Cooper Supp.* 11. Chelsea College, 1861; *Britten.*
First record: *Merrett*, 1666; also the earliest notice of the plant as British. The Acton plant is *R. parvulus*, L, a small form of *hirsutus.*

18. R. arvensis, *L.* *Corn Crowfoot. Hedgehogs.*
R. arv. echinatus, C. B. Pin. (Blackst.).
Cyb. Br. i. 91; Comp. 85. Curt. F. L. f. 6.
Cornfields and waste ground; common. A. June—August.
I. Cornfields, Harefield, frequent; *Blackst. Fasc.* 84. Wood Hall, Pinner, abundant. Harrow Weald Common. Near Elstree. Uxbridge; *Cole.*
II. Near Staines, 1841; *Herb. Young.* A single plant, Staines, 1866. Stanwell Moor, 1867.
III. Cornfields near Hounslow; *Hemsley.* A few plants by the roadside near Heston. Isleworth; *Cole.*
IV. Stanmore, 1858; Pinner Drive; *Melvill,* 3. Fields by Hampstead Wood; *Cooper,* 101. Harrow Weald.
V. Greenford, abundant; *Melvill,* 3.
VI. Edmonton. Enfield.
VII. [Near St. John's Wood Chapel, 1815: *Herb. G. & R.*] Kentish Town; *Herb. Hardw.* Waste ground, St. Mark's, Chelsea !; *Newb.*

* Curtis's plate does not figure the characteristic achenia.

First record: *Blackstone*, 1737.　This is usually only a roadside plant, sprung from stray seed, but occasionally it becomes a cornfield pest.

19.　R. parviflorus, *L.*
R. hirsutus annuus fl. minimo, Ray (Pet. Blackst.).
Cyb. Br. i. 90; Comp. 35.　Syme E. B. i. t. 37.

Dry places and cornfields; very rare.　A.　May, June.
I. Several places near Harefield, plentifully; *Blackst. Fasc.* 84.　Page's Lane, Uxbridge, 1839, H. Kingsley; *Herb. Young.*
III. In a lane near Thistleworth (=Isleworth), Mr. Doody; *Pet. Midd.*
VII. [About Hackney; *Blackst. Spec.* 80.]　[Kentish Town; *Mart. App. P. C.* 72.]

First record: *Doody*, 1695.　Not very recently met with.

CALTHA, *Linn.*

20.　C. palustris, *L.*　　*Marsh Marigold.*
C. palustris major, Ger. em. (Blackst.).
Cyb. Br. i. 91; iii. 374; Comp. 85.　Syme E. B. i. t. 40.

Banks of streams, wet fields, and marshes; common.　P.　March—May.
I. Harefield!; *Blackst. Fasc.* 13.　Ruislip Moor.　Pinner, 'introduced,' W. M. H.; *Melvill*, 4.　Stream at Yewsley!; *Newb.*
II. Staines.　Bet. Hampton and Sunbury!; *Newb.*　Teddington.
III. By the Thames, Twickenham.　By the Cran, Hounslow Heath.　Duke's River, Isleworth.
IV. Hampstead Heath!; *Johns. Eric.*　Hendon; Harlesden Green; *Cole.*　Bog in Turner's Wood.　Near Finchley!; *Newb.* (the small-flowered plant).
V. Greenford; *Cooper*, 108.　Hammersmith; *Cole.*
VI. Edmonton.　The Alders, Whetstone (a small-flowered plant, but not *C. Guerangerii*, Bor.).
VII. Eel-brook Common, Parson's Green.　Ken Wood pastures, abundant.　Side of London Canal, Hackney; *Cherry.*

First record: *Johnson*, 1629.　*With double flowers;* Harefield; *Blackst. Fasc.* 13.　*C. riparia, Don*; on the banks of the Thames in marshes; *Don Dichlam. Plants.* i. 44.　Referred to var. β. *Guerangerii*, by Syme (*E. B.* i. 51.) Does not differ greatly from the type. We have not met with it in Middlesex.

HELLEBORUS, *Linn.*

21.　H. viridis, *L.*　　*Bearsfoot.*
Cyb. Br. i. 94; iii. 376; Comp. 86.　Curt. F. L. f. 6 (probably drawn from a Finchley plant).

Woods and copses; very rare.　P.　February—April.

I. Near Harefield, 1791, Miss Jane Baynes; *Herb. Linn. Soc. & Fl. Brit.*
ii. 598.

III. Down Barn Hill, near Harrow, J. Woods; *B. G.* 406.

IV. In Caper's Wood bet. the 7 and 8 mile stones in the Mill Hill
Road, 1765; *MS. note by Michael Collinson in his father's (Peter
Collinson) copy of Blackst. Fasc.* In a small wood near Finchley,
discovered by Mr. Jacob Rayer; *Curt. F. L.* (probably the same
locality).

(VII. Ken-wood, Hunter; *Park Hampst.* 30.)

First record: *Michael Collinson*, 1765. Perhaps originally planted in its
few localities.

[AQUILEGIA, *Linn.*

22. A. vulgaris, *L.* *Columbine.*
A. sylvestris, C. B. P. (Blackst.).
Cyb. Br. i. 96; iii. 376; Comp. 86. Syme E. B. i. t. 46.

Woods and thickets; very rare. P. June.

I. About Harefield, rarely; *Blackst. Fasc.* 7.

No other record, but no reason to doubt the accuracy of this one.]

Delphinium Consolida, *L.* *Larkspur.* Cyb. Br. i. 97; iii. 377 (includes
D. Ajacis, Reich.). Syme E. B. i. t. 47 (small figure). VII. Waste
ground about Chelsea and Brompton, frequent; *Britten.* From gardens, probably.

Aconitum Napellus, *L.* *Monkshood.* Cyb. Br. i. 97; Comp. 87. Syme
E. B. i. t. 48. III. In a plantation adjoining Hounslow Heath, 1862;
Hemsley (*v. s.*). No doubt of garden origin.

BERBERIDACEÆ.

BERBERIS, *Linn.*

23. B. vulgaris, *L.* *Barberry.*
Spina acida sive Oxyacantha (Ger.).
Cyb. Br. i. 391; Comp. 87. Syme E. B. i. t. 51.

Hedges, &c.; rather common. Shrub. May—June.

I. ? At Iver . . . most of the hedges are nothing else but Barberry
bushes; *Ger.* 1144. (Iver village is in Bucks.)

II. Road from Staines to Hampton, near Ashford.

III. Many places about Twickenham; Marsh Farm, North End. Side of
Duke's River, near Isleworth.

IV. Hampstead Heath; *Mart. App. P. C.* 66.

V. Gunnersbury Lane, Ealing, 1821; *Herb. G. & R.* Perivale; *Lees.*

VI. Footpath from East Barnet to Whetstone; *Fl. Herts Supp* 1.
VII. Bet. Hampstead Heath and Belsize, E. H. Button; *Cooper Supp.* 11.
[Hedges, Primrose Hill; *Lond. Flor.* 186.] Kentish Town, 1845;
Herb. Hardw.

First record : *Gerarde*, 1597. The above localities are those in which the
plant seems most likely to be native. It occurs frequently, evidently
introduced, in plantations, churchyards, garden hedges, &c.

NYMPHÆACEÆ.

NYMPHÆA, *Linn*

24. N. alba, *L.*   *White Water-Lily.*
N. a. major vulgaris, Park. (Blackst.).
Cyb. Br. i. 100; iii. 378; Comp. 88.   Syme E. B. i. t. 53.
Ponds, and still parts of streams ; rather common.   P.   July.
I. In Windsor Lake, on Uxbridge Moor, plentifully ; *Blackst. Fasc.* 65.
In Harefield river, plentifully ; *Blackst. Spec.* 60.
II. Staines! ; *Newb.* Thames, off Laleham, 1815 ; *Herb. G. and R.*   Pond
by Walton Bridge. In the waters of Hampton Court, very luxuriant.
In a large piece of water adjoining Staines Road, near Twickenham
Common.
III. Ornamental water, Twickenham Park.   Pond at Roxeth, doubtfully
wild ; *Melvill,* 4.
V. In the rivulet at Brent Bridge, in the Uxbridge Road * ; *Blackst.*
Spec. 60. Canal bet. Brentford and Hanwell ; *Hemsley.* Northolt ;
Melvill, 4.   Greenford ; *Cooper,* 108.   Ornamental water, Sion
Park.
VI. Ponds, Enfield Chase ; *Phyt. N. S.* vi. 301.   Ponds, Edmonton.
VII. In the moat at Fulham ; *Mart. App. P. C.* 66.   Ponds, Ken wood,
Hunter ; *Park Hampst.* 30.
First record : *Blackstone,* 1737.   No doubt a native, but planted in very
many of its localities for ornament.

NUPHAR, *Linn.*

25. N. lutea, *Sm.*   *Yellow Water-Lily.*   *Brandy-Bottle.*
Nymphæa lutea, Ger. em. (Blackst.).
Cyb. Br. i. 101; Comp. 88.   Syme E. B. i. t. 54.
Rivers and ponds ; rather common.   P.   July.
I. Uxbridge Moor, Harefield river, and elsewhere ; *Blackst. Fasc.* 65.
II. Colne, at Stanwell and Staines.   Thames bet. Twickenham and

---

* This may be the place where Lobel observed a small form.   (See *Lob. Adv.* 257.)

Teddington, and bet. Kingston and Hampton. Pond on common by
Walton Bridge.   Hampton Court Gardens.

III. Cran, in several places.   Duke's River, Isleworth, Twickenham, &c.

IV. Brent, at Willesden ; *Herb. Hardw.*

V. Brent, at Greenford ; *Cooper*, 108.   Canal, Twyford! ; *Newb.*   Brent
Bridge ; *Forst. Midd.*   Canal bet. Brentford and Hanwell; *Hemsley.*
Brent, at Hanwell! ; *Newb.*

VI. Ponds, Enfield Chase ; *Phyt. N. S.* vi. 301.   Lea, near Stratford.
Freeman ; *Fl. Essex*, 13.   Edmonton.   Pond at Whetstone.

VII. Moat at Fulham, near the garden gate ; *Pet. Midd.*   Ken Wood ponds.

First record : *Petiver*, 1695.   Often planted, but undoubtedly a native.

PAPAVERACEÆ.

PAPAVER, *Linn.*

26. P. Argemone, *L.*

Argemone capitulo longiore (Johns.).   *Pap. laciniato fol. capitulo hispido
longiore, R. Syn.* (Blackst.).

Cyb. Br. i. 103 ; Comp. 89. Curt. F. L. f. 5.

Cornfields and waste ground ; rare. A. June—August.

I. Harefield ! ; *Blackst. Fasc.* 71.   Bet. Harefield and Uxbridge ; *Cole.*

II. Bet. W. Drayton and Yewsley! ; *Newb.*   Sunbury.

V. Sandy field, Chiswick !; Acton! ; *Newb.*

VII. [Chelsea and Hammersmith fields, Mr. Robert Lorkin and I ; *Johns.
Ger.* 373.   Ibid. copiously ; *Morison*, 2, 279.   Cornfields about
Chelsea ; *Pet. Midd.*]

First record : *Johnson*, 1633.

27. P. hybridum, *L.*

Argemone capitulo torulo (Johns.).   *P. laciniato fol. capit. hispido rotundiore, R. Syn.* (Blackst.).

Cyb. Br. i 103 ; Comp. 88.   Syme E. B. i. t. 62.

Cornfields and walls ; very rare. A. June—August.

I. Frequent among corn, Harefield ; *Blackst. Fasc.* 71.

III. About 6 plants on a wall, S. side of road bet. Sion House and Wood
Lane, Isleworth, 1866 ; *Cole.*

VII. [Chelsea and Hammersmith fields, Mr. Robert Lorkin and I ; *Johns.
Ger.* 373.   Bet. London and Chelsea ; *Morison*, 2, 279.   Cornfields
about Chelsea ; *Pet. Midd.*]

First record : *Johnson*, 1633.

28. P. Rhœas, *L.*   *Common Poppy.   Corn Rose.*

Cyb. Br. i. 105 ; Comp. 89.   Curt. F. L. f. 3.

Cornfields, roadsides and walls ; very common. A. June—August.

Occurs in all the districts, and is generally abundant.

VII. On the South Heath, Hampstead.

First record: *Blackstone,* 1737.

29. P. dubium, *L.*

Pap. laciniato fol. capit. longiore glabro, R. Syn. (Pet.).

Cyb. Br. i. 104; iii. 373; Comp. 89. Curt. F. L. f. 5 (drawn from a London specimen).

Cultivated and waste ground, and tops of walls; common. A. May—July.

I. Wall at Pinner. Sandpits, Pinner; *Melvill,* 5.

II. Bet. Kingston Bridge and Hampton Court.

III. Very abundant on walls at Isleworth. Waste ground at Twickenham. Wall at Harrow.

V. Walls about Chiswick, Turnham Green, and Brentford, abundant. Gravel-pit near Hanwell!; *Newb.*

VI. Walls at Enfield.

VII. Maiden Lane, Islington; *Herb. Hardw.* Cornfields about Chelsea; *Pet. Midd.*

Var. β P. Lecoquii, Lamotte. Syme E. B. i. t. 60.

I. not uncommonly about Pinner; *Hind.*; III. garden at Roxeth (a single plant) W. M. H.; *Melvill,* 5. We have seen no specimens.

First record: *Petiver,* 1695. The tops of the numerous high brick walls in the W. suburbs of town bear a large crop of this plant annually. It grows to a height of 3 feet, but some specimens are but an inch or two. All that we have seen must be referred to the variety *a. P. Lamottei,* Bor. of *Syme E. B.* i. 89.

P. somniferum, *L. White or Garden Poppy.* Cyb. Br. i. 106. Syme E. B. i. t. 57. I. Island in the Colne, near Iver, 1848; *Morris (v. s.).* III. About Isleworth and Twickenham, in many places. IV. Hampstead; *Irv. MSS.* VII. Parson's Green; *Britten.* Near Cremorne. Back of Adelaide Road, N.W. Both the forms, *P. hortense* and *P. officinale,* have been noticed. The plant seems only a straggler from the garden, and does not become a cornfield weed, as in parts of Kent and Surrey.

Meconopsis cambrica, *Vig.* Syme E. B. i. t. 63. I. Questionably naturalised at Pinner, Hind; *Melvill,* 5. Was this in the Parsonage garden? A plant of Western Europe (v. *Comp.* 89).

Glaucium luteum, *Scop.* Comp. 90. Syme E. B. i. t. 66. VII. On the site of the International Exhibition of 1862, in 1865. A seashore plant.

Glaucium phœniceum, *Crantz.* Glaucium corniculatum, *Curt.* (Syme E. B.) E. B. 1433, reproduced in Syme E. B. i. t. 65 (drawn from a

Middlesex specimen). I. Spontaneously in the Parsonage garden, Pinner, 1865; *Hind.* VII. In Chelsea garden it has, from time immemorial, come up every year as a weed; *E. B.*

CHELIDONIUM, *Linn.*

30. C. majus, *L.* *Celandine.*

Cyb. Br. i. 107; Comp. 90. Syme E. B. i. t. 67.

Hedges and old walls, also a garden weed; common. P. May—August.

I. Harefield!; *Blackst. Fasc.* 18. Pinner; *Melvill,* 5. Near Elstree. Harrow Weald Common, probably introduced.

II. Common in this district. Sunbury, Hampton, Bushey Park, Teddington.

III. Isleworth!; *Herb. G. & R.* About Twickenham, frequent.

IV. Garden hedge at the Spaniards, Hampstead Heath; *Burnett,* 2, 86. Stanmore.

V. Turnham Green. Greenford; *Melvill,* 5. Perivale; *Lees.*

VI. Crouch End, Hornsey, 1833; *Herb. Young.*

VII. Wall of kitchen garden, Ken Wood. Near Downshire Hill; *Burnett,* 2, 86. Bishop's Walk, Fulham; *Britten.* Parson's Green.

First record: *Blackstone,* 1737. In some of the above stations it has doubtless sprung from the rubbish of neighbouring gardens. *With semi-double flowers.* (V.) Hedge at Turnham Green, with the usual form.

FUMARIACEÆ.

CORYDALIS, *Cand.*

31. * C. solida, *Hook.*

Cyb. Br. i. 110; iii. 379. Syme E. B. i. t. 68.

I. Naturalised . . . at Uxbridge; *Syme, E. B.* i. 101.

Also occurs 'perfectly naturalised' in a wood (Puget's Wood) on the north side of Totteridge Green, Herts, W. P., A. I., and J. R. M.; *Phyt. N. S.* i. 391, and ii. 157.

32. * C. lutea, *DC.*

Cyb. Br. i. 110. Syme E. B. i. t. 69.

On old garden walls. P. May—September.

I. Harrow Weald, *Varenne.* Uxbridge, for several years; *Cole.*

II. Harmondsworth. Sunbury.

IV. Miss Hill's garden wall, Harrow; *Melvill,* 6.

V. Rev. R. Lindsey's garden wall, Ealing; *Lond. Flor.* 174.

VI. Warren Lodge, Edmonton.

First record: *Varenne,* 1827. Scarcely naturalised.

[C. claviculata, *DC. Fumaria alba latifolia, Park.* (Blackst.). Cyb.
Br. i. 109 ; iii. 379 ; Comp. 90. Syme E. B. i. t. 70. VII. In the
hedges near Bonner's Row, Bethnal Green ; *MS. note (Alchorne's)
quoted in Phyt.* iii. 166. The only record. Possibly a rampant
Fumaria mistaken for it. It does not occur in Herts.]

FUMARIA, *Linn.*

33. F. capreolata, *L.*
Cyb. Br. i. 111 ; iii. 379 ; Comp. 90.

1. *F. pallidiflora, Jord.* Syme E. B. i. t. 71.
Hedges and bushy places ; very rare. A. July.
III. By the Duke's River at Chase Bridge, Whitton. Near Harrow,
W. M. H. ; *Melvill,* 6.
IV. Near Kingsbury Reservoir ; *Farrar.*

2. *F. Boræi, Jord. F. pallidiflora, var. β.* (Bab. Man.). *F. capreolata,*
Curt. F. L. f. 6.
VI. Sparingly near Edmonton ; *Curt. F. L.*
Curtis very likely confounded other forms with *Boræi,* which, however,
his plate represents well.

3. *F. muralis, Sond.* Syme E. B. i. t. 74.
Hedges on a sandy soil ; rare. A. July—September.
II. Bet. Teddington Station and Bushey Park. Fulwell.
III. Twickenham, near the railway station.
V. Bet. Turnham Green and Brentford, on hedgebanks in the market
gardens ; *Newb.* Probably *F. muralis.*
This subspecies is the most frequent in the neighbourhood of London.
First record: *Curtis,* about 1798.

34. F. officinalis, *L.*   *Fumitory.*
F. vulgaris (Johns). *F. officinarum et Dioscoridis, C. B. Pin.* (Blackst.).
Cyb. Br. i. 111 ; Comp. 91. Curt. F. L. f. 2.

Cultivated land and hedgebanks, especially on a light sandy soil ; com-
mon. A. May—September.
I. Harefield ! ; *Blackst. Fasc.* 30.
II. Hampton. Teddington.
III. Isleworth. Heston. Hounslow. Twickenham.
IV. Clay-pits at Harrow Weald ; *Melvill,* 6. Hampstead Heath ; *Johns.
Enum.*
V. Common in this district, especially in the market gardens.
VI. Edmonton.
VII. [Marylebone fields, 1815 ; *Herb. G. & R.*]

First record: *Johnson*, 1632. The larger rampant form is, probably, most common. On waste ground near the King's Road, Chelsea, we noticed a small-flowered plant, which is perhaps *F. Wirtgeni*, Koch.

35. F. micrantha, *Lagasca.*
    Cyb. Br. i. 112; iii. 315; Comp. 91. Syme E. B. i. t. 75.

Cultivated land; rare. A. June—September.
II. Near Hampton, in garden ground!; *Newb.*
V. Near Shepherd's Bush Ry. Station!; bet. Acton and Turnham Green!; near Ealing Cemetery!; hedgebank in the market gardens bet. Turnham Green and Brentford!; *Newb.*

First record: *Newbould*, 1866.

CRUCIFERÆ.

CHEIRANTHUS, *Linn.*

36. *C. Cheiri, L.*   *Wallflower.*
Leucoium luteum, C. B. Pin. (Blackst.).
    Cyb. Br. i. 155; iii. 385. Syme E. B. i. t. 106.

Old garden walls. P. April—June.
I. Harefield; *Blackst. Fasc.* 51.
III. Walls, Isleworth. Harrow, evident escape; *Melvill*, 6.
V. Chiswick; *Cole.* Perhaps naturalized; *Newb.*
VI. Wood Green; *Cole.* Hornsey Lane; *Herb. Hardw.* [Bury St., Edmonton.]
VII. Hampstead; *Irv. MSS.*

First record: *Blackstone*, 1737. We cannot point out a locality where this grows wild in Middlesex; it is scarcely naturalised.

NASTURTIUM, *R. Br.*

37. N. officinale, *R. Br.*   *Water Cresses.*
N. aquaticum supinum, C. B. Pin. (Blackst.). *Sisymbrium nasturtium, L.* (Curt.).
    Cyb. Br. i. 147; iii. 384; Comp. 100. Curt. F. L. f. 6.

Ditches, ponds, and streams; very common. P. June—September.
Generally distributed through all the districts.
VII. In Hyde Park, 1817; *Herb. G. & R.* Primrose Hill, 1845; *Morris*, (*v. s.*). Homerton; *Cherry.* South Heath, Hampstead, and other places.

First record: *Blackstone*, 1737. A very large form, perhaps the variety *siifolium* of *Syme E. B.* i. 177, was noticed (II.) in the streams on Stanwell Moor, and (III.) by the Thames, near Richmond Bridge. The latter plant had a stem more than 3 ft. long and ¾ in. thick.

This plant is extensively cultivated in the northern and eastern suburbs, in shallow pits of large size, intersected and divided into rectangular portions, by narrow dikes. It is gathered in early morning, and cried about the streets for sale, mostly by children.

38. N. sylvestre, *R. Br.*

Eruca aquatica, Ger. (Pet.). *Sisymbrium sylvestre, L.* (Mart.).

Cyb. Br. i. 148 ; iii. 384 ; Comp. 101. Curt. F. L. f. 3 (drawn from a London specimen).

Wet roadsides and river banks ; rather common. P. June—September.

I. Ruislip, Hind ; *Melvill,* 8. Eastcott. Moss Lane, Pinner.

II. Thames side, Teddington; *Newton MSS.* By the Thames side, generally, from Sunbury to Teddington.

III. River side, Twickenham.

IV. Stanmore. Brick field, Burgess Hill, Hampstead.

V. Chiswick.

VII. [In a ditch in the road bet. Whitechapel and Mile End ; *Pet. Midd.*] [In the ditches about Tothill Fields, Westminster ; *Blackst. Spec.* 20. Very abundant there ; *Curt. F. L.* and *Herb. Forst.*] Hammersmith ; *E. B.* 2324. By the Thames at Fulham ; *Britten.* Bet. Kensington Square and Chelsea, 1820 ; *Pamplin.* Inside of Kensington Garden railings, S.E. corner. South Heath, Hampstead.

First record : *Petiver,* 1695.

39. N. palustre, *DC.*

Sisymbrium terrestre, Sm. (Curt.). *N. terrestre, R. Br.* (Melv.).

Cyb. Br. i. 147 ; iii. 384 ; Comp. 101. E. B. 1747 (drawn from a Tothill Fields specimen); but Curt. F. L. f. 5 is a better figure.

Damp ground, waste or cultivated, old walls, &c. ; common. P. June—September.

I. Near Eastcott ; near Ruislip Common ; Woodready, Pinner ; *Melvill,* 7. Stanmore Heath. South Mims.

II. Staines Moor. Feltham Green.

III. Headstone ; *Melvill,* 7. Hounslow Heath. Whitton. Hatton. Twickenham. Isleworth.

IV. Kingsbury, 1827–30 ; *Varenne.* Harrow Weald; *Melvill,* 7. North Heath, Hampstead.

V. Hanwell, 1845 ; *Morris* (*v. s.*). Ealing Common ! ; *Newb.* Brentford.

VI. Edmonton.

VII. [Tothill Fields ; *Curt. F. L.* Where the Penitentiary now stands ; *Pamplin.*] Kensington ; *Pamplin.* [On Blackfriars Bridge, 1815] ; Uxbridge Road, 1816 ; *Herb. G. & R.* Eel-brook common, abundant, 1862. Chelsea. Near Buckingham Palace. South Heath, Hampstead. Kentish Town ; *Herb. Hardw.* [Thames Embankment, 1866, and formerly on the river wall of Temple Gardens.] Isle of Dogs ! ; *Newb.* Hackney Wick.

CRUCIFERÆ.

First record: *Curtis*, about 1780. There are specimens in Buddle's Herbarium, v. 123, f. 11. He distinguished it from *N. amphibium*, with which Linnæus afterwards combined it.

BARBAREA, *R. Br.*

40. B. vulgaris, *R. Br.* B. eu-vulgaris (Syme). *Winter Cress. Yellow Rocket.*

Barbarea, *Ger.* (Johns. Park.).

Cyb. Br. i. 145; Comp. 100. Syme E. B. i. t. 120.

Hedgebanks, roadsides, meadows; common. B. or P. May—July.

I. Harefield!; *Blackst. Fasc.* 9. Elstree reservoir. Wood Hall, Pinner (a very slender form). South Mims.

II. Staines. Bet. Kingston Bridge and Hampton Court, by the river.

III. Hatch End, near Pinner. Hounslow. Twickenham.

IV. Bet. Whetstone and Totteridge. Near Finchley!; *Newb.*

V. Acton Green!; *Newb.* Hanwell!; *Warren.*

VI. Edmonton. Whetstone. Hadley!; *Warren.* Near Enfield; *Tucker.*

VII. [In the next pasture to the conduit head behind Grayo's Inne that bringeth water to Mr. Lambe's Conduit in Holborne; *Park. Theat.* 820.] Bet. Kentish Town and Hampstead; *Johns. Eric.* Highgate Cemetery. Regent's Park. Hackney Wick.

First record: *Johnson*, 1629. The form with slender arched pods, *divaricata*, Lond. Cat., occurs (II.) near Hampton Court, by the Thames. It is often mistaken for *B. arcuata*, Reich., and is probably the plant intended by that name found at (III.) Roxeth; *Melvill*, 6; and (IV.) in the brook dividing Middlesex and Herts, near Totteridge; *Fl. Herts Supp.* 6.

B. stricta, *Andrz.* Syme E. B. ii. t. 122. II. By the towing-path bet. Kingston Bridge and Hampton Court, in seed, 1867; *Bloxam.* Requires confirmation.

41.* B. præcox, *R. Br.* *American Cress.*

Cyb. Br. i. 146. Syme E. B. i. t. 124.

Waste ground and roadsides; also a garden weed; rather common. B. May, June.

I. Ruislip. Pinner; *Melvill*, 7.

III. Hedgebank, Bath road, near 11-mile stone. In clover, Marsh Farm, Twickenham. Roxeth; *Melvill*, 7.

IV. Harrow Weald Churchyard. Hampstead; *Irv. MSS.*

V. Apperton. Northolt; *Melvill*, 7. About Brentford; *Irv., H. B. P.*, 694. Turnham Green.

VI. Near Trent Park, Southgate; *Cole.*

VII. Near St. Mark's College!; *Newb.* Site of International Exhibition, abundant.

First record: *Irvine*, about 1830. Originally cultivated, now quite naturalised.

TURRITIS, *Linn.*

42. T. glabra, *L.* Arabis perfoliata, *Lam.* (Syme E. B.). *Tower Mustard.*
Turritis (Ger.).

Cyb. Br. i. 144 ; iii. 383; Comp. 100. Curt. F. L. f. 4.

Dry roadsides and hedgebanks; very rare. B. June, July.

III. In a lane near Thistleworth, Mr. Doody; *Pet. Midd.*

 V. A few specimens out of flower bet. Turnham Green and Brentford,
  about 1860 ; *Newb.*

 VI. At Pyms, by a village called Edmonton ; *Ger.* 213.

First record: *Gerarde*, 1597 *;* also the earliest for the plant as British.
Occurs just out of the county bet. Harefield and Rickmansworth, Herts.

[Arabis turrita, *L.* Syme E. B. i. t. 118. VII. ' I sowed seeds on the
 north wall of Kensington Gardens, east of Bayswater Gate, on the
 side next the gardens,' J. Denson; *Mag. Nat. Hist.* ix. 90.]

CARDAMINE, *Linn.*

C. impatiens, *L.*   *Nasturtium minimum annuum fl. albo* (Morison).
Card. impatiens, vulgo Sium minus impatiens, Ger. em. (Pet.).   Cyb.
Br. i. 139; iii. 383; Comp. 99. Syme E. B. i. t. 12.

VII. [Near London, by the first stone going from S. Ægidius' Church *
 towards Hampstead; *Morison*, 2, 222.] On the moat-sides near the
 garden-gate at Fulham ; *Pet. Midd.* By the Thames side, near the
 Physic Garden, Chelsea, plentifully; Dr. Watson ; *Hill*, 334 : copied
 in *Mart., App. P. C.* 65. By the ditch-sides in Hell (=Eel) Brook, at
 Parson's Green, plentifully ; *MS. note (by Stanesby Alchorne) in a
 copy of Blackst. Spec. quoted in Phyt.* iii. 166.

First record: *Morison*, 1680. It is not unlikely that some other plant
was mistaken for this in the above stations ; probably *C. sylvatica*, as
well as, perhaps, *C. amara* ; Johnson, however, speaks of it (*Johns.
Ger.* 260) as cultivated in his time, and it may have been a garden
escape. There is no recent record of *C. impatiens* in Middlesex, and
it is marked as doubtfully native in the N. Thames sub-province of
Cyb. Br. Supp. It occurs, however, in Surrey.

43. C. hirsuta, *L.*   *Small Lady's Smock.*
Comp. 98.

 1. *C. sylvatica*, Link.

C. impatiens altera hirsutior, R. Hist. (Newt., Pet., Blackst.). *C. hirsuta*
(Curt.).   Cyb. Br. iii. 318. Curt. F. L. f. 4 (drawn from a Middlesex
specimen).

Ditch banks, wet roadsides, and woods ; common. A., B. or P. April—
September.

     * St. Giles-in-the-Fields.

I. Scarlet Spring, and Gutter's Dean Woods, near Harefield; *Blackst. Fasc.* 14. Near Yewsley!; *Newb.* Eastcott. Pinner, &c.
II. Near Hampton Court; *Newb.* Near Charlton.
III. Hatton.
IV. Near Finchley!; *Newb.* Stanmore. Scratch Wood. Bishop's Wood.
V. Hanwell!; near Willesden Junction!; *Warren.*
VI. Enfield. Edmonton.
VII. Ditch bet. Cane Wood and the close newly stubbed up; by the water a little beyond Pancras towards Kentish Town; *Newt. MSS.* New River banks bet. Canberry House and Newington; *Pet. Midd.* [Lane opp. Mother Red-Cap's, Dr. Watson; *Hill,* 334.] Chelsea Water-works; about Highgate and Hampstead; *Curt. F. L.* Walham Green; *Morris* (*v. s.*). Millfield Lane.

2. *C. hirsuta,* L. (Bab.). C. eu-hirsuta (Syme E. B.).
Cyb. Br. iii. 317. Syme E. B. i. t. 110.

Dry places, garden-walks, walls, &c.; rather rare. A. April—August.
I. Pinner. Ruislip; *Melvill,* 7.
IV. Near Stanmore Church.
V. Near Willesden Junction!; *Warren.* Ealing Common; canal-bank, Twyford!; *Newb.*
VI. Hadley!; *Warren.*
Certainly less common than *C. sylvatica.* As a garden weed abundant at Totteridge, Herts, just beyond our bounds.
First record: *Petiver,* 1695.

44. C. pratensis, *L.* *Lady's Smock. Cuckoo Flower.*
Cardamine, Ger. em. (Blackst.). *C. altera, Lob.* (Johns.).
Cyb. Br. i. 138; iii. 383; Comp. 98. Syme E. B. i. t. 109.

Rich, moist meadows and heaths; very common. P. May.
Occurs in all the districts, and is abundant in I., IV., and VI.; rarer in the other districts, and rather scarce in V., on light soil.
VII. Primrose Hill. West End Lane. Kentish Town. Hackney Marshes, &c.
First record: *Johnson,* 1629.

45. C. amara, *L.* *Bitter Cress.*
C. fl. majore elatior, J. B. (Blackst.). *C. major et ramosior* (R. Syn.). *Nasturtium aquaticum annuum, Park.* (Pet.).
Cyb. Br. i. 137; iii. 382; Comp. 98. Curt. F. L. f. 3.

Wet places in woods and river banks; rare. P. May—July.
I. River-side near Harefield; plentifully about Uxbridge; *Blackst. Fasc.* 14. Neighbourhood of Uxbridge; *Curt. F. L.* Uxbridge; *Cole.*
II. Thames-side, bet. Hampton and Hampton Court!; *Newb.*

IV. Wood near the 'Spaniards,' Hampstead ; *Irv. MSS.* Turner's Wood, abundant in the bog. Kingsbury, by the brook, 1827–30; *Varenne.*

V. Canal-side, Greenford, Hind; *Melvill*, 7. Island in the Thames opposite Kew Palace, 1837 ; *Herb. Young.*

VII. Osier-holts by the Thames-side, over against my Lord Bishop of London's garden at Fulham, W. Sherard; *R. Syn.* i. 238. Mr. Sweet told me he used to find it there; *Pamplin.* [Banks of Thames bet. Peterborough House and Chelsea; *Pet. Midd.*] [Isle of Dogs, J. Woods; *B. G.* 408.]

First record: *W. Sherard*, 1690.

DENTARIA, *Linn.*

46. D. bulbifera, *L.* Cardamine bulbifera (Syme E. B.) *Coral Wort.* Cyb. Br. i. 136; Comp. 98. Brit. Ent. 144 (drawn from a Harefield specimen.)

Woods and groves ; very rare. P. May.

I. In the old Park Wood near Harefield, abundantly!; *Blackst. Fasc.*, 23. In a grove by Harefield Church, 1853 ; *Herb. Hardw.* 1866! *Cole and Griffiths.* Garret Wood, plentiful (partly in Herts), 1855; *Phyt. N. S.* i. 62.

First record: *Blackstone*, '1734' (see MS. note in his copy of *Johns. Merc. Bot.*). It had been previously noticed in Sussex (v. *Park. Th. Bot.* 620); but Blackstone (*Spec.* 17) thought that Parkinson ' mistook the plant' there. Since then, however, it has been found in several parts of that county. It has been constantly gathered at Harefield since its first discovery. In 1752, Mr. W. Watson exhibited specimens collected there to the Royal Society. (See *Phil. Trans.* 47, 428.) We have also seen specimens dated 1819, *J. J. Bennett*; 1840, *H. Kingsley*; 1842, *Dr. Bromfield*; 1843, *Mr. Henfrey.* It still grows there in profusion, amongst wild hyacinths and dog's mercury. E. B. t. 309 was drawn from a plant in Dr. Goodenough's garden at Ealing, where it was ' perfectly naturalised.'

Hesperis matronalis, *L. Dames' Violet.* Cyb. Br. i. 157. Syme E. B. i. t. 103. IV. Near Harrow Weald; *Melvill*, 8. We have not noticed this anywhere wild. It is common in cottage gardens.

SISYMBRIUM, *Linn.*

47. S. officinale, *Scop.* *Hedge Mustard.* Eruca hirsuta siliqua caule adpressa, Erysimum dicta, R. Syn. (Blackst.). Cyb. Br. i. 149 ; Comp. 101. Curt. F. L. f. 6.

Roadsides and waste places, and on walls; very common. A. or B. June—August.

Common almost everywhere, even in London itself, on pieces of waste land. Haller mentions see *Sm. E. F.* iii. 196) that it springs up wherever houses have been burnt.

First record: *Blackstone*, 1737.

[**48. S. Irio,** *L.*   *London Rocket.*
Irio lævis apula, Col. (Merrett). *Erysimum latifolium neapolitanum,*
   Park. (Ray). *Erysimum latifolium majus glabrum, C. B. P.* (Morison).
   Cyb. Br. i. 150; iii. 384; Comp. 102. Curt. F. L. f. 5 (drawn from a
   London plant).

On walls and dry waste ground; very rare. A. or B. July, August.

VII. Almost everywhere in the suburbs of London; *Merrett,* 66. Especially on earth mounds bet. the City and Kensington; in 1667 and
   1668, after the City was burnt, it grew very abundantly on the ruins
   round St. Paul's; *R. Cat.* i. 104. Copiously about Chelsea; *Morison,* 2, 219; where, and also in the *Præludia* of the same author,
   p. 498, is an interesting account of the growth of this species after
   the great fire. Plentifully on the Lord Cheney's wall at Chelsea;
   Pet. Midd. Bet. Brick Lane and Islington; *Pet. Bot. Lond.* 291.
   At the end of Goswell Street; *Hill,* 338. Frequent enough about
   London; *Curt. F. L.* In Chelsea Garden and all that neighbourhood a troublesome weed; *E. B.* 1631. Brompton, Mr. Borrer;
   about Haggerstone, and near Chelsea, E. Forster; opposite Shoreditch Workhouse, L. W. Dillwyn; *B. G.* 408. Growing in 1832
   beneath brick walls by the side of a then new road leading from
   Earl's Court to the new church near Walham Green, which road
   passes the north boundary of the cemetery, not very plentifully. . . .
   Mr. Haworth told me that when he first came to live at Chelsea,
   about 1790-95, it used to grow in great abundance in various places
   by the roadside bet. Little Chelsea and Hyde Park Corner;
   Pamplin (*v. s.*). See also *New B. G.* 97.

First record: *Merrett,* before 1666; also the first notice as British. We
   have seen no specimens collected since 1832, nor ever met with it ourselves, though, no doubt, it was formerly very abundant, as the above
   localities are confirmed by specimens in all the older herbaria collected near London.]

[**49. * S. Sophia,** *L.*   *Flixweed.*
Sophia chirurgorum (Morison). *Nasturtium sylv. tenuissime divisum,*
   C. B. P. (Blackst.).

Cyb. Br. i. 151; Comp. 102. Syme E. B. i. t. 98.

Waste ground and rubbish; very rare. A. June—August.

I. Harefield, not unfrequent; *Blackst. Fasc.* 64.

VII. On the site of a hovel (which was burnt down after the inhabitants
   had died of the plague) not far from S. James's Palace, close to
   Barkshire House, in vast quantity; *Morison,* 2, 219.

First record: *Morison,* 1680. No modern authority. A scarce plant
   round London; it is said to have grown in Battersea Fields, Surrey,
   some twenty years back, but has not recently been seen there.
   Morison considers that this and similar weeds arise spontaneously
   without seed; see also the *Præludia* of the same author.]

D

34   CRUCIFERÆ.

50. S. thalianum, *Gaud.* Arabis thaliana, *L.* (Syme E. B.). *Wall Cress. Thale Cress.*

Pilosella siliquata (Merrett). *P. siliquata Thalii* (Blackst.).

Cyb. Br. i. 140; iii. 383; Comp. 99. Syme E. B. i. t. 115.

Walls and dry ground; common. A. May, June.

I. Harefield!; *Blackst. Fasc.* 77. Ruislip. Bet. Uxbridge and Hillingdon; *Cole.*

II. Staines, in many places. Road from Sunbury to Walton Bridge.

III. Farmyard wall, Headstone. Isleworth. Heston. Hounslow Heath.

IV. Hampstead Heath; *Irv. MSS.,* and *Cooper,* 99. Hedgebank bet. Bishop's Wood and Finchley; *Davies.*

V. Turnham Green. Chiswick, field by the river. Osterley Park walls; *Masters.* Hanwell!; *Newb.* Acton; *Tucker.*

VI. Colney Hatch. Edmonton. Enfield.

VII. [On the ditch sides in the way to Marybone; and in a close on the left hand of the lane from Islington to Kingsland; *Merrett,* 93.]

First record: *Merrett,* 1666.

ALLIARIA, *Adans.*

51. A. officinalis, *Andrz.* Sisymbrium Alliaria, *Scop.* (Syme E. B.). Erysimum Alliaria, *L.* (L. Cat.). *Jack-by-the-Hedge. Sauce-alone. Alliaria, Matth.* (Johns.).

Cyb. Br. i. 154; Comp. 102. Syme E. B. i. t. 100.

Hedges; very common. B. May, June.

Generally distributed throughout the county, and usually abundant.

VII. Occurs in many of the northern suburbs, as Blackstock Lane, Green Lanes, Highgate, &c.

First record: *Johnson,* 1629.

ERYSIMUM, *Linn.*

52. E. cheiranthoides, *L.*   *Worm-seed.*

Cyb. Br. i. 152; iii. 384; Comp. 102. Syme E. B. i. t. 102.

On river-banks, a weed in garden ground, also on rubbish, and in waste land; common. A. or B. June—September.

I. Banks of Colne, near Harefield Mill; *Fl. Herts,* 27.

II. Riverside, Staines, 1841; *Herb. Young.* Gravel bed of Thames, shortly below Hampton Court Bridge, Watson; *New B. G.* 588. West Drayton!; *Newb.* Teddington, abundant. Hampton, abundant. Many places about Staines.

III. About Twickenham, abundant. Cornfields at Wood End, Hind; *Melvill,* 9.

IV. Brick field, Burgess Hill.

V. Shepherd's Bush; Acton; Turnham Green!; *Newb.*

VI. Edmonton.

VII. Camden Town; *Irv. MSS.* Hampstead, 1850, Syme; *Phyt.* iv. 46. Chelsea; *Britten.* Sandy End, Fulham. Site of International Exhibition, South Kensington. Green Park, a weed in gardens and flower-beds. Kilburn, 1862; *Bromwich.*

First record: *Irvine*, about 1830. It is especially abundant in land near the Thames.

BRASSICA, *Linn.*

53. * **B. polymorpha,** *Syme.*
Comp. 103.

1. *B. napus, L.* *Navew. Winter Rape. Coleseed.*
Napus sylvestris, C. B. Pin. (Blackst.).
Cyb. Br. i. 160; iii. 385. Syme E. B. i. t. 88.
River banks and cultivated land; common. A. or B. May—July.
I. Harefield; *Blackst. Fasc.* 63.
II. By the Thames, not unfrequent in this district. By the Thames, near Hampton, 1829; *Winch MSS.* ('*B. campestris*'). Bet. Sunbury and Walton Bridge. Bet. Hampton Court and Kingston Bridge.
III. Near Richmond bridge. Cornfields at Wood End, Hind; *Melvill*, 10.
V. Chiswick; *Fox.* Very abundant on banks of Thames from Putney upwards, 1852, Syme; *Phyt.* iv. 859.
VII. Marylebone Fields, 1815; *Herb. G. & R.* Kentish Town; *Herb. Hardw.* Side of Lea, bet. Lea Bridge and Tottenham; *Cherry.*

2. *B. campestris, L.* *Summer Rape. Colza. Swedish Turnip.*
Cyb. Br. i. 158; iii. 385. Syme E. B. i. t. 89.
Sides of Fields and road-sides; rare. A. or B. May—July.
III. Twickenham, Mr. Quekett; *Varenne.*
VII. Bet. Kentish Town and Hampstead; behind Primrose Hill; *Herb. Hardw.* [All the ditch-banks about Islington; *Pet. Bot. Lond.* 289.]
Hardly to be distinguished from *B. napus*, but a much less common weed in Middlesex. We have not observed it ourselves.

3. *B. Rapa, L.* *B. campestris var. β* (Bab.). *Turnip.*
Rapistrum aliud sylvestre non bulbosum (Park).? *Rapum minus* (Ger.).
Cyb. Br. i. 160; iii. 385. Syme E. B. i. t. 90.
Fields and waysides; rare. A. or B. May—July.
III. Commonly at Harrow; *Melvill*, 10.
VI. Edmonton.
VII. [Going from Shoreditch by Bethnal Green to Hackney; *Park. Theat.* 864.] Hampstead; *Herb. Hardw.* Bet. Kentish Town and Hampstead.
This is usually found in fields where it has been cultivated, and can scarcely be said to be naturalised. It was cultivated in Gerarde's time near

Hackney, ' and brought to the Crosse in Cheapside by the women of that village, to be sold'; *Ger.* 178.

First record: *Parkinson*, 1640.

B. Cheiranthus, *Vill.* *B. monensis, var. β* (Bab. Man.). *Sinapis Cheiranthus*, Koch (L. Cat.). Syme E. B. i. t. 92. VII. A stray plant at Chelsea in 1852; *Irv. H. B. P.* 704.

SINAPIS, *Linn.*

54. S. nigra, *L.* Brassica nigra, *Koch* (Syme E. B.). *Black Mustard.*
Sinapi sativum secundum, Ger. (Johns.)
Cyb. Br. i. 162; Comp. 104. Syme E. B. i. t. 85.
River-sides, fields, and waste places; rather common. A. June—August.
II. Fields near Queen's River, Hampton.
III. By the Railway, Hounslow Heath; *Warren.*
IV. Bet. Whitchurch and Stanmore.
V. Twyford!; Shepherd's Bush!; *Newb.*
VI. Bet. Tottenham and Edmonton; *Cherry.*
VII. [Banks at back of Old Street; and in the way to Islington; *Johns. Ger.* 245.] On ditch sides in many places about London; *Merrett*, 114. . Highgate; Gospel Oak Fields; *Herb. Hardw.* Camden Town, 1818; *Bennett (v. s.).* Chelsea, 1815; *Herb. G. & R.* Upper Clapton; *Cherry (v. s.).* Near St. Mark's College; *Newb.* South Heath, Hampstead.

First record: *Johnson*, 1633.

55. S. arvensis, *L.* Brassica Sinapistrum, *Bois.*(Syme). *Charlock.*
Rapistrum arvorum, Ger. em. (Blackst.).
Cyb. Br. i. 161; Comp. 104. Curt. F. L. f. 5.
Cornfields and cultivated ground; very common. A. June—September.
Common in all the districts.
VII. Marylebone Park, 1823; *Herb. G. & R.* Primrose Hill, 1827; *Varenne.* Kentish Town; *Herb. Hardw.* I. of Dogs.

First record: *Blackstone*, 1737. Mr. Irvine notices a ' *var. glabra* ' at Chelsea; *H. B. B.* 703.

56. * S. alba, *L.* Brassica alba, *Bois.* (Syme). *White Mustard.*
Sinapi album (Johns.).
Cyb. Br. i. 161; iii. 385; Comp. 104. Curt. F. L. f. 5.
Cornfields, but usually stray plants on waste ground and roadsides; common. A. July—September.
I. Stanmore Heath.
II. Stanwell Moor. Tangley Park. Teddington. Near Hampton Court.
III. Waste ground about Twickenham; isolated specimens. Harrow; *Herb. Harr.*

IV. Harrow Weald. Bet. Whitchurch and Stanmore. Near Bishop's Wood.

V. By canal, Greenford; *Melvill*, 10. Near Ealing; *Britten.*

VI. Colney Hatch; *Herb. Hardw.* Edmonton.

VII. [Banks at the back of Old Street; and in the way to Islington; *Johns. Ger.* 245.] Upper Clapton; *Cherry.* Near South Heath, Hampstead. Roadside, Kensington Gore, one plant.

First record: *Johnson,* 1633; also the first notice as British. This species is grown for 'small salad;' hence, no doubt, its origin in some of the above stations.

DIPLOTAXIS, *Cand.*

57. D. tenuifolia, *DC.* Brassica tenuifolia, *Bois.* (Syme). Sinapis tenuifolia, '*Br.*' (L. Cat.).

Eruca sylvestris (Ger.). *Sinapi sylvestre minus Bursæ-pastoris folio, Lob.* (Johns.). *E. sylv. major lutea caule aspera* (Buddle). *Brassica Erucastrum, L.* (Huds. I.). *Brassica muralis* (Huds. II.). *Sisymbrium tenuifolium, L.* (Smith.)

Cyb. Br. i. 163; iii. 386; Comp. 104. Curt. F. L. f. 3.

On old walls and rubbish; rare. P. July—September.

VII. Most brick and stone walls about London covered with it; *Ger.* 192. [Hampstead Heath; *Johns. Enum.*] [On the walls of the City in great plenty; *Mill. Bot. Off.* 189.] [Chelsea, Tower Ditch, &c.; *Newton MSS.*] [On London Wall, bet. Cripplegate and Bishopsgate; wall of Charterhouse; *Pet. Midd., Bot. Lond.* and *Herb. Pet.* 152, 10.] [Walls round the Tower; back of Bedlam; near Hyde Park; *Curt. Fl. Lond.*] [London Bridge, Mr. Jones; *With.* iii. 593; and *Herb. Linn. Soc.*] [Wall near Hyde Park Corner, 1817; *Herb. G. & R.* We have also seen plants from this wall collected in 1821.] [Walls in Church Lane and Silver Street, Kensington; *Mag. Nat. Hist.* ix. 90.] Roadside bet. Kentish Town and the floor-cloth manufactory; *Irv. MSS.* Westminster School wall!; *Macreight,* 22. It still grows there, in front of Lord John Thynne's house in the Schoolyard, 1867. Upper Clapton; *Cherry.* Near Parson's Green; *Britten.* Waste ground at Hammersmith. Bet. Chelsea Hospital and the river.

First record: *Gerarde,* 1597; also the first notice as a British plant. It is mentioned as growing abundantly on the walls of London by almost all subsequent authors on British plants, and there are many specimens in the old herbaria. Though now much less common it is still to be found on a few old walls in town.

58. *D. muralis, DC.* Brassica muralis, *Bois.* (Syme). Sinapis muralis, '*Br.*' (L. Cat.).

Cyb. Br. i. 164; iii. 386; Comp. 105. Syme E. B. i. t. 94.

Waste ground, roadsides, on a sandy soil, and walls; rather common. A. B. or P.? June—October.

II. Very sparingly on the gravel bed of the Thames, shortly below Hampton Court Bridge, Watson; *New B. G. Supp.* 588. Hampton, abundant. Bet. Kingston Bridge and Hampton Court in plenty. Staines common. Near Yewsley!; *Newb.*

III. Common about Twickenham. Isleworth. Bet. Hounslow and Lampton.

V. Near Shepherd's Bush!; Chiswick!; Turnham Green; *Newb.* On Hammersmith Bridge; *Warren.*

VII. Parsons Green. Brompton Cemetery. Chelsea College; *Britten.* Near Cremorne; *Fox.* Upper Clapton; *Cherry.* North End, Fulham. Brook Green. Entrance to Grosvenor Canal. South Heath Hampstead. Behind York Cottages in plenty, 1862; *Bromwich.*

First record: *Watson,* 1837. This plant was not distinguished from the preceding by British botanists previous to Smith; there are no specimens in the herbaria of Buddle or Petiver; nor is it given in Ray's Synopsis. It is not improbably a comparatively recent introduction First noticed in England near Ramsgate, Kent, by Mr. Dillwyn in 1801 (*Linn. Trans.* vi. 389),* where it was said to have been introduced with foreign oats. It quickly became a troublesome cornfield weed there, and is so still throughout the I. of Thanet. The larger form, β *Babingtonii* (Syme) appears to be the more common about London.

Lunaria biennis, *Mönch. Honesty. White-Satin. (Viola Lunaria s. Bolbonac.* Hath been found wild in the woods about Pinner and Harrow-on-the-Hill; *Ger.* 378.) No doubt an error; the plant is common in gardens, and single plants sometimes occur on waste ground, as in (V.) a newly made road at Chiswick.

Königa maritima, *R. Br. Alyssum mar.,* Lam. (Syme). Cyb. Br. i. 134. Syme E. B. i. t. 140. IV. [Amongst rubbish on Hampstead Heath; *Irv. MSS.*] VII. [Near St. John's Wood Chapel, 1815; *Herb. G. & R.*] Ditch at Shepherd's Bush, 1845; *Morris (v. s.).* Escaped from gardens.

DRABA, *Linn.*

D. muralis, *L.* Cyb. Br. i. 132; Comp. 97. Syme E. B. i. t. 135. VII. Walls of Chelsea garden, J. E. Smith; *Herb. Linn. Soc.* A garden escape, or planted there.

* It had previously been detected by Dr. S. F. Gray of the British Museum (where, we are not told), but was considered a variety of *D. tenuifolia* by him and Smith. (See under *Sisymbrium Irio,* E. B. 1631.)

59. D. verna, *L.*   *Whitlow Grass.*

Paronychia vulgaris, Dod. (Ger.).

Cyb. Br. i. 133 ; iii. 382 ; Comp. 97. Curt. F. L. f. 1.

Wall-tops and dry ground ; common. A. March—May.

I. Wood Hall, Pinner, in a ploughed field. Harefield ; *Blackst. Fasc.* 72.

II. Teddington (in flower Jan. 3, 1854); *Phyt. N. S.* ii. 279. Road bet. Staines and Hampton.

III. Isleworth, &c., abundant.

IV. Neasdon ; Kingsbury ; *Cole.* Bet. Whetstone and Totteridge. Hampstead Heath ; *Johns. Enum.* Stanmore.

V. Horsington ; *Melvill,* 8. Greenford ; *Herb. Young.* Osterley Park walls ; *Masters.* Turnham Green.

VI. Muswell Hill ; *Cole.* Edmonton. Enfield.

VII. [Plentifully on the brick wall in Chauncerie Lane, belonging to the Earl of Southampton ; *Ger.* 500.] Marylebone, 1827 ; *Varenne.* Kensal Green Cemetery. Haverstock Hill.

First record : *Gerarde,* 1597 ; also the first notice as British.

[COCHLEARIA, *Linn.*

60. C. anglica, *L.*   *Scurvy Grass.*

Cyb. Br. i. 128 ; iii. 381 ; Comp. 97. Syme E. B. i. t. 133.

VII. Isle of Dogs ; *Mart. App. P. C.* 65.

It does not now grow on the Middlesex shore, though abundant on both sides of the Thames as high as Woolwich, and occurring also near Greenwich (opposite the above locality), whence the specimen from which Syme's figure was drawn was obtained.]

ARMORACIA, *Rupp.*

61. *A. rusticana,* *Rupp.* Cochlearia Armoracia, *L.* (Syme). *Horse Radish.*

Raphanus rusticanus (Ger.).

Cyb. Br. i. 129 ; iii. 381. Syme E. B. i. t. 129.

Railway banks, waste places, sides of fields and ditches ; common. P. May—July.

I. Stanmore Heath. Pinner Chalk-pits. Uxbridge ! ; Harefield ! ; *Newb.*

II. Staines, abundant.

III. About Isleworth, common. Hounslow. Grove Road, Harrow.

IV. Peterborough Road, Harrow ; *Melvill,* 8.

V. Side of stream, Apperton ; *Melvill,* 8. Horsington Hill. Brentford. Ealing ! ; *Newb.* Acton ; *Tucker.*

VI. Colney Hatch. Church fields, Edmonton.

VII. Bishop's Walk, Fulham ; *Britten.* [At a small village called Hogsdon (=Hoxton), in the field next to a farm-house leading to

Kingsland, *Ger.* 187.]   Banks of the Lea; Forster, *Herb. Brit. Mus.*
Isle of Dogs!; *Newb.* Sandy End, Fulham. Hackney Wick.

First record: *Gerarde*, 1597; also first notice in Britain. No doubt
derived from neighbouring gardens. It is very persistent in ground
where it has once been introduced, and must be considered thoroughly
established in Middlesex, though it does not produce seed.

62. A. amphibia, *Koch.*   Nasturtium amph., *R. Br.* (Syme).
          Great Water Radish.
Raphanus aquaticus, (Ger.).   *Sisymbrium amphibium, L.* (Sm.)
Cyb. Br. i. 149; Comp. 101.   Syme E. B. i. t. 128.

Sides of rivers, streams, and ditches; rather common. P. June—
September.
I. Colne, at Harefield!, and at Yewsley!; *Newb.*
II. Staines Common. Common by the Thames at Sunbury, Walton
    Bridge, Halliford, Hampton, Hampton Court, Kingston Bridge and
    Teddington.
III. By the river at Twickenham. By the Cran on Hounslow Heath and
    other parts of its course. Duke's River, Isleworth.
V. Continues down the course of the Thames at Brentford, Strand-on-the-
    Green, and Chiswick. Canal at Greenford; *Melvill*, 7. Twyford!;
    Newb. Hanwell!; *Cherry.*
VII. Bishop's Walk, Fulham; *Britten.* River bank at Cremorne. [In the
    chinks of a stone wall upon the river Thames by the Savoy in
    London; *Ger.* 186.] Brick wall of a house fronting the river (now the
    embankment) at bottom of Surrey Street, Strand, 1867. Isle of Dogs,
    1817; *Herb. G. and R.* [Westbourne Green, 1827—30; *Varenne.*]
    Hackney Marshes.

First record: *Gerarde*, 1597; also first as a British plant. A common
species by the Thames, and by most of the streams flowing into it.

THLASPI, *L.*

63. *T. arvense, *Linn.*   *Penny Cress.*
Cyb. Br. i. 117; iii. 380; Comp. 93. Curt. F. L. f. 6.

Waste places and roadsides; rather rare. A. June—September.
I. A single plant in a garden at Harefield!; *Newb.*
II. Isolated plants at Hampton Station, and near Kingston Bridge.
III. Single specimens in several places about Twickenham; in plenty near
    Twickenham Ry. Station. Field near Hounslow; *Newb.*
IV. Near the Spaniards, Hampstead, a few specimens; *Curt. F. L.*
V. In a field bet. Acton and Turnham Green!; *Newb.*
VII. [Tothill fields;] bet. Tottenham and Newington, J. Woods; *B. G.* 407.
    Newington Common, about 50 plants, 1865; *Grugeon* (*v. s.*). Chel-
    sea; *Britten.* Prince Albert Road, S. Kensington, plentiful in 1855,

E. I.; *Phyt. N. S.* i. 439. Adelaide Road; *Newb.* Green Park, sown with grass on the site of the old reservoir.

First record: *Curtis*, about 1798. Certainly introduced in all its localities in the county, and not permanent in them.

Camelina sativa, *Crantz. Alyssum sativum, Sm.* (B. G.). Cyb. Br. i. 134. Syme E. B. i. t. 141. I. One plant in a corn-field at Pinner; *Melvill*, 9. IV. Bank of the Midland Ry. near Edgwarebury. VII. Stoke Newington; Highgate; Isle of Dogs, J. Woods; *B. G.* 407. Hampstead Heath, Hunter; *Cooper*, 99. Came up in Kensington Gardens, with grass sown for turf, 1834; *Mag. Nat. Hist.* viii. 389. South Heath, Hampstead. Site of International Exhibition of 1862. Introduced with foreign seed, and not thoroughly naturalised.

TEESDALIA, *R. Br.*

64. T. nudicaulis, *R. Br.*

Nasturtium petræum, Tab. (R. Syn.). *Iberis nudicaulis, L.* (Curt.). *Teesdalia Iberis, DC.* (Cooper).

Cyb. Br. i. 121; iii. 380; Comp. 94. Curt. F. L. f. 6.

Sandy and gravelly commons; rare. A. April—June.

II. In Hampton Court Park abundantly, and in the fields thereabouts, Sam. Doody; *R. Syn.* ii. 344. Teddington field, abundantly; *Blackst. Fasc.* 64.

III. Hounslow Heath; *Dicks. H. S.* and *Curt. F. L.* Mr. Gosling's waste ground, Whitton, G. Francis; *Cooper Supp.* 12. Near Twickenham, 1842; *Twining (v. s.).*

First record: *Doody*, 1696. We have no very recent authority for its growth in Middlesex, though it is probably still to be found about Teddington and Hounslow.

Iberis amara, *L. Bitter Candytuft.* Cyb. Br. i. 122; iii. 381; Comp. 94. Syme E. B. i. t. 149. VII. Near Highgate; *Lond. Fl.* 162. Lane opposite Kilburn Wells, 1816; *Herb. G. & R.* A garden escape always.

LEPIDIUM, *Linn.*

65. * L. Draba, *L.*

Cyb. Br. i. 124; iii. 381. Syme E. B. i. t. 158.

Waste ground; very rare. P. May, June.

VII. Abundant for several years on the south bank of the N. London Ry. at Edgware Road Station, 1866; *Newb.*

It has not been noticed elsewhere, and seems of much less frequent occurrence on the north than on the south side of London, where it is quite naturalised in several places in North Surrey.

66. L. campestre, *R. Br. Pepperwort.*

Thlaspi (Turn.). *Thl. vulgatissimum* (Ger.). *Thl. campestre, L.* (Curt.). Cyb. Br. i. 124; Comp. 95. Curt. F. L. f. 5.

Hedgebanks, fields, and waste places; rather common. B. June—August.
Among the crops, Harefield; *Blackst. Fasc.* 98. Bet. Harefield and
Ruislip. Bet. Pinner Ry. Station and Harrow Weald Common. Old
chalk-pit, Harefield!; *Cole.*
IV. Hampstead; *Irv. MSS.* Pinner Drive; *Melvill,* 9.
V. Sion; *Turn. Names.* Gunnersbury Lane, Ealing; *Curt. F. L.* Near
Apperton!; *Newb.*
VI. Near Enfield. By the path from Hornsey unto Waltham Cross;
Ger. 206.*
VII. [In London it groweth in Maister Riche's garden and in Maister
Morgannne's also; *Turn.* ii. 152. Perhaps cultivated there.]
First record: *Turner,* 1548; the first notice as British.

67. L. Smithii, *Hook.*
Cyb. Br. i. 124; Comp. 95. Syme E. B. i. t. 157.
Hedgebanks; rare. P. June—August.
II. Road bet. Staines and Hampton. Near Charlton, in plenty.
III. Near Hounslow, on the Hanworth Road, and on the same road near
Hanworth.
First record: *the Authors,* 1866. Appears to be confined to the
sandy S.W. of the county.

L. sativum, *L. Garden Cress.* Cyb. Br. iii. 317. Syme E. B. i. t. 155.
II. Road bet. Teddington and Twickenham. IV. Harrow Weald.
V. Apperton; *Herb. Harr.* VII. Parson's Green; *Britten.* Back of
Adelaide Road. Extensively grown in kitchen gardens, whence it is
cast out when in seed, and so easily gets to roadsides and waste ground.

68. * L. ruderale, *L.*
Cyb. Br. i. 125; Comp. 95. Syme E. B. i. t. 154.
Waste ground; rare. A. June—September.
II. In some quantity bet. the Ry. station and Grand Junction Canal at
West Drayton!; *Newb.*
III. Southall; *Dr. Cobbold,* who showed specimens at the Linn. Society's
Meeting, June 6th, 1867.
V. [A few plants on Turnham Green, about 1860;] in some abundance
by the canal at Apperton!; *Newb.*
VII. Rubbish near Highgate Archway; *Irv. MSS.* and *Jewitt* (*v.s.*).
Brickfield by Sir G. Duckett's Canal; *Cherry* (*v. s.*). Site of Inter-
national Exhibition at S. Kensington.
First record: *Irvine,* about 1830. Probably not native in Middlesex.
Four of the above stations are by the course of the canal which crosses
the county, and to such places the seeds might be easily conveyed with
merchandise. In the other three localities the plant is an evident
introduction.

* This locality may perhaps refer to *Thlaspi Discoridis,* Ger.= *Thlaspi arvense,* L.

CAPSELLA, *Vent.*

69. C. Bursa-pastoris, *DC. Shepherd's Purse. Mother's Heart.*
Bursa-Pastoris, Ger. (Blackst.).
Cyb. Br. i. 120 ; Comp. 93. Syme E. B. i. t. 152.

On and under walls, by roadsides, waste places, and fields ; very common.
A. March—September.

This well-known weed occurs almost everywhere throughout the county.
In London itself, it grows abundantly in the squares and less frequented
streets, being one of the few plants which can resist the adverse
influences of the metropolis.

First record : *Robert Turner*, 1664.

SENEBIERA, *Pers.*

70. S. Coronopus, *Poiret.* Coronopus Ruellii, *Gaertn.* (L. Cat.).
Swine Cress.
Nasturtium supinum capsulis verrucosis, Ray Meth. em. (Blackst.).
Cyb. Br. i. 116 ; Comp. 92. Syme E. B. i. t. 160.

Roadsides, foot of walls, waste places ; very common. A. June—Sep-
tember.
Throughout all the districts.
VII. [Tothill Fields; *Ger.* 347.] Kensington Gardens (*v. s.*); by Trinity
Church, Albany St., N.W. ; *Warren.* Fulham. Green Park.
Kentish Town. Highbury. Hackney Wick.

First record : *Gerarde*, 1597.

71. * S. didyma, *Pers.* Coronopus didyma, *Sm.* (Lond. Cat.).
Lepidium didymum, L. (Smith).
Cyb. Br. i. 115 ; iii. 379 ; Comp. 92. Syme E. B. i. t. 160.

Road-sides, on a sandy soil ; rare. A. June—September.
II. Road by Strawberry Hill House, a few plants.
III. Road bet. Hounslow and Hanworth, by the Drilling Ground ! ; *Newb.*
Several places on the Staines Road, near the 11-mile stone, especially
in a gravel-pit on the S. side of the road.
VI. Southgate, and lane bet. Southgate and Colney Hatch, 'well esta-
blished ;' *Phyt. N.S.* vi. 303.
VII. Highgate Archway; *Irv. MSS.* and *Lond. Flora*, 163. In a man-
ner naturalised in Chelsea Gardens; *E. B.* 248. Chelsea College,
1861 ; *Britten.* Roadside at Parson's Green, 1862, a few specimens.
Abundant there a few years before ; *Irvine.* Still there in 1867 ! ;
Newb. In some plenty in the Isle of Dogs ; *Cherry* (*v. s.*).

First record : *J. E. Smith*, 1795. This has much the look of a native
in III., but in the other districts is, no doubt, an introduction. It
has been said to be of American origin.

RAPHANUS, *Linn.*

72. R. Rhaphanistrum, *L.*   *Wild Radish. Jointed Charlock.*
Rapistrum album articulatum, Park. (Blackst.).
Cyb. Br. i. 166 ; iii. 387 ; Comp. 105. Curt. F. L. f. 4.

Fields, cultivated and waste ground; common. A. June—August.
I. Harefield! ; *Blackst. Fasc.* 86. Pinner chalkpits. South Mims.
II. Near W. Drayton! ; *Newb.* Staines. Stanwell. Sunbury. Hampton.
III. Twickenham, Hounslow.
IV. Bet. Hendon and Finchley Road; *Morris (v. s.).* Side of N.W. Ry.
     near Harrow Station ; *Melv.* 10.
V. Lampton. Near Shepherd's Bush ! ; *Newb.* Near Hanwell! ; *Cherry.*
VI. Colney Hatch. Edmonton.
VII. Side of London Canal; *Cherry (v. s.)*

First record: *Blackstone,* 1737. The *white-flowered plant,* though not
so common as that with yellow petals, is by no means unfrequent in
the county.

RESEDACEÆ.

RESEDA, *Linn.*

73. R. lutea, *L.*   *Wild Mignonette.*
Cyb. Br. i. 169 ; iii. 387 ; Comp. 106. Syme E. B. ii. t. 162.

Open places on chalk; very rare. B. or P. June—August.
I. Harefield, chalkpits, 1868! ; *Newb.*
Has not been seen elsewhere in the county.

74. *R. suffruticulosa, *L.*   (Including *R. alba,* L.)
R. fruticulosa, Auct.
Cyb. Br. i. 170 ; iii. 387. Syme E. B. ii. t. 163.

I. A weed for two years in the Parsonage garden, Pinner ; *Hind.*
VII. Chelsea, T. Moore ; *Herb. Brit. Mus.* Frequently noticed in that
     neighbourhood, 1866 ; *Fox (v.s.).*

First record : *T. Moore,* 1851. Barely naturalised. Commonly cultivated
in gardens.

75. R. luteola, *L.*   *Weld. Dyer's Weed.*
Luteola herba salicis folio, C. B. Pin. (Blackst.).
Cyb. Br. i. 168 ; iii. 387 ; Comp. 106. Syme E. B. ii. t. 164.

Railway banks and waste places ; rather rare. B. June—August.
I. Harefield! ; *Blackst. Fasc.* 54. Stanmore Heath, a single plant.
II. Teddington. Bet. Staines and Hampton. Bet. Long Ditton Ferry
     and Hampton Bridge, by the river.

III. Many places about Twickenham. In abundance, and growing to a great height at corner of Ailsa Road, Twickenham Park. Near Hounslow!; *Newb.*

IV. Opp. gate of Kingsbury Churchyard; *Farrar.*

V. Near Lampton. Kew Bridge Ry. Station!; *Newb.* Brentford; *Lees.*

VII. [Wall, S. side of Hyde Park, 1815; *Herb. G. & R.*] [St. Pancras, 1816; *E. T. Bennett, (v.s.)*.] Love Lane, Kentish Town; *Irv. MSS.* Parson's Green, *Irvine*; Craven Hill. King's Road, Chelsea. Site of International Exhibition, S. Kensington.

First record: *Blackstone*, 1737.

R. odorata, *L.*, the *sweet Mignonette* of the gardens, seems to be getting established on waste ground by Cremorne.

VIOLACEÆ.

VIOLA, *Linn.*

76. V. palustris, *L.*

Cyb. Br. i. 174; Comp. 107. Curt. F. L. f. 3.

Spongy bogs; very rare. P. May—July.

IV. Hampstead Heath!; *Huds.* i. 330. We have seen specimens collected there in 1815 and 1817; *Herb. G. & R.*, and in 1821; *Bennett.* It still grows in good quantity in the bog behind Jack Straw's Castle, but flowers now very sparingly.

First record: *Hudson*, 1762. It has not been seen elsewhere in Middlesex.

77. *V. odorata, *L.* Sweet Violet.

V. martia alba et purpurea (Blackst.).

Cyb. Br. i. 174; Comp. 107. Curt. F. L. f. 1.

Ditches, hedgebanks, and shrubberies; rather common. P. March, April.

I. Harefield, frequent!; *Blackst. Fasc.* 109. Road bet. Wood Hall and Pinner.

II. By the river bet. Kingston Bridge and Hampton Court, abundant.

III. Roxeth; *Melv.* 11. Worton. Path leading to river-side by Orleans House, Twickenham.

IV. Harrow Park; *Melv.* 11.

V. Apperton; *Cole.* Dry ditch under garden wall, Sion Park.

VI. Edmonton, in a field.

VII. Wall near Hampstead, 1819; *Herb. G. & R.*

First record: *Blackstone*, 1737. Of garden origin in most, if not all, of the above stations. The white-flowered plant has been noticed at Harefield; *Blackst.*, and Harrow Park; *Melv.* 11.

78. V. hirta, *L.*

V. martia hirsuta major inodora, Plot. (Blackst.).

Cyb. Br. i. 176; iii. 388; Comp. 108. Curt. F. L. f. 1.

Hedgebanks and woods; very rare. P. April, May.

I. By the sides of the old Park Wood, and in the chalkpit, Harefield!;
    Blackst. Fasc. 110. A mile from Harefield on the Uxbridge Road,
    1855; *Phyt. N. S.* i. 62.

First record : *Blackstone*, 1737. Confined to the chalk.

79. V. sylvatica, *Fries.*   *Dog Violet.*   ·

V. canina sylvestris, Ger., and *V. canina minor, R. Syn.* (Blackst.*).
V. canina, L., and *V. pumila, Vill.* (Melv.†).
Cyb. Br. i. 177 ; iii. 389 ; Comp. 108. Curt. F. L. f. 2.

Woods, hedges, and heaths; common. P. April—June.

I. Harefield!; and Uxbridge; *Blackst. Fasc.* 110. Ruislip and Harrow
    Weald Commons; *Melvill,* 11. Ruislip Wood. Pinner Wood.
    Elstree. South Mims.
II. Road bet. Staines and Hampton.
III. Whitton. Gravel pit near Hounslow.
IV. Harrow district, common ; *Melv.* 11. Deacon's Hill. Bentley Priory.
    Bishop's and Turner's Woods.
VI. Hadley!; *Warren.* Edmonton. Winchmore Hill Wood.

First record : *Merrett,* 1666. *With white flowers.* In Hampstead Wood,
    on that side the chestnut walk where the two ways meet ; *Merrett,* 125.
    All we have seen belongs to the sub-species *V. Riviniana,* Reich.
    (*sylvatica var.* β of *Bab. Man.*); but Mr. Hind says (*Melv.* 11),
    V. Reichenbachiana, Bor., also occurs about Harrow. The small
    heath form—*V. flavicornis,* Forst., not Smith—is found (I.) on Stan-
    more Heath, (IV.) North Heath and Bishop's Wood, and (VII.)
    South Heath, Hampstead.

80. V. canina, *L.* (Bab.).

Cyb. Br. iii. 319 ; Comp. 108. Syme E. B. ii. t. 175.

Dry sandy places ; rare. P. April—June.
I. Near Harefield, by the Ruislip road, a few specimens.
II. Near Staines, on the road to Hampton, abundant.
III. Drilling-ground, Hounslow. Near Hatton. Heston gravel-pits.
First certain record : *the Authors,* 1866. The small form of *V. sylvatica*
    is frequently thus named.

81. V. tricolor, *L.*   *Wild Pansy. Heartsease. Love-in-idleness.*

Cyb. Br. i. 180 (with *V. arvensis*); Comp. 109. Curt. F. L. f. 1.

Weed in cultivated ground, especially gardens ; rather rare. A. May—
    September.
I. Harefield ; *Cole.*
II. Teddington, G. Francis ; *Cooper Supp.* 12. Meadow, near Hampton
    Court Park ; *Blackst. Fasc.* 110.

---

* Blackstone's plant may be 80. *V. canina.*   † On authority of *Herb. Harr.*

III. Cottage garden, Twickenham. Isleworth, 1815; *Herb. G. & R.*
IV. Cornfields, Hampstead; *Irv. MSS.*
VI. Garden weed, Edmonton.
VII. Victoria Street, Westminster.
Scarcely wild.

Var. β. V. arvensis, Murr. *Corn Pansy. V. bicolor arvensis, C. B. P.*
(Blackst.); Comp. 109. Syme E. B. ii. t. 179.

A cornfield weed; also in waste places; common. A. (occasionally P.)
June—October.

I. Harefield!; *Blackst. Fasc.* 110; abundant, 1867. Harrow Weald
Common. Ruislip, Eastcott, and Pinner; *Melv.* 11.
II. Teddington. Fulwell. Near Strawbery Hill. Tangley Park.
III. Twickenham allotments.
IV. Hampstead; *Irv. MSS.*
V. Chiswick!; *Newb.* Acton; *Tucker (v. s.).*
VI. Colney Hatch. Edmonton.
VII. Near Paddington Cemetery; *Warren.* Parson's Green.
First record: *Blackstone*, 1737.

DROSERACEÆ.

DROSERA, *Linn.*

82. D. rotundifolia, *L.*   *Sundew.*
Ros solis, Dod. (Johns.). *R. s. fol. rotundo, C. B. Pin.* (Blackst.).
Cyb. Br. i. 183; Comp. 110. Syme E. B. ii. t. 182.
Spongy bogs; rare. P. July, August.

I. Battles-well, near Harefield; *Blackst. Fasc.* 87. [Ruislip Heath;
M. & G. 473.] Bushey Heath; *Fl. Herts. Supp.* 6. Harefield
common, abundant. Low parts of Harrow Weald Common, in
plenty.
IV. Hampstead Heath!; *Johns. Eric., and many subsequent authors.* In
some plenty, amongst *Sphagnum*, in the bog behind Jack Straw's
Castle.
First record: *Johnson*, 1629.

(D. intermedia, *Hayne.* D. longifolia (M. & G.). Cyb. Br. i. 184;
iii. 389; Comp. 110. Syme E. B. ii. t. 184. Bogs on Ruislip and
Harrow Heaths, in great abundance; on a common (=Harrow Weald
Common?) betwixt Pinner and Stanmore; *M. & G.* 474. There is no
other record of this as a Middlesex species, and the authors above
quoted are not alone sufficiently worthy of trust to stand as satisfactory vouchers for its occurrence. The plant, however, grows in
Bucks, near the Middlesex boundary, and is not uncommon in North
Surrey.)

POLYGALACEÆ.

POLYGALA, *Linn.*

83. P. vulgaris, *L.*   *Milkwort.*

P. repens, Ger. em. (Blackst.).

Cyb. Br. i. 186; iii. 390; Comp. 111.   Syme E. B. ii. tt. 185 and 187.

Heaths and commons; rather rare.   P.   June—August.

I. Harefield!; *Blackst. Fasc.* 79.   Ruislip and Harrow Weald Commons; *Melv.* 12.   Stanmore Heath.

III. Drilling ground, Hounslow.

IV. Deacon's Hill, near Elstree.   North Heath, Hampstead.

VI. Edmonton.

VII. South Heath, Hampstead.

First record: *Blackstone,* 1737.   With the exception of the Harefield plant, which is typical '*eu-vulgaris*' of Syme, all the Middlesex specimens are to be referred to *P. depressa,* Wenderoth (*Syme E. B.* t. 187). A variety noticed under the name of *P. myrtifolia palustris humilis et ramosior* was found (IV.) in some boggy meadows near Highwood Hill, beyond Hendon; *Blackst. Spec.* 77.

CARYOPHYLLACEÆ.

DIANTHUS, *Linn.*

[**84. D. prolifer,** *L.*   '*Childing Sweet Williams.*'

Armeria prolifera, Ger. (How).   *Caryophyllus sylvestris prolifer, C. B. P.* (R. Syn. iii.).

Cyb. Br. i. 189; Comp. 112.   Syme E. B. ii. t. 196.

Sandy and gravelly places; very rare.   A.   July.

In the grounds 'twixt Hampton Court and Tuddington; *How,* 10, and *Merrett,* 10.   Hampton Court; *Macreight,* 28 (probably only copied from the former authors).

First record: *How,* 1650; also first as a British plant.   Perhaps, as suggested by Mr. H. C. Watson, this may have been only *D. deltoides* (q. v.); but the locality is not an unlikely one, especially as there is satisfactory evidence of the occurrence of the plant near Windsor.]

[**85. D. Armeria,** *L.*   *Deptford Pink.*

Caryophyllus pratensis noster major (Park.).   *Caryoph. latifl. barbatus minor annuus fl. minore* (Newton).   *Caryoph. pratensis, Ger. em.* (Merrett   ) .

Cyb. Br. i. 190; Comp. 112.   Syme E. B. ii. t. 191.

Dry, grassy places; very rare. A. or B. July, August.

II. In Tuddington field; *Merrett*, 22.

VII. A lesser sort of this among the thicke grasse towards Totnam Court, near London; *Park. Theat.*, 1338. In a little wood cut down on the right hand of the road, a little beyond the bottom of the hill beyond Highgate; *Newton MSS*.

First record: *Parkinson*, 1640. Last: *Newton*, about 1680.]

[D. plumarius, *L.* Common Pink. *Caryophyllus simplex fl. minore pallide rubente, C. B. P.* (R. Syn. iii.). Cyb. Br. i. 191. Syme E. B. ii. t. 195. III. A variety of this is found near Twickenham; *Dill. in R. Syn.* iii. 336. Probably on a garden wall, semi-naturalised.]

86. D. deltoides, *L.* *Maiden Pink.*
Armeriæ species fl. summo caule singulari (Pet.). *Caryophyllus minor repens nostras* (R. Syn. ii.).
Cyb. Br. i. 192; iii. 391; Comp. 113. Syme E. B. ii. t. 192.

Dry pastures; very rare. P. June—September.

II. In the park at Hampton Court; *Pet. Midd.* (a small form; see *Herb. Budd.* 124, fol. 1). In Hampton Court Park, abundantly, and in the fields thereabouts; *Doody MSS.* In Hampton Court Park, but not plentifully; *Blackst. MSS.* Teddington meadows, 1838, D. Cooper; *Herb. Young.* Hampton; *Lond. Fl.* 168.

IV. [On Hampstead Heath; *Huds.* i. 162.]

First record: *Petiver*, 1695. Also occurs at Totteridge Green, Herts; *Phyt. N. S.* ii. 157.

SAPONARIA, *Linn.*

87. *S. officinalis,* *L.* *Soapwort.*
Cyb. Br. i. 193; Comp. 113. Curt. F. L. f. 2.
Roadsides; rare. P. August.

I. In a lane near Uxbridge Churchyard, plentifully; *Blackst. Fasc.* 90.

VII. [Near Kingsland Turnpike; *MS. note (Alchorne's) quoted in Phyt.*, iii. 166.] The Lots near Cremorne, Britten; *Bot. Chron.* 21.

First record: *Blackstone*, 1737. An escape from cultivation, scarcely naturalised.

S. Vaccaria, *L.* Cyb. Br. i. 194; iii. 391. I. A single plant at Pinner lime-pits, Hind; *Melvill*, 13.

CUCUBALUS, *Linn.*

[**88. C. bacciferus,** *Linn.*
Cyb. Br. i. 194. Brit. Ent. vol. xvi. t. 761 (drawn from an Isle of Dogs specimen).
Very rare. P. August.

VII. On the banks of the ditch on the left-hand side of the road leading down the middle line of the Isle of Dogs, from Blackwall to the

Ferry House, rather nearer the latter than the docks; discovered there August, 1837, by G. Luxford, A.L.S.; who read a note on the plant before the Linn. Soc. on November 21, 1837, in which he says, 'If not truly indigenous it is at least perfectly naturalised. I . . . passed it . . . the first time I saw it (in June. 1837) . . . thinking it to be merely Cerastium aquaticum;' *Trans. Linn. Soc.* xviii. p. 687. See also *Mag. Nat. Hist.*, N.S. i. 45. There are specimens, collected by Luxford in 1837, in *Herb. Linn Soc.*; by H. M. Colman in 1840, and J. Freeman in 1841, in *Herb. Young*, and by Brewer in 1846, in *Herb. Brit. Mus.* In 1852, Mr. Thomas Westcombe found the plant 'growing in considerable abundance, and thought there was no probability of its extermination;' *Phyt.* iv. 605. Mr. J. B. Syme collected it in 1853. since which date we believe it has not been seen. The part, including that where it grew, is now covered by new docks, a railway, &c.

This plant has not occurred elsewhere in Great Britain. The locality in Anglesea (see *R. Syn.* iii. 267, and *E. B.* 1577), was an error, as appears from a letter from Mr. Foulkes, the supposed discoverer, given in Smith's '*Linnæan Correspondence*,' vol. ii. p. 171.

Mr. Syme considers it 'almost certainly introduced,' and Messrs. Watson and Babington as 'not native;' it is, however, found in France, and throughout Germany, and formerly grew in Belgium, in localities quite similar to ours.]

SILENE, *Linn.*

S. anglica, *L.* Cyb. Br. i. 197; Comp. 114. Syme E. B. ii. t. 202. II. A single plant by a farm road, Tangley Park; a casual straggler, doubtless introduced.

89. S. inflata, *Sm.*   *Bladder Campion.   Spatling Poppy.*
Lychnis sylvestris quæ Ben album vulgò, C. B. P. (Blackst.).
Cyb. Br. i. 195; Comp. 113. Syme E. B. ii. t. 199.

Roadsides and cornfields; rather rare. P. June—August.

I. Harefield, frequent!; *Blackst. Fasc.* 54. Road south of Harrow Weald Common.

II. Staines. Bet. Sunbury and Walton Bridge. Teddington, abundant. Bet. Kingston Bridge and Hampton Court.

III. Roxeth; *Melv.* 12. Hounslow. Twickenham (a var.); *Doody MSS.*

IV. Pinner Drive; *Melv.* 12. Edgware, 1827–30; *Varenne.* Bishop's Wood; *Lawson.* Stanmore.

First record: *Doody*, about 1696.

90. *S. noctiflora, L.*
Cyb. Br. i. 291; iii. 391; Comp. 115. Syme E. B. ii. t. 209.

Waste ground on a sandy soil; very rare. A. July, August.

III. A single plant on the edge of Hounslow Heath, near the road, 1866.

VII. Lane from Walham Green to Hammersmith, 1848; *Morris* (*v.s.*). Parson's Green, 1864; waste ground in Chelsea College, 1858–1863; *Britten.*

First record: *Morris*, 1848. Scarcely naturalised.

LYCHNIS, *Linn.*

91. L. Flos-cuculi, *L.*   *Ragged Robin.*   *Meadow Pink.* *Jagged Pink* (Pet.).

Cyb. Br. i. 205; Comp. 116. Curt. F. L. f. 1.

Wet meadows and heaths; very common. P. May—August. Generally distributed through all the districts.

VII. Site of Exhibition at South Kensington!; *Naylor.* Millfield Lane, Ken Wood, and South Heath, Hampstead.

First record: *Petiver*, 1715.

92. L. vespertina, *Sibth.* Silene pratensis, *Gren. & Godr.* (Syme). *White Campion.*

L. sylvestris alba simplex, C. B. P. (Blackst.). *L. dioïca fl. albo, Sm.* (Herb. G. & R.).

Cyb. Br. i. 206; Comp. 116. Syme E. B. ii. t. 210.

Waste ground, hedges, and clover-fields; common. B. or P. May—Sept.
I. Harefield, frequent; *Blackst. Fasc.* 55. Ruislip; *Farrar.*
II. Common in this district. Staines. Stanwell Moor. Sunbury. Teddington.
III. Hounslow. Hatton. Whitton. Twickenham.
IV. Near Hampstead; *Lawson.*
V. Near Hanwell; *Hemsley.* Chiswick, common. Railway bank, Kew Bridge Station.
VI. Edmonton.
VII. [Marylebone fields, 1817; *Herb. G. & R.*] Craven Hill. Green Park. Constitution Hill. Site of International Exhibition. Blackstock Lane.

First record: *Blackstone*, 1737. This species seems to prefer a light soil. being less common on the stiff clayey parts of the county than on the sands and gravels.

93. L. diurna, *Sibth.* Silene diurna, *Gren. & Godr.* (Syme). *Red Campion.*

L. sylvestris minus hirsuta fl. rubello simplici (Morison). *L. sylvestris rubello flore, Ger. em.* (Blackst.). *L. dioica fl. rubro, Sm.* (Herb. G. & R.).

Cyb. Br. i. 206; Comp. 116. Curt. F. Lond. f. 2.

Woods and thick hedges; very common. P. May—September. Occurs in all the districts; especially abundant in IV.

VII. [Marylebone fields, 1817; *Herb. G. & R.*] Common about the northern outskirts of town, as Kentish Town, &c.

First record : *Morison*, 1682. Most abundant on the clay, and much less frequent in the SW. part of the county. With *white flowers* (I.) Old Park Wood, Harefield. Also (IV.) in Bentley Priory woods ; *Melv.* 13.

94. ***L. Githago,** *Lam.*   *Corn Cockle.*

L. segetum major, C. B. P. (Blackst.).

Cyb. Br. i. 207 ; Comp. 117. Syme E. B. ii. t. 215.

A cornfield weed, and a straggler in waste places ; rather rare. A. June —August.

I. Harefield, not frequent ; *Blackst. Fasc.* 55. Ruislip ; *Melv.* 13.

II. Bet. Kingston Bridge and Hampton Court, on earth brought from elsewhere.

III. Wood End ; *Melv.* 13. A single plant near Twickenham Ry. Station.

IV. Stanmore, 1815 ; *Varenne.* Midland Railway, near Edgware, a single specimen.

V. Near Ealing! ; near Acton! ; *Newb.* By canal, Apperton, only one plant.

VI. Edmonton.

First record : *Blackstone*, 1737. This, like many cornfield plants, is scarce in Middlesex, from corn being only cultivated in a small part of the county. In other parts it becomes a wayside weed of uncertain appearance.

(Buffonia annua, *DC. Alsine polygonoides tenuifolia, flosculis ad longitudinem caulis velut in spicam dispositis, Pluk.* (R. Syn. ii.). *B. tenuifolia, Sm.* (M. & G.). Cyb. Br. i. 234. E. B. t. 1313. III. ' On Hounslow Heath ;' *MS. note by Doody in his copy of R. Syn.* ii. *opp. p.* 210, ' 7,' and copied by Dillenius in *R. Syn.* iii. 346. We found it . . . . on Hounslow Heath, where Mr. Doody observed it ; *M. & G.* 199. Mr. Watson suggests that *Moenchia erecta* was mistaken for it. It seems to us more probable that Doody accidentally wrote his note of the locality opposite the wrong species, intending perhaps to refer to ' An. 8 ' = *Spergula.* Milne and Gordon may have found a tall-growing plant of *Juncus bufonius.* The Eng. Bot. figure was drawn from a garden specimen, and the continental distribution of the species does not favour the probability of its having occurred here.)

SAGINA, *Linn.*

95. **S. procumbens,** *L.*   *Pearlwort.*

Saxifraga pusilla graminea fl. parvo herbido, et muscoso (Merrett).
Saxifraga anglica alsinefolia, Ger. em. (Blackst.).

Cyb. Br. i. 209 ; iii. 392 ; Comp. 118. Syme E. B. ii. t. 248.

Between paving-stones, on damp walls, gravel paths, and wet sandy places; common. P. May—September.

I. Harefield!; *Blackst. Fasc.* 91. Eastcott. Near Potter's Bar.

II. Sunbury. Teddington Station.

III. Bet. Ashford and Hounslow Heath; *Curt. F. L.* f. 3. Hounslow Heath, abundant. Walls at Harrow; *Herb. Harr.*

IV. North Heath, Hampstead.

V. Walls at Cheswick; *Merrett,* 109. Brick tombs in Chiswick churchyard. Acton; *Tucker (v. s.).*

VI. Edmonton. Hadley!; *Warren.*

VII. [River wall of Temple Gardens.] Hampstead. [Courtyard of Burlington House.] Kensington Gardens, near the Scotch Firs, 1868; *Warren (v. s.).*

First record: *Merrett,* 1666.

96. S. apetala, *L.*

Saxifraga anglica alsinefolia annua, Plot. (R. Syn. iii.).

Cyb. Br. i. 210; Comp. 117. Syme E. B. ii. t. 246.

On walls and dry ground, especially gravel walks; common. A. May—September.

I. Eastcott; *Melv.* 13.

II. About Sunbury. Hampton.

III. Twickenham.

IV. Near Harrow Station; *Melv.* 13. Harrow Weald churchyard. Edgware and Whitchurch churchyards.

V. Near Ealing!; Hanwell churchyard!; *Newb.* Chiswick.

VI. Edmonton.

VII. Common in nearly all garden-walks about London; *Dill. in R. Syn.* iii. 478.

First record: *Dillenius,* 1724. A variety is recorded by Mr. Hind from Ruislip (v. *Melv.* 13).

97. S. ciliata, *Fries.*

Cyb. Br. Supp. 52, 81; Comp. 118. Syme E. B. ii. t. 147.

Roadsides and dry places, on a sandy soil; rare. A. June—August.

I. South side of Harrow Weald Common.

II. Road from Staines to Hampton, in several places.

III. Hounslow Heath, abundant. Drilling Ground, Hounslow.

VII. Chelsea Hospital; *Irv. H. B. P.* 769.

First record: *Irvine,* 1858.

98. S. subulata, *Wimm.*

Spergula saginoides, L. (Curt.). *Spergula subulata, Swartz* (Sm.).

Cyb. Br. i. 212; iii. 393; Comp. 118. Curt. F. L. f. 4.

Damp sandy heaths; very rare. P. June—August.

I. Mr. Lightfoot showed it me . . . . on Uxbridge Moor; *Curt. F. L.*

VII. A weed in Chelsea Botanic Garden, 1825; *Winch MSS.* No doubt
introduced there.

First record: *Curtis*, about 1790. We have seen no Middlesex
specimens.

99. S. nodosa, *E. Meyer.*
Spergula nodosa, L. (Sm.). *Alsine palustris fol. tenuissimis, Park.*
(Blackst.).
Cyb. Br. i. 213; Comp. 119. Curt. F. L. f. 4 (drawn from a Middle-
sex specimen).

Damp places on heaths; very rare. P. June—August.
I. Uxbridge Moor, abundantly, Mr. Hill, 1736; *Blackst. MSS.* Hare-
field Moor; *Blackst. Spec.* 3.
III. Hounslow Heath; *Curt. F. L.*
IV. On Hampstead Heath; *Huds.* i. 178.

First record: *Hill*, 1736. Not met with recently, but there is no other
reason to believe it extinct in all the above stations.

ALSINE, *Wahl.*

100. A. tenuifolia, *Wahl.* Arenaria tenuifolia, *L.* (Lond. Cat.).
Cyb. Br. i. 217; iii. 394; Comp. 121. Syme E. B. ii. t. 243.

Dry sandy places; very rare. A. May, June.
II. Hampton Court; *Mr. G. Francis (v. s.).*
No other locality recorded. We have not met with it.

ARENARIA, *Linn.*

101. A. trinervis, *L.*
Alsine plantaginis folio, J. B. (Blackst.).
Cyb. Br. i. 215; Comp. 122. Curt. F. L. f. 4.

Woods and hedges; common. A. May, June.
I. Harefield; *Blackst. Fasc.* 5. Pinner. Road from Elstree to Penni-
wells. South Mims.
II. Staines. Near Hampton Court!; *Newb.*
III. Twickenham, 1843, T. Twining; *Herb. Brit. Mus.*
IV. Harrow Weald; Harrow Park, &c.; *Melv.* 14. Deacon's Hill.
Moist parts of woods about Highgate and Hampstead; *Pet. Midd.*
Bishop's Wood and North Heath, Hampstead.
V. Horsington Hill. Near Willesden Station!; *Warren.*
VI. Whetstone. Near Enfield. Hadley!; *Warren.*
VII. [Marylebone fields, 1817; *Herb. G. & R.*]

First record: *Petiver*, 1695.

102. A. serpyllifolia, *L.*
Alsine minor, Tab. (Johns.). *Alsine minor multicaulis, C. B. P.* (Blackst.).
Cyb. Br. i. 217; iii. 394; Comp. 120.

Walls and dry fields; common. A. May—October.

1. *A. serpyllifolia*, L. (Bab.). Syme E. B. ii. 235.

I. Uxbridge, 1839, H. Kingsley; *Herb. Brit. Mus.* Harefield; *Blackst. Fasc.* 5. Iu Harefield chalk-pits.

II. Teddington.

III. Twickenham, abundant. Isleworth. Harrow; *Melv.* 14.

IV. Hampstead Heath; *Johns. Enum.*

V. Turnham Green, abundant. Hanwell!; *Warren.*

VI. Edmonton.

VII. Hampstead. Clapton, 1846; *Morris (v. s.).*

2. *A. leptoclados*, Guss. Syme E. B. ii. t. 236.

I. Harefield!; *Newb.*

II. Bet. W. Drayton and Yewsley!; *Newb.* Staines.

III. Field near Hounslow!; *Newb.* Twickenham. Isleworth. Harrow; *Herb. Harr.*

V. Acton Green!; *Newb.* South wall of Chiswick House.

VI. Colney Hatch!; *Newb.* Clapton; *Morris (v. s.).*

First record: *Johnson*, 1632. *A. leptoclados* will probably turn out to be the commoner Middlesex plant.

STELLARIA, *Linn.*

103. S. media, *Wither.*   *Chickweed.*

Alsine media, C. B. P. (Blackst.).

Cyb. Br. i. 222; Comp. 122. Syme E. B. ii. t. 229.

Cultivated and waste land; very common. A. February—October.

Abundant throughout all the districts, but we have few notes of its occurrence in I.

VII. In London itself it is a common weed, growing in the squares, and on pieces of waste ground everywhere.

The apetalous form (*var. β*, Bab.), called *S. Boræana* by Jordan, was met with (V.) in a sandy field at Chiswick. (VII.) Isle of Dogs!; *Newb.*

First record: *Blackstone*, 1737.

104. S. Holostea, *L.*   *Great Stitchwort. All-bones.*

Cyb. Br. i. 223; Comp. 123. Syme E. B. ii. t. 230.

Woods and hedges, among grass; common. P. April—June.

I. Potter's Bar. South Mims. Harefield!; *Newb.*

II. Staines.

III. Hatch End. Heston.

IV. Stanmore. Harrow. Hampstead; *Pet. H. B. Cat.* Bishop's and Turner's Woods, and North Heath, Hampstead.

V. Near Hanwell!; *Newb.* Acton; *Tucker (v. s.).* Near Brentford; *Hemsley.* Near Willesden Station; *Warren.*

VI. Colney Hatch. Whetstone. Edmonton. Near Enfield; *Tucker.*

VII. Marylebone fields, 1815; *Herb. G. & R.* Near Lea Bridge; *Cherry.* Hornsey Wood.

First record: *Petiver*, 1715. This is not included in *Blackst. Fasc.*

105. S. glauca, *Withering.*

Cyb. Br. i. 223; iii. 395; Comp. 123.   Syme E. B. ii. t. 231.

Shallow pools and marshes; rare.   P.   June—August.

II. Ditch by the road leading from Buckinghamshire to Staines, 1860; *Phyt. N.S.* iv. 263.

III. Marshy ground by the Cran near Hatton.

IV. Beside the Brent, near Stonebridge; *Hind.*   Ditch near plantation by the N.W. Ry. bridge over Brent, H. J. Wharton; *Farrar.*

First record: '*J. W. T. and A. J.*,' 1860.

106. S. graminea, *L.*   *Small Stitchwort.*

Holosteum Ruellii; gramen leucanthemum minus (Johns.).

Cyb. Br. i. 224; Comp. 123.   Syme E. B. ii. t. 232.

Roadsides, heaths, and hedgebanks. especially on a gravelly and sandy soil; common.   P.   May—August.

I. Harefield!; *Newb.*   Harrow Weald Common; Ruislip; *Melv.* 14. Elstree.   Potter's Bar.   South Mims, abundant.

II. Roadsides about Staines, Ashford, &c.; abundant.   Bet. Kingston Bridge and Hampton Court.

III. Hounslow Heath.   Hatton.   Twickenham.

IV. Near Bishop's Wood.   Football field, Harrow; *Melv.* 14.

V. Apperton.   Chiswick.   Near Brentford; *Hemsley.*

VI. Edmonton.   Hadley!; *Warren.*

VII. [Marylebone fields, 1815; *Herb. G. & R.*]   Bet. Kentish Town and Hampstead; *Johns. Eric.*   Kensington Gardens, 1845; *Morris* (*v. s.*).   Hornsey; *Cherry.*   South Heath, Hampstead.   Hackney Wick.

First record: *Johnson*, 1629.

107. S. uliginosa, *Murr.*

Alsine long. ulig. prov. locis, J. B. (Pet.). *Alsine aquatica media, C. B. P.* (Blackst.).

Cyb. Br. i. 225; Comp. 123.   Syme E. B. ii. 233.

In shallow water, marshes, &c.; rather common.   A.   May—July.

I. Harefield, rarely; *Blackst. Fasc.* 4.   Ruislip; *Melvill,* 14.   Eastcott. Harrow Weald Common.   Elstree.

III. Gravel pit near Hounslow.

IV. Sheepcote Farm, near Harrow; *Melv.* 14.   Wetter parts of Hampstead woods; *Pet. Midd.*   Bishop's and Turner's Woods, and North Heath, Hampstead.   Deacon's Hill.   Near Finchley!; *Newb.*

VI. Copse near Whetstone.   Edmonton.   Hadley!; *Warren.*

VII. Kensington Gardens, 1817; *Herb. G. & R.*   South Heath, Hampstead.

First record: *Petiver*, 1695.

108. S. aquatica, *Scop.* Malachium aquat., *Fries* (Bab.). Cerastium aquat., *L.* (L. Cat.). *Water Chickweed.*
Alsine cochleariæ longæ facie (Merrett). *Alsine aquatica major, C. B. P.* (Blackst.).
Cyb. Br. i. 226; iii. 396; Comp. 124. Curt. F. L. f. 1.

Ditches and damp waste ground; common. P. July—September.
I. Harefield!; *Blackst. Fasc.* 4. Elstree Reservoir; *Fl. Herts.* 358. Colne, near Yewsley!; *Newb.*
II. West Drayton!; *Newb.* Staines and Stanwell Moors. Near Strawberry Hill. Between Kingston Bridge and Hampton Court, on made ground.
III. Waste ground near Twickenham Ry. Station. By the river near Twickenham Church, on made ground.
IV. Harrow Weald; *Melvill,* 15. Stanmore. Edgware; *Varenne.* Near Neesdon, common; *Farrar.*
V. Bet. Acton and Turnham Green!; *Newb.* Brent, near Hanwell!; *Warren.*
VI. About Edmonton, in several places.
VII. Divers places near London; *Merrett,* 4. Bet. Chelsea Hospital and the river, 1861. Hackney Wick, in several places.

First record: *Merrett,* 1666. Appears to be less common by the Thames, than by its tributaries.

CERASTIUM, *Linn.*

109. C. glomeratum, *Thuill.* *Mouse-ear Chickweed.*
Alsine myositis humilior et rotundo fol. (Merr.). *Alsine hirsuta magno flore, C. B. P.* (Blackst.). *C. vulgatum, L.* (Sm.). *C. viscosum, L.* (Curt.).
Cyb. Br. i. 227; Comp. 124. Curt. F. L. f. 2.

Heaths, dry roadsides, and walls; common. A. April—September.
I. Harefield!; *Blackst. Fasc.* 4. Near Uxbridge, 1839, D. Cooper; *Herb. Young.*
II. Near Staines, on Hampton Road.
III. Hounslow Heath, abundant. Hatton. Twickenham.
IV. Sudbury; *Melvill,* 14. Near Whetstone. Hampstead Heath; *Warren.*
V. Apperton; *Melvill,* 14. Chiswick. Hanwell!; *Newb.*
VI. Colney Hatch. Edmonton. Hadley!; *Newb.*
VII. Near Hampstead Church and in Hyde Park; *Merrett,* 6.
First record: *Merrett,* 1666.

110. C. triviale, *Link.*
Alsine myositis procerior et longiore fol. (Merr.). *Alsine hirsuta altera viscosa, C. B. P.* (Blackst.). *C. viscosum, L.* (Sm.). *C. vulgatum, L.* (Curt.).
Cyb. Br. i. 228; Comp. 124. Curt. F. L. f. 2.

Meadows, and rather damp places; very common. B. or P. April—September.

Found in all the districts, and generally abundant.

VII. Common. Primrose Hill. Kensington Gore. Kensington Gardens. Hackney Wick. Isle of Dogs, &c.

First record: *Merrett*, 1666. Small specimens are often named *C. semidecandrum*; this is the case with the Sudbury plant in *Herb. Harr.* recorded at *Melv.* 15.

111. C. semidecandrum, *L.*
C. hirsutum minus parvo flore, Cat. Giss. (R. Syn. iii.).
Cyb. Br. iii. 327 ; Comp. 125. Curt. F. L. f. 2.
Dry, sandy places and wall-tops ; rather rare. A. April—June.
II. Hampton Court ; *Newb.*
III. Hounslow Heath, abundant. Wall, Twickenham.
IV. Some parts of Hampstead Heath, abundant! ; *Warren.*
V. Abundant in a sandy field at Chiswick.
VII. Frequent about London ; *Dill. in R. Syn.* iii. 348. About Hackney particularly ; *Curt. F. L.* Kensington, 1863 ; *Newb.*
First record : *Dillenius*, 1724 ; also the first record as British.

112. C. arvense, *L.*   *Field Chickweed.*
Cyb. Br. i. 230 ; ii. 396 ; Comp. 125. Curt. F. L. f. 6.
Dry banks and fields ; rare. P. May—July.
II. By the Thames below Hampton Court Bridge, H. C. W. ; *New B. G.* 98. By the towing-path near Hampton Court, abundant. Probably same locality.
IV. Near Hampstead, 1845 ; *Morris (v. s.).* Side of N. W. Railway, bet. Harrow Station and Kenton Bridge, in small quantity ; *Melvill,* 15 (*v. s.*).
V. Abundant in a sandy field by the Thames at Chiswick.
First record : *H. C. Watson*, 1835.

MOENCHIA, *Ehrh.*

113. M. erecta, *Sm.* Cerastium quaternellum, *Fenzl.* (Syme E. B.).
Holosteum minimum tetrapetalon, &c., Ray (Blackst.). *Sagina erecta, L.* (Sm.). *Moenchia glauca, Pers.* (New B. G.).
Cyb. Br. i. 208 ; iii. 392 ; Comp. 117. Curt. F. L. f. 2.
Dry, sandy places ; rare. A. April, May.
I. On Harefield and Uxbridge Commons, abundantly ; *Blackst. Fasc.* 42. Road bet. Harefield and Ruislip.
III. Hounslow Heath, abundant. Near Hatton.
IV. About Highwood Hill ; *Blackst. Spec.* 38. On Hampstead Heath! ; *Pet. Midd.*
VII. [Hyde Park ; *Dicks. H. S.* In the dry part north of the Magazine,

1820; *Bennett* (*v. s.*).] [Old wall in the King's Road, Chelsea, Pamplin; *New B. G.* 98.]

First record : *Petiver*, 1695.

Polycarpon tetraphyllum, *L.* Comp. 176. Syme E. B. ii. t. 258. Garden at Harrow, occurring by accident, Hind; *Melvill*, 31.

LEPIGONUM, *Fries.*

114. L. rubrum, *Fr.* Spergularia rubra, *Fenzl.* (Syme and L. Cat.). *Red Spurrey.*

Alsine spergulæ facie (Johns.). *Spergula purpurea* (R. Syn.). *Arenaria rubra, L.* (Sm.).

Cyb. Br. i. 220 ; iii. 394 ; Comp. 120. Syme E. B. ii. f. 254.

Heaths, commons, and dry roadsides ; common. A. or B. May—September.

I. Harrow Weald Common. Stanmore Heath. Harefield! ; *Newb.*
II. Near Teddington Station.
III. Whitton ; *Twining* (*v. s.*). Hounslow Heath. Drilling Ground. Hatton.
IV. Peterboro' Road, Harrow ; *Melv.* 31. Hampstead Heath. Stanmore.
V. Gravel pits near Hanwell! ; *Newb.*
VI. About Highgate; *Cooper*, 104. Near the Hyde, Edmonton. Hadley! ; *Warren.*
VII. [Tothill Fields, nigh Westminster ; *Johns. Ger.* 1125.] Hyde Park ; *Blackst. MSS.* South Heath, Hampstead. Kensington Gardens, 1866.

First record : *Johnson*, 1633.

SPERGULA, *Linn.*

115. S. arvensis, *Linn.* *Corn Spurrey.*

Cyb. Br. i. 214; Comp. 119.

Sandy fields, heaths, and roadsides ; common. A. June—September.

Var. a, sativa. Curt. F. L. f. 5.
II. Teddington Station. Strawberry Hill. Bet. Kingston Bridge and Hampton Court.
III. Near Hospital Bridge. Near Hounslow.
IV. Hampstead Heath, and neighbourhood of 'The Spaniards' ; *Curt. F. L.*
V. Chiswick. Apperton.
VI. Edmouton. Near Enfield ; *Tucker.*
VII. Cremorne. Kensington Gore. Shepherd's Bush. Green Park. Adelaide Road. Thames Embankment. I. of Dogs.

Var. β, vulgaris. Syme E. B. ii. t. 253.
I. Stanmore Heath. Harrow Weald Common. Harefield ! ; *Newb.*

II. Near Teddington Station. Road bet. Staines and Hampton.
III. Hounslow Heath. Twickenham Park.
IV. Field in Bishop's Wood.
VII. Cromorne. Shepherd's Bush. Kensington Gore. South Heath, Hampstead. [Court-yard of Burlington House, Piccadilly, 1868.]

First record: *Curtis*, about 1790. The two forms frequently grow together, but *var.* β seems rather less frequent.

MALVACEÆ.

MALVA, *Linn.*

116. M. moschata, *L.*   *Musk Mallow.*
Malva verbenaca (Ger.).   *Alcea* vulgaris, Dod.* (Johns.).
Cyb. Br. i. 238; Comp. 127. Curt. F. L. f. 4.

Fields and roadsides, on a dry soil; rather rare. P. July—September.
I. Harefield!; *Blackst. Fasc.* 3.   Ruislip; Pinner Hill; *Melv.* 16.
Woodready. Stanmore Heath. Road bet. Penniwells and Elstree.
Harefield Common, 1866.
II. One plant near the church, Twickenham Common.
IV. Harrow; Kingsbury; *Varenne.* Field bet. Turner's Wood and North
End, J. Bliss; *Park Hampst.* 30. Hampstead Heath; *Johns. Enum.*
Stanmore, near the Heath. Stone Bridge; *Farrar.*
V. Horsington Hill. G. W. Ry. bank, near Hanwell Station. Wood in
Osterley Park; *Irvine.* Perivale; *Lees.*
VI. Wood Green, Munby: *Naturalist,* 1867, 181. Edmonton.
VII. [On the left hand of Tyborne; bet. London and a bathing-place called
the Old Foorde; in the bushes as you go to Hackney; *Ger.* 786.]
[In a field as you go from Hamsteed Church to the town; *Park.
Theatr.* 306.] [Meadow near Highgate on the left from London;
Blackst. MSS.] [New churchyard, Hampstead, *E. H. Button; Coop.
Supp.* 11.]

First record: *Gerarde,* 1597; also first notice as British.

117. M. sylvestris, *L.*   *Mallow.*
Cyb. Br. i. 238; Comp. 127. Curt. F. L. f. 2.

Roadsides and waste places; common. P. June—September.
I. Harefield!; *Blackst. Fasc.* 58. South Mims.
II. Staines. Hampton Court!; *Newb.*
III. Twickenham. Harrow; *Melv.* 16.
IV. Stanmore churchyard. Stanmore village.
V. Greenford; *Melv.* 16. Apperton. Hanwell; *Newb.*
VI. Enfield. Edmonton. Southgate; *Tucker.*

* Misprinted '*Aloca*' in Ralph's reprint.

VII. Hampstead; *Irv. MSS.* Old St. Pancras Churchyard. Near Chelsea Hospital. Thames Embankment. I. of Dogs. Hyde Park; *Newb.*
First record : *Blackstone,* 1737. No doubt overlooked in other places.

118. M. rotundifolia, *L.*   *Small Mallow.*
M. sylvestris pumila, Ger. em. (Blackst.).
Cyb. Br. i. 239 ; Comp. 128. Curt. F. L. f. 3.
Waste places, commons, and roadsides ; common. P. June—September.

I. Harefield! ; *Blackst. Fasc.* 58. Harrow Weald Common. Stanmore Heath.
II. Staines. Sunbury, in many places. Hanworth. Bet. Hampton Court and Kingston Bridge.
III. Twickenham. Isleworth. Hounslow. Harrow; Whitton ; *Hemsley.* Roxeth ; Headstone ; *Mclv.* 16.
IV. Stanmore, in several places. North Heath, Hampstead. Peterboro' Road, Harrow ; *Melv.* 16. Neesdon ; *Farrar.*
V. Brentford. Norwood. Acton! ; *Newb.*
VI. Bury Street, Edmonton.
VII. Kensal Green ; *Cole.* Chelsea College ; *Britten.* South Heath, Hampstead.

First record : *Blackstone,* 1737.

M. parviflora, L. Herb.? V. Waste ground by the canal, Apperton. This seems different from *M. parviflora* of Hudson—*M. pusilla* of *E. B.* 241, reproduced in *Syme E. B.* ii. t. 283 as *M. borealis,* Wallm. ; it also differs from the figure in *Jacq. Vindob.* 39. Smith does not allude to the fruit characters in the description of his *M. pusilla* ; nor are the remarkable carpels figured in his plate ; there is no specimen in his herbarium. The fruit added in Syme's plate is quite that of our plant. The same species occurs at Mitcham (v. *Seem. J. of Bot.* iv. 149), where it was introduced with foreign corn; it is a tall, upright plant, thus differing from *M. parviflora* of the *Linn. Herb.*

? *M. crispa, L.* Site of Exhibition of 1862. A doubtful species, not uncommon in gardens.

ALTHEA, *Linn.*

[119. A. officinalis, *L.*   *Marsh Mallow.*
Cyb. Br. i. 240 ; iii. 397 ; Comp. 128. Syme E. B. ii. t. 278.
VII. By the Thames side at the I. of Dogs just before you come to the ferry for Greenwich ; *MS. note (Alchorne's) quoted in Phyt.* iii. 166. I once (1737) gathered it by the side of Chelsea Waterworks ; *Blackst. Spec.* 4. Naturalised in a lane leading up to the Wood House, Hornsey, from Newington ; *Lond. Fl.* 181.

First record: *Blackstone*, 1737. In the first of these stations it was probably a native, in the other two a naturalised plant.]

Hibiscus Trionum, *L.*   VII. Site of Exhibition of 1862, S. Kensington,
1867.

___

TILIACEÆ.

Tilia europæa, *L.*   *T. intermedia, DC.* (Syme, and L. Cat.).   *Lime.* Cyb.
Br. i. 244.   Syme E. B. ii. 286.   Tree.   July.   Probably always
planted in the places where it is found in this county —parks, pleasure-
grounds, and garden hedges.   There are fine trees in the great avenue
of Bushey Park, and it grows by the towing-path bet. Kingston Bridge
and Hampton Court.   Loudon states that at Sion are trees supposed
to be 80 years old, 75 feet high; and at Ken Wood 90 ft. high, planted
90 years; the oldest tree he knew of is at Fulham Palace, and has
a trunk 7 or 8 ft. in diameter; *Arb. et Frut. Brit.* i. 371.   Doody
notices a variety ' on Enfield Green going to the Chase;' *Dill. in R.
Syn.* iii. 474.

T. parvifolia, *Ehrh.*   Cyb. Br. i. 243, iii. 398; Comp. 128.   Syme E. B.
ii. t. 287.   Tree.   August.   This is said to be truly native in old woods
in England; in Middlesex, however, it is everywhere planted, and is
found in Bushey Park, at Sion, and in many of the London parks, as
St. James's, with the other species.

T. grandifolia, *Ehrh.*   Cyb. Br. i. 245; Comp. 129.   Syme E. B. ii. t. 285.
Tree.   June, July.   Planted with the other species at Sion, and at
Hampton Court; *Sm. E. Fl.* iii. 20.   At Hampstead ; *Lond. Fl.* 181.

Some North American species are also frequently planted ; *Bentham*, 93.

___

HYPERICACEÆ.

HYPERICUM, *Linn.*

120. * H. calycinum, *L.*

Cyb. Br. i. 253.   Syme E. B. ii. t. 267.

Shrubberies and hedges.   P.   July—October.

IV. Julians, Harrow ; *Hind.*   Lane by Lord Southampton's Park, Highgate, Mrs. Lowe ; *Cooper*, 110.

V. In great plenty in the grounds of Chiswick House.

Long cultivated as an ornamental plant, and now semi-wild in the places
where it has been originally planted.

121. H. Androsæmum, *L.*   *Tutsan.*

Clymenon Italorum (Ger.).   *Androsæmum vulgare, Park.* (Merr.).

Cyb. Br. i. 245; iii. 398; Comp. 129.   Curt. F. L. f. 3.

Bushy places and shady lanes ; very rare.   Small shrub.   June—
August.

I. In a thicket near Harefield Church; also near Bacher Heath;
Blackst. Fasc. 44. Harefield; *J. F. Young MSS.*

IV. [Hampstead Heath; *Johns. Eric.*] [Hampstead Wood; *Ger.,* 435,
Park. Theat., 577, and many subsequent writers. In the drier parts
of Hampstead and Highgate Woods, *Pet. Midd.*; common till these
were grubbed up and the land subjected to cultivation; *Loud. Arb.
et Frut. Brit.* i. 403.] [Stanmore, 1827-30; *Varenne.* Hedge on
the left hand of the lane leading from Stanmore to Stanmore Heath,
1858.]

VI. Side of road bet. Highgate and Muswell Hill, 1745; *Blackst. Spec.* 38.
Near Enfield, collected by Mr. Irvine; *Cole.*

First record: *Gerarde,* 1597. Formerly more frequent in the county.

122. H. quadrangulum, *L.* H. tetrapterum, *Fries* (Syme and L. Cat.).
St. Peter's Wort.

Ascyron (Turn.). *Hyp. Ascyron dict. caule quadrangulo, J. B.* (Blackst.).
Cyb. Br. i. 248; Comp. 130. Curt. F. L. f. 4.

Ditches and sides of streams; common. P. July, August.

I. Harefield!; *Blackst. Fasc.* 44. Pinner. Stanmore Heath. Woodready. Elstree.

II. Staines Moor. Bet. Hampton and Hampton Court!; *Newb.* Hanworth Park.

III. By the Cran on Hounslow Heath. Whitton. Twickenham. Roxeth;
Herb. Harrow.

IV. Mill Hill, Miss Atkins; *Herb. Brit. Mus.* Bet. Hampstead and new
road to Barnet, Mann; *Cooper,* 116. Stanmore.

V. Sion Park; *Turn. Names.* Canal near Hanwell; *Hemsley.* Near
Lampton. Perivale; *Lees.*

VI. Highgate Wood; *Johns. Eric.* Wood Green, Munby; *Nat.* 1867, 181.
Edmonton.

VII. South Heath, Hampstead. Near Temple Mills; *Cherry.*

First record: *Turner,* 1548; also first record as British.

123. H. perforatum, *L.* *St. John's Wort.*
Cyb. Br. i. 215; iii. 398; Comp. 129. Curt. F. L. f. 1.

Hedgebanks and roadsides; common. P. July, August.

I. Harefield Common. Harrow Weald Common. Pinner. Elstree.
Potter's Bar. South Mims.

II. Near Charlton. Staines!; *Newb.* Bet. Kingston Bridge and Hampton
Court.

III. Hounslow. Twickenham. Harrow; *Herb. Harrow.*

IV. Hendon; *Cole.* Harrow Weald. Deacon's Hill. North Heath, Hampstead. Sudbury Station; *G. Johnson.* Finchley Road; *Newb.*

V. Near Ealing!; *Newb.* Horsington Hill. Hanwell!; *Warren.*

VI. Whetstone. Edmonton.

VII. Edgeware Road, 1815; *Herb. G. & R.*

First record: *Goodger and Rozea*, 1815. A form with very narrow strapshaped leaves occurs (II.) near Staines on the Hampton Road. *H. lineolatum*, Jord., if we understand that plant, seems more frequent than the type.

124. ? H. dubium, *Leers.*
Cyb. Br. i. 247; iii. 399 ; Comp. 129.   Syme E. B. ii. t. 269.
Sides of ditches, &c.; very rare.   P.   July.
II. One plant under the hedge by the towing-path by the river, bet. Kingston Bridge and Hampton Court, nearly opposite Thames Ditton, 1867 ; *Bloxam.*
III. ? Mr. Hind thinks he gathered it somewhere near Harrow before 1860, but he has kept no specimen; nor are there any in *Herb. Harrow.*
Requires confirmation. Said (*Fl. Herts.* 51) to grow at North Mims, just beyond the Middlesex boundary. (See also *Fl. Brit.* 1404.)

125. H. humifusum, *L.*
H. minus supinum, C. B. P. (Blackst.).
Cyb. Br. i. 249; Comp. 130.   Curt. F. L. f. 3.
Heaths and dry roadsides ; rather common.   P.   July, August.
I. Harefield! ; *Blackst. Fasc.* 44.   Stanmore Heath.   Ruislip Common. Harrow Weald Common ; *Melvill,* 17.
II. Fulwell.   Tangley Park.   Near Staines, on road to Hampton.
III. Near Hounslow.   Near Hatton.
IV. Deacon's Hill.
V. Hanwell, 1846 ; *Morris* (*v.s.*).
VI. Hadley! ; *Warren.*
VII. [Hyde Park, 1815 ; *Herb. G. & R.*]   South Heath, Hampstead.
First record: *Blackstone,* 1737.

126. H. hirsutum, *L.*   *Tutsan St. John's Wort.*
H. Androsæmum dictum, J. B. (Blackst.).
Cyb. Br. i. 251 ; iii. 399 ; Comp. 131.   Curt. F. L. f. 3.
Woods and hedges ; rather rare.   P.   July.
I. Harefield, frequent! ; *Blackst. Fasc.* 43.   Pinner.
II. Staines.
IV. Common in this district.   Hampstead; *Pet. H. B. Cat.*   Bet. Golder's Green and Finchley, 1861 ; *Phyt. N. S.* v. 319.   Green Hill! ; near Roxborough Turnpike ; Station Road, Harrow! ; *Melvill,* 17. Bentley Priory.   Scratch Wood, and fields near it.   Harrow Weald.   Edgware, abundant.
V. Hedge by Horsington Hill.   Near the Brent, Hanwell! ; *Warren.*
VII. [Duval's Lane, bet. Islington and Hornsey, 1816 ; *Bennett* (*v.s.*).] [Regent's Park, 1827–30 ; *Varenne.*] [Field bet. Kentish Town and South End, Hampstead ; *Herb. Hardw.*]
First record: *Petiver,* 1715.   Almost restricted to the chalk and clay of the county.

127. H. pulchrum, *L.*

H. pulchrum Tragi (Blackst.).

Cyb. Br. i. 250; Comp. 131. Syme E. B. ii. t. 273.

Dry heaths and roadsides, also dry parts of woods ; rather common. P. June—August.

I. Old Park and other woods at Harefield!; *Blackst. Fasc.* 44. Ruislip; Pinner; *Melvill*, 17. Harefield Common. Harrow Weald Common. Stanmore Heath. Long Lane, Hillingdon; *Warren*.

II. Abundant by roadsides bet. Staines and Hampton.

III. Hounslow Heath. Drilling Ground, &c.

IV. Woods about Hampstead and Highwood Hill; *Blackst. Spec.* 38. North Heath, Hampstead. Deacon's Hill.

VI. Colney Hatch. Edmonton. About Highgate; *Cooper*, 104. Near Trent Park, Southgate; *Cole*. Hadley!; *Warren*. Near Enfield; *Tucker*.

VII. [In St. John's Wood* and other places ; *Johns. Ger.* 540.] Hornsey Wood, 1815; *Herb. G. & R.* Ken Wood ; *Burnett*, 105.

First record : *Johnson*, 1633.

[128. H. elodes, *L.*

H. palustre supinum tomentosum (Blackst.).

Cyb. Br. i. 253 ; iii. 399 ; Comp. 131. Syme E. B. ii. t. 276.

Boggy places on heaths ; very rare. P. July, August.

III. On Hounslow Heath, where *Millefolium galericulatum minus* (= *Utricularia minor*) grows ; *Doody MSS.* and *Mart. Tourn.* ii. 32.

IV. On the great bog on Hampstead Heath ; *Blackst. Spec.* 39, *and many subsequent writers.*

First record : *Doody*, 1696. Last : *Hudson*, 1762.]

ACERACEÆ.

ACER, *L.*

129. A. campestre, *L.*　　　*Maple.*

Acer minus, Ger. (Blackst.).

Cyb. Br. i. 254; iii. 399 ; Comp. 132. Syme E. B. ii. 321.

Hedges and woods; very common. Tree. May.

Throughout the county. Appears scarce about Twickenham in III.

VII. About the northern outskirts of London; Kentish Town, Hornsey Wood, Green Lanes, &c.

First record : *Blackstone*, 1737. There are some very old trees (IV.) on North Heath, Hampstead.

---

* Probably Little St. John's Wood at Highbury ; see note to **151.** Great St. John's Wood was in Marylebone.

130.   * A. pseudo-Platanus, *L.*   *Sycamore.*

Cyb. Br. i. 255; iii. 400.   Syme E. B. ii. t. 320.

Hedges and woods; common.   Tree.   May.

An ornamental tree, commonly planted in parks and woods.   Though
not native, it is naturalised, and many existing trees are no doubt
self-sown, as those in the hedges about (I.) South Mims, and
(III.) Hounslow, Hatton, &c.

Several American species are also cultivated in plantations.

GERANIACEÆ.

GERANIUM, *L.*

131.   G. pratense, *L.*   *Meadow Crane's-Bill.*

Cyb. Br. i. 260; iii. 401; Comp. 133.   Curt. F. L. f. 4.

Sides of streams and meadows; rare.   P.   June—September.

I. Pinner Hill, Hind; *Melvill*, 18.

II. Near the Thames, in this district frequent.   Bet. Chertsey Bridge and
Sheperton, E. Forster; *B. G.* 409.   Bet. Sunbury and Hampton!;
Newb.   Bet. Hampton Court and Kingston Bridge, abundant.   Bet.
Strawberry Hill and Teddington.   Staines Moor.

III. By the Cran, Marsh Farm, Twickenham.   [At Harrow; *Sm. E. B.*,
404.]

IV. Kingsbury Churchyard, one plant; *Farrar.*

First record: *J. E. Smith*, 1797.

G. Phæum, *L.*   Cyb. Br. i. 259; iii. 400.   Syme E. B. ii. t. 294.
IV. Field near Mrs. Butler's garden, Julian Hill, near Harrow;
Melvill, 18.   Doubtless an escape from the garden.

G. striatum, *L.*   *Pencilled Geranium.*   Cyb. Br. i. 258.   I. Well established on a hedgebank in Pinner Hill grounds;* *Hind.*   Commonly
cultivated in gardens.

132.   G. pyrenaicum, *L.*

G. perenne (Huds.).

Cyb. Br. i. 261; iii. 401; Comp. 133.   Curt. F. L. f. 3.

Waste ground; rather rare.   P.   May—July.

I. Plentiful at Eastcott; *Hind.*   Harefield!; near the church; *Cole.*

III. Twickenham, 1838; *Herb. Young.*

V. Turnham Green.   About Chiswick, abundant; the railway bank bet.
the bridge over the Thames at Barnes and Chiswick Station is
covered with it.   Acton; *Tucker* (*v. s.*).

* The plant in this locality is erroneously referred to *G. pyrenaicum* in *Melv.* 18.

VI. Near Enfield; *Huds.* i. 265. In and about Enfield Churchyard. Dillwyn; *B. G.* 409. In a hedge by a footpath near the old palace at Enfield, E. Forster; *Herb. Brit. Mus.*

VII. Bet. Hyde Park and Little Chelsea; *Huds.* i. 265. About Brompton; *Huds.* ii. 303. A weed in Chelsea Gardens; *Curt. F. L.* Chelsea; *Irv. H. B. P.* 750. By St. Mark's College !; *Newb.* West London Cemetery, Brompton.

First record: *Hudson*, 1762; also first notice as British. There do not seem to be very good grounds for considering this species an introduction, though it is certainly remarkable that it was not noticed by the numerous good observers before Hudson. Curtis noticed it *with white flowers* in Chelsea Gardens.

133. G. pusillum, *L.*

G. columbinum majus, fl. minore cæruleo (Dill.). *G. parviflorum* (Curt.). Cyb. Br. i. 263; iii. 402; Comp. 134. Curt. F. L. f. 6.

Roadsides and dry places; rather common. A. May—September.

I. Harefield !; *Cole.* Harrow Weald Common. Bet. Yewsley and Iver Bridge !; *Newb.*

II. Road from Staines to Hampton. Near Sunbury. Bet. Hampton and Hampton Court!; *Newb.* Bet. Hampton Court and Kingston Bridge. Near Teddington Railway Station.

III. Near Hounslow. Near Twickenham. Near Richmond Bridge.

V. Near Brentford; *Hemsley.* Chiswick, near the river, abundant.

VI. Colney Hatch.

VII. West side of London, in neglected gardens and fallow fields near Little Chelsea, quite a weed; *Curt. F. L.* Paddington; *Herb. G. & R.* South Heath, Hampstead; *Lawson.* Parson's Green. West London Cemetery, Brompton. Isle of Dogs, abundant.

First record: *Dillenius,* 1724. *Var. fl. albo.*—Near the river at Hackney; *Dill. in R. Syn.* iii. 359. The ' *G. pusillum* ' from Greenford of *Herb. Harr.* (v. *Melv.* 19) is *G. molle.*

134. G. dissectum, *L.*

G. columbinum majus dissectis fol., Ger. em. (Blackst.). Cyb. Br. i. 264; Comp. 134. Curt. F. L. f. 6.

Roadsides, fields, and waste places; very common. A. or B. May—September.

In all the districts generally.

VII. Hackney Wick; Isle of Dogs, &c.

First record: *Blackstone,* 1737.

135. G. columbinum, *L.*

G. columb. dissect. fol. pediculis florum longissimis, Ray (Blackst.). Cyb. Br. i. 265; iii. 402; Comp. 134. Syme E. B. ii. t. 303.

Dry places; very rare. A. or B. June, July.

F 2

I. In a field near Harefield Common, in a gravel pit plentifully; *Blackst. Fasc.* 32. Near Uxbridge, Miss Gawler; *Cooper*, 117.

VII. [Marylebone Park, 1753; *Hill*, 353.]

First record: *Blackstone*, 1737.

136. G. rotundifolium, *L.*

G. columb. majus. fl. etiam majori purpureo (Buddle).

Cyb. Br. i. 262; Comp. 133. Syme E. B. ii. t. 301 (drawn from a Hackney specimen).

Dry hedgebanks, fields, and waysides; rare. A. or B. May—July.

III. Garden at Roxeth, introduced by Mr. Hind; *Melvill*, 19.

V. Field by the river and waysides near Chiswick. A weed in a garden bet. Turnham Green and Brentford; *Newb.*

VII. Various places about London; *Budd. MSS.* [Bet. Islington and Canbury House; *Mart. App. P. C.* 72.] [Lane by the 'Adam and Eve' at Islington, J. Woods; *B. G.* 109.] Near Holloway, 1831; *Varenne.* In the Lea Bridge Road, Upper Clapton; *Wollaston.* Under the east walls of the college at Hackney, T. F. Forster; *B. G.* 109. By the side of the footpath leading across the fields from Hackney Church to Brooksby's Walk, Homerton, 1839; *Herb. Hardw.* Hackney, 1865; *Reeves.* On walls at Chelsea; *Irvine.*

First record: *Buddle*, about 1712.

137. G. molle, *L.* *Dove's Foot. Crane's-bill.*

Ger. secundum, Diosc., Pes columbinus (Johns.). *G. columbinum, Ger.* (Blackst.).

Cyb. Br. i. 263; Comp. 134. Curt. F. L. f. 2.

Meadows, roadsides, and waste ground; very common. A. or B. April—August.

In all the districts.

VII. Frequently occurs in waste places, as Parson's Green, Brompton Cemetery, Victoria St. Westminster, Kensington Gardens, &c.

First record: *Johnson*, 1629.

138. G. lucidum, *L.*

G. saxatile, Ger. (Merr.). *G. lucidum saxatile, Ger. em.* (Blackst.).

Cyb. Br. i. 265; iii. 402; Comp. 135. Syme E. B. ii. t. 304.

Dry hedgebanks and walls; rare. A. or B. May—August.

I. In the lane leading to Harefield Mill, plentifully; *Blackst. Fasc.* 32. Hedgebank near Harefield Church!; *Newb.*

III. Abundant under the hedge in a road leading from Twickenham Ry. Station to Twickenham Park just beyond Amyand House, and in a footpath leading from this road. Bet. Worton and Whitton. Near Whitton; *Hemsley.*

V. Betwixt Branford and Sion House, plentifully; *Merrett*, 46. Sunken wall on Greenford Road about 100 yards from the Church; *Melv.* 18. Norwood; *Masters.* Bet. Turnham Green and Ealing; *Newb.*

VII. [On a wall on Newington Green, Buddle; *Newton MSS.*]
First record: *Merrett*, 1666.

139. G. robertianum, *L.*   *Herb Robert.*
Cyb. Br. i. 266; Comp. 135.   Syme E. B. ii. t. 305.
Hedgebanks and shady places; very common.   A. or B.   May—September.
Frequent in all the districts except VII. where we have not observed it.

First record: *Johnson,* 1629.   Blackstone noticed it *with white flowers* at Harefield in 1735 and 1745 (*Fasc.* 33 and *Spec.* 27).   The *seashore variety* is mentioned by Smith as being 'a weed in Chelsea Gardens;' *E. Fl.* iii. 236.

ERODIUM, *L'Hérit.*

140. E. cicutarium, *Sm.*   *Stork's-bill.*
Geranium arvense, Tab., Myrrhida Plinii (Johns.).   *Geranium cicutarium* (Curt.).
Cyb. Br. i. 257; iii. 400; Comp. 132.   Curt. F. L. f. 1.
Roadsides and waste ground, on sandy soil; rare.   A. or B.   June—September.
I. Pinner.   Uxbridge; *Cole.*
II. A few plants on the road from Staines to Hampton.
V. Chiswick; *Hemsley.*   Kew Bridge Station, N. Lond. Ry.; *Newb.*
VI. Edmonton.
VII. Bet. Kentish Town and Hampstead; *Johns. Eric.*   [Hyde Park wall, 1817; *Herb. G. & R.*]   Side of canal, Homerton, a single specimen; *Cherry.*   Kensington Gardens, 1868, a single plant; *Warren (v. s.).*

First record: *Johnson,* 1629.   Remarkably scarce in Middlesex.   The old botanists (as some moderns do still) divided this into several species.   Buddle mentions a *large-flowered plant* found by Doody and himself near Chelsea College, and also *Ger. inodorum cicutæ foliis angustissimis sectis* shown him by Doody in several places near London. From the specimen in *Herb. Budd.* v. 11, f. 18, the latter seems a small form, and is probably that mentioned in *R. Syn.* iii. 357, as found by Doody near Chelsea.   *Ger. pimpinellæ folio* of Dillenius found by him near Hackney (*R. Syn.* iii. 358) may have perhaps some claims to notice as a variety.   Mr. Woodward says (*With.* ii. 724), 'I have cultivated it for several years without observing any variation in it, and Mr. Curtis . . . is satisfied it is a distinct species.'   It is the *E. commixtum* of Jordan.

141. E. moschatum, *Sm.*
Cyb. Br. i. 257; iii. 400; Comp. 132.   Syme E. B. ii. t. 308.
Waste places and roadsides; very rare.   A. or B.   May—September.
II. Laleham, 1815; *Herb. G. & R.*   Mr. Varenne assures us that this is

correctly named. It has not been met with elsewhere in the county; perhaps escaped from a garden.

E. littoreum, *Willd.* VII. Near the Chelsea Old Waterworks, Pimlico ; *Irv. H. B. P.* 752.

LINACEÆ.

LINUM, *Linn.*

142. *L. usitatissimum, *L.* *Flax.*
Cyb. Br. i. 235. Syme E. B. ii. t. 292.

Waste ground ; rare. A. July—September.
IV. Harrow ; Pinner Drive; *Melv.* 15.
V. Greenford ; *Melv.* 15.
VII. [Near St. John's Wood burying chapel, 1817; *Herb. G. & R.*] Green lane, Paddington, W. Chatterley; *Cooper*, 104. Maiden Lane, G. E. Dennes ; *Cooper Supp.* 11. Bet. Highgate and Holloway ; *Herb. Young.* South Heath, Hampstead. Cremorne. Hackney Wick.
First record: *Goodger and Rozea*, 1817. Probably in most cases springs from the sweepings of bird-cages ; also spread by bird-catchers.

143. L. catharticum, *L.* *Purging Flax.*
L. sylvestre cathart., Ger. em. (Blackst.).
Cyb. Br. i. 237 ; Comp. 127. Syme E. B. ii. t. 291.

Heaths and dry places; rare. A. June—August.
I. Harefield ; *Blackst. Fasc.* 53. Stanmore Heath. Harrow Weald Common. Pinner Chalk-pits.
IV. Railway embankment near Harrow ; *Melvill*, 15 (*v. s.*). Hampstead ; *Irv. MSS.*
First record : *Blackstone,* 1737. Chiefly found in the north of the county.

RADIOLA, *Gmel.*

144. R. Millegrana, *Sm.* *All-seed. Flax-seed.*
Millegrana minima, Ger. em. (Blackst.).
Cyb. Br. i. 237 ; Comp. 127. Syme E. B. ii. t. 288.

Damp sandy places on heaths ; very rare. A. July, August.
I. Harefield Common ! ; Hillingdon Heath ; *Blackst. Fasc.* 61. Still grows in the former abundantly.
III. Hounslow Heath ; *Pet. Midd.*
First record : *Petiver,* 1695.

BALSAMINACEÆ.

IMPATIENS, *Linn.*

I. Noli-me-tangere, *L.*   Cyb. Br. i. 268 ; iii. 402 ; Comp. 135.   Syme E. B.
ii. t. 313.   I. Several places near Pinner Marsh, *Hind.*   A garden
escape.

145. * **I. fulva,** *Nutt.*

' *I. Noli-me-tangere* ' (Lond. Fl.).

Cyb. Br. i. 268.   Syme E. B. ii. t. 314.

Sides of streams and ditches ; rather rare.   A.   July—September.

I. Side of Grand Junction Canal, Harefield.

II. Thames near Hampton Court, T. Ralph ; *Lond. Fl.* 171.   Bet. Sunbury and Hampton ! ; Bushey Park ! ; *Newb.*   Thames side by
Hampton Court Palace.

III. Near Isleworth on road to Twickenham ; sides of some ponds near
the mills, Hounslow, A. Williamson ; *Phyt.* i. 814.   Twickenham,
by sides of ditches connected with the Thames, at Orleans House,
and bet. Marble Hill and Richmond Bridge ; and on waste ground
near the church.   Ditch behind Kneller Park.   Ditches about
Whitton Dean.   Sides of the Cran from Baber Bridge downwards,
and in the ditches connected with it.   Duke's River, near Worton.

V. Near Kew Bridge ; *Jewitt.*

VII. [Sparingly by a ditch near the Vale of Health pond, Hampstead, before
1866 ; *Melvill.*]

First record : *Ralph*, 1838.   An American species, so thoroughly and
perfectly naturalised as to give no suspicion of its exotic origin.   It
almost certainly originated from the gardens of Albury Park, Surrey ;
a small stream, the Tillingbourne, flows through these gardens and
runs into the Wey above Guildford, and this in turn flows into the
Thames a little above Shepperton.   In this way the seeds have been
carried by the water-current and by barges, &c., throughout the
Thames valley district.   The first notice we have met with of it is in
Phyt. i. 40, where Mr. John Stuart Mill states that he saw it near
Albury in 1822.   It was first figured in England in *Brit. Ent.* vol.
xvi. t. 747 (1839) from a Surrey specimen.   (It is stated in *Phyt.
N. S.* i. 166, that ' a Balsam twice as tall as *I. fulva* grows on the Colne
bet. Harefield and Denham.'   Of this we can learn nothing.)

146. * **I. parviflora,** *DC.*

Syme E. B. ii. t. 315.

Waste ground and a garden weed ; rare.   A.   June—September.

II. On ground laid out for building at Fulwell, bet. Strawberry Hill and
the Hanworth road ; in abundance, especially on the tops of the
stacks of bricks, which must have remained undisturbed for some
years.

VII. [Chelsea College, 1858–61, destroyed in the alterations, J. Britten;
Bot. Chron. 57.] Abundant weed in the gardens of a new square at
Notting Hill, 1866 and 1867, *Naylor* (*v. s.*). Weed in the garden of
70 Adelaide Road, N.W., 1857; *Syme* (*v. s.*).

First record: *Britten*, 1858. Well naturalised, and getting more completely so.

OXALIDACEÆ.

OXALIS, *Linn.*

147. O. acetosella, *L.*   *Wood Sorrel.*

Trifolium acetosum Oxys Plinii (Johns.). *Trif. acet. vulgare, C. B. P.*
(Blackst.).

Cyb. Br. i. 270; Comp. 136. Syme E. B. ii. t. 310.

Woods and shady hedgebanks, and under bushes on heaths; rather rare.
P. April, May.

I. Harefield!; *Blackst. Fasc.* 101. In the Old Park Wood. Pinner;
Harrow Weald Common; *Melv.* 19.

IV. Wood at Highgate; *Johns. Eric.* Bishop's and Turner's Woods, and
North Heath, Hampstead. Deacon's Hill, Finchley; *Newb.*

VI. Winchmore Hill Wood. Hadley!; *Newb.*

VII. Under the garden wall of Kenwood, at the entrance to Millfield Lane.

First record: *Johnson*, 1629.

O. corniculata, *L.* Cyb. Br. i. 271; iii. 402. Syme E. B. 311. VII. Extremely abundant, with *O. stricta*, on waste ground in Hollywood School
playground, Brompton, 1861; in 1864 *O. stricta* alone remained;
Britten. Waste ground about Chelsea College, 1862, J. Britten;
Phyt. N. S. vi. 349. An introduction in these localities.

148. * O. stricta, *L.*

O. corniculata (Cooper).

Cyb. Br. i. 272; iii. 403. Syme E. B. ii. 312.

A garden weed; rather rare. A. or P. July—September.

II. Weed in gardens at Sunbury, B. G.; *Cooper*, 114.

III. Garden ground, Twickenham.

IV. Under a garden wall by the roadside near the church, Stanmore.

V. Garden at Turnham Green; *Newb.*

VI. Edmonton.

VII. Parson's Green. Garden of 70 Adelaide Road. Hollywood
School playground, abundant; waste ground about Chelsea College;
Britten.

First record: *Cooper*, 1836. Also introduced, but better established
about London than *O. corniculata*, for which it is often mistaken.

CELASTRACEÆ.

Staphylea pinnata, *L.* Cyb. Br. i. 273 ; iii. 404. VII. Kenwood, Hunter ; *Park Hampst.* 29. A cultivated shrub.

EUONYMUS, *Linn.*

149. E. europæus, *L.* *Spindle-tree.*

E. Theophrasti, Ger. em. (Blackst.).

Cyb. Br. i. 272 ; iii. 404 ; Comp. 136. Syme E. B. ii. t. 317.

Hedges and woods ; rather rare. Shrub. May, June.

I. Harefield! ; *Blackst. Fasc.* 27. Old Park Wood. Pinner. Near Yewsley! ; *Newb.*

II. Bet. Hampton Court and Kingston Bridge, in several places. Staines! ; *Newb.*

IV. Harrow Grove, abundant ; *Melv.* 20. Bishop's Wood ; *Lawson.*

V. Apperton ; Greenford ; *Melv.* 20. Horsington Hill.

First record : *Blackstone,* 1737. Loudon mentions (*Arb. et Frut. Brit.* 497) a bush of this in Kensington Gardens, 15 feet high.

RHAMNACEÆ.

RHAMNUS, *Linn.*

150. R. catharticus, *L.* *Buckthorn.*

Cyb. Br. i. 273 ; Comp. 136. Syme E. B. ii. t. 318.

Hedges ; rare. Shrub or small tree. June, July.

I. In some places near Harefield, sparingly ; *Blackst. Fasc.* 86. Near Harefield, but just out of Middlesex! ; *Newb.* By Elstree Reservoir.

II. Near Staines. Road from Hampton Court to Sunbury, H. C. W. ; *New B. G.* 98. Bet. Sunbury and Walton Bridge. Bet. Hampton and Hampton Court! ; *Newb.*

IV. By the Brent in the meadows bet. Hendon and Hampstead ; *Irv. MSS.* By the Brent, near Neasdon ; *Farrar.*

V. Near Heston ; *Masters.* By the Brent, about Perivale ; *Lees.*

VII. [Lane bet. Paddington and Notting Hill, 1820 ; *Bennett* (*v. s.*).]

First record : *Blackstone,* 1737.

151. R. Frangula, *L.* *Black* or *Berry-bearing Alder.*

Alnus nigra baccifera, Lugd. (Johns.).

Cyb. Br. i. 274 ; iii. 404 ; Comp. 136. Syme E. B. ii. t. 319.

Woods and bushy places ; rather rare. Shrub. May, June.

I. In Whiteheath Wood on Harefield Common, abundantly ; *Blackst. Fasc.* 4. Pinner ; *Melv.* 20. Side of road from Harrow Weald Common to Burnt Oak Farm.

IV. Harrow Park ; *Melv.* 20. Hampstead Heath ; *Johns. Enum.* Woods at Hampstead ; *Ger.* 1286. North Heath and Bishop's Wood, plentiful. By the Brent near Kingsbury ; *Farrar.*

V. Horsington Wood ; *Lees.*

VI. Highgate Wood, 1827–30; *Varenne.* Colney Hatch!; *Newb.* Winchmore Hill Wood.

VII. [Great plentie in a wood* a mile from Islington towards Harnsey; *Ger.* 1286.] [In St. John's Wood by Hornsey ; *Park. Theat.* 240. In a wood † against the boarded river ; *Pet. Midd.*] [Hornsey Wood ; *Miller,* 25 ; and many subsequent authors.]

First record: *Gerarde,* 1597. Chiefly found on the clay.

LEGUMINIFERÆ.

ULEX, *Linn.*

152. U. europæus, *L.* *Furze. Gorse. Whin.*
Genista spinosa vulgaris (Johns.).
Cyb. Br. i. 275 ; Comp. 137. Syme E. B. iii. t. 323.
Heaths, roadsides, and railway banks; very common. Shrub. February—June.
Occurs in many places in all the districts.
VII. South Heath, Hampstead.
First record: *Johnson,* 1632.

153. U. nanus, *Forst.*
Genista spinosa minor, Park. (Budd.).
Cyb. Br. i. 277 ; iii. 404 ; Comp. 137. Syme E. B. iii. t. 325.
Heaths and roadsides ; rather rare. Shrub. July—September.
I. Lane near Pinner Hill, Mrs. Tooke; *Hind.* Gravel-pits by the high road opp. Hillingdon Place; *Warren.* Harefield Common, abundant.
II. Hampton Common, 1808 ; *Winch. MSS.* Waste ground, Fulwell.
III. Hounslow, Planchon; *Bot. Gaz.* i. 289. Hounslow Heath, 1820 ; *Bennett* (*v. s.*). Drilling ground, abundant. Roadside from drilling ground to cemetery.
IV. On Hampstead Heath, abundantly; *Budd. MSS.* Ibid., 1809 ; *Winch. MSS.*
V. Wyke Green ; *Masters.* Old Oak Common; *Britten.*
First record : *Buddle,* about 1710. Our plant is the restricted *U. nanus* of Planchon ; *U. Gallii,* Planch., does not seem to occur.

GENISTA, *Linn.*

154. G. tinctoria, *L.* *Dyer's Green-Weed.*
Genistella tinct., Ger. em. (Blackst.).
Cyb. Br. i. 278 ; Comp. 137. Syme E. B. iii. t. 228.

---

* Highbury Wood, which began to be destroyed in 1650 ; Cream Hall, in Highbury Vale, was a farmhouse built on the site of the wood.

† Little St. John's Wood, only separated by a field from Highbury Wood, abutted on the New River. (See Tomlin's *Islington,* p. 198.)

Rough pastures ; rare. Small shrub. July—September.

I. Harefield ; *Blackst. Fasc.* 31. Field by Elstree Reservoir. Harrow Weald Common, W. M. H. ; *Melv.* 21.

IV. Near Wood House, Kingsbury; *Irv. MSS.* Hampstead Heath ; *Johns. Enum.* Meadow bet. Finchley and Bishop's Wood, E. H. Button ; *Cooper Supp.* 12. Near Scratch Wood, Edgewarebury, abundant. S.E. side of the Brent towards the reservoir ; *Farrar.*

V. Northolt ; *Melv.* 21.

First record: *Johnson*, 1632. Chiefly on the stiff clay. In both the localities where we have seen it, it grows with *Trifolium medium*; and a similar companionship is noticed by the authors of the *Fl. Herts.*

155. G. anglica, *L.* *Needle Furze.*
Genistella, Trag. (Johns.). *G. aculeata, Ger. em.* (Blackst.).
Cyb. Br. i. 279 ; iii. 405; Comp. 138. Syme E. B. iii. t. 326.

Damp parts of heaths ; rather common. Shrub. June, July.

I. Harefield ! ; *Blackst. Fasc.* 31. Ruislip Common ; *Melv.* 21. Harrow Weald Common, scarce. Stanmore Heath.

III. Hounslow Heath. Drilling ground.

IV. Hampsteed Heath ! ; *Ger.* 1139.

V. Old Oak Common ; *Britten.* Near Willesden Junction ; *Newb.*

VI. Hadley ! ; *Warren.* Enfield.

VII. South Heath, Hampstead.

First record : *Gerarde*, 1597.

SAROTHAMNUS, *Wimm.*

156. S. scoparius, *Koch.* *Broom.*
Genista vulgaris, Ger. em. (Blackst.). *Spartium Scoparium, L.* (Curt.).
Cyb. Br. i. 274; iii. 404; Comp. 137. Curt. F. L. f. 5.

Heaths, waysides, and railway banks; very common. Shrub. May—July.

In all the districts, preferring a dry gravelly soil.

VII. Poplar ; *Newb.* Bank of North London Railway, near Dalston Junction. South Heath, Hampstead. Kilburn.

First record : *Blackstone*, 1737.

ONONIS, *Linn.*

157. O. arvensis, *Fries.*
Anonis non spinosa purpurea, Ger. (Blackst.).
Cyb. Br. i. 280 ; Comp. 138. Syme E. B. iii. t. 331.

Roadsides, on sand ; rare. P. June—August.

I. In the lane near Harefield Mill ; *Blackst. Fasc.* 7. Abundant on a bank by the paper mills : perhaps the same locality. Another place towards Rickmansworth ! ; *Newb.* (Harrow Weald and Ruislip Commons ; *Melv.* 21; probably *O. campestris* was intended.)

II. Staines!; bet. Hampton and Sunbury!; *Newb.* By the towing-path
bet. Hampton Court and Kingston Bridge, nearer the latter. Near
Teddington.

IV. Hampstead; *Irv. MSS.*

First record: *Blackstone,* 1737.

158. O. campestris, *Koch.* O. spinosa, *L.* (L. Cat.). *Rest Harrow.*
Anonis spinosa fl. purpureo, Ger. (Blackst.).

Cyb. Br. i. 281; iii. 405; Comp. 138. Syme E. B. iii. t. 330.

Roadsides, commons and fields; rather common. P. June—September.

I. In the lane near Harefield Mill; *Blackst. Fasc.* 6. Pinner Hill;
Hind. Near Yewsley!; *Newb.* Road from Harrow Weald Common
to Burnt Oak Farm.

II. About Staines in several places. By the towing-path bet. Kingston
Bridge and Hampton Court.

IV. Hampstead; *Irv. MSS.* Near Wembley Green; near Brent Reservoir;
Farrar. Roadside by Harrow Weald Church. Golder's Green.

V. Bet. Perivale and Ealing; *Lees.*

VII. South Heath, Hampstead; abundant.

First record: *Gerarde,* 1597. Gerarde (p. 1141) noticed it *with white
flowers* in our London pastures; and Newton (*MSS.*), bet. Battle
Bridge * and Hygate.

MEDICAGO, *Linn.*

M. sativa, *L. Lucerne.* Cyb. Br. i. 283. Syme E. B. iii. t. 334. Of
rare occurrence. Not indigenous, and but little cultivated in the
county. Has been observed about (I.) Pinner; *Melv.* 21; (V.) near
Willesden Ry. Station; and (VII.) on the site of the Exhibition of
1862 at South Kensington. It is fond of railway banks.

159. M. lupulina, *L. Trefoil. Nonsuch. Yellow Clover.*
Trifolium luteum minimum, Ger. em. (Blackst.).

Cyb. Br. i. 284; iii. 406; Comp. 139. Curt. F. L. f. 2.

Dry fields, roadsides and waste ground; very common. A. or B. May—
August.

Through all the districts.

VII. South Heath, Hampstead. Green Lanes, Newington. Kensington
Gore. Craven Hill. Hackney Wick. Isle of Dogs!; *Newb.*

First record: *Blackstone,* 1737. More often introduced than really wild;
much cultivated for fodder under the name of *Trefoil.*

160. M. maculata, *Sibth.*
Medica arabica, Cam. and *M. minor fruct. cochl. asp.* (Johns.). *Tri-
folium cochleat. fol. cordato maculato, C. B. P.* (Blackst.). *M. polymorpha,*
L. (Curt.).

Cyb. Br. i. 285; iii. 406; Comp. 139. Curt. F. L. f. 6.

_____

* A bridge over the River Fleet, near where King's Cross Station is now.

Roadsides, pastures, and waste places, on a light soil; rather common.
A. June—August.

I. On Uxbridge Common, plentifully; *Blackst. Fasc.* 101.
II. Bet. Hampton and Hampton Court!, H. C. W.; *New B. G.* 98. Bet.
Sunbury and Walton Bridge. Towing-path bet. Hampton Court and
Kingston Bridge, abundant. Near Teddington Station.
III. Several places about Twickenham. Isleworth. Field near Richmond
Bridge, abundant.
IV. Hampstead Heath; *Johns. Enum.*
V. Near Hanwell!; *Warren.* Near Kew Bridge, *Jiwitt.* About Chis-
wick Ry. Station. Abundant in the field by the river, opp. Mortlake.
VII. [Site of old reservoir in Green Park, sown with grass seed.]

First record: *Johnson,* 1632.

161. * M. denticulata, *Willd.*

Cyb. Br. i. 286; iii. 406; Comp. 140. Syme E. B. iii. t. 338.

Waste ground; rare. A. May—August.

I. One plant on a manure heap, Down Fields, Pinner, W. M. H.;
Melv. 21.
V. Rather plentifully in a field bet. Turnham Green and Acton, Mr. J. W.
Lawrence; *Phyt.* ii. 811.
VII. Parson's Green, 1856, A. J.; *Phyt. N. S.* ii. 168.

Var β. M. apiculata, Willd. *Medica polycarpos fructu minore com-
presso scabro* (Doody).

VI. Behind Ponder's End, near Enfield, and in several places thereabouts;
Doody MSS.
VII. [In the cornfields near Paddington; *Doody, loc. cit.* In an open field
bet. Paddington and Kensington, plentifully; *Budd. MSS.*]

First record: *Doody,* about 1700. Not very recently met with. *M. maculata* may have been mistaken for it in some of the above localities.

MELILOTUS, *Linn.*

162. * M. officinalis, *Willd.* *Melilot.*
M. germanica, Ger. em. (Blackst.). *Trifolium Melilotus-officinalis, L.*
(Cooper).
Cyb. Br. i. 287; iii. 407; Comp. 140. Syme E. B. iii. t. 341.

Waste ground; rather common. B. or P. June—August.

I. Cornfields, Harefield; *Blackst. Fasc.* 59. In great quantity at the
lime-works, Harefield. Pinner Cemetery; *Melv.* 22.
II. Staines!; *Newb.* Near Kingston Bridge.
IV. Hampstead Heath; *Cooper,* 99. Railway bank near Harrow Station;
Melv. 22.
V. Northolt; Greenford, F. W. Longman; *Melv.* 22.
VI. Enfield.

VII. Homerton; Upper Clapton; *Cherry.* Near Chelsea Hospital, 1861.
[Near the Victoria Tower, Westminster.] By the side of the Great
North Road; *Burnett,* 105.

First record: *Blackstone,* 1737.

163. * **M. arvensis,** *Wallr.*
Cyb. Br. iii. 332.   Syme E. B. iii. t. 343.
Waste ground; rare.   B.   June—September.
I. Harrow Weald Common.
II. Railway bank at Fulwell Station.
VII. Brook Green.   Site of Exhibition of 1862, S. Kensington.   Side of
Hackney Canal.
First record: *the Authors,* 1866.   Occurs under the same circumstances
as *M. officinalis,* but is less frequent.

164. * **M. vulgaris,** *Willd.*   M. alba, *Lam.* (Syme).
Cyb. Br. i. 288; iii. 407;   Comp. 140.   Syme E. B. iii. t. 342.
Waste places; rare.   B.   June—September.
II. Near Strawberry Hill.   By the towing-path bet. Kingston Bridge and
Hampton Court; *Bloxam.*
V. Chiswick; *Fox.*
VII. Shepherd's Bush.   Near Cremorne.   Site of Exhibition of 1862, S.
Kensington.   Chelsea College; *Britten.*
First record: *Britten,* 1858.

M. sulcata, *Desf.* VII. Site of Exhibition of 1862, S. Kensington.   Introduced.

M. messanensis, *Desf.* VII. Chelsea College; *Britten.*   Introduced.

M. cærulea, *Lam.* VII. Site of Exhibition of 1862, S. Kensington.   Of
foreign origin.

M. parviflora, *Desf.* Cyb. Br. iii. 332.   Syme E. B. iii. t. 344.   VII.
Chelsea College, 1861; Parson's Green; *Britten.*   Not native.

TRIGONELLA, *Linn.*

[**165.** **T. ornithopodioides,** *DC.*   Falcatula ornith., *Brotero* (Bab.).
Trifol. siliquis ornithopodii, R. Syn. (Doody).   *Fœnugræcum humile
repens, Ornithopodii siliquis brevibus erectis* (Dill.).   *Trifolium orthitho-
podioides,* L. (Curt.).
Cyb. Br. i. 290; iii. 407; Comp. 141.   Curt. F. L. f. 2.
Gravelly and sandy places; very rare.   A. or B.   June, July.
IV. Hampstead Heath, J. Woods; *B. G.* 109.
V. Hanwell Heath, Dr. Goodenough; *Cooper,* 108.
VII. In Tuthill fields, Doody; *R. Syn.* i. 246.   In Tuttle fields, plen-
tifully; *Budd. MSS. and Herb., and subsequent writers.*   Curtis
found it there abundantly.
First notice: *Doody,* 1690.   Last: about 1805.]

TRIFOLIUM, *Linn.*

166. T. pratense, *L.*   *Red Clover.*   *Cow-Grass.*

Cyb. Br. i. 295 ; Comp. 142.   Syme E. B. iii. t. 347.

Fields, roadsides and waste places ; very common. P. or B. May—August.

Found in all the districts. Though often the result of former cultivation, it is probably native ; the smaller procumbent form seems to be the original state, which has been 'improved' by long culture.

First record : *Blackstone*, 1737.

167. T. medium, *L.*

T. purp. majus fol. longiore et angustiore fl. saturatior., Ray (Budd.).

Cyb. Br. i. 296 ; Comp. 142.   Syme E. B. iii. t. 348.

Pastures, heaths and roadsides ; rather rare. P.   June—August.

I. Wayside at Pinner Hill, W. M. H. ; *Melv.* 22. Stanmore Heath. Field near Elstree Reservoir. Near South Mims.

II. River bank bet. Richmond Bridge and Hampton Court ! ; *Newb.*

IV. Meadow near Scratch Wood. Burgess Hill.

V. Field bet. Acton and Turnham Green ! ; *Newb.*

VI. Bet. Colney Hatch and Whetstone.

VII. About London ; *Budd. MSS.* South Heath, Hampstead, where *Ononis* grows. Hedgebank E. side of Finchley Road, near New West End.

First record : *Buddle*, about 1710.

[168. ? T. ochroleucum, *L.*

Cyb. Br. i. 293 ; iii. 408 ; Comp. 141. Curt. F. L. f. 6.

VI. Sparingly about Barnet ; *Curt. F. L.*

Perhaps in Herts. The locality has not been corroborated, but is likely to have been correct, as the plant has occurred in neighbouring parts of Herts, and is 'abundant in N.W. Essex.']

T. incarnatum, *L. Scarlet Clover.* Cyb. Br. i. 294. Syme E. B. iii. t. 352. An occasional straggler from cultivation. V. Field near Chiswick ! ; *Newb.* VII. Site of Exhibition of 1862, S. Kensington ! ; *Naylor.*

169. T. arvense, *L.*   *Haresfoot.*

T. arv. humile spicatum, sive Lagopus, C. B. P. (Blackst.).

Cyb. Br. i. 298 ; iii. 409 ; Comp. 143. Curt. F. L. f. 6.

Dry sandy ground ; rare. A. June—September.

I. Harefield, in corn ; *Blackst. Fasc.* 102.

II. Field near Teddington Ry. Station, abundant. Teddington Park. One plant by roadside bet. Staines and Hampton.

IV. Hampstead Heath ; *Cooper*, 99.

V. Gravel-pits near Hanwell ! ; *Warren.*

VII. Chelsea College, 1860, Britten ; *Bot. Chron.* 58.

First record : *Blackstone,* 1737. Curtis (*loc. cit.*) says, ' Very common about London ; ' yet in Middlesex, II. is the only district in which we have seen it growing.

170. T. striatum, *L.*

T. parv. hirsut., fl. parv. dilute purp. in glomerulis mollior. et oblong., semine majus, Ray (Blackst.).

Cyb. Brit. i. 299 ; iii. 409 ; Comp. 143. Syme E. B. iii. t. 356.

Dry sandy ground ; rather common. A. May—July.

I. On Oliver's Mount on Uxbridge Moor, plentifully ; *Blackst. Fasc.* 103.

II. Roadsides bet. Staines and Hampton, about Ashford and Charlton, abundant.

III. Near Twickenham Ry. Station ; *Syme.* Hounslow Heath. Near Hatton.

V. Sandy field by the river near Chiswick Ry. Station, abundant.

VII. [Once found on a bank in the first field from Southampton Row in the road to Hampstead ; *Blackst. Spec.* 100.]

First record : *Blackstone,* 1737.

[171. T. scabrum, *L.*

Cyb. Br. i. 298 ; iii. 409 ; Comp. 143. Syme E. B. iii. t. 357.

VII. Waste gravelly ground bet. Bayswater and Paddington, 1834 ; *Irv. MSS.* and *Lond. Fl.* 178.

Requires confirmation ; we have seen no specimens.]

172. T. subterraneum, *L.*

T. pumilum supinum flosc. long. albis, P. B. (Pet.).

Cyb. Br. i. 292 ; iii. 408 ; Comp. 141. Curt. F. L. f. 2.

Sandy ground ; rather rare. A. May, June.

I. Harefield and Uxbridge Commons, plentifully ; *Blackst. Fasc.* 102.

II. Roadside bet. Hampton and Hampton Court, H. C. W. ; *New. B. G.* 98. Roadside bet. Staines and Ashford. Common by Walton Bridge, abundant.

III. Hounslow Heath, abundant. Near Hatton.

V. Field by the river, Chiswick.

VII. [Tothill fields ; *Pet. Midd.* and *Budd. MSS.*] [Hyde Park 1780, Smith ; *Herb. Linn. Soc.* and *E. B.* 1048.]

First record : *Plukenet,* 1690. (See *Ray's Letters,* p. 231.)

[173. T. glomeratum, *L.*

T. supinum c. glomerulis ad caulium nodos. globosis, fl. purpurascentibus (R. Hist.).

Cyb. Br. i. 301 ; iii. 409 ; Comp. 144. Curt. F. L. f. 4.

Gravelly and sandy commons. A. May, June.

IV. Hampstead Common, 1809 ; *Winch. MSS.*

V. Hanwell Heath, Dr. Goodenough ; *Sm. Fl. Brit.* ii. 789.

Thomas Willisel first gathered this in 1686, ' about London ' (*R. Hist.* i.

948); but whether in this county is not known. If not, the first
record is *Goodenough*, about 1800. It has not been recently collected
in Middlesex.]

174. T. fragiferum, *L.* *Strawberry Trefoil.*
Cyb. Br. i. 303; Comp. 144. Curt. F. L. f. 2 (drawn from a Middlesex
specimen).
Wet meadows, sides of streams, &c.; rather common. P. July—September.
I. Harefield; *Blackst. Fasc.* 101.
II. Bet. Hampton and Hampton Court!; *Newb.* By the Thames bet.
Hampton Court and Kingston Bridge. About Staines.
III. Thames bank bet. Twickenham and Richmond Bridge. Roxeth;
Hind.
IV. Near the clay-pits, Harrow Weald. Near Harrow Ry. Station. Bet.
Wilsdon and Neesdon; *Farrar.*
VII. [Hyde Park; *Dicks. H. S.*] Lanes about Hornsey and near Pancras;
Curt. F. L. South Heath, Hampstead. Eel Brook Meadow, Parson's Green.
First record: *Curtis,* about 1780.

175. * T. hybridum, *L.* *Alsike Clover.*
T. elegans, Savi. Cyb. Br. iii. 332. Syme E. B. iii. t. 361.
Waste ground and roadsides. P. June—September.
I. Harefield.
II. Towing-path bet. Kingston Bridge and Hampton Court; *Bloxam.*
V. Chiswick; *Newb.* Apperton.
VII. Green lanes, Newington!; *Newb.* About Little Chelsea and Parson's
Green, common. Site of Exhibition of 1862, S. Kensington.
First record: *the Authors,* 1866. Cultivated as a fodder plant, and hence
widely disseminated.

176. T. repens, *L.* *White or Dutch Clover.*
T. pratense album, C. B. P. (Blackst.).
Cyb. Br. i. 291; Comp. 141. Curt. F. L. f. 3.
Waysides, fields, waste land, &c.; very common. P. April—September.
In all the districts.
VII. In London itself it frequently occurs, e.g. in the parks and squares,
and most pieces of waste ground.
First record: *Ray,* 1670. A monstrosity in which the parts of the flower
are leaflike and the pedicels longer is frequently met with, and is re-
corded under the name of *T. alb. umbella siliquosa* as found 'about
London' by Ray; *Cat.* i. 304. (See also *Ray's Letters,* p. 222.) 'In
the exceptionally hot summer of 1868, many plants in the turf of
Regent's Park became almost the plant noticed by Townsend in the
Scilly Isles; see *Seem. J. of Bot.* ii. 1, and t. 13;' *Newb.*

T. resupinatum, *L.*   Cyb. Br. i. 301 ; iii. 409.   Syme E. B. iii. t. 364.   VII.
Site of Exhibition of 1862, S. Kensington.   An exotic species.

177.   T. procumbens, *L.*   *Hop Trefoil.*

T. luteum lupinum, Ger. em. (Blackst.).   *T. agrarium, L.*   (Curt.).
Cyb. Br. i. 304 ; Comp. 145.   Curt. F. L. f. 3.

Dry waste places and roadsides ; rather common.   A.   June—August.
   I. In corn, Harefield ; *Blackst. Fasc.* 101.   Cornfields, Wood End
      Melv. 22.   Pinner chalkpits, abundant.   South Mims.
   II. Near Staines.   Hampton.   Tangley Park.
   III. Twickenham Park.
   IV. Harrow Weald churchyard.
   V. Boston Road, Brentford ; *Cherry.*   Kew Bridge Ry. Station.
   VI. Edmonton.   Enfield.
VII. Green Lanes, Newington ! ; *Newb.*

First record : *Blackstone,* 1737.

178.   T. minus, *Relhan.*   *Small Yellow Clover.*

T. procumbens, L. (Curt.).
Cyb. Br. iii. 334 ; Comp. 145.   Curt. F. L. f. 5.

Fields, waste places and roadsides ; very common.   A.   May—August.
Through all the districts.
VII. South Heath, Hampstead.

First record : *Irvine,* about 1830.

179.   T. filiforme, *L.*

T. lupulinum minimum, Inst. R. H. (Budd., Blackst.).
Cyb. Br. i. 304 ; iii. 334 ; Comp. 145.   Syme E. B. iii. t. 367.

Sandy heaths, commons and roadsides ; rather common.   A.   May—
   July.
   I. On Harefield Common, plentifully ; *Blackst. Fasc.* 102.   Roadside
      bet. Harefield and Ruislip.   Embankment of Ruislip Reservoir,
      W. M. H. ; *Melv.* 22.   Harrow Weald Common.   Bet. South Mims
      and Potter's Bar.
   II. Roadsides about Ashford and Staines in several places.   Bet. Sun-
      bury and Walton Bridge, and on a common by Walton Bridge,
      abundant.
   III. Hounslow Heath.
   V. Field by the Thames opposite Mortlake.
VII. [In Tuttle fields, Westminster ; *Budd. MSS.* and *Herb.* vol. cxix. fol.
      39.] [Hyde Park, 1815 ; *Herb. G. & R.*] Railway bank, Hackney
      Marsh ; *Grugeon* (*v. s.*).

First record : *Buddle,* about 1710, who carefully distinguished it from
   small *T. minus* ; also first notice as a British plant.   One or two of
   the *quoted* localities may produce only *T. minus,* but the true plant is
   common enough in the county.

LOTUS, *Linn.*

180. L. corniculatus, *L.*   *Bird's-foot Trefoil.*
Trif. siliquosum minus, Ger. em. (Blackst.).
Cyb. Br. i. 305 ; Comp. 145.   Syme E B. iii. t. 368.

Meadows, heaths and fields ; common.   P.   June—August.

   I. Harefield ! ; *Blackst. Fasc.* 103.   By Elstree Reservoir.   South Mims.
  II. Near Teddington Station.
 III. Hounslow Heath.   Twickenham.
 IV. Hampstead Heath ! ; *Irv. MSS.*   Stanmore.
  V. Ealing Common ! ; Hanwell ! ; *Newb.*
 VI. Near Enfield ; *Tucker.*   Edmonton.
VII. South Heath, Hampstead.

  Var. β. L. tenuis, Kit.   Cyb. Br. i. 305 ; iii. 410 ; Comp. 146.   Syme
    E. B. iii. t. 369.

  V. Near Chiswick Railway Station ! ; *Newb.*
VII. Site of Exhibition of 1862, S. Kensington.

First record : *Blackstone,* 1737.

181. L. major, *Scop.*
Trifolii siliquosi var. major, Ger. em. (Blackst.).
Cyb. Br. i. 306 ; Comp. 146.   Syme E. B. iii. t. 370.

Ditch-banks and wet places ; common.   P.   July—September.

   I. Harefield ; *Blackst. Fasc.* 103.   Elstree.   South Mims.
  II. Bet. Hampton and Hampton Court ! ; *Newb.*   Staines Moor.   Fulwell.
 III. Common at Harrow ; *Melv.* 22.   Twickenham.
 IV. North Heath, Hampstead.   Sudbury.
  V. Near Twyford ! ; Ealing Common ! ; bet. Acton and Turnham Green ! ;
    Newb.
 VI. Edmonton.   Enfield.
VII. Belsize ; *Irv. MSS.*   Kilburn ; *Varenne.*   Finchley Road ; *Newb.*
    South Heath, Hampstead.

First record : *Buddle,* about 1710.   The more or less glabrous forms
    were distinguished as species by the older botanists ; Buddle notices
    L. pentaphyllos major alter glaber near Chelsea ; and Doody (*R. Syn.*
    iii. 334) *L. pentaphyllus medius pilosus* in the fields behind Mother
    Huffs.

(Anthyllis Vulneraria, *L. A. leguminosa* (Ger.).   Cyb. Br. i. 282 ; Comp.
    139.   Syme E. B. iii. t. 333.   On Hampstead Heath, right against the
    beacon on the right hand going from London ; *Ger.* 1061, who,
    however, may have only intended **192,** *Ornithopus perpusillus.*   The
    locality is improbable.)

182. V. hirsuta, *Koch.*   *Hairy Tare.*
Cracca minor (Johns.). *Vicia segetum c. siliquis plurimis hirsutis,*
C. B. P. (Blackst.). *Ervum hirsutum, L.* (Curt.).
Cyb. Br. i. 321 ; Comp. 151. Curt. F. L. f. 1.
Cultivated and waste ground, meadows and hedges; very common. A.
June—August.
Throughout all the districts.
VII. South Heath, Hampstead. Thames Embankment. Site of Exhibition
of 1862, S. Kensington.
First record : *Johnson,* 1629.

183. V. tetrasperma, *Moench.*   *Smooth Tare.*
V. segetum siliquis singularibus glabris, C. B. P. (Blackst.). *Ervum
tetrasp., L.* (Curt.).
Cyb. Br. i. 321 ; iii. 413 ; Comp. 151. Curt. F. L. f. 1.
Cornfields, meadows and waste ground; rather common. A. June—
September.
  I. Harefield, frequent; *Blackst. Fasc.* 109. Pinner chalkpits.
  II. Bet. Hampton and Hampton Court!; *Newb.*
  III. Near Hounslow; *Hemsley.* Twickenham.
  IV. Pinner Drive; Railway bank; *Melv.* 24. Near Scratch Wood,
Edgwarebury.
  V. Field by Thames, opp. Mortlake; *Fox.* Horsington Hill.
VII. Behind Adelaide Road, N.W. South Heath, Hampstead, a single
plant.
First record : *Blackstone,* 1737.

  V. gracilis, *Lois.* Cyb. Br. Supp. 55; Comp. 152. Syme E. B. iii. t.
384. III. Meadows, Roxeth ; *Hind in Phyt. N. S.* v. 199, and *Melvill*
24. Mr. Hind has some doubt about this. Requires confirmation
as a Middlesex plant.

184. V. Cracca, *L.*
V. multiflora C. B. P. (Blackst.).
Cyb. Br. i. 316 ; Comp. 149. Syme E. B. iii. 385.
Hedges and bushy places; common. P. June—August.
  I. Harefield!; *Blackst. Fasc.* 109. Pinner; *Melv.* 23. Stanmore
Heath. Elstree. South Mims.
  II. Staines Moor. Hampton. Teddington. Bet. Sunbury and Walton
Bridge.
  III. Isleworth. Twickenham. Near Hatton.
  IV. Railway bank by Kenton Bridge; *Melv.* 23. Hendon, *Colc.* Stan-
more ; *Varenne.* Near Scratch Wood. Bishop's Wood.
  V. Bet. Acton and Turnham Green!; near Twyford!; *Newb.*

VI. Edmonton. Whetstone.

VII. [Marylebone fields, 1815; *Herb. G. &. R.*]

First record: *Blackstone*, 1737.

185. V. sepium, *L.*

V. maxima dumetorum, Ger. em. (Blackst.).

Cyb. Br. i. 319; Comp. 151. Syme E. B. iii. t. 388.

Woods and hedgebanks; rather common. P. June—August.

I. Harefield!; *Blackst. Fasc.* 108. Pinner. South Mims.

III. Near Worton.

IV. Railway embankment near Harrow; *Melv.* 23. Deacon's Hill. Bishop's Wood, Hampstead. Mill Hill; *Cole.* Near Finchley!; *Newb.*

V. Horsington Hill. Near Lampton.

VI. Hadley!; *Warrin.* Edmonton. Enfield.

VII. Hornsey Wood; *Herb. Hardw.* Millfield Lane.

First record: *Blackstone*, 1737. Noticed *with pure white flowers* near the Vicarage, Hendon; *Irv. Lond. Fl.* 175.

186. V. sativa, *L.* *Vetch.*

1. *V. sativa*, Smith.

Comp. 150. Syme E. B. iii. t. 392.

Borders of fields and roadsides; always the result of cultivation. Very little grown in Middlesex.

2 *V. angustifolia*, Roth. *Cracca major* (Johns.). *V. sylvestris*, Ger. em. (Blackst.).

Cyb. Br. iii. 335; Comp. 150. Syme E. B. iii. t. 393.

Fields, waste places, and roadsides, chiefly on a light sandy soil; common. A. May—July.

I. Harefield!; *Blackst. Fasc.* 108. Elstree. South Mims.

II. Staines. Sunbury. Teddington.

III. Near Hounslow. Near Hatton. Near Hospital Bridge.

IV. Harrow Weald; *Hind.* Renter's Lane, Hendon; *Davies.*

V. Apperton; *Hind.* Near Ealing!; near Twyford!; Hanwell; *Newb.* Turnham Green. Near Chiswick Railway Station. Field by Thames just opp. Mortlake, abundant.

VI. Gt. Northern Ry. bank, Colney Hatch.

VII. Bet. Kentish Town and Hampstead; *Johns. Eric.* [Hyde Park *Dicks. H. S.*] W. Lond. Cemetery, Brompton, abundant. Isle of Dogs.

Under this is included *V. Bobartii*, Forst.(*Syme E. B.* iii. t. 594), extreme states of which look very different from the type. Such forms are found at Sunbury, in the field by the Thames opposite Mortlake and in the W. London Cemetery. The plant near Chiswick Ry. Station was as large as *V. sativa*, Sm., but preserved its distinctive characters. Dickson's Hyde Park plant is labelled '*V. lathyroides*,' and the

locality is given under that species in *With.* eds. iii. to vii. and in *B. G.*
409; Smith however (*E. Fl.* iii. 282) quotes it rightly under *V. angusti-*
folia. Buddle records several varieties (he did not consider them
species) about Hampstead.

First record: *Johnson*, 1629.

187. V. lathyroides, *L.*
Cyb. Br. i. 317; Comp. 150. Syme E. B. iii. t. 395.
Dry sandy places; very rare. A. May—June.
I. Waste ground at Ruislip; *Hind in Phyt. N. S. v.* 199 and *Melv.* 23.
V. Plentiful in a sandy field by the Thames just opposite Mortlake,
   growing mixed with a small form of *V. angustifolia.*
VII. [In Hyde Park; *Smith E. Fl.* iii. 283.]
First record: *J. E. Smith,* 1825. There is some doubt about the Ruislip
   plant.

LATHYRUS, *Linn.*
[**188. L. Aphaca,** *L.*
Cyb. Br. i. 322; iii. 413; Comp. 152. Curt. F. L. f. 5.
Dry cultivated places; very rare. A. May—August.
III. Near Twitnam (=Twickenham) Park; *Doody MSS.*
VI. About Enfield; *Doody, loc. cit.* In the old camp near Enfield called
   Oldbury field; *Forst. Midd.* Frequently about Tottenham and
   Enfield; *Curt. F. L.*
First notice: *Doody,* about 1700. Last: Curtis, about 1790.]

189. L. Nissolia, *L.*
Ervum sylvestre, Dod. (Johns.). *Vicia fol. gramineo siliqua porrectis-*
sima (Merrett). *Catanance leguminosa quorundam, J. B.* (Blackst.
Fasc., Ray). *Nissolia vulgaris Tournefortii* (Blackst. Spec.).
Cyb. Br. i. 323; iii. 413; Comp. 152. Curt. F. L. f. 6.
Sides of fields, hedges, bushy places and woods; rather rare. A. June—
   July.
I. In a meadow near Harefield Church, plentifully; *Blackst. Fasc.* 16.
IV. Hampstead Heath; *Johns. Enum.* Kenton Brook, sparingly; hedgerow
   near Kenton, plentifully, F. W. Longman; *Melv.* 24. New Road,
   from Finchley to Regent's Park, E. H. Button; *Cooper, Supp.* 12.
V. Near the canal, Apperton; *Farrar.*
VI. Side of a wood behind the 'Green Man,' Muswell Hill; in a meadow
   near Tottenham High Cross, plentifully *Blackst. Spec.* 59. By the
   path side bet. White Hart Lane and Tottenham New River, J. Woods;
   B. G. 109. Edmonton, over against the Church; *Mart. App. P. C.*
   72. At the south entrance to the G. N. Ry. tunnel bet. Wood Green
   and Colney Hatch, sparingly, G. Munby; *Naturalist,* 1867, p. 180.
VII. [Pasture and medow ground about Pancridge (=Pancras) Church;
   Johns. Ger. 1250.] Fields towards Highgate; *Park. Theat.* 1079.

[About Tyburn and Maribone, Dr. Dale; *Merrett, 125.*] [Bushes about Pancras Church; *R. Cat.* i. 61.] [On the borders of the field going to Pancras Church and the Tile-hill; *Dill. in R. Syn.* iii. 325.] [Ken Wood, Hunter; *Park. Hampst.* 30.] [Near the Mitre public-house by the Paddington Canal, 1816; *Herb. G. & R.*]

First record: *Johnson*, 1632. Seems to have been formerly common in the northern suburbs. Easily overlooked.

190. L. pratensis, *L.*

L. sylvestris fl. luteo, Ger. em. (Blackst.).

Cyb. Br. i. 324; Comp. 153. Curt. F. L. f. 3.

Meadows and pastures; common. P. June—September.

I. Harefield!; *Blackst. Fasc.* 51. Elstree. South Mims.

II. Near Hampton Court!; *Newb.* Staines.

III. Hounslow Heath. Twickenham.

IV. About Harrow, common; *Melv.* 24. Sudbury. Stanmore.

V. Brentford; *Hemsley.* Near Twyford!; Hanwell; *Newb.*

VI. Edmonton. Whetstone.

VII. Millfield Lane.

First record: *Blackstone*, 1737.

(L. sylvestris, *L.* Hampstead Wood, Hunter; *Cooper*, 100. This has never been corroborated.)

191. L. macrorrhizus, *Wimm.*

Terræ glandes (Ger.).* *Astragalus sylvaticus, Thal.* (Johns.). *Lathyrus sylvestris lignosior* (Park.). *Orobus tuberosus, L.* (Herb. G. & R.).

Cyb. Br. i. 329; Comp. 154; Syme E. B. iii. t. 406.

Woods and heaths; rather common. P. May—July.

I. Harefield!; *Blackst. Fasc.* 8. Near Ruislip reservoir; *Melv.* 24. Near Elstree reservoir. Stanmore Heath; *Fl. Herts. Supp.* 3.

IV. In Hampsteede Wood; *Ger.* 1057. Hampstead Heath!; *Johns. Eric.* Bishop's Wood. Scratch Wood. Railway bank near Harrow; *Melv.* 24.

V. Wormholt Scrubs, 1815; *Herb. G. & R.*

VI. Barnet; *Herb. Hardw.* Winchmore Hill Wood. The Alders, near Whetstone. Hadley; *Newb.*

First record: *Gerarde*, 1597. The narrow-leaved form, *O. tenuifolius*, Roth, is found in (I.) Pinner Woods; *Melv.* 24, and (IV.) Bishop's Wood.

ORNITHOPUS, *Linn.*

192. O. perpusillus, *L.* *Bird's-foot.*

Ornithopodium majus et minus (Ger.). *Ornithopodium perpus., Lob.* (Johns.). *Ornith. radice nodosa, Parkins.* (Blackst.).

Cyb. Br. i. 311; iii. 411; Comp. 148. Syme E. B. iii. t. 378.

* Gerarde's figure represents *Lathyrus tuberosus*; Johnson (*Ger. em.* 1237) corrects him.

Sandy heaths, commons and roadsides; common. A. June—August.
I. Harefield!; *Blackst. Fasc.* 70. Ruislip Common; *Melv.* 23. Harrow
  Weald Common.
II. Towing-path bet. Kingston Bridge and Hampton; *Bloxam.* Near
  Teddington Ry. Station. Common by Walton Bridge. Roadside
  bet. Staines and Hampton, abundant.
III. Hounslow Heath. Drilling ground. Near Hatton, &c.
IV. On Hampstead Heath!; *Ger.* 1061, *Johns. Eric.*, and many subsequent
  writers. Near the Brent reservoir; *Farrar.* Stanmore.
V. Field near Wyke House Lane, Brentford; *Cherry.* Hanwell;
  Warren.
VI. Hadley!; *Warren.*
VII. [Hyde Park beyond the Spring, 1790; *E. B.* 369, and *Herb. Linn. Soc.*
  Ibid., 1816; *Herb. G. & R.*]

First record: *Gerarde,* 1597.

ONOBRYCHIS, *Gaertn.*

193. O. sativa, *Lam.* 　　　*Saintfoin.*
O. sive Caput Gallinaceum, Ger. em. (Blackst.).
Cyb. Br. i. 313; iii. 411; Comp. 149. Syme E. B. iii. t. 381.

Borders of fields on the chalk; very rare. P. June—August.
I. In a meadow near Harefield chalkpit; *Blackst. Fasc.* 66. Abundant
  in the fields above the canal at Harefield. Railway bank at Pinner;
  Melv. 97.
VII. Chelsea College, 1860, Britten; *Bot. Chron.* 58.

First record: *Blackstone,* 1737. Only native on the chalk.

Coronilla varia, *L.* VII. Chelsea; *Irv. H. B. P.* 684. Introduced.

ROSACEÆ.

PRUNUS, *Linn.*

194. P. communis, *Huds.*
1. *P. spinosa, L.* 　　　*Blackthorn. Sloe.*
P. sylvestris, Matth. (Johns.).
Cyb. Br. i. 330; iii. 415; Comp. 154. Syme E. B. iii. t. 408.
Hedges, and on heaths; very common. Shrub. March—May.
In all the districts.
VII. West End Lane. Finchley Road, &c.
There are some apparently very old bushes on North Heath, Hampstead.

2. *P. insititia, L.* 　　　*Bullace.*
P. sylv. fol. latioribus fructu majore (Johns.).
Cyb. Br. i. 330; Comp. 155. Syme E. B. iii. t. 409.

Hedges; rather rare. Large shrub. April, May.
I. Hedge near Pinner Wood. Harefield!; *Newb.*
II. Bet. Kingston Bridge and Hampton Court. Tangley Park.
III. Harrow; Roxeth; *Melv.* 24. Twickenham.
IV. Hampstead Heath; *Johns. Enum.* Near Sudbury Ry. Station. Dollis Hill; *Fox.*
VII. About Primrose Hill; *Irv. MSS.* Kentish Town.

3. *P. *domestica,* L.   *Wild Plum.*
Cyb. Br. i. 330. Syme E. B. iii. t. 410.
Hedges; rare. Small tree. April, May.
I. Road bet. Harefield and Ruislip; *Fox (v.s.).*
III. Roxeth; *Hind.*
IV. Neesden, near Willesden, 1863; *Morris (v.s.).*
Not a native.
First record: *Johnson,* 1632. The three sub-species are not separated by well-defined characters.

195. *P. Padus, *L.*   *Bird Cherry.*
Cyb. Br. i. 331; iii. 415; Comp. 155. Syme E. B. iii. t. 413.
Woods and hedges. Large shrub. May.
IV. Woods about Highgate; *Irv. MSS. and Lond. Fl.* 186. Hampstead Heath; *Cooper,* 99. Bishop's Wood; *Cooper,* 103. Harrow Park, probably introduced, W. M. H.; *Melv.* 25.
First record: *Irvine,* about 1830. Barely naturalised; often planted in parks and shrubberies. London mentions a tree at Syon House, 36 ft. high, and with a trunk 11 in. in diameter.

196. P. Avium, *L.*   *Wild Cherry.*
Cerasus sylvestris fructu rubro, J. B. (Blackst.). *C. s. fructu nigro, J. B.* (J. Mart.). *P. Cerasus,* L. (Mart. and Irv.).
Cyb. Br. i. 333; iii. 416; Comp. 155. Syme E. B. iii. t. 411.
Woods; rare. Tree. April, May.
I. Harefield; *Blackst. Fasc.* 17. Ibid., 1855; *Phyt. N. S.* i. 62.
III. Harrow Grove; *Melv.* 24. Twickenham.
IV. Bishop's Wood!; *Mart. App. P. C.* 68. Harrow Park; *Melv.* 24. Turner's Wood. Bentley Priory. *With black fruit*: Hampstead Heath and Bishop's Wood; *Mart. Tourn.* i. 167.
First record: *J. Martyn,* 1732.

197. P. Cerasus, *L.*
Cyb. Br. i. 332; iii. 415; Comp. 155. Syme E. B. iii. t. 412.
Hedges; rare. Shrub. May.
I. Near Harrow Weald Common. Harefield!; *Newb.* Deacon's Hill?
II. Tangley Park, near Hampton.
III. Near Isleworth.

IV. Edgeware; *Varenne.* Near Finchley; *Newb.*

V. Wood near Ealing, probably planted; *Hemsley.*

First record: *Varenne*, about 1830.

SPIRÆA, *Linn.*

198. S. Ulmaria, *L.*   *Meadow-Sweet.*

Ulmaria vulgaris, Park. (Blackst.).

Cyb. Br. i. 334; Comp. 156. Curt. F. L. f. 5.

Sides of streams and ditches, and in wet meadows; common. P.   June — September.

I. Harefield!; *Blackst. Fasc.* 111.   Eastcott.   Elstree.   Near Potter's Bar.   South Mims.

II. Staines.   Teddington.

III. Twickenham.   By the Cran, abundant.   Hounslow Heath.

IV. Frequent about Harrow; *Melv.* 25.   Sudbury.   Stanmore.   Near Finchley!; *Newb.*

V. Greenford; *Cooper*, 108.   Twyford!; Hanwell!; *Newb.*

VI. Edmonton.   Whetstone.

VII. Isle of Dogs; *Cooper*, 115.   Thames side, near Fulham.

First record: *Blackstone*, 1737.

199. S. Filipendula, *L.*   *Dropwort*

Œnanthe (Turn.).   *Filipendula* (Ger.).

Cyb. Br. i. 335; iii. 416; Comp. 156.   Syme E. B. iii. t. 416.

Meadows, banks and roadsides; rare. P.   June—August.

II. Bet. Hampton and Sunbury!; *Newb.*   In some plenty under bushes by the roadside near Charlton.   Hedgebank by road on north side of Bushy Park.

III. Meadow on south side of Richmond Bridge, abundant.

V. Great plenty . . . in a field adjoining to Sion House . . . on the side of a meadow called Sion Meadow; *Turn. Names, Ger.* 902.

First record: *Turner*, 1548; also first as a British plant. Seems confined in Middlesex to the precincts of the Thames.

SANGUISORBA, *Linn.*

200. S. officinalis, *L.*   *Great Burnet.*

Bipennella italica (Turn.).   *Pimpinella sylvestris* (Ger.).   *P. major vulgaris* (Park.).

Cyb. Br. i. 350; iii. 422; Comp. 165.   Syme E. B. iii. t. 421.

Damp meadows, on clay; rare. P.   June—August.

IV. In a field 5 or 6 miles from London on the road to Harrow, 1817; *Herb. G. & R.*   Pastures about Whitchurch and Stanmore, abundant, 1827–30; *Varenne.*   Meadow on the left hand of the road to Finchley, 6 or 7 miles down it, 1845; *Morris* (v. s.).   Near Stone-

bridge in a willow copse by the Brent, H. J. Wharton; *Farrar.*
Field at Kingsbury.
V. Muche about Sion; *Turn. Names.* Meadow by the Brent opp. a copse
called the Hundred Oaks; *Lees.*
VII. [Upon the side of a cawsey which crosseth . . . a field . . . bet.
Paddington and Lysson Greene; *Ger.* 889.] [By Pancras Church;
in two or three fields nigh unto Boobies Barn; *Park. Theat.* 583.]
(Ken Wood, Hunter; *Park Hampst.* 28.)
First record: *Turner,* 1548; also first as a British plant.

POTERIUM, *Linn.*

201. P. Sanguisorba, *L.*   *Salad Burnet.*
Pimpinella septima, C. B. P. (Blackst.).
Cyb. Br. i. 361; iii. 422; Comp. 165. Curt. F. L. f. 2.
Fields, dry pastures, &c.; chiefly on a dry or chalky soil; rather rare.
P. June—August.
I. In Harefield chalk-pit!; *Blackst. Fasc.* 77, and other places about
Harefield.
II. Bet. Sunbury and Hampton!; *Newb.* Meadows by the Thames; at
Hampton, bet. Hampton Court and Kingston Bridge, at Teddington
abundantly, and between Teddington and Strawberry Hill.
III. River side near Richmond Bridge.
IV. Bishop's Wood; *Lawson.*
VII. Brompton Cemetery, 1861; Kensal Green Cemetery, 1862, Britten;
Bot. Chron. 58.
First record: *Blackstone,* 1737.

AGRIMONIA, *Linn.*

202. A. Eupatoria, *L.*   *Agrimony.*
Cyb. Br. i. 339; Comp. 165. Syme E. B. iii. t. 417.
Fields, roadsides and waste places; rather common. P. June—
August.
I. Harefield!; *Blackst. Fasc.* 3. Pinner chalk-pits. Elstree.
II. Road bet. Staines and Hampton, not common. Towing-path bet.
Kingston and Hampton Court.
III. Twickenham.
IV. Common about Harrow; *Melv.* 27. Wembley; *Cole.* Bet. Whit-
church and Stanmore.
V. Greenford; *Cooper,* 108. Hanwell!; *Newb.* Apperton. Lampton.
VI. Edmonton. Enfield.
First record: *Blackstone,* 1737.

ALCHEMILLA, *Linn.*

203. A. vulgaris, *Linn.*
Cyb. Br. i. 361; iii. 423; Comp. 166. Syme E. B. iii. t. 423.

Meadows and pastures; rare. P. May—August.

I. In a field bet. Ruislip reservoir and the road to Harefield, 1866!; *Griffiths.*

IV. Harrow; not an uncommon plant in hilly pastures about Stanmore in 1827–30; *Varenne.*

VI. Among brushwood about Colney Hatch, near Barnet; *M. & G.* 193.

First record: *Milne and Gordon,* 1793. Totteridge Park, Herts, very large, J. R. M., *Phyt. N. S.* ii. 158.

204. A. arvensis, *Scop.*   *Parsley Piert.*
Percepier anglorum, Lob. (Johns.). *Polygonum selinoides* (Parkins.).
Cyb. Br. i. 363; Comp. 166. Syme E. B. iii. t. 422.

Cultivated fields, roadsides and waste ground; rather common. A. May—August.

I. Harefield; *Blackst. Fasc.* 74. Pinner; Ruislip Common; *Melv.* 27. Road bet. Harefield and Ruislip. Harrow Weald Common.

II. Staines. Teddington. Tangley Park, Hampton. Near Strawberry Hill.

III. Near Whitton, 1842; *Twining* (*v.s.*). Hounslow Heath. Near Hatton.

IV. Hampsteede Heath by the footpathes; *Park. Theat.* 449, and *Irv. MSS.* Stanmore; *Varenne.*

V. Near Ealing!; *Newb.*

VI. Colney Hatch. Edmonton. Near Enfield; *Tucker.*

VII. [Hide Park, Tuthill fields, &c.; *Johns. Ger.* 1594.]

First record: *Johnson,* 1633.

POTENTILLA, *Linn.*

205. P. anserina, *L.*   *Silver Weed.*
Pentaphylloides argentina dicta, R. Syn. (Blackst.).
Cyb. Br. i. 342; iii. 417; Comp. 158. Curt. F. L. f. 3.

Damp roadsides and waste places; very common. P. June—August. In all the districts.

VII. Tottenham; *Cherry.* Isle of Dogs. Thames Embankment.

First record; *Blackstone,* 1737.

P. hirta, *L.* VII. Rubbish at Parson's Green, 1856, A. J.; *Phyt. N. S.* ii. 168.

P. recta, *L.* VII. Parson's Green; *Irv. H. B. P.* 624. Ibid., 1862, J. Britten; *Seem. J. of Bot.* i. 375. Side of Basin, West India Docks; *Cherry* (*v.s.*).

206. P. argentea, *L.*
Pentaphyllum rectum minus, Park. (Blackst.).
Cyb. Br. i. 342; Comp. 158. Syme E. B. iii. t. 435.

Sandy roadsides and fields ; rather rare. P. June—August.

I. In a field near Harefield Common, sparingly ; *Blackst. Fasc.* 74. Lane bet. Potter's Bar Station and North Mims Wood ; *Phyt. N. S.* i. 407.

II. Under the wall of Ashford churchyard, a single specimen; *Phyt. N. S.* iv. 264. Several places on the road from Staines to Hampton. Near Teddington Ry. Station, one plant. Roadside near Teddington in some plenty. Near Strawberry Hill. In abundance on the bank bet. the towing-path and road, near Hampton Court.

IV. Hampstead Heath ; *Cooper*, 99.

V. Field by the Thames opp. Mortlake; *Fox.* Gravel-pits near Hanwell! ; *Warren.*

First record : *Blackstone,* 1737.

207. P. reptans, *L.*   *Cinquefoil.*

Pentaphyllum vulgatissimum, Park. (Blackst.).

Cyb. Br. i. 345 ; Comp. 159. Curt. F. L. f. 1.

Fields, roadsides, and waste places; very common. P. June—September.

Throughout all the districts.

VII. Hyde Park, 1817 ; *Herb. G. & R.* Near Cumberland Gate, 1868 ; *Warren.* Lea Bridge ; *Cherry.* Kensington Gardens. By Chelsea Hospital. Kentish Town. Primrose Hill. Isle of Dogs. West End, Hampstead, &c.

First record : *Blackstone,* 1737.

208. P. Tormentilla, *Nestl.*   *Tormentil.*

Tormentilla, Ger. em. (Blackst.). *T. officinalis, L.* (Curt.).

Cyb. Br. i. 345 ; iii. 417 ; Comp. 159. Curt. F. L. f. 5.

Heaths, roadsides, and dry parts of woods; rather common. P. July —September.

I. Harefield ! ; *Blackst. Fasc.* 99. Bet. Ruislip and Harefield. Harrow Weald Common ; *Melv.* 27. Stanmore Heath. Uxbridge ! ; *Newb.*

II. Abundant on roadsides about Ashford, Sunbury, &c. Bushey Park. Fulwell.

III. Drilling Ground. Hounslow Heath, &c.

IV. Highgate Wood ; *Johns. Eric.* Bishop's Wood. North Heath, Hampstead.

VI. Enfield. Colney Hatch.

VII. South Heath, Hampstead.

Var. β. P. procumbens, Sibth. *Tormentilla reptans alat. fol. profund. serratis, Plot.* (Pet.). *T. reptans, L.* (Mart.). Comp. 159. Syme E. B. iii. t. 431.

IV. Child's Hill Lane, Hampstead.

V. Horsington Hill.

VI. Winchmore Hill Wood.

VII. [In a ditch bet. the boarded river* and Islington Road ; *Pet. Midd.*]
First record : *Johnson*, 1629.

209. P. Fragariastrum, *Ehrh.*   *Barren Strawberry.*
Cyb. Br. i. 346 ; iii. 418 ; Comp. 160.   Syme E. B. iii. t. 427.
Hedgebanks, woods and heaths ; rather common.   P.   April, May.
    I. Stanmore Heath.   Near Potter's Bar.
III. Whitton, by the church.
  IV. Hampstead ! ; *Pet. H. B. Cat.*   Bishop's Wood. North Heath. Very
      common round Harrow ; *Melv.* 27.   Mill Hill ; *Herb. Hardw.*
  V. Near Lampton.
  VI. Colney Hatch.   Edmonton.   Hadley ! ; *Warren.*
VII. West End Lane.
First record ; *Petiver*, 1713.   No doubt overlooked elsewhere.

[Comarum palustre, *L.*   *Potentilla Comarum, Nestl.* (Syme). Cyb. Br. i.
    348 ; Comp. 160.   Syme E. B. iii. t. 437.   IV. It thrives very well at
    the head of the lesser bog on Hampstead Heath, where it was planted
    some years ago by Mr. Rand ; *Blackst. Spec.* 70.   A few plants are
    growing upon a bog at Hampstead ; *Mart. Mill. Dict.* It must have
    died out soon after Martyn saw it.]

FRAGARIA, *Linn.*

210. F. vesca, *L.*   *Wild Strawberry.*
F. vulgaris, C. B. P. (Blackst.).
Cyb. Br. i. 349 ; iii. 419 ; Comp. 160.   Syme E. B. iii. t. 438.
Woods, hedgebanks and shady places ; rather common.   P.   May—June.
    I. Harefield ! ; *Blackst. Fasc.* 29.   In old park woods.   Elstree.   South
     Mims.
  II. Hampton Court walls.   Waste ground near Strawberry Hill.
III. Harrow Grove ; *Melv.* 26.
  IV. Harrow Park ; *Melv.* 26.   Stanmore ! ; *Varenne.* Wembley Park ;
     Farrar. Deacon's Hill.   Bishop's Wood, Hampstead.
  V. Wormholt Scrubs, 1815 ; *Herb. G. & R.*
  VI. Winchmore Hill Wood.   Hadley ! ; *Warren.*   Near Enfield ; *Tucker.*
First record : *Merrett*, 1666.   Merrett mentions *Fragaria fructu hispido,*
    Ger. (IV.) in Hampstead Wood ; and (VII.) [In Hide-park] ; *Merr.*
    39.   This Ray considered a monstrosity.

? F. elatior, *Ehrh.*   *Hautboy Strawberry.*   Cyb. Br. i. 349.   Syme E. B.
    iii. t. 439.   I. Pinner Cemetery ; *Hind.*   IV. Hedgebank near Harrow
    Weald, abundant.   Railway embankment near the iron bridge, far
    from houses ; *Herb. Harr.*   The name of this garden escape is

---

* The New River was thus called in that part of its course between Hornsey and Islington where it was carried over Highbury Vale in a wooden trough lined with lead, and
462 feet long. It was demolished in 1776.

uncertain. There may be several species in cultivation, and it appears that *F. elatior* is not now in much repute with gardeners.

RUBUS, *Linn.*

211. R. idæus, *L.*   *Raspberry.*

Cyb. Br. i. 354; iii. 420; Comp. 161. Syme E. B. iii. t. 442.

Heaths and woods; rather rare.   Shrub.   June—July.

I. In a lane on Mr. Austin's farm on Harefield Moor; *Blackst. Fasc.* 88. Woods by Pinner Lane; *Fl. Herts Supp.* 9. Stanmore Heath. Harrow Weald Common. Steep bank round Bentley Priory, abundant.

II. Shaklegate Lane, Teddington, G. Francis; *Cooper Supp.* 12.

III. Gravel-pit nr. Hounslow. Twickenham Park. Inclosure, Whitton Park.

IV. Harrow Park; hedge in London Road; *Melv.* 25.   Near the great bog, Hampstead, J. Bliss; *Park Hampst.* 30.   Hampstead Heath; *W. S. Coleman.*

V. Twyford; *Lees.*

VI. Wood near Colney Hatch, doubtfully wild, 1821; *Bennett* (*v. s.*). Copse near Whetstone.   Edmonton.

First record: *Blackstone*, 1737. Certainly native in I., but no doubt the result of cultivation in many localities.

212. R. fruticosus, *L.**   *Bramble. Blackberry.*

R. vulgaris (Johns. Enum.).

Cyb. Br. i. 353; iii. 336–46; Comp. 161–2.

Hedges, thickets, woods and heaths.   Shrub.   B.   June—August.

1. *R. affinis*, W. & N.

Cyb. Br. iii. 340.

V. Horsenton Wood; *Lees in Phyt.* iii. 400.

Is not '*R. affinis*' of the Home counties, *R. calvatus*?; *Warren.*

2. *R. Lindleianus*, Lees.

R. leucostachys (Lindl. Syn. II.).   *R. nitidus* (Coop. Melv.).   Cyb. Br. iii. 339.

I. Harrow Weald Common, 1861; *Melv.* 26.†

II. Shacklegate Lane, Teddington, G. Francis; *Coop. Supp.* 12.

III. Hanworth Rd. nr. Hounslow, 1866!; *Newb.* Hospital Bridge; *Bloxam.*

---

* We much regret the imperfection of the list of Middlesex Brambles. It has resulted from the illness of one, and absence from the county of the other, of the Authors during the summer of 1868, which was to have been occupied by an investigation of the subject. We are indebted to the Hon. J. Leicester Warren for many parts of the list, which is made up of his observations and those of Mr. Hind, Mr. Lees, Mr. Meehan, and the authors of the *Flora Hertfordiensis*, with a few other, and some original, localities. It is, however, probable that further investigation would not add more than three or four recognised names to the list, though it would greatly improve the account of their distribution through the county. We have followed Mr. Syme in making the named forms, considered species in *Bab. Man.* vi., sub-species.

† All the fruticose *Rubi* (17), recorded in *Melv.* 25, 26, were found by the Rev. W. M. Hind, and named by Professor Babington.

IV. Willesden Lane, near Kilburn ; *Warren.*
V. Horsenton Wood ; *Lees.*
No doubt fairly general ; *Warren.*

3. *R. rhamnifolius,* W. & N.
Cyb. Br. iii. 340.   Syme E. B. iii. t. 446.
I. Harrow Weald Common (*R. cordifolius,* W. & N.).
II. Shacklegate Lane, Teddington, G. Francis ; *Coop. Supp.* 12.
IV. Bishop's Wood ! ; *Warren.*
V. Horsenton Hill ; *Lees.*
VII. Bet. Kensington Gore and Chelsea, E. F. ; *Herb. Mus. Brit.*
Frequent in N. Surrey ; *Warren.*

4. *R. discolor,* W. & N.
R. fruticosus, Sm. (Auct.).   Cyb. Br. iii. 341.   Syme E. B. iii. t. 447.
I. Uxbridge ; Hillingdon, very common ; *Warren.*
II. Towing-path bet. Hampton Court and Kingston Bridge ; *Bloxam.*
Bet. Hampton and Hampton Court ! ; *Newb.*
III. Hedges at Harrow ; *Melv.* 25.   Hanworth Road, near Hounslow ! ;
Newb.   Near Twickenham Park.
IV. Bishop's Wood, &c.
V. Hanwell ; *Warren.*   Ealing Common ! ; *Newb.*   Horsenton ; *Lees.*
VI. Near Barnet ; *Warren.*
VII. Kilburn, near Hampstead, &c.   Isle of Dogs ! ; *Newb.*

No doubt universally distributed.   The large South-England form,
which seems near *R. speciosus* of continental authors, is abundant
(I.) at Hillingdon ; (IV.) in Bishop's Wood ; and (VI.) in Hadley
Wood, and is prevalent in the metropolitan circuit, though I have
never observed it in the north of England ; *Warren.*

5. *R. thyrsoideus,* Wimm.
Cyb. Br. iii. 340.
III. Hedges at Roxeth, 1860 ; *Melv.* 25.

6. *R. leucostachys,* Sm.
Cyb. Br. iii. 341.   Syme E. B. iii. t. 448.
III. Hedges at Harrow, 1860 ; *Melv.* 25.
IV. Child's Hill Lane.
V. Ealing, 1846–47 ; *Meehan in Phyt.* iii. 9.   Hedge in the lane from
Brentford Ferry, 1866 ; *Baker.*

Var. β.   R. vestitus, Weihe.
III. Shady places at Harrow Park, 1860 ; *Melv.* 25.   Hospital Bridge ;
Bloxam.
V. Horsenton Wood ; *Lees.*
A plant found (V.) at Horsenton Hill by Mr. Lees has been supposed
to be a var. of *R. tomentosus,* Borkh. ; but Professor Babington con-
siders it probably *R. leucostachys* (see *Syme E. B.* iii. 261).

7. *R. carpinifolius*, W. & N.
   Cyb. Br. iii. 341.
I. Woods near Pinner Lane ; *Fl. Herts. Supp.* 9.
II. Hedges, Hampton, near the place where the Queen's River crosses the
   Thames Valley Railway! ; *Bloxam.*
   Mr. Bloxam's *carpinifolius* is probably the *R. umbrosus* of Bab. Man. ;
   Warren.

8. *R. villicaulis*, W. & N.
   Cyb. Br. iii. 341.
IV. Bet. Willesden Lane and Edgware Road Ry. Station, a bush or two ;
   Warren, who says that this is the *R. sylvaticus* of Bloxam and con-
   tinental authors and that it and the foregoing *R. speciosus* run near
   each other at times.

9. *R. macrophyllus*, Weihe.
   Cyb. Br. iii. 342. Syme E. B. iii. t. 450.
Var. R. macrophyllus, W. & N.
V. Ealing ; Chiswick ; *Mcehan in Phyt.* iii. 9. Horsenton Hill.

Var. R. Schlechtendalii, W. & N.
I. Woods near Pinner Lane ; *Fl. Herts. Supp.* 9.
Vars. amplificatus and umbrosus are almost certain to occur ; *Warren.*

10. *R. Bloxamii*, Lees.
   Cyb. Br. iii. 342.
IV. A bush or two bet. Willesden Lane and Edgware Road Ry. Station ;
   so named by Mr. Bloxam, but the plant is not quite like the Twy-
   cross *Bloxamii*, which I have seen in a growing state ; *Warren.*
   Middlesex ; *Lees in Phyt.* iv. 917.

11. *R. Hystrix*, Weihe.
   Cyb. Br. iii. 343.
I. Woods near Pinner Lane ; *Fl. Herts. Supp.* 9.
III. Shady places at Harrow Park, 1859 ; *Melv.* 25.
IV. Bishop's Wood, Hampstead, abundant ! ; *Warren.*
   A common bramble in the home counties ; frequent in Herts and
   Surrey ; *Warren.*

12. *R. rosaceus*, Weihe.
   Cyb. Br. iii. 344.
I. Pinner Hill, 1859 ; *Melv.* 25.
III. Wood End ; *Melv.* 25.
   Little more than a name ; I have never seen a specimen so labelled
   that might not be referred to either *scaber* or *Hystrix* ; *Warren.*

13. *R. pygmæus*, Weihe.
I. Edge of Pinner Wood, 1861 ; *Melv.* 26. Near Pinner ; *Syme E. B.*
   iii. 182.
   A scarce form.

14. *R. scaber*, Weihe.
   Cyb. Br. iii. 345.
III. Harrow, 1860; *Melv.* 25.
   V. Horsenton Hill; *Lees in Phyt.* iv. 917.
   Middlesex, Bloxam; *Cyb. Br.* iii. 345. Dried specimens of *R. Hystrix* and *R. scaber* when not well selected are difficult to discriminate; I suspect the distribution of these two, as Mr. Bloxam understands the names, will be found to supplement one another, and that *Hystrix* is the prevalent form in the counties about London where *scaber* is either absent or very rare; *Warren.*

15. *R. rudis*, Weihe.
   Cyb. Br. iii. 343.
   I. Woods near Pinner Lane; *Fl. Herts. Supp.* 9.
   V. Horsenton Wood; *Lees.*

16. *R. Radula*, Weihe.
   Cyb. Br. iii. 343. Syme E. B. iii. t. 452.
   I. Woods near Pinner Lane; *Fl. Herts. Supp.* 9. Wood at Swakeleys and Long Lane, Hillingdon; *Warren.*
III. Shady places at Wood End, 1859; *Melv.* 25.
IV. Hedge N. of Bishop's Wood, Hampstead; *Warren.*
VII. Kensal Green Lane; about Kilburn in several places; *Warren.*
   A general bramble of the home counties; *Warren.*

17. *R. Köhleri*, Weihe.
   Cyb. Br. iii. 343. Syme E. B. iii. t. 453.
   II. Shacklegate Lane, Teddington, G. Francis; *Coop. Supp.* 12.
III. Near Hounslow Ry. Station! ; *Newb.*
IV. Bishop's Wood, Hampstead, Sowerby; *Herb. Mus. Brit.*
   V. Horsenton Wood; *Lees.*

Var. R. pallidus, Weihe.
   I. Harrow Weald Common. Swakeleys and near Hillingdon; *Warren.*
VII. Side of road bet. Cricklewood and Edgware Road Railway Station probably general; *Warren.*

18. *R. fusco-ater*, Weihe.
   Cyb. Br. iii. 344.
   I. Woods by Pinner Lane; *Fl. Herts. Supp.* 9.
   V. Acton; *Meehan in Phyt.* iv. 917.
   Requires confirmation; *Warren.*

19. *R. Lejeunii*, Weihe.
   Cyb. Br. iii. 344.
   V. Horsenton Wood, Lees: I have a specimen so labelled authenticated by Mr. Bloxam; *Warren.*
   Middlesex, Bloxam; *Cyb. Br.* iii. 344. Edge of wood, Oxhey, Herts, 1861; *Melv.* 26.

20. *R. Guntheri*, Weihe.
 Cyb. Br. iii. 344.
 I. Woods near Pinner Lane ; *Fl. Herts. Supp.* 9.

[21. *R. humifusus*, W. & N.
 VII. Shady lane at Bellsize House, Hampstead.; *Irv. Brit. Bot.* 193.
 Weak woodland forms of *R. pallidus* are often thus misnamed ; *Warren.*]

22. *R. glandulosus*, Bell.
 R. Billardi, Weihe (Lees). Syme E. B. iii. t. 454.
 II. Shacklegate Lane, Teddington, G. Francis ; *Coop. Supp.* 12.
 V. Horsenton Hill ; *Lees.*

 Var. β. R. hirtus, W. & K. *R. fuscus* (Lees).
 I. Woods near Pinner Lane ; *Fl. Herts. Supp.* 9.

23. *R. Balfourianus*, Blox.
 Cyb. Br. iii. 242.
 I. Pinner ; *Melv.* 26.
III. Wood End ; *Melv.* 26.
 V. Lane from Castle Bar, Ealing, to Twyford, in some quantity; *Lees in Phyt.* iv. 917. Banks of the Brent ; *Lees.*
See Warren's note on *R. althæifolius.*

24. *R. corylifolius*, Sm.
 Cyb. Br. iii. 345 ; Comp. 161. Syme E. B. iii. t. 455.
 No doubt universally distributed.
 I. Hedges at Pinner, 1860 ; *Melv.* 26. Bet. Yewsley and Iver Bridge ! ; *Newb.* Hillingdon ; *Warren.*
 II. Towing-path bet. Hampton Court and Kingston Bridge ; Teddington Park ; *Bloxam.* Bet. W. Drayton and Yewsley! ; *Newb.*
 III. Harrow, 1860 ; *Melv.* 26. Near Hatton. Hospital Bridge ; *Bloxam.*
 V. Footpath bet. Brentford and Ealing ; *Meehan in Phyt.* iii. 917.
 VII. In nearly every hedge about London, e.g. near the Finchley Road Station; *Warren.* Isle of Dogs ! ; *Newb.*

 Var. β, conjungens.
III. Roxeth, 1860 ; *Melv.* 26.

 Var. γ, purpureus.
III. Harrow, 1860 ; *Melv.* 26. I have never observed this about London ; *Warren.*

25. *R. althæifolius*, Hort.
 I. Pinner, 1860 ; *Melv.* 26.
III. Harrow ; *Melv.* 26.
 Middlesex ; *Syme E. B.* iii. 194. See Mr. Warren's note on *R. tuberculatus.*
 Specimens departing from *corylifolius* in the direction of *nemorosus* are generally labelled *althæifolius.* There are, roughly speaking, three

H 2

groups in this section, *corylifolius, nemorosus,* and *cæsius*; and whilst *althæifolius* is a botanical dust-bin between the first and second, *Balfourianus* is a similar receptacle for intermediates between the second and third ; *Warren.*

26.　*R. tuberculatus,* Bab.
'*R. dumetorum*' (Blox.).　Cyb. Br. iii. 345.
I. Long Lane, bet. Swakeleys and Hillingdon ; *Warren.*
III. Harrow ; *Herb. Harr.*
IV. Willesden Lane ; *Warren.*
V. Horsenton Hill.
VII. A bramble of this group in the hedge bounding Kensington Gardens from the Palace northwards! ; *Warren.*

A common hedge bramble about London.　Our plant must certainly be ncluded under this somewhat large 'species' of Babington ; it is a less intense form than the northern *nemorosus* δ, *ferox,* Leight., and is certainly distinct from the *nemorosus var. pilosus* of continental authors, typical plants of which grow in a hedge near some old fir-trees on Barnet Common.*　Probably, some of the weakest specimens of the metropolitan form of '*tuberculatus*' would be placed among the forms of *R. althæifolius* by many botanists, the limits of which 'species' I am unable to exactly determine ; *Warren.*

27.　*R. Cæsius,* L.　　　*Dewberry.*
R. repens fructu cæsio, Ger. em. (Blackst.).　Cyb. Br. i. 352; iii. 419 and 346; Comp. 161.　Syme E. B. iii. t. 456.
I. Harefield ; *Blackst. Fasc.* 88.　Bet. Uxbridge and Hillingdon ; *Warren.*　Chapel Lane, Pinner ; *Hind.*
II. Bet. Sunbury and Hampton! ; *Newb.*　Staines Moor.　By the towing-path bet. Hampton Court and Kingston Bridge, plentiful.
III. By the Cran, Twickenham.
V. Brentford ; *Cherry.*　Perivale ; *Lees.*

---

* The group of *Nemorosi* in this genus requires complete rearrangement ; they may be roughly defined as all setose brambles of Babington's *Cæsii* which are not *R. cæsius,* L., *R. corylifolius* being for all practical purposes an unsetose bramble.　The following is suggested as a provisional settlement, one by no means exhaustive, new forms being almost certain to be found :—

　　R. nemorosus.
a. pilosus=R. tuberculatus, *Bab.*　Distinguished from *ferox* by the more equal armature of the stem, and by the more downy leaflets and longer panicle.　Leaflets flat.
β. intermedius.　The weakest and least setose of the group.　The barren stem is less prickly than in *pilosus,* the panicle with a few large distant prickles, and few much smaller ones between.　Common about London.
γ. ferox, Blox.　Strong and intense.　Corresponds to Babington's *diversifolius.*　Prickles of barren stem and panicle abundant and unequal, panicle very short, leaflets very variable but always rugose.　The common hedge-bramble of Cheshire, York, and Shropshire, but I have never seen it quite typical south of these counties.
δ. diversifolius, Blox.　By this I understand a very intense form of var. *γ,* with a long open panicle.　An extremely prickly plant ; much more local than *ferox,* but hardly worth a letter to itself. (*Warren.*)

VII. Hackney Marshes.

Not uncommon about London, but much less general than *R. corylifolius*; the usual form is probably *α agrestis*, but *β tenuis* occurs at Swakeleys: Hillingdon; *Warren.*

First record: *Johnson,* 1632.

GEUM, *Linn.*

213. G. urbanum, *L.*   *Avens.   Herb Bennet.*

Caryophyllata, Ger. em. (Blackst.).

Cyb. Br. i. 337; iii. 416; Comp. 156.   Syme E. B. iii. t. 457.

Shady hedgebanks and woods; very common.   P.   June—August.

Generally distributed through all the districts.

VII. Seen in this district north of London.

First record: *Blackstone,* 1737.

ROSA, *Linn.*

214. R. spinosissima, *L.*   *Burnet Rose.*

R. pimpinella (Ger.).   *R. sylvestris pomifera* (Johns.).   *R. pimpinellæ-folio, Ger. em.* (Blackst.).

Cyb. Br. i. 355; Comp. 162.   Syme E. B. iii. t. 461.

Open sandy heaths; very rare.   Small shrub.   May, June.

II. In Tuddington Field, abundantly; *Blackst. Fasc.* 87.

III. On Twittenham Common among the furze, plentifully; *Waring.* Hounslow, and other commons near Twickenham, R. Castles; *New B. G. Supp.* 588.

IV. [Hampstead Heath; *Johns. Enum.*]

VII. [In a pasture bet. Knight's bridge and Fulham; *Ger.* 1088.]

First record: *Gerarde,* 1597.   We have not seen this in Middlesex.

215. R. Sabini, *Woods.*

Cyb. Br. Supp. 56, 86; Comp. 162.   Syme E. B. iii. t. 465.

III. Near Blackmoor Farm, near Twickenham, R. Castles; * *New B. G. Supp.* 588.   'R. gracilis,' near Twickenham, 1843, T. Twining *Herb. Brit. Mus.*

First record: *Castles,* 1837.

216. R. villosa, *L.*   R. mollissima, *Fries* (Syme E. B. and L. Cat.).

Cyb. Br. Supp. 56 and 86; Comp. 163.   Syme E. B. iii. t. 466.

Hedges; very rare.   Shrub.   June, July.

III. Hedge at Roxeth, by the Harrow Gasworks, W. M. Hind; *Melv.* 28.

IV. Field near the Railway Station, at Greenhill, W. M. Hind; *Melv.* 28.

V. Hedges near Twyford; bet. Twyford and Apperton; *Lees.*

---

* 'Mr. C. has favoured me with cultivated specimens of the several roses mentioned here on his authority.'—*H. C. Watson* in *New B. G. Supp.* 588.

First record: *Hind*, 1860. Specimens named *R. villosa*, gathered by Mr. Hind, were distributed in 1860 by the ' Thirsk Botanical Exchange Club,' so that the name is probably correct; *R. tomentosa* is, however, said also to grow in both Mr. Hind's stations.

217. R. tomentosa, *Sm.*

R. villosa, var β (Huds.).

Cyb. Br. Supp. 57, 86 ; Comp. 163. Syme E. B. iii. t. 467.

Hedges; rather rare. Shrub. June, July.

I. ' *Var. η* (resembling *R. Borreri* and *R. micrantha*),' near Potter's Bar, 1814, J. Woods; *Linn. Trans.* xii. 202. There are specimens collected by Woods in 1815, in *Herb. Linn. Soc.* Hedges between Potter's Bar and South Mims in several places.

III. Roxeth, Hind ; *Melv.* 28. ·

IV. Hampstead ; *Irv. MSS.* Hedge at Greenhill; *Melv.* 28 (*v. s.*).

V. Near Apperton!; *Newb.*

VI. Highgate, 1827–30 ; *Varenne.*

VII. Hedges bet. Stoke Newington Church and Stamford Hill, J. Woods; *B. G.* 406. ' *Var. ε*,' Stoke Newington, J. Woods; *Linn. Trans.* xii. 201. Keb Wood, Hunter; *Park Hampst.* 30.

First record : *Woods*, 1805. Mr. Woods (*loc. cit.*) divided this into very numerous varieties ; besides *vars. ε* and *η*, he found his *vars. ν* and *ξ* in Middlesex.

218. R. Borreri, *Woods.* R. inodora, *Fries* (Bab.).

Cyb. Br. Supp. 57 and 86; Comp. 163. Syme E. B. iii. t. 471.

Hedges; very rare. Shrub. June, July.

IV. Hampstead ; *Irv. MSS.*

VI. Near Southgate, 1814, J. Woods ; *Linn. Trans.* xii. 211.

First record: *Woods*, 1814.

219. R. micrantha, *Sm.*

Cyb. Br. Supp. 57 and 86 ; Comp. 163. Syme E. B. iii. t. 469.

Hedges; very rare. Shrub. July, August.

I. Potter's Bar, probably Mr. Woods; *Herb. Kew* (*fide Newb.*).

II. By the (Hampton Court) Park pales bet. Ditton Ferry and Kingston, R. Castles ; *New B. G. Supp.* 588.

First record: *Woods*, about 1814.

220. R. rubiginosa, *L.* *Sweet Briar.*

R. sylvestris odora, Ger. em. (Blackst.).

Cyb. Br. Supp. 57, 86 ; Comp. 164. Syme E. B. iii. t. 468.

Heaths and hedges; rather rare. Shrub. June, July.

I. Harefield, sparingly; *Blackst. Fasc.* 87. Bet. Rickmansworth and Harefield on the common moor; on Stanmore Heath!, Mrs. Shute; *Fl. Herts. Supp.* 10.

III. Roxeth; *Melv.* 27.

IV. Hedge near football field, Harrow; *Melv.* 27. Stanmore; *Varenne.* Hampstead; *Irv. MSS.*

V. Greenford Road near Green Lane; *Melv.* 27. Osterley Park; *Masters.*

VII. Stoke Newington; *Herb. of King's Coll. Lond.*

First record: *Blackstone*, 1737. No doubt planted in some of its localities.

221. R. sepium, *Thuil.* R. rubiginosa, var. β? (L. Cat.).

Cyb. Br. Supp. 57. Syme E. B. iii. t. 470.

VII. Green Lanes, Middlesex, 1818, J. Woods; *Herb. Linn. Soc.*

Given in *Cyb. Br. Supp.* as a plant of the North Thames sub-province, probably on the above authority, as it is not included in Essex or Herts Floras.

222. R. canina, *L.* *Dog Rose.*

R. can. inodora, Ger. em. (Blackst.).

Cyb. Br. i. 357; Comp. 164. Curt. F. L. f. 5 (represents *R. dumalis,* Bechst.?).

Hedges and borders of woods; very common. Shrub. June, July.

Through all the districts. Most of the forms observed fall under Mr. Baker's group of *Eu-caninæ*; *R. lutetiana,* Leman, and *R. dumalis,* Bechst. (vars. α and β of Bab. Man.), are the most frequent; *R. urbica,* Leman (var. ε. Bab. Man.) is also common. *R. surculosa,* Woods (var. γ of Bab. Man.), recorded from Harrow; *Melv.* 28, and Hampstead; *Irv. MSS.*, is also probably to be referred to this group.

In the group *Hispidæ, R. andevagensis,* Bastard, was found by Woods ('*canina, γ glandulifera*') near Potter's Bar; *Linn. Trans.* xii. 225, and *R. verticillacantha,* Merat, by Mr. Warren in Willesden Lane.

Of the *Sub-rubiginosæ, R. tomentella,* Leman, was noticed in Willesden Lane by Mr Warren. '*Var. β,*' J. Woods, a single bush observed near Stoke Newington by Mr. Woods; *Linn. Trans.* xii. 228.

First record: *Blackstone,* 1737.

223. R. systyla, *Woods.*

R. collina (Sm. E. B.). *R. stylosa,* Desv.

Cyb. Br. i. 358; iii. 421; Comp. 164. Syme E. B. iii. t. 475.

Hedges; rare. Shrub. June, July.

I. Near Pinner Ry. Station.

IV. Hampstead, in a hedge near Turner's Wood; *Syme (v. s.).*

VII. Clapton, E. Forster; near Hornsey, Woods; *Linn. Trans.* xii. 230. Welham Green, 1815, Woods; *Herb. Linn. Soc.* Stoke Newington, 1820; *Herb. Mus. Brit.*

First record: *Woods,* 1815. A specimen in *Herb. Harr.* named *systyla* is probably a garden rose. Mr. Syme's plant is referred by Baker to the *R. fastigiata* of Deseglise.

224. R. arvensis, *Huds.*

Cyb. Br. i. 358; iii. 422; Comp. 164. Syme E. B. iii. t. 476.

Heaths, borders of fields and hedges; common. Shrub. June, July.

I. Harrow Weald Common. Stanmore Heath. South Mims. Harefield!; *Newb.*

II. West Drayton!; *Newb.* Tangley Park.

III. Near Isleworth. Bet. Twickenham and Worton.

IV. Hampstead; *Irv. MSS.* Bishop's Wood!; *Warren.*

V. Bet. Turnham Green and Acton!; near Twyford!; *Newb.* Near Lampton.

VI. Edmonton. Hadley!; *Warren.*

VII. [Marylebone fields, 1815; *Herb. G. & R.*] South Heath, Hampstead.

First record: *Merrett*, 1666. *Rosa canina sylvestris unico flore et fructu,* found in the fields bet. Hackney and London; *Merrett*, 105, and in Bishop's Wood, Mr. Martyn; *R. Syn.* iii. *Indic. plant. dub.* is referred by Hudson to this species.

CRATÆGUS, *Linn.*

225. C. Oxyacantha, *L.* *Hawthorn.* *May.* *White Thorn.*
Oxyacanthus, Ger. em. (Blackst.).
Cyb. Br. i. 364; iii. 424; Comp. 167.

1. *C. oxyacanthoides,* Thuill.
C. obtusata, DC. (Macreight). Syme E. B. iii. t. 479.
Woods and hedges; rather rare. Small tree. May.

I. Near Ruislip.

II. Hedge near Staines.

IV. Abundant in Bishop's Wood, Hampstead.

V. Near Apperton!; Hanwell!; *Newb.* Horsenton Hill.

VI. About Tottenham; *Macreight*, 74. Plentiful in Enfield Chase; *Syme E. B.* iii. 236. Near Hadley in several places!; *Warren.*

2. *C. monogyna,* Jacq.
Syme E. B. iii. t. 480.

Heaths, woods and hedges; very common. Tree or shrub. May, June.

Common in all the districts, but generally planted as a quickset hedge. There are however, very old thorns in (II.) Bushey Park, on (III.) Hounslow Heath, and by the river side bet. Twickenham and Richmond Bridge, and (IV.) on Hampstead North Heath; which are doubtless of spontaneous growth.

First record: *Johnson*, 1632. A form with the leaves more deeply and narrowly cut was noticed by Dillenius about London; *R. Syn.* iii. 454. A large-leaved form is recorded by Buddle as *Oxyacanthus folio et fructu majore* lately observed by Mr. Petiver not far from the boarded river; *Budd. MSS.* *With dark pink flowers;* near the canal bridge,

Apperton; *Melv.* 28. There are some fine hawthorns now in the Regent's Park, which are probably the same as those mentioned with admiration by Pepys in his Diary.

MESPILUS, *Linn.*

[226. *M. germanica, *L.*   *Medlar.*
M. sylvestris spinosa* (Merrett).
Cyb. Br. i. 364 ; iii. 424. Syme E. B. iii. t. 478.
Hedges; very rare. Tree. May, June.
III. Harrow Grove ; *Melv.* 28.
IV. In the hedges betwixt Hampsted Heath and Highgate; *Merrett,* 77.
First record: *Merrett,* 1666.   Merrett's station may have produced the wild plant, but it has not been corroborated by recent observers.]

PYRUS, *Linn.*

227. P. communis, *L.*   *Wild Pear.*
P. sylvestris, Ger. em.* (Blackst.).
Cyb. Br. i. 366; iii. 424; Comp. 167. Syme E. B. iii. t. 488.
Hedges and woods ; rather rare. Tree. April, May.
I. Harefield, frequent; *Blackst. Fasc.* 82.
III. Hedge near Twickenham.
IV. Two fields S. of road bet. Neesdon and Blackfoot Hill ; *Farrar.* Near the path across the meadows from Child's Hill to Hendon; *Irv. Lond. Fl.* 187. In hedge-wastes to the north of Finchley; *Loud. Arb. et Frut.* 882.
VI. Highgate Wood, 1850; *Herb. Hardw.* Hadley!; *Warren.*
VII. [A solitary tree in field bet. Primrose Hill and Adelaide Road, 1862, perhaps remains of a garden.]

First record: *Blackstone,* 1737. Loudon (*loc. cit.* 888) says that the oldest pear trees in the neighbourhood of London are at Twickenham, where are some 50 to 60 ft. high, and with trunks 18 in. to 3 ft. in diameter. No doubt often not native, but accidentally sown from orchard pears. The specimen in *Herb. Harr.* labelled *P. communis* seems a foreign *Pyrus* allied to *Aria.*

228. P. Malus, *L.*   *Crab.*
Cyb. Br. i. 366 ; Comp. 167. Syme E. B. iii. t. 489.
Hedges and heaths; rather common. Tree or Shrub. May.
I. Harefield! ; *Blackst. Fasc.* 58. Harrow Weald Common.
II. Staines.
III. Harrow; *Melv.* 29.
IV. Kingsbury ; *Cole.* North Heath and Bishop's Wood, Hampstead. Finchley ; *Newb.*

V. Wormholt Scrubs; *Britten.* Near Kew Bridge Ry. Station!; *Newb.*

VI. Colney Hatch. Edmonton.

VII. [Marylebone fields, 1818; *Herb. G. & R.*]

Var. β tomentosa, Koch. *P. mitis,* Syme E. B. iii. t. 490. *Wild Apple.*

I. Bet. Ruislip and Harefield.

II. Tangley Park.

III. Harrow; *Herb. Harr.*

First notice: *Blackstone,* 1737. *Var. β,* though probably native, is often of garden origin.

(P. domestica, *Sm.* *Service Tree. Sorbus, Ger.* (Merrett). Cyb. Br. i. 369. Syme E. B. iii. t. 487. VII. In the pine-walks by Hampstead; *Merrett,* 115. Ken Wood, Hampstead, Hunter; *Park Hampst.* 30. Probably *P. Aucuparia* was the species intended.) Loudon, *Arb. and Frut.* 924, mentions a planted tree in a field (formerly part of the nursery) adjoining Brompton Park Nursery. It is 40 feet high, and the diameter of the trunk 18 inches. He adds that it bears abundantly most years, but not every year.

229. P. Aucuparia, *Gaertn.* *Mountain Ash. Rowan.*

Sorbus sylvestris alpina, Lob. (Johns.). *Sorbus seu Fraxinus sylvestris* (Park.). *S. sylv. fol. domesticæ similis* (Budd.).

Cyb. Br. i. 366; Comp. 168. Syme E. B. iii. t. 486.

Woods; rather rare. Tree. May, June.

I. Harefield Common!; *Newb.*

III. Harrow Grove; *Melv.* 29. Inclosure, Whitton Park.

IV. Harrow Park, &c.; *Melv.* 29. Hampstead Heath and Wood!; *Johns. Eric. and many subsequent authors.* Woods by Highgate, &c.; *Park. Theat.* 1418. Bishop's Wood, abundant.

VI. Winchmore Hill Wood. Whetstone.

First notice: *Johnson,* 1629. Planted for ornament in many places. Wild in districts I. IV. and VI.

230. P. Aria, *Sm.* P. eu-Aria (Syme E. B.). *White Beam.*

Aria Theophrasti effigie Alni, Lob. (Johns.). *Sorbus sylvestris Aria Theophrasti dicta* (Park.). *Sorbus sylvestris anglicus* (How). *Sorbus alpina, Chabr.* (Blackst.). *Cratægus Aria, L.* (Mart.).

Cyb. Br. i. 367; iii. 425; Comp. 168. Syme E. B. iii. 482.

Hedges, heaths and woods; rare. Tree. May, June.

I. On Uxbridge Common and in Harefield chalkpit, *Blackst. Fasc.* 96. Uxbridge Common near the Warren House; *Blackst. MSS.*

IV. Hampstead Heath; *Johns. Eric. &c.* One tree on Hampstead Heath . . . . on the left hand of the highway as you go to Hendon; *Park. Theat.* 1421. Bishop's Wood; *Mart. App. P. C.* 68. Hedge by the road bet. Hendon and Finchley!; *Irv. Lond. Fl.* 187.

V. Small trees in hedges near Apperton ! ; *Newb.*

VI. Hornsey, sp. Sowerby ; *Winch. MSS.*

First record : *Johnson,* 1629. Native on the chalk in District I. Perhaps planted in most or all of the other localities.

231. P. torminalis, *Ehrh.*   *Wild Service-tree.*

Sorbus torminalis (Ger.).

Cyb. Br. ii. 367 ; iii. 425 ; Comp. 168. Syme E. B. iii. t. 481.

Woods and hedges ; rather rare. Tree. May.

I. Harefield ; *Blackst. Fasc.* 96. Just out of the county in Herts ! ; *Newb.* Woods by Pinner Lane ; *Fl. Herts. Supp.* 10.

IV. Frequent in this district. Hampstead Heath ; *Johns. Eric.* Woods adjacent to Hampstead Heath ; *MSS. Budd.* Bishop's Wood ! ; *Mart. App. P. C.* 68. Hedge in foot-ball field, Harrow ; *Melv.* 29. Scratch Wood, Edgwarebury, abundant.

V. Bet. Perivale and Greenford, one tree ; *Lees.*

VII. Formerly common. [Many small trees in a little wood a mile beyond Islington ; *Ger.* 1290.] In the pine-walks by Hampstead ; *Merrett,* 115. Cane Wood ; *Mart. App. P. C.* 68. [Hedges bet. Regent's Park and Hampstead, 1818 ; *Bennett (v.s.).*] [Hedge on north side of Primrose Hill near Swiss Cottage ; *Herb. Hardw.*] Millfield Lane. Hornsey Wood.

First record : *Gerarde,* 1597. The trees in Bishop's and Scratch Woods are small, but there are two fine ones in Millfield Lane. Almost confined to the clay.

LYTHRACEÆ.

LYTHRUM, *Linn.*

232. L. Salicaria, *L.*   *Purple Loosestrife.*

Lysimachia purpurea spicata, Ger. em. (Blackst.).

Cyb. Br. i. 384 ; Comp. 174. Curt. F. L. f. 3.

Sides of streams and ditches ; very common. P. July—September.

In all the districts.

VII. Fulham. Hackney Marshes.

First record : *Blackstone,* 1737. [Blackstone mentions a state, *Lysimachia purp. trifol. caule hexagono* (VII.) by the sides of the canals at Chelsea waterworks ; *Blackst. Spec.* 50.]

[233. L. Hyssopifolia, *L.*

Gratiola angustifolia, Ger. em. (Merr., Blackst.).

Cyb. Br. i. 383 ; iii. 430 ; Comp. 174. Syme E. B. iv. t. 492.

Damp places in fields ; very rare. A. June—September.

II. Marsby field by the road bet. Stanes and Lalam (= Laleham), Mr. Nichols ; *Blackst. Spec.* 33.

III. On Hounslow Heath ; *Huds.* ii. 206.
V. ? Bet. Acton and Uxbridge, amongst the corn ; *Merrett,* 59.
    First record : *Merrett,* 1666. Last : *Hudson,* 1778.]

PEPLIS, *Linn.*

234. P. Portula, *L.*
Alsine rotundifol. sive Portulaca aquatica (Johns.).
Cyb. Br. i. 384 ; iii. 431 ; Comp. 174. Curt. F. L. f. 4.

Wet places, especially on heaths and commons ; rather common. A. ?
    July, August.
  I. Harefield ! ; *Blackst. Fasc.* 5. Harrow Weald and Ruislip Commons ;
      Melv. 30. Stanmore Heath. South Mims.
 II. Feltham Green.
III. Hounslow Heath. Near Hatton.
 IV. North Heath and Turner's Wood, Hampstead.
 VI. Edmonton. Hadley ! ; *Warren.*
VII. Bet. Kentish Towne and Hampstead ; *Johns. Ger.* 615.

First record : *Johnson,* 1633 ; also the first notice of the plant as British.
    Johnson figured it (fig. 4) in the plate affixed to the *Enum. Plant.*
    Hampst. under the name *Alsine aquat. fol. rotundioribus.*

ONAGRACEÆ.

EPILOBIUM, *Linn.*

235. E. angustifolium, *L.*   *Rose Bay. French Willow.*
Cyb. Br. i. 369 ; Comp. 168. Syme E. B. iv. t. 495.
Gravelly banks and woods ; rare. P. July, August.
  I. Harrow Weald, introduced, W. M. H. ; *Melv.* 29.
 II. Bank by the pier of the aqueduct which carries the water of the
      Queen's River across the Thames Valley Ry. not far from Hampton
      Station, in some abundance. Solitary plants in two places in the
      unused roads, Fulwell.
 VI. Near East Barnet, Herts, J. Woods ; *B. G.* 404.
VII. Cane Wood, Hampstead ; *Hill,* 201. Hedgebank in Lord Mansfield's
      premises, J. Bliss ; *Park Hampst.* 30. On soil brought from else-
      where near Paddington Cemetery, where it has held its ground for
      four years (1868) ; *Warren.*

First record : *Hill,* 1760. There is some doubt as to the nativity of this
    in Middlesex. In Hill's locality however it was probably wild, and
    the plant found in District II. is the form *E. macrocarpum,* Steph.,
    which is not that usually cultivated in gardens.

236. E. hirsutum, *L.*   *Great Willow Herb. Codlins and Cream.*
Lysimachia hirsuta siliquosa magno flore, C. B. P. (Blackst.).
Cyb. Br. i. 370; iii. 426 ; Comp. 169.   Curt. F. L. f. 2.
Sides of streams and ditches, and in damp ground ; very common. P.
June—September.
In all the districts.
VII. [Marylebone fields, 1815; *Herb. G. & R.*] Victoria Park; *Cherry.*
Kentish Town. Regent's Park, by Hanover Gate, &c. By Kilburn
Ry. Station.
First record : *Blackstone,* 1737.   A more or less glabrous state is
not uncommon in the county.

237. E. parviflorum, *Schreb.*
Lysimachia siliquosa hirsuta parvo flore, C. B. P. (Blackst.).   *E. villosum*
(Curt.).
Cyb. Br. i. 371; Comp. 169.   Curt. F. L. f. 2.
Wet places, sides of ditches, &c.; very common. P. June—August.
In all the districts.
VII. Regent's Park, 1830; *Varenne.* Eel-brook Meadow, Parson's Green.
Hackney Marsh.
First notice: *Buddle,* about 1710. The subglabrous form, *E. rivulare,*
Wahl., is nearly as frequent as the type ; Buddle records it as
Chamænerion glabrum parvo flore, by the Thames side and in
several other places about London ; *Budd. MSS.* and *Herb. Budd.*
vol. x. f. 37.   The specimens in *Herb. Harr.* are *tetragonum.*

238. E. montanum, *L.*
Lysimachia siliquosa glabra major, C. B. P. (Blackst.).
Cyb. Br. i. 371 ; Comp. 169.   Curt. F. L. f. 3.
Hedgebanks, old walls and woods ; very common. P. June—September.
In all the districts.
VII. Chelsea, 1815; *Herb. G. & R.*   Kentish Town ; *Herb. Hardw.*
Parson's Green. Hampstead.
First record : *Blackstone,* 1737.

239. E. roseum, *Schreb.*
Cyb. Br. i. 372 ; iii. 426 ; Comp. 170. Syme E. B. iv. t. 501.
Damp waste places, a weed in garden ground; rare. P. July—Sep-
tember.
I. Moss Lane, Pinner, on the hedgebanks.
IV. Totteridge, Herts, J. Woods; *B. G.* 332.
VI. Garden weed, Edmonton.
VII. Rather common. Marylebone Infirmary garden ; in 1830 common in
the little gardens in front of houses in the New Road, especially
tl at part near Tottenham Court Road; *Varenne.* Hyde Park, opp.
Bayswater Road ; *Irv. H. B. P.* 610. Grafton St., Holloway ; *Newb.*

Gardens, Pentonville Road. Devonshire Hill, Hampstead. 70 Adelaide Road. Under walls of Henry VII.'s Chapel, Poet's Corner, Westminster.

First record : *Varenne*, 1830.

240. E. tetragonum, L. (*Bab.*).

Cyb. Br. i. 373 ; Comp. 170. Curt. F. L. f. 2 (drawn from a Middlesex specimen).

Sides of ditches, roadsides, &c.; rather common. P. July—September.
I. Harrow Weald Common. Moss Lane, Pinner.
II. Teddington. Near Strawberry Hill.
III. Near Hounslow. About Twickenham.
IV. About Harrow, very common; *Melv.* 30. North Heath, Hampstead; *Lawson* (*v. s.*). Brent side near Willesden; *Herb. Hard.* Harrow Weald. Stanmore, on a wall. Near Sudbury Ry. Station. Brick-field at Burgess Hill, Hampstead.
V. Near Brentford; *Hemsley.* Acton !; near Shepherd's Bush Station ! ; *Newb.* Ditch by the line, Kew Bridge Ry. Station. Chiswick.
VII. Lane from Newington to Hornsey; *Curt. F. L.* Regent's Park, 1830 ; *Varenne.* Sandy End, Fulham. Site of Exhibition of 1862.

First record : *Curtis*, about 1780.

241. E. obscurum, *Schreb.*

Cyb. Br. Supp. 57. Syme E. B. iv. t. 503.

Boggy places, sides of ditches, &c. ; common. P. June—August.
I. By the canal, Harefield. Harrow Weald Common. Stanmore Heath. Moss Lane, Pinner. Elstree. Potter's Bar.
II. Near Sipson. Charlton. Near Strawberry Hill.
III. About Hounslow. Twickenham.
IV. Mill Hill, A. Atkins; *Herb. Mus. Brit.* Field near Edgwarebury. North Heath and Bishop's Wood, Hampstead. Finchley Road at Burgess Hill.
V. Recorded for this district.
VI. Colney Hatch. Edmonton.
VII. South Heath, Hampstead, by the ponds. Finchley Road.

First record : *Atkins*, 1837. Probably this is *E. virgatum*, Fries, of Mr. Hind's catalogue of Harrow plants in *Phyt. N. S.* v. 200 ; but neither that name nor *E. obscurum* is given in the ' Harrow Flora.'

242. E. palustre, *L.*

Cyb. Br. i. 372 ; iii. 426 ; Comp. 170. Syme E. B. iv. t. 504.

Boggy places, chiefly on heaths ; rare. P. July—September.
II. Bet. Hampton and Hampton Court !; *Newb.*
III. Hounslow Heath, by a little tributary to the Cran.
IV. North Heath, Hampstead !; *Varenne.*

VI. Edmonton.

First record: *Varenne*, 1827–30.

Œnothera biennis, *L.* *Evening Primrose.* Cyb. Br. i. 375. Syme E. B. iv. t. 508. II. Roadside near Twickenham Common Church, one plant. III. Twickenham, near the railway junction. IV. Fields at Hendon, naturalised; *Irv. Lond. Fl.* 199. VII. Chelsea, 1864; *Britten.* Cremorne; *Fox.* Near the Paddington Cemetery!; *Warren.* Shepherd's Bush. Victoria St., Westminster. Thames Embankment. Site of Exhibition of 1862, S. Kensington. Perhaps naturalised; usually a garden escape.

Œ. odorata, *Jacq.* Syme E. B. iv. t. 509. III. Clifden Road, Twickenham; site of an old garden. Cultivated.

Œ. pumila, *L.* VII. Spontaneously in a garden at Chelsea, 1861; *Phyt. N. S.* vi. 23. Cultivated.

CIRCÆA, *Linn.*

243. C. lutetiana, *L.* *Enchanter's Nightshade.*
Cyb. Br. i. 376; Comp. 171. Syme E. B. iv. t. 511.

Woods, shrubberies, and shady ditch-banks; rather common. P. June —August.

I. Harefield!; *Blackst. Fasc.* 18. Moss Lane, Pinner.
II. Bet. Hampton and Tangley Park. Harrow Grove; *Melv.* 30.
IV. Ditch on further side of Hampstead Heath; *Blackst. MSS.* Golder's Green; *Herb. Hardw.* Near Hendon Church, Button; *Cooper Supp.* 11. Harrow Park; *Melv.* 30. Child's Hill Lane, Hampstead. Stanmore.
V. Sion Park. Road bet. Isleworth and Brentford.
VI. Hornsey Lane, Button; *Cooper Supp.* 11. Bury Street, Edmonton.
VII. Mill Farm Lane, Highgate, G. E. Dennes; *Cooper Supp.* 11. Near Fulham Church, 1815; *Herb. G. & R.* Highgate Hill, plentiful.

First record: *Blackstone*, 1737.

HALORAGACEÆ.

MYRIOPHYLLUM, *Linn.*

244. M. verticillatum, *L.*
Cyb. Br. i. 377; iii. 428; Comp. 172. Syme E. B. iv. t. 513.

River and streams; rare. P. July, August.

I. Near Uxbridge; *Lightfoot MSS.*
II. Canal, West Drayton!; *Newb.* Pond on common by Walton Bridge.
V. Greenford; *Cooper*, 108. Ditch near Greenford; *Farrar.*
VII. By where the New River runs underground at Stoke Newington; and

in a little ditch at the west end of the village, J. Woods ; *B. G.* 412. Floating in the Hackney Canal.

First record : *Woods,* 1805.

245. M. spicatum, *L.*

Millefolium aquaticum pennatum spicatum, C. B. (Pet.).

Cyb. Br. Supp. 57, 87 ; Comp. 172. Syme E. B. iv. t. 514 (drawn from an Isle of Dogs specimen).

Ponds and ditches ; common. P. June—September.

II. In the canal at Hampton Court; *Pet. Midd.* Bet. Teddington and Bushey Park ! ; *Newb.* Round Pond, Bushey Park. River bet. Hampton Court and Kingston Bridge.

IV. Pond in Bentley Priory grounds.

V. (Canal at Greenford ; *Melv.* 30.) Ditches in Market Gardens, Fulham ; *Britten.* Sion Park.

VI. Edmonton.

VII. In a slow rivulet near Poplar; *Pet. Midd.* Isle of Dogs ! ; *E. B.* 83. River Lea, Forster ; *Fl. Essex,* 117. Round Pond, Kensington Gardens, 1868 ; *Warren* (*v. s.*). Walham Green, 1845 ; *Morris* (*v.s.*). Ponds at the head of the Serpentine, Kensington Gardens. Reservoir, S. Heath, Hampstead.

First record : *Petiver,* 1695. There is some doubt about the species in several of the above localities. The specimen in *Herb. Harr.* from Ruislip (see *Melv.* 30) is *Ranunculus floribundus* (submerged leaves).

246. M. alterniflorum, *DC.*

Potamogeiton sive millefolium aquat. pinnatum minus et ramosius (Budd.). *Millefolium aquat. pennat. minus fol. sing. &c., H. Ox.* (R. Syn iii.). *Potamog. pennis tenuissimis, Pet.* (Blackst.).

Cyb. Br. Supp. 57, 87 ; Comp. 172. Syme E. B. iv. t. 515.

Streams and ponds ; rare. P. June—September.

I. In a field called Innins joining to Harefield Common, in a pond ; *Blackst. Fasc.* 81.

III. In the river on Hounslow Heath ; *Doody MSS. and Budd. MSS. & Herb.*

First record : *Doody,* about 1700 ; also the first record as British. There is a poor figure in *Pet. H. B. Cat.* t. 6, f. 6. We have not met with this species in Middlesex, but perhaps one or two of the stations given for *M. spicatum* produce it. Occurs on Totteridge Green, Herts (v. *Fl. Herts.* 100).

HIPPURIS, *Linn.*

247. H. vulgaris, *L.*   *Mare's-Tail.*

Equisetum palustre, brev. fol. polyspermon, C. B. P. (Blackst.).

Cyb. Br. i. 377 ; iii. 428 ; Comp. 172. Curt. F. L. f. 4.

Streams, ditches, and boggy places; rare. P. June, July.
- I. In Harefield River; in a bog on Uxbridge Moor, plentifully; *Blackst. Fasc.* 26. Water near Uxbridge; *Dicks. H. S.* Bet. Yewsley and Iver Bridge!; *Newb.*
- IV. About Hampstead; *Cooper,* 100.
- VI.? In the Lea, but perhaps not so low down as Middlesex; *Pamplin.*
- VII. [Near Hornsey; *Huds.* i. 2. In the New River at Hornsey; *Curt. F. L.*] [Near Stoke Newington and Highgate; *M. & G.* 8.] [In a ditch which was formerly the course of the New River bet. Stamford Hill and the Green Lane, Hornsey, E. Forster; *Herb. Mus. Brit.*]

First record: *Blackstone,* 1737.

CUCURBITACEÆ.

BRYONIA, *Linn.*

248. B. dioica, *L.*   *White Bryony.*
B. alba, Ger. em. (Blackst.).
Cyb. Br. i. 385; Comp. 175. Syme E. B. iii. t. 517.

Hedges; very common. P. May—July.
Distributed through all the districts.
- VII. Marylebone, 1815; *Herb. G. & R.* Islington; Hornsey Wood, 1839; *Herb. Hardw.* Primrose Hill, 1845; *Morris (v.s.).* Kentish Town, &c., common.

First record: *Blackstone,* 1737.

PORTULACEÆ.

MONTIA, *Linn.*

249. M. fontana, *L.*   *Blinks.*
Alsine parv. pal. tricoccos, Portul. aquat. similis (Ray, Blackst.). *Alsine flosculis conniventibus* (Merrett).
Cyb. Br. i. 386; Comp. 175. Curt. F. L. f. 3.

Wet places on heaths and commons; rather common. A. B. or P. April —August.
- I. On Harefield Common, plentifully!; *Blackst. Fasc.* 5. Ruislip and Harrow Weald Commons; *Melv.* 31. Stanmore Heath.
- II. Common by Walton Bridge.
- III. Whitton, G. Francis; *Coop. Supp.* 12. Near Hatton.
- IV. Hampstead; *Pet. H. B. Cat.* North Heath, Hampstead.
- VI. Hadley!; *Warren.*

VII. [Frequent in Hide Park ; *Merrett*, 5.] Bet. Kentish Town and Hampstead ; *R. Cat.* i. 17. Chelsea Common ; *Herb. Mus. Brit.* South Heath, Hampstead.

First record : *Merrett*, 1666. *M. minor*, Gm. is the more common variety in Middlesex ; but the large floating plant, *M. rivularis*, Gm. occurs at (I.) Stanmore Heath, and probably in other places.

Claytonia perfoliata, *Don.* Syme E. B. ii. 260. VII. The late Mr. Anderson introduced it into the Botanic Garden at Chelsea, where it soon became a most troublesome weed, and remains so at the present day (1853), coming up spontaneously by thousands in various parts of the garden ; *E. Newman in Phyt.* iv. 983. Mr. T. Moore informs us that it is still (1869) an abundant weed there.

PARONYCHIACEÆ.

HERNIARIA, *Linn.*

[250. H. glabra, *L.*

H. hirsuta, L. (Huds.).

Cyb. Br. i. 388 ; iii. 431 ; Comp. 176. Syme E. B. vii. t. 1171.

Dry fields and commons. P.? July.

VI. In meadows at Colney Hatch ; *Huds.* i. 94. We found it in a field at Finchley, and likewise at Colney Hatch ; *M. & G.* 455. Finchley Common, 1795, Mr. Dickson ; *C. C. Babington in E. B. S.* iii. 2857.

First record : *Hudson*, 1760 ; last, 1795. Likely to be refound. Mr. Babington possesses a specimen of Dickson's Finchley Common plant, marked *H. hirsuta*, of which he says, ‘ It is not that plant, being exactly *H. glabra*’ (*loc. cit.*) ; and again, ‘ The Finchley Common plant was *H. glabra*’ (*Bab. Man.* vi. 136).]

[251. *H. hirsuta, *L.*

Cyb. Br. i. 389. E. B. t. 1379.

VI. ? Near Highgate ; *Dicks. H. S.* fasc. 19 (published in 1802).

The specimen in the British Museum copy appears to be correctly named *H. hirsuta.* It was probably an introduction at Highgate ; and, though it has not since been met with in Middlesex, it has occurred at Hartlepool, and is stated in *Syme E. B.* vii. 183, to have ‘ recently occurred near Coventry, no doubt accidentally introduced.’ Mr. Syme considers that this and *H. glabra* should be combined, each standing as a sub-species, whilst *H. ciliata* is kept distinct ; and in L. Cat. *H. hirsuta* is entered in italics as var. β of *H. glabra*.]

SCLERANTHUS, *Linn.*

252. S. annuus, *L.*   *Knawel.*

Polygonum selinoides seu Knawel Germanorum, Ger. em. (Blackst.).

Cyb. Br. i. 139 ; iii. 431 ; Comp. 176. Syme E. B. vii. t. 1175.

Dry sandy ground, roadsides, and cornfields; rather rare. A or B. June—September.

I. Harefield!; *Blackst. Fasc.* 79. Pinner, W. M. H.; *Melv.* 65. Harrow Weald Common. Hillingdon!; *Warren.*

II. Bet. Teddington Ry. Station and Bushey Park. Teddington Park. Tangley Park.

III. About Hounslow. Near Hospital Bridge.

IV. Kingsbury; *Varenne.* Hampstead; *Irv. MSS.*

Var β. S. biennis, Reuter. Syme E. B. vii. t. 1176. Frequent in the vicinity of London; *Syme E. B.* vii. 182. The Harefield and Hillingdon plants seem to be this, and it may be the more general form.

First record: *Blackstone,* 1737.

(? Ortegia hispanica, *L. Juncaria salmaticensis*; on the right hand of Bradford Bridge, at the lower end of Gray's Inn Lane, by London, near the watercourse that passes along thereby; *Park. Theat.* 453. Parkinson, perhaps, mistook *Scleranthus annuus* for this.)

CRASSULACEÆ.

Tillæa muscosa, *L.* Cyb. Br. i. 395; iii. 434; Comp. 178. Syme E. B. iv. t. 524. IV. An old farm wall near Kingsbury, Mr. Wharton; *Farrar.* A troublesome weed in gravel walks .... near London; *Hook. Br. Fl.* iii. 80. Mr. H. C. Watson remarks: ' Through cultivation as a botanical curiosity, it has become naturalised in places near London' (*loc. cit.*); but we have never seen it under such circumstances in Middlesex.

SEDUM, *Linn.*

253. *S. Telephium, *L.* (Syme and L. Cat.). *Orpine. Live-long.* Telephium vulgare, C. B. P.* (Blackst.).

Cyb. Br. i. 396; Comp. 178. Syme E. B. iv. tt. 526 and 527.

Hedgebanks and woods; rare. P. July, August.

I. Harefield; *Blackst. Fasc.* 97. North Mims, wood by the track leading to the keeper's lodge, 1856; *Phyt. N. S.* i. 408. Hedgebank bet. Potter's Bar and South Mims. (*S. Fabaria,* Koch.?)

IV. Stanmore, 1827–30; *Varenne.* Hampstead Wood; *Cooper* 102, and *Irv. MSS.* Meadow opposite the Swan Inn bet. Hampstead and Hendon, E. H. Button; *Cooper Supp.* 11.

VII. Kenwood, Hunter; *Park Hampst.* 30.

First record: *Blackstone,* 1737. Probably usually arises from garden plants.

254. *S. album, *L.* *White Stonecrop.* S. minus fol. longiusculo tereti, fl. alb. offic., J. B.* (Blackst.).

Cyb. Br. i. 399. Curt. F. L. f. 1 (drawn from a Middlesex specimen).

Walls; rare. P. July, August.

I. Old wall at Cowley, 1855, 'probably *album*'; *Phyt. N.S.* i. 65.

III. Wall at Twickenham, Mr. Borrer; *B. G.* 405.

IV. Stanmore, probably not wild, 1827–30; *Varenne.*

V. Old wall at Ealing; *Lees.* Abundant on the high wall of Sion House grounds on the left hand of the high road beyond Brentford.

VII. [Kentish Town, on a wall just beyond the chapel, abundantly; *Blackst. Spec.* 91, and *Curt. F. L.*] [Wall left-hand side leading from Bromley to Bromley Hall; *Curt. F. L.*] Fulham, &c., W. Pamplin; *New B. G.* 99, and *Cooper*, 113.

First record: *Blackstone*, 1746. Often, if not always, planted.

255.   * S. dasyphyllum, *L.*

Cyb. Br. i. 398.   Curt. F. L. f. 3 (drawn from a Middlesex specimen).

Walls; rare.   P.   June, July.

III. On walls at Twickenham; *Irv. Lond. Fl.* 171.

V. Old wall at Chiswick, 1843; *Herb. Hardw.*

VII. [Near Hammersmith; *Huds.* ii. 197.] [Walls of the Botanic Gardens, Chelsea; *Mart. App. P. C.* 65.] [Wall near Chelsea Hospital, on left-hand side of the horse-road on turning the corner out of Paradise Row; wall on left-hand side of the lane from Kensington Gravel-pits to Acton; *Curt. F. L.* Wall in the lane leading from Kensington Church to the gravel-pits, E. Forster; *Herb. Mus. Brit.* and *B. G.* 405.]

First record: *Hudson*, 1778. Probably a result of cultivation in all the above localities.

256.   S. acre, *L.*   *Yellow Stonecrop.   Wall Pepper.*

S. minus vermiculatum acre, C. B. P. (Blackst.).

Cyb. Br. i. 400; iii. 435; Comp. 179.   Syme E. B. iv. t. 532.

Walls and sandy ground; common.   P.   June, July.

I. Harefield; *Blackst. Fasc.* 92.

II. Staines!; *Newb.*   Among stones on the river bank near Hampton Court.   Bet. Twickenham and Teddington.

III. Grove Wall, Harrow; garden wall, Roxeth; *Melv.* 32.   Isleworth. Twickenham, in many places.   Towing-path, Hampton Court!; *Newb.*

IV. Stanmore, Near Whetstone.   Hampstead Heath; *Irv. MSS.*

V. Brentford.   Turnham Green.   Field by the river opp. Mortlake. Chiswick.   Drayton Green; *Cole.*   Hanwell; *Newb.*

VI. Edmonton, abundant.

VII. Kilburn; *Cole.*   Haverstock Hill, N.W.   King's Road, Chelsea. Brompton.   Parson's Green.

First record: *Blackstone*, 1737.

(*S. sexangulare, L.*   Cyb. Br. i. 401; iii. 435.   Syme E. B. iv. t. 533.

IV. Hampstead Heath; *Cooper*, 102.   Probably *S. acre* was intended.)

257. *** S. reflexum,** *L.*

S. minus hæmatodes, Ger. em. (Blackst.).

Cyb. Br. i. 401 ; iii. 435. Syme E. B. iv. t. 534.

Roofs and walls ; rare. P. July, August.

I. Harefield ; *Blackst. Fasc.* 92. Near Elstree ; *Herb. Rudge.*

III. Roxeth ; *Melv.* 32.

IV. Mill Hill, 1841 ; *Herb. Hardw.* Stanmore, 1830 ; *Varenne.*

VI. Highgate, 1819 ; *Herb. G. & R.*

First record: *Blackstone*, 1737. No doubt originally planted in all its
 localities.

Sempervivum tectorum, *L.* *House-leek.* Cyb. Br. i. 403 ; iii. 435.
 Curt. F. L. f. 3. Walls and roofs. P. July. II. Wall by the river,
 Sunbury. III. Whitton. Wall opposite Cole's Bridge, Twickenham.
 IV. About Harrow, not unfrequent ; *Melv.* 32. Hampstead ; *Irv.
 MSS.* VI. Edmonton. VII. Kentish Town ; *Herb. Hardw.* Always
 planted.

[COTYLEDON, *Linn.*

258. *** C. Umbilicus,** *L.*

Umbilicus Veneris (Ger.).

Cyb. Br. i. 403 ; Comp. 180. Syme E. B. iv. t. 539.

Walls and hedgebanks. P. June—August.

VII. Upon Westminster Abbey over the doore that leadeth from Chaucer
 his tombe to the old palace ; *Ger.* 424. In this place it is not now
 to be found ; *Johns. Ger.* 529. In an old gravel-pit near Highberry
 barn ; *Mart. App. P. C.* 71.

First record: *Gerarde*, 1597 ; last, *Martyn*, 1763. Possibly planted in
 the above localities, for it does not occur in Essex, Herts or Bucks,
 and in but one spot in Surrey, on its western border.]

GROSSULARIACEÆ.

RIBES, *Linn.*

R. Grossularia, *L.* *Gooseberry.* Cyb. Br. i. 394 ; iii. 433 ; Comp.
 178. Syme E. B. iv. t. 518. Hedges. Shrub. April, May. III.
 Grove Wall, Harrow ; *Melv.* 33. IV. Hampstead ; *Irv. MSS.*
 Harrow Park ; *Melv.* 33. Hedge near Dirt House on the Barnet
 Road ; *Fox.* VI. Edmonton. VII. Bet. Belsize Park and Hampstead.
 Only where it has been cultivated.

259. *** R. nigrum,** *L.* *Black Currant.*

R. fructu nigro, Ger. em. (Blackst.).

Cyb. Br. i. 392 ; iii. 432 ; Comp. 177. Syme E. B. iv. t. 523.

Wet meadows and sides of streams; very rare. Shrub. April, May.
  I. In a meadow near the Warren Pond at Breakspears, plentifully;
    Blackst. Fasc. 87.
III. One bush by the Cran on Hounslow Heath.

First record: *Blackstone*, 1737.

260. * R. rubrum, *L.* *Red Currant.*
  Cyb. Br. i. 393; iii. 432; Comp. 177. Syme E. B. iv. t. 520.
  Woods, &c.; rare. Shrub. April, May.
  I. Sparingly in some coppices near Harefield; *Blackst. Spec.* 83. Chalk-
    pit in Old Park, Harefield.
  IV. Harrow Park, doubtfully wild; *Melv.* 33. Bishop's Wood, Hamp-
    stead.
  VI. Copse near Warren Lodge, Edmonton.

First record: *Blackstone*, 1737. The Bishop's Wood plant is *Ribes
  sativum* of *Syme E. B.* iv. 42, which that botanist thinks probably
  always the produce of seed from the garden. It seems, however,
  well naturalised.

SAXIFRAGACEÆ.

SAXIFRAGA, *Linn.*

261. S. tridactylites, *L.*
Paronychia rutaceo folio (Ger.).
  Cyb. Br. i. 415; iii. 437; Comp. 182. Syme E. B. iv. t. 552.

  Walls; common. A. April—June.
  I. Harefield; *Blackst. Fasc.* 72. Pinner; Eastcott; *Melv.* 33. Ruislip.
  II. Staines. Halliford. Sunbury.
  III. Grove Wall, Harrow; *Melv.* 33. Heston House. Abundant about
    Isleworth.
  IV. Hampstead Heath; *Johns. Enum.* Harrow Park, &c.; *Melv.* 33.
    Hampstead, by Jack Straw's Castle. Stanmore. Whetstone.
  V. Osterley Park; *Masters.* Drayton Green; *Cole.* Walls at Brent-
    ford, Turnham Green, &c., abundant.
  VI. Enfield. Edmonton.
  VII. [Plentifully on the brick wall in Chauncerie Lane belonging to the
    Earl of Southampton; *Ger.* 500.] Fulham; Hammersmith; *Britten.*
    Haverstock Hill, 1866. Shepherd's Bush.

First record; *Gerarde*, 1597.

262. S. granulata, *L.*
S. alba (Ger.). *S. alba vulgaris* (Park.).
  Cyb. Br. i. 413; iii. 436; Comp. 181. Syme E. B. iv. t. 555.

# # # #

Meadows and open hilly places; very rare. P. May, June.

I. In a moist meadow near Moor Hall, Harefield; *Blackst. Fasc.* 90.

IV. In a hilly meadow above Child's Hill, Hampstead; *Irv. MSS.*

V. Sparingly on a ditch-bank by the roadside which skirts the south of Chiswick House Grounds, 1869!; *Warren.*

VII. [In the great field called the Mantells,* by Islington; *Ger.* 693.] [On the back side of Graye's Inne, where Mr. Lambe's Conduit heade standeth; *Park. Theat.* 423.]

First record: *Gerarde,* 1597. This is likely to occur in other places, as it is abundant in some neighbouring parts of Surrey.

CHRYSOSPLENIUM, *Linn.*

263. C. oppositifolium, *L.*

Saxifraga aurea Dodonæi (Blackst.).

Cyb. Br. i. 418; Comp. 183. Syme E. B. iv. t. 563.

Wet places in woods, and shady sides of ditches; rare. P. April.

I. Side of ditch in a meadow just below Coney's Farm at Harefield, plentifully; *Blackst. Spec.* 89.

IV. On Hampstead Heath; *Huds.* i. 156. Bank of a ditch by the roadside on the west side of the Heath, 1824; *Pamplin.* Turner's Wood, below the 'Spaniards;' *Irv. MSS.* In a shady ditch that runs across Ken or Bishop's Wood; *Burnett,* 105; *Macreight,* 93. By the stream in the south part of Bishop's Wood, abundant, 1869.

First record; *Blackstone,* 1746.

PARNASSIA, *Linn.*

264. P. palustris, *L.*   *Grass of Parnassus.*

P. vulgaris & palustris, Inst. R. H. (Blackst.).

Cyb. Br. i. 419; iii. 438; Comp. 184. Syme E. B. iv. t. 565.

Wet meadows; very rare. P. July—September.

I. In the moist meadows near Harefield Mill, particularly in a bog in the meadow next Mr. Ashby's fishing-house; *Blackst. Fasc.* 72. 'Observed by Mr. Ligo, who communicated (it) to me 1735;' *Blackst. MSS.* Meadows bet. Harefield and Rickmansworth, 1868!; *Newb.*

First record: *Blackstone,* 1735.

* Usually called the Commandry Mantells. Part of the manor of the priory of St. John of Jerusalem, Clerkenwell. It covered sixty-six acres, comprising the greater part of modern Pentonville, and extending southward to the New River head.

UMBELLIFERÆ.

HYDROCOTYLE, *Linn.*

265. H. vulgaris, *L.*   *Penny-wort.*
Cotyledon palustris (Ger.).
Cyb. Br. i. 423 ; Comp. 185.   Syme E. B. iv. t. 566.

Bogs and marshes, especially on heaths ; rather rare.   P.   July—September.

I. Harefield Common !; *Blackst. Fasc.* 21.   Ruislip Common ; wet ground round Ruislip Reservoir ; *Melv.* 33.   Harrow Weald Common.   Stanmore Heath.

II. Bushey Park ! ; *Newb.*

III. Hounslow Heath, sparingly.   Whitton Park inclosure.

IV. Hampstead Heath ! ; *Johns. Eric. and many subsequent writers.*

VII. South Heath, Hampstead.

First record ; *Johnson,* 1629.

SANICULA, *Linn.*

266. S. europæa, *L.*   *Wood Sanicle.*
S. sive Diapensia, Ger. em. (Blackst.).
Cyb. Br. i. 424 ; iii. 438 ; Comp. 185.   Syme E. B. iv. t. 568.

Woods ; rather rare.   P.   May—July.

I. Harefield ; *Blackst. Fasc.* 90, and *Young MSS.* Pinner Wood ; *Melv.* 33.

IV. Hampstead ; *Pet. H. B. Cat.* Bishop's Wood, common.   Woods at Highwood Hill ; *Blackst. Spec.* 89.   Bentley Priory Woods ; *Melv.* 33.   Stanmore ; *Varenne.*

VI. Highgate Woods ; *Cooper,* 104.   Hadley Wood ! ; *Newb.*

VII. Cane Wood ; *Mart. App. P. C.* 68.

First record : *Petiver,* 1713.

[CICUTA, *Linn.*

267. C. virosa, *L.*   *Water Hemlock.*
Sium alterum Olusatri facie, Lob. (Ray, Blackst.).
Cyb. Br. i. 428 ; iii. 439 ; Comp. 187.   Syme E. B. iv. t. 571.

Ponds and river-sides ; rare.   P.   July, August.

I. In the river by Mercer's Mill, near Cowley . . . . found by Dr. Wilmer, 1746 ; *Blackst. MSS.* Uxbridge, on the river banks ; *Mart. App. P. C.* 74.

II. Near Staines Bridge and in the road to Uxbridge ; *Doody MSS.* About Drayton ; *Blackst. MSS.* In the river Colne near Colnbrook ; *Huds.* ii. 122.

III. In a shallow pool of water by the highway side on Hounslow Heath
near the town's end, and in pools of water about Thistleworth;
R. Cat. i. 285. In one of the ponds near the road at Hayes, three
miles from Uxbridge, gathered once by Dr. Wilmer; *Blackst.
Spec.* 92.

First record: *Ray,* 1670; last, 1763. Confined to the western side of
the county in the Colne and Cran districts, where it seems to have
died out. Can some other plant have been mistaken for it?]

APIUM, *Linn.*

268. A. graveolens, *L.*   *Wild Celery. Smallage.*
Cyb. Br. i. 429; Comp. 187. Syme E. B. iv. t. 572.

Ditches where the water is brackish, also on walls washed by such water;
very rare. P. June—September.
VII. Edge of ditch in the Isle of Dogs, near the timber-dock, 1866.
Ibid. 1869!; *Warren.* In Tower ditch, and on the walls; on a
wharf-wall at Millbank; (in the crevices of the tombs in Pancras
churchyard;) *M. & G.* 429.

First record: *Milne and Gordon,* 1793. One of our few semi-maritime
species. The last locality appears improbable. Mr. Lees records it
(V.) at Greenford, and bet. Perivale and Harrow, but it must there be
of garden origin.

PETROSELINUM, *Hoffm.*

P. sativum, *Hoffm.* Cyb. Br. i. 429. Syme E. B. iv. t. 576. III. Old
garden wall at rear of houses in Hog Lane (=Crown St.), Harrow,
now nearly extinct; *Melv.* 33. A garden escape.

269. P. segetum, *Koch.*
Sium arvense seu segetum, J. R. H. (Blackst.). *Sison segetum, L.* (Forst.).
Cyb. Br. i. 430; iii. 440; Comp. 187. Syme E. B. iv. t. 577.

Damp fields; very rare. A or B. August, September.
VII. [In the brickfield adjoining to Tyburn turnpike, 1744; *Blackst.
Spec.* 92.] [About Hampstead Heath, Hyde Park and Tothill Fields;
Forst. Midd.] Isle of Dogs!, 1836; *Herb. Young.* Abundant and
very large by the side of a ditch near the timber-dock in the Isle
of Dogs, 1866. Ibid. 1869!; *Warren.*

First record: *Blackstone,* 1744. Perhaps *Sison Amomum* was mistaken
for this in the stations given in Forster's Middlesex list.

HELOSCIADIUM, *Koch.*

270. H. nodiflorum, *Koch.*
Sium umbellatum repens, Ger. em. (Blackst.). *Sium nodiflorum, L.*
(G. & R.).
Cyb. Br. i. 431; Comp. 188. Syme E. B. iv. t. 573.
Ditches and sides of streams; common. P. July—September.

I. Harefield!; *Blackst. Fasc.* 94. Pinner.

II. Staines. Near Sipson. Sunbury (a very large form). Towing-path, Hampton Court!; *Newb.*

III. Cran on Hounslow Heath. Twickenham.

IV. Bet. Finchley Road Ry. Station and Finchley; *Newb.* Golder's Green.

V. Canal, Greenford; *Melv.* 34. Hanwell!; *Warren.*

VI. Edmonton.

VII. Almost in every watery place about London; *Johns. Ger.* 257. Paddington Canal, 1815; *Herb. G. & R.* Clapton, 1830; *Varenne.* South Heath, Hampstead. Millfield Lane. Blackstock Lane, Highbury. Ditch bet. Hyde Park and Kensington Gardens. Eelbrook Meadow, Parson's Green. Hackney Wick. Isle of Dogs.

Var. β. H. repens, Koch. *Sium repens*, Jacq. Syme E. B. iv. t. 574,

II. Towing-path bet. Kingston Bridge and Hampton Court; *Bloxam.*

VI. [Finchley Common, Woods; *Winch MSS.*]

VII. [Tothill Fields, E. Forster; *Herb. Mus. Brit.* and *B. G.* 402.] South Heath, Hampstead; *Irv. MSS.* Eel-brook Meadow, J. A.; *Phyt. N. S.* i. 464.

Forster's specimens in *Herb. Mus. Brit.* are true *repens*; but small examples of *H. nodiflorum* are frequently called by that name.

First record: *Johnson,* 1633.

271. H. inundatum, *Koch.*

Sium minimum (Pet.). *Sium pusillum fol. variis, N. D.* (R. Syn. ii.). *S. minimum umbellatum fol. variis, Pluk.* (Blackst.). *Sison inundatum, L.* (Huds.).

Cyb. Br. i. 432; Comp. 188. Syme E. B. iv. t. 575.

Overflowed places, or in water; rather common. P. June—September.

I. In the bogs on Harefield Common, plentifully; *Blackst. Fasc.* 94. Harrow Weald Common; Pinner Hill; *Melv.* 34. Stanmore Heath.

II. Bushey Park.

III. Several ponds on Hounslow Heath; *Pet. Midd.* Hounslow Heath towards Hampton, S. Doody; *R. Syn.* ii. 344 (specimens in *Herb. Rudge* and *Dicks. H. S.*). Near Hatton.

IV. Bet. Kingsbury and Harrow; *M. & G.* 411. Hampstead Heath; *Forst. Midd.* Golder's Green; *Irv. MSS.*

V. Greenford; *Cooper,* 108. Apperton; *Melv.* 34.

VII. Bet. Newington and Hornsey; *Huds.* i. 104. In several parts of the New River; *M. & G.* 411. Eel-brook Meadow, Parson's Green, 1867.

First record: *Petiver,* 1695.

SISON, *Linn.*

272. S. Amomum, *L.*

Sison (Morison). *Sium aromaticum Sison, Inst. R. H.* (Blackst.).
Cyb. Br. i. 433; Comp. 188. Syme E. B. iv. t. 578.

Hedges and roadsides; very common. B. July—September.
In all the districts.

VII. Ditch-banks round London; *Moris. Umb.* 15. Marylebone Fields,
1815; *Herb. G. & R.* Islington; *Herb. Hardw.* Kentish Town,
abundant. Near Hornsey Wood. Kilburn.

First record: *Morison,* 1672.* Mr. Watson (*Comp.* 188) seems to doubt
the nativity of this species.

BUNIUM, *DC.*

273. B. flexuosum, *L.* *Earth-Nut.*

Bulbo-castanum et B. altera (Ger.). *Bulbocastanum minus, Ger. em.*
(Blackst.). *B. majus, Ger. em.* (R. Syn. iii.). *Nucula terrestris
septentrionalium, Lob.* (Johns.). *Bunium Bulbocastanum, L.* (Curtis
and others).
Cyb. Br. i. 435; Comp. 189. Curt. F. L. f. 4.

Fields, heaths and woods; rather common. P. May—July.

I. Harefield!; *Blackst. Fasc.* 12. Pinner and Stanmore Woods;
Melv. 34.

IV. Hampstead Heath!; *Johns. Enum.* Harrow Park; *Melv.* 34.
Bishop's Wood, Hampstead. Near Finchley!; *Newb.*

V. Near Brentford; *Hemsley.* Chiswick House grounds; *Fox.*

VI. Field adjoining Highgate; *Ger.* 906. Field bet. Hornsey Wood and
Old Fall, Mr. Martyn; *R. Syn.* iii. 209. Hadley!; *Warren.*
Colney Hatch. Enfield.

VII. [Field next the conduit heads by Maribone; *Ger.* 906. Near Maribone Park; *Merrett,* 17.] Kensington; Paddington; *Brit. Phys.*
Kensington Gardens, abundant, but stunted in growth.

First record: *Gerarde,* 1597. The Kensington Gardens plant was
recorded by Smith, *Fl. Brit.* i. 301, as *B. Bulbocastanum, L.* on the
authority of Mr. W. Wood. In *B. G.* 402, however, is a note stating
that the latter botanist was then 'of opinion the plants he gathered
were only large specimens of *B. flexuosum.*' See also *E. Fl.* ii. 55.
Martyn's Hornsey plant has been often referred to *B. Bulbocastanum*
(see *Cyb. Br.* i. 436), and we were at first inclined to consider it
that species, though all the places at present known to produce it
are on a chalk soil, in consequence of some evident care shown in the
record. But, from a note by Thomas Martyn in vol. i. (article

---

* 'A kinde of *Seseli* growing everywhere in the pastures about London,' described in
Ger. 894 may be this. Johnson, *Ger. em.* 1051, says, 'I am ignorant what our author
means by this description.'

Bunium) of his edition of *Mill. Gard. Dict.*, it seems evident that the plant found by his father was but a young and luxuriant example. of *B. flexuosum* with the leaf-segments somewhat broader than usual, and not the true *B. Bulbocastanum*, L., which, at that time, though well known on the Continent, was not understood by English botanists.

CARUM, *Linn.*

274. *C. Carui, *L.*   *Caraway.*

C. officinale (Pet.).

Cyb. Br. i. 434; iii. 440.   Syme E. B. iv. t. 582.

Waste places; rare. B. May—September.

I. Railway-bank, Pinner; *Farrar.*

II. Waste ground by the canal, West Drayton!; *Newb.*

VI. For some distance along the top of the Great Northern Ry. bank, by the footpath north of Colney Hatch Station.

First record: *Petiver,* 1695. This I have found more than once about London; *Pet. Midd.* I have sometimes found it in the fields about the town; *Mill. Bot. Off.* 120. Not a native; cultivated for its fruit in some parts of England, and hence escaped, or originates from 'caraway seeds' accidentally dropped by persons eating them.

ÆGOPODIUM, *Linn.*

275. Æ. Podagraria, *L.*   *Gout-Weed. Herb Gerarde.*

Angelica sylvestris minor, C. B. P. (Blackst.).

Cyb. Br. i. 433; iii. 440; Comp. 188. Syme E. B. iv. t. 580.

Hedgebanks, shrubberies, shady places, and a garden weed; very common. P. June—August.

In all the districts.

VII. [Limehouse, near the horseferry; *Forst. Midd.*] Weed in Horticultural Gardens, South Kensington. Highgate Cemetery. About Kentish Town. Hackney Wick. Grafton Road, Holloway; *Newb.*

First record: *Blackstone,* 1737.

PIMPINELLA, *Linn.*

276. P. magna, *L.*

P. saxifraga, Ger. (R. Syn. ii.). *P. major* (M. & G.).

Cyb. Br. i. 437; iii. 441; Comp. 190. Syme E. B. iv. t. 586.

Hedges and fields; very rare. P. June—August.

IV. In a meadow on the left hand on this side Brent-bridge near Kingsbury, in Harrow-on-the-Hill Road, Mr. Vardy; *Doody MSS.* (On Hampstead Heath; *M. & G.* 413.) Bet. Totteridge Green and Finchley; *Irv. MSS.*

VI. Hadley!; *Newb.*

First record, *Doody,* about 1700. Seems to be common in immediately adjacent parts of Herts. (See *Fl. Herts.* 119.)

277. P. Saxifraga, *L.*   *Burnet Saxifrage.*
P. sax. minor fol. sanguisorbæ, R. Syn. iii. (Blackst.).   *P. minor*
(M. & G.).
Cyb. Br. i. 437 ; Comp. 189. Syme E. B. iv. t. 585.
Fields, meadows, and roadsides ; very common. P. July—September.
Throughout all the districts.
VII. South Heath, Hampstead. Hackney Wick ; *Cherry.*
First record : *Blackstone,* 1737.

SIUM, *Linn.*

278. S. latifolium, *L.*
Sium Dioscoridis seu Pastinaca aquatica major, Park. (Blackst.).
Cyb. Br. i. 438 ; iii. 441 ; Comp. 190. Syme E. B. iv. 587.
River-sides ; rare. P. July, August.
I. In several parts of Harefield River ; *Blackst. Spec.* 92 ; but
Messrs. Webb and Coleman say : ' We have searched for it below
Rickmansworth in vain ; ' *Fl. Herts.* 120.
II. In several places by the Thames. Bet. Hampton Court and Hampton
Village, H. C. W. ; *New B. G. Supp.* In some plenty about a
quarter of a mile above Kingston Bridge. A large patch a little
below Teddington Lock.
III. By the Thames, Twickenham ; *M. & G.* 399. Sparingly about mid-
way bet. Twickenham and Richmond Bridge.
VII. (In the brickfield near Tyburn turnpike ; *Hill,* 147. Hyde Park ;
Cockfield, 14.) [By the Thames near the Botanic Garden, Chelsea,
1825 ; *Herb. G. & R.*]
First record : *Blackstone,* 1746. With the exception of Blackstone's
station, all the trustworthy localities are on the Thames bank.

279. S. angustifolium, *L.*
S. majus angustifolium, Ger. em. (Blackst.).
Cyb. Br. i. 439 ; iii. 441 ; Comp. 190. Syme E. B. iv. t. 588.
Sides of streams and ditches ; rather rare. P. July—September.
I. Harefield ! ; *Blackst. Fasc.* 94. Colne, near Yewsley ! ; *Newb.* East-
cott ; *Farrar.*
II. Canal, West Drayton ! ; *Newb.* Staines Moor.
IV. Edgware Road, 1830 ; *Varenne.* Bet. Hampstead and Golder's
Green ; *Irv. MSS.* Finchley Road ; *Newb.* Roadside by the clay-
pits, Harrow Weald. All along the Brent ; *Farrar.*
V. Canal at Greenford ; *Melv.* 35. Apperton ; *Lees.*
VII. Isle of Dogs ; *Young, MSS.* Ditch on South Heath, Hampstead,
abundant.
First record : *Blackstone,* 1737.

BUPLEURUM, *Linn.*

280. B. tenuissimum, *L.*

B. angustifolium monspeliense, Ger. (Merrett).

Cyb. Br. i. 439; iii. 441; Comp. 190. Syme E. B. iv. t. 591.

Fields; very rare. A. August, September.

V. On Ealing Common, profusely; *Lees* (*v. s.*).

VI. A field north of Highgate where water had stood, growing with *Polygonum aviculare*, 1841, W. Mitten; *Phyt.* i. 204. A specimen was presented to the Botanical Society of London by the discoverer; *loc. cit.* 268.

VII. [At Paddington, beyond the bridge in the way to Harrow-upon-the-Hill; † *Merrett*, 17.]

First record: *Merrett*, 1666.

[281. *B. rotundifolium, *L.* *Thorough-Wax.*

Perfoliata vulgaris, Ger. em. (Blackst.).

Cyb. Br. i. 441; Comp. 191. Syme E. B. iv. t. 589.

Cornfields; very rare. A. August.

I. In 1735, in a field of corn near Harefield Mill, but not plentifully; *Blackst. Fasc.* 75.

The only record. Considered 'native' by Mr. Watson, but surely liable to the same doubts as other cornfield weeds.]

B. protractum, *Link.* II. Stray bits in garden ground bet. Hampton and Sunbury!; *Newb.* VI. A few plants by the roadside bet. Muswell Hill and Hornsey. Probably introduced with foreign seed.

Ammi majus, *L.* VII. Side of West India Dock Basin, near the water, a few specimens; *Cherry* (*v. s.*).

ŒNANTHE, *Linn.*

282. Œ. fistulosa, *L.*

Filipendula aquatica (Ger.). *Œ. aquatica, C. B. P.* and *Œ. aquatica triflora caulibus fistulosis, H. Ox.* (Blackst.).

Cyb. Br. i. 441; iii. 441; Comp. 191. Syme E. B. iv. t. 593.

Wet pastures and heaths; rather rare. P. June, July.

I. Wet meadows and in the river near Harefield Mill; *Blackst. Fasc.* 66. Ruislip; *Hind.*

II. Staines Common. Common by Walton Bridge.

III. Hounslow Heath.

IV. Hampstead; *Irv. MSS.*

V. Greenford; *Lees.*

VII. Neare the river about the Bishop of London's house at Fulham; *Ger.* 902. About Blackwall; *M. & G.* 384. [Marylebone Fields,

† This locality of Merrett's is quoted in *E. B. Supp.* 2763, for *B. falcatum.*

1815; *Herb. G. & R.*] Near Homerton; by the London Canal, nearly opposite the gas-works; *Cherry* (*v. s.*). [Eel-brook Meadow, Parson's Green.]

First record: *Gerarde*, 1597.

283. Œ. Lachenalii, *Gmel.*
Cyb. Br. i. 442; Comp. 192. Syme E. B. iv. t. 596.
Ditches; very rare. P. July—September.
II. Ditch by the road, four miles from Staines towards Hampton Court, 1847; also in the cross road thence to Sunbury; *H. C. Watson.*
The only record.

284. Œ. silaifolia, *Bieb.?*
Cyb. Br. i. 443; iii. 442; Comp. 192. Syme E. B. iv. t. 595.
Wet meadows; very rare. P. June, July.
IV. Pool in a little plantation nearly opposite Woodford House, Kingsbury; *Farrar.*
V. Banks of the Brent near Greenford, abundant in 1863, but, we fear, now extirpated, from the bank being cut away; *Melv.* 97.
VII. [Marylebone Fields, 1815; *Herb. G. & R.*] Mr. Varenne, who has the original specimen, believes it to be *silaifolia.*
First record: *Goodger and Rozea*, 1815. We have seen no specimens. It is possible that Gerarde and Blackstone may have intended this species by the name *Filipendula aquatica*, quoted under *Œ. fistulosa.* Indeed, the imperfect specimen in *Herb. Budd.* bearing Gerarde's name, looks more like *silaifolia* than any other species; but the two plants· were confounded by the early botanists, and are even now sometimes mistaken for each other.

285. Œ. crocata, *L.* *Hemlock Water-Dropwort.*
Filipendula cicutæ folio (Johns.). *Œnanthe maxima* (Morison). *Œnanthe cicutæ facie Lobelii, Park.* (Dill.).
Cyb. Br. i. 444; iii. 442; Comp. 192. Syme E. B. iv. t. 597.
River-banks and ponds; rather common. P. July, August.
I. Harefield; *Cole.*
II. By the Thames, bet. Strawberry Hill and Teddington, abundant.
III. Abundant on the river-bank bet. Twickenham and Richmond Bridge; also bet. Richmond Bridge and the railway bridge.
V. Abundant about Hanwell, Southall, &c.; *Masters.* By the Thames, Chiswick.
VII. By the Thames in many places. [In great abundance among the oysiers against Yorke House a little above the horse ferry against Lambeth; *Johns. Ger.* 1060.] Thames side, plentifully; *Herb.* and *MSS. Budd.* Ditches in the Isle of Dogs, abundant; *M. & G.* 386. Near Cremorne, plentiful.
First record: *Johnson*, 1633. Almost confined to the bank of the

Thames, where it is a conspicuous species. 'Very rare' in Herts, a single specimen only having been gathered near Rickmansworth, close to our boundary.

286. Œ. Phellandrium, *Lam.*

Phellandrium aquaticum (M. & G.).

Cyb. Br. i. 445 ; Comp. 193.   Syme E. B. iv. t. 598.

Sides of ponds and streams, preferring still water ; rare.   B. or P. July, August.

V. Banks of the canal bet. Hanwell and Brentford ; *Hemsley.*   In the Brent at Apperton, abundant ; *Lees.*

VII. [At Milbank among the piles ; *M. & G.* 435.]   Pool near Hornsey, 1815 ; *Herb. G. & R.*   Pond at Kentish Town, abundant ; *Barnett,* i. 10.   [Copenhagen Fields, 1841 ; *Herb. Hardw.*]   Marshy ground near Hammersmith, 1850, W. Wing ; *Herb. Mus. Brit.*   Seven Sisters Road, Holloway, 1864, G. Munby ; *Naturalist,* 1867, 179.

First record: *Milne and Gordon,* 1793.   Chiefly in the metropolitan district.   The next species, *Œ. fluviatilis,* is frequently misnamed *Œ. Phellandrium.*

287. Œ. fluviatilis, *Coleman.*

Millefolium aquaticum, Matth. (Dillen.).   *Phellandrium, J. B.* (Blackst.).

Cyb. Br. i. 445 ; iii. 354 and 443 ; Comp. 193.   Syme E. B. iv. t. 599.

In running water ; rather common.   B. or P.   July, August.

I. In Harefield River (= Colne), plentifully ! ; *Blackst. Spec.* 72.   Grand Junction Canal, Harefield (torn up and floating).

II. River Colne near Staines and Stanwell, abundant.   River Thames from Sunbury Lock to Walton Bridge, Surrey ; *Watson.*

III. In the Cran, abundant throughout its course.   Duke's River near Isleworth.

VII. In Hackney River (= Lea), abundantly ; *Dill. in R. Syn.* iii. 216.   Ponds at the entrance of Hornsey Wood, ditches in the Isle of Dogs, Mr. Hurlock ; *Blackst. Spec.* 72.*   [New River Head, Clerkenwell ; *Irv. H. B. P.* 592.]   Ditch near Victoria Park, 1867 ; *Cherry* (*v. s.*).   Lea at Temple Mills, and below that point.

First record: *Dillenius,* 1724.   This rarely flowers in the open stream, being there quite under water, but in the quieter sheltered places it rises above water and flowers, as in the Cran at Hospital Bridge.

ÆTHUSA, *Linn.*

288. Æ. Cynapium, *L.*   *Fool's Parsley.*

Cicutaria tenuifolia, Ger. em. (Blackst.).

Cyb. Br. i. 446 ; Comp. 193.   Curt. F. L. f. 1.

Cultivated ground, especially gardens and waste places ; very common.   A.   June—August.

---

* Mr. Hurlock's localities more probably refer to **286.**

Throughout all the districts.

VII. Very frequent. [St. Martin-in-the-Fields Churchyard, 1826; *Burnett*, i. 8.] Hampstead. Green Park. Parson's Green. Craven Hill. Kensington Gore. Montague St. Guildford St. Euston Road. Hackney Wick. Isle of Dogs. A weed in many of the squares.

First record: *Blackstone*, 1737.

FŒNICULUM, *Hoffm.*

289. ***F.* officinale,** *All.* F. vulgare, *Gaertn.* (Syme E. B. and L. Cat.). *Fennel.*

F. sylvestre, Park. (Waring).

Cyb. Br. i. 447; iii. 443; Comp. 193. Syme E. B. iv. t. 601.

Cultivated and waste land; rather rare. P. July—September.

II. In an oat-field near Strawberry Hill. Near Twickenham-Common Church.

III. Queen St., Twickenham, abundant in an old orchard.

V. Great Western Ry. bank near Hanwell Station.

VII. [About the gravel-pit at Hyde Park Corner; *Waring*.] New Finchley Road, 1840; *Herb. Hardw.* Highgate, G. E. Dennes; *New B. G. Supp.* Shepherd's Bush. Victoria St., Westminster.

First record: *Waring*, 1770. Always the result of cultivation, and not well established.

SILAUS, *Besser.*

290. S. pratensis, *Bess.*

Saxifraga anglicana facie Seseli pratensis, Lob. (Johns., Blackst.).

Cyb. Br. i. 449; iii. 444; Comp. 194. Syme E. B. iv. t. 604.

Meadows and commons; rather common. P. August, September.

I. Harefield!; *Blackst. Fasc.* 91. Stanmore Heath. Near Elstree Reservoir. Woodhall, Pinner.

II. Stanwell Moor. Near Charlton. Common by Walton Bridge.

III. Harrow; *Melv.* 35, and *Hemsley.*

IV. Fields bet. Child's Hill and Hendon; *Irv. MSS.* Pinner Drive; *Melv.* 35. Bentley Priory meadows, abundant.

V. Ealing Common; near Twyford; *Newb.* Horsington Hill.

VII. Bet. Kentish Town and Hampstead Town; *Johns. Eric.*, 1815; *Herb. G. & R.*, 1819; *Bennett (v. s.)*, 1845; *Herb. Hardw.*

First record: *Johnson*, 1629.

ANGELICA, *Linn.*

291. A. sylvestris, *L.*

A. sylvestris major, C. B. Pin. (Blackst.).

Cyb. Br. i. 451; Comp. 195. Syme E. B. iv. t. 607.

Sides of streams and ditches, and other wet places; common. P. July, August.

K

I. Banks of Harefield River!; *Blackst. Fasc.* 6. Pinner Wood; *Melv.* 35. Elstree. South Mims.

II. Canal, West Drayton!; *Newb.* Staines. Bet. Kingston Bridge and Hampton Court.

III. Twickenham. Hounslow Heath. Hatton.

IV. Highgate Wood; *Johns. Eric.* Road by bathing-place, Harrow; *Melv.* 35. Near Finchley!; *Newb.* Bentley Priory. Bishop's Wood. Mill Hill, 1837 ; *Herb. Mus. Brit.*

V. By the Brent near Hanwell!; *Newb.*

VI. Finchley, 1815; *Herb. G. & R.* Edmonton.

VII. Near Kentish Town ; *Park. Theat.* 940. Duckett's Canal, 1867 ; *Cherry.* Sandy End, Fulham.

First record : *Johnson,* 1629.

ARCHANGELICA, *Hoffm.*

292. * **A. officinalis,** *Hoffm.* Angelica arch. *L.* (Syme E. B.). *Garden Angelica.*

Angelica sativa, C. B. Pin. (Dill.).

Cyb. Br. i. 451 ; iii. 444.   Syme E. B. iv. t. 608.

Wet ground by streams ; very rare.   P.   July—September.

VII. [Frequent about the Tower of London and on the banks of the moats ; *Doody MSS.,* quoted in *Mart. App. P. C.,* &c.]   Wild in many places by the Thames side ; *Waring.* Banks of the Thames ; *Macreight,* 102.   Osier-holt, surrounded by ditches which communicate with the Thames, Sandy End, Fulham, 1867, thoroughly established.

First record : *Doody,* about 1700.   Doody doubted whether this was native about the Tower ; and it may be the remains of former cultivation in all its stations.   It has been planted in many of the squares and gardens in London for ornament, and in Lincoln's Inn Fields now comes up from self-sown seed in abundance.

PASTINACA, *Linn.*

293. **P. sativa,** *L.*   *Parsnep.*

P. sylvestris latifolia, C. B. Pin. (Blackst.).

Cyb. Br. i. 454 ; iii. 444; Comp. 196.   Syme E. B. iv. t. 612.

Chalky banks, also waste ground and railway banks; rather common. B.   July, August.

I. Harefield chalkpit!; *Blackst. Fasc.* 73.   Pinner chalkpits.

II. By the river bet. Kingston Bridge and Hampton Court, abundant. Railway bank, Sunbury, abundant.

III. Roxeth ; *Melv.* 35.   Twickenham.   Near Whitton.   Hampton Court!; *Newb.*

IV. Railway embankment, Harrow ; *Melv.* 35.

V. Near Ealing !; *Newb.*

VI. Railway bank, Colney Hatch. Pickard Lock, Edmonton. Side of Lea bet. Edmonton and Tottenham ; *Cherry (v. s.).*

VII. Copenhagen Fields ; *Herb. Hardw.* Isle of Dogs! ; *Newb.* Railway bank, Hampstead.

First record : *Blackstone,* 1737. The majority of specimens in Middlesex are the remains of cultivation. Wild on the chalk.

HERACLEUM, *L.*

294. H. Sphondylium, *L.* *Cow Parsnep. Hog-Weed.*
Sphondylium vulgare, Ger. em. (Blackst.).
Cyb. Br. i. 454 ; Comp. 196. Syme E. B. iv. t. 613.

Hedges ; very common. P. July.
Throughout all the districts.

VII. Marylebone Fields, 1815 ; *Herb. G. & R.* Sandy End.

First record : *Blackstone,* 1737. Blackstone noticed the state (*var. β angustifolium,* Bab.) with laciniated leaves (figured in *Pluk. Phyt.* t. 63) in several places near Harefield ; *Blackst. Fasc.* 97. Mr. Irvine mentions a ' *var. majus,*' found at Chelsea Old Waterworks ; *Irv. H. B. P.* 598.

TORDYLIUM, *L.*

295. T. maximum, *L.* *Great Hartwort.*
T. semine minus hirto et limbo quasi lævi seu parum granulato (Morison). *Tordylium sive Seseli creticum minus, Park.* (Ray, Buddle). *Small Hartwort* (Petiver). *Tordylium maximum, Inst. R. H. sive Seseli creticum majus, Park.* (Dillenius).
Cyb. Br. i. 455; Comp. 196. Syme E. B. iv. t. 614. Hook. Curt. vol. v. (from an Oxford specimen).

Hedgebanks and waste ground. A. July.

III. Frequent about London ; *Morison, Umb.* 40. Found by Mr. Doody about Thistleworth (= Isleworth) ; *R. Syn.* i. 63. I have found it in various adjacent localities there for many years ; *Doody MSS.* Mr. Doody showed me this growing in two or three places about Thistleworth; *MSS. Budd.* (There are specimens in *Herb. Budd.* vol. cxx. fol. 3, and also in *Herb. Pet.* vol. cli. fol. 147.) Not now to be found in Mr. Doody's locality, Mr. Clements ; *Hill,* 137. Mr. Francis gave me seeds of this plant gathered near Twickenham in 1827; the plants raised from them did (if I mistake not) supply the figure in Baxter's *British Plants,* vi. 443 ; *Pamplin.* Near Twickenham, July, 1837, W. Wilson ; bet. Twickenham and Isleworth, in a hedgebank, W. Borrer; *Herb. Mus. Brit.* In a hedge on the right-hand side of a lane leading from St. Stephen to the Marquis of Aylesbury's Lodge, about midway, within Twickenham parish, but close to Isleworth, R. Castles ; *Watson.*

132 UMBELLIFERÆ.

First record: *Morison*, 1672; also first as a British plant. Probably now extinct from the great increase of building about Isleworth and Twickenham. There is no date to Borrer's or Castles' specimens, so that the last year in which the plant was found is, so far as we have been able to ascertain, 1837. Ray (*Syn.* ii. 102) suspected this to have a garden origin. Mr. Watson classes it with his 'denizens.']

(T. officinale, *L.* E. B. 2440. About London, Petiver; *Smith Fl. Br.* i. 295. There is no reason to suppose that this was ever found. Smith was doubtless misled by Petiver's figure (*H. B. Cat.* xxiv. 6); he sets himself right in *Eng. Fl.* ii. 105. The specimens in *Herb. Pet.* are *T. maximum.* Morison (*Hist. Ox.* iii. 316) points out that confusion in the synonymy of the two plants was probably due to the examination of specimens with unripe fruit.)

DAUCUS, *Linn.*

296. D. Carota, *L.*   *Wild Carrot.*
D. sylvestris tenuifolia, Ger. em. (Blackst.).
Cyb. Br. i. 456; Comp. 196. Syme E. B. iv. t. 615.

Fields, waste ground, and railway banks; rather common. B. June—August.
I. Harefield!; *Blackst. Fasc.* 22. Pinner chalkpits. Roadside by Pinner Ry. Station.
II. Staines!; bet. Hampton and Hampton Court!; *Newb.* Bet. Hampton Court and Richmond Bridge, abundant. Teddington. Road from Sunbury to Hampton, scarce.
III. Near Drilling Ground, Hounslow. Field by Hospital Bridge. Railway bank, Feltham, abundant. Twickenham Park inclosure.
IV. Railway embankment, Harrow; *Melv.* 37. Fields by Scratch Wood, Edgwarebury.
VII. [Paddington, 1827–30; *Varenne.*] Bet. Belsize House and New Finchley Road; *Irv. Lond. Fl.* 196. Railway bank, Hampstead. Stray plants by the roadside at Kensington Gore and Constitution Hill.

First record: *Blackstone*, 1737.

TORILIS, *Adans.*

297. T. Anthriscus, *Gaertn.* Caucalis A., *Huds.* (Syme E. B.). *Hedge Parsley.*
Caucalis minor flosculis rubentibus, Ger. em. (Blackst.).
Cyb. Br. i. 458; Comp. 197. Curt. F. L. f. 6.

Hedges; very common. A. July, August.
Throughout all the districts.
VII. Side of Hackney Canal; *Cherry* (*v. s.*). Hackney Wick.

First record: *Blackstone*, 1737.

298. T. infesta, *Spreng.* Caucalis infesta, *Curtis* (Syme E. B.).
Caucalis segetum minor, Anthrisco hispido similis, R. Hist. (Blackst.).
Cyb. Br. i. 459; Comp. 197. Curt. F. L. f. 6.

Cornfields; rare. A. July—September.
I. Harefield, common!; *Blackst. Fasc.* 16. Near Pinner Ry. Station.
II. Near Sunbury, H. C. W.; *New B. G.* 99.
III. Near Hounslow!; *Newb.*
IV. About Hampstead; *Irv. MSS.*
VII. [Paddington, 1830; *Varenne.*]

First record: *Blackstone,* 1737.

299. T. nodosa, *Gaertn.* Caucalis nodosa, *Huds.* (Syme E. B.).
Caucalis nod. echinato semine (Johns.).
Cyb. Br. i. 459; iii. 446; Comp. 197. Syme E. B. iv. t. 621.

Dry fields and banks; rare. A. June—September.
IV. About Hampstead; *Irv. MSS.*
V. A few stray plants seen in this district; *Newb.*
VI. Hedgebank near the 'Green Man' Inn, Finchley Common, 1826
     Herb. Hardw.
VII. [On the bankes about S. James and Pickadilla; *Johns. Ger.* 1023.]
     Isle of Dogs near the timber-dock, 1866. Ibid, 1869!; *Warren.*

First notice: *Johnson,* 1633.

SCANDIX, *Linn.*

300. S. Pecten-Veneris, *L.* *Shepherd's Needle.*
Pecten Veneris, Ger. em. (Blackst.).
Cyb. Br. i. 460; Comp. 198. Syme E. B. iv. t. 627.

Cornfields; rare. A. June—September.
I. Harefield!; *Blackst. Fasc.* 73; and *Cole.*
III. Wood End, W. M. H.; *Melv.* 36. Near Hounslow!; *Newb.*
IV. About Hampstead; *Irv. MSS.* Kenton, W. M. H.; *Melv.* 36.
V. Near Ealing!; Acton; *Newb.*
VI. Finchley Common; *Herb. Hardw.*
VII. [Marylebone; *Herb. G. & R.*] Kentish Town; *Herb. Hardw.*

First record: *Blackstone,* 1737.

ANTHRISCUS, *Hoffm.*

301. A. sylvestris, *Hoffm.* Chærophyllum sylvestre, *L.* (Syme E. B.).
     Wild Chervil.
Myrrhis sylvest. seminibus lævibus, C. B. P. (Blackst.). *Chærophyllum
sylvestre, Sm.* (Auct.).
Cyb. Br. i. 461; iii. 446; Comp. 198. Curt. F. L. f. 4.

Sides of hedges, meadows and shady places; very common. P. April—
June.

In all the districts.

VII. Kensington Gardens, 1868; *Warren* (*v. s.*). Hampstead. Kentish
 Town. Regent's Park. Primrose Hill. Hornsey Wood. Hammersmith. Victoria Park.

First record: *Blackstone*, 1737.

A. Cerefolium, *Hoffm.* Chærophyllum sativum, *Lam.* (Syme E. B.). *Garden
 Chervil.   Scandix Cerefolium* (Forst.). Cyb. Br. i. 462; iii. 446.
 Syme E. B. iv. t. 623.   VII. [In Clapton Field, E. Forster; *Herb. Mus.
 Brit.*] [Near Bellsize House, Hampstead; *Irv. MSS.*] Kentish Town,
 1850; *Herb. Hardw.*   An escape from neighbouring gardens.

302.   A. vulgaris, *Pers.* Chærophyllum Anthriscus, *Lam.* (Syme E. B.).
 Myrrhis sylvestris æquicolorum, Col. (Johns.).   *Myrrhis sylvestris seminibus asperis, C. B. Pin.* (Blackst.).   *Chærophyllum sylv. sem. brevibus
 hirsutis* (Budd.).   *Scandix Anthriscus, L.* (Huds.).

Cyb. Br. i. 461; iii. 446; Comp. 198. Curt. F. L. f. 1.

Hedgebanks and waste ground; rare.   A.   April—June and September.

I. Harefield; *Blackst. Fasc.* 63.

IV. Hampstead Heath; *Johns. Enum.*

V. Hedgebank opposite south wall of Chiswick House.

VI. Road bet. Edmonton and Ponder's End; *Cherry* (*v. s.*).

VII. Hackney Marshes, 1864.

First record: *Johnson*, 1632.   Not now common in Middlesex; yet
 Buddle mentions it as general 'about London,' and Hudson says
 'round London everywhere.'

CHÆROPHYLLUM, *Linn.*

303.   C. temulum, *L.*   *Rough Chervil.*
 Cerefolium sylvestre, Ger. em. (Blackst.).
 Cyb. Br. i. 462; Comp. 198. Curt. F. L. f. 6.

Hedgebanks; rather common.   P.   June, July.

I. Harefield!; *Blackst. Fasc.* 17.   Bet. Harefield and Ruislip.   Elstree.
 South Mims.

II. Harmondsworth.   Teddington, abundant.

III. Isleworth Lane, 1815; *Herb. G. & R.*   Twickenham.

IV. Harrow, very common; *Melv.* 36.   Near Finchley!; *Newb.* Hampstead; *Irv. MSS.*

V. Hanwell!; *Warren.*   Near Ealing!; *Newb.*

VI. Edmonton.

VII. Upper Clapton; bet. Lea Bridge and Tottenham; *Cherry* (*v. s.*).

First record: *Blackstone*, 1737.

CONIUM, *Linn.*

304.   C. maculatum, *L.*   *Hemlock.*
 Cicuta, Ger. em. (Blackst.).
 Cyb. Br. i. 426; Comp. 186. Curt. F. L. f. 1.

Hedges and waste places; rare. B. June—August.

I. Harefield; *Blackst. Fasc.* 18. Road bet. Uxbridge and West Drayton, sparingly, 1855; *Phyt. N. S.* i. 62.

III. About Whitton, G. Francis; *Coop. Supp.* 12. Field by Roxborough Turnpike, and hedges there; fields near Rifle Range; Headstone; *Melv.* 36.

V. Horsington!; *Melv.* 36.

VII. [Marylebone, 1815; *Herb. G. & R.*] Hedges bet. Haverstock Hill and Kentish Town; *Irv. MSS.* and *Lond. Fl.* Gospel Oak and Five-acre Fields, bet. Kentish Town and Hampstead; *Herb. Hardw.* Site of Exhibition of 1862, South Kensington.

First record: *Blackstone*, 1737.

SMYRNIUM, *Linn.*

305. *S. Olusatrum, L.*   *Alexanders.*

Hipposelinum, Ger. em. (Blackst.).

Cyb. Br. i. 427; iii. 439; Comp. 186. Syme E. B. iv. t. 631.

Waste land and hedges; rare. B. April—June.

I. In a lane and some fields about Cowley; *Blackst. Fasc.* 42. Bet. Uxbridge and West Drayton, 1½ mile from the latter, 1855; *Phyt. N. S.* i. 62.

II. About Hampton Court, G. Francis; *Cooper Supp.* 12. Hampton Court, under the wall by the side of the river!, W. Pamplin; *Lond. Flora*, 197. Still there in some quantity.

IV. (Among the ruins of old Stanmore Church, W. F. Longman; *Melv.* 36.)

VII. [At Milbank by the side of the river; and one plant . . . on the wall at Whitehall Stairs; *M. & G.* 431.]

First record: *Blackstone*, 1737. Perhaps formerly cultivated. There are no specimens in *Herb. Harr.*

Coriandrum sativum, *L.* Cyb. Br. i. 464. Syme E. B. iv. t. 632. VII. Thames side about Chelsea; *Irv. H. B. P.* 606. Chelsea College, 1859, J. Britten; *Bot. Chron.* 58. Not naturalised.

ARALIACEÆ.

ADOXA, *Linn.*

306. A. Moschatellina, *L.*

Radix cava minima fl. viridi, Ger. em. (Pet., Blackst.).

Cyb. Br. i. 420; iii. 438; Comp. 184. Syme E. B. iv. t. 636.

Shady hedgebanks and woods; rather rare. P. April, May.

I. Old Park Wood, and many other places near Harefield; *Blackst. Fasc.* 83. Grove near Harefield Church, 1853; *Herb. Hard.* Harefield, 1867; *Fox & Cole.* Roadside bet. Uxbridge and West Drayton, 1855; *Phyt. N. S.* i. 62.

IV. On the top of Hampstead Hill, among the bushes near the Wood; *Mart. App. P. C.* 67. Kingsbury; *Cole.* Reuter's Lane bet. Hendon and Edgware Road; *Davies.* Fields by Blackpot Farm, near the Brent, 1865; *Farrar.*

VI. Side of brook at Maiden Bridge, Enfield; *Forst. Midd.* Near Edmonton, Enfield, &c.; *Macreight,* 108. Abundant in a copse near Warren Lodge, Edmonton. Southgate, F. W.; *Phyt. N. S.* vi. 349.

VII. [On the moatside as you enter into Jack Straw's Castle*; *Pet. Midd.* At Highbury Barn†; *Forst. Midd.*] Hornsey Wood; Lane at Tottenham; *Blackst. Spec.* 55. Field adjoining Mill Farm (= Millfield or Milford) Lane, east side of Ken Wood grounds!, G. E. Dennes; *Coop. Supp.* 11; abundant in 1866.

First record: *Petiver,* 1695.

HEDERA, *Linn.*

307. H. Helix, *Linn.* *Ivy.*

Cyb. Br. i. 421; Comp. 184. Syme E. B. iv. t. 633.

In woods, hedges, on old walls, &c.; adhering to the trunks of trees or masonry, or creeping on the ground; common. Shrub. October.

I. South Mims. Harefield!; *Newb.*
II. Near Hampton Court!; *Newb.* Staines.
III. Hounslow Heath. Hatton.
IV. Harrow district, abundant; *Melv.* 36. Stanmore; *Varenne.* North Heath, Hampstead.
V. Near Willesden Junction!; Hanwell!; *Newb.*
VI. Hadley!; *Warren.* Edmonton.
VII. Ken Wood. West End, Hampstead.

First record: *Varenne,* 1827. Not mentioned in *Blackst. Fasc.*, and often passed over. Much planted for ornament.

CORNACEÆ.

CORNUS, *Linn.*

308. C. sanguinea, *L.* *Dogwood. Wild Cornel.*
Cornus fœmina, Ger. em. (Blackst.).
Cyb. Br. i. 421; iii. 438; Comp. 185. Syme E. B. iv. t. 635.
Woods, thickets and hedges; very common. Shrub. June, July.
In all the districts.

---

* This must not be confounded with the tavern so called now existing on Hampstead Heath (v. 76). Jack Straw's Castle was the popular name of the area enclosed by the moat which surrounded the old Manor-house or Castle of Highbury; this mansion, then occupied by the Prior of the Knights Hospitallers of St. John, was in 1381 destroyed by the mob under Jack Straw in Wat Tyler's insurrection. The present house was built on the site in 1781.

† Highbury Barn occupies the site of a barn belonging and adjacent to the Manor-house.

VII. [Marylebone fields, 1815; *Herb. G. & R.*] Hornsey Wood. South Heath, Hampstead.

First record: *Blackstone*, 1737.

LORANTHACEÆ.

VISCUM, *Linn.*

309. V. album, *L.*   *Mistletoe.*

Viscum, Ger. (Pet., Blackst.).

Cyb. Br. i. 5; iii. 447; Comp. 199. Syme E. B. iv. t. 635.

Parasitic on the branches of various trees; rare. P. March, April.

I. Harefield, on various trees; *Blackst. Fasc.* 111.

II. Plentifully on *Hawthorn* in Bushey Park. Mr. Jesse has seen it in Bushey Park and grounds of Hampton Court on the following trees: *Tilia sp., Acer Opulus, A. rubrum, Robinia pseudacacia, Laburnum, Pyrus Aucuparia, Æsculus Hippocastanum,* * *Ulmus montana; Gentleman's Mag.* new series, vol. i. (Jan. 1866), p. 72.

IV. On *Lime-trees* . . . at Cannons, . . . Edgeware; . . . on a *Virginia walnut-tree* growing in our fields at Mill Hill, Thos. Knowlton; *Peter Collinson, quoted by Dillwyn in Hortus Collinsonianus.* p. 57.

VI. On the *Lime-trees* at Bone (=Bohunt) Gate, Thos. Knowlton; *Peter Collinson, loc. cit.*

VII. [On some trees at Clarendon House, St. James's; *Pet. Midd.*]

First record: *Petiver*, 1695.

CAPRIFOLIACEÆ.

SAMBUCUS, *Linn.*

310. S. Ebulus, *L.*   *Dane-wort.*

Ebulus (Ger.). *E. sive Sambucus humilis, Ger. em.* (Blackst.).

Cyb. Br. ii. 7; Comp. 199. Curt. F. L. f. 3.

Hedges; rare. P. June, July.

I. Meadow near Breakspears; Uxbridge Moor, plentifully; *Blackst. Fasc.* 24. Uxbridge churchyard; *M. & G.* 464.

III. Scrutage Lane, Heston; *Masters.* Marsh Farm, Twickenham, near the railway.

V. [In the highway at old Brainford (Brentford) Towne's end; *Ger.* 1238.] Near Chiswick Ry. Station; *Naylor.* Near Greenford; *Lees.*

VI. Enfield Chase; *Phyt. N. S.* vi. 301.

VII. [Plentifully in the lane at Kilburn Abbey; *Ger.* 1238.] [Wild in the Bishop of Ely's garden in Holborn, Dr. Watson; *Hill,* 163.]

---

* 'As far as I can ascertain, a solitary instance of its being found on that tree;' *Jesse, loc. cit.*

First record : *Gerarde*, 1597. Said to be frequently introduced into its stations; it has the look of a native plant at Twickenham.

311. S. nigra, *L.*   *Elder.*
Sambucus (Johns., Blackst.).
Cyb. Br. ii. 6; iii. 447; Comp. 199. Syme E. B. iv. t. 637.
Hedges; very common. Small Tree. June, July.
Generally distributed through all the districts.
VII. Old trees in many parts of London; in some cases probably the remains of former hedgerows. There are, or were, large trees in Old St. Pancras Churchyard. Kensington Gardens, 1868; *Warren* (*v. s.*).
First record : *Johnson*, 1632. No doubt often planted. The form or variety with lanceolate cut leaflets (β *laciniata*, Bab.) occurs (III.) in a hedge by the roadside bet. Twickenham and Isleworth near the latter place. *With whitish-green berries*, in a hedge near Perivale Church ; *Lees.*

VIBURNUM, *Linn.*
312. V. Lantana, *L.*   *Wayfaring Tree.*
Viburnum, C. B. Pin. (Blackst.).
Cyb. Br. ii. 8; iii. 447; Comp. 200. Syme E. B. iv. t. 640.
Hedges ; rare. Shrub. May, June.
I. Harefield, chiefly on calcareous soil; *Blackst. Fasc.* 108.
II. Hammonds, near Staines.
IV. Hendon, 1823, Mr. Maurice; *Varenne.* Hampstead ; *Irv. MSS.* Harrow Park ; *Melv.* 36.
First record : *Blackstone*, 1737.

313. V. Opulus, *L.*   *Wild Guelder Rose.*
Sambucus aquatilis sive palustris, Ger. em. (Pet., Blackst.).
Cyb. Br. ii. 8; Comp. 200. Syme E. B. iv. t. 639.
Woods and hedges ; rather rare. Shrub. May—July.
I. Harefield!; *Blackst. Fasc.* 90. Bet. Yewsley and Iver Bridge!; *Newb.*
II. Staines.
III. Wood End ; *Melv.* 37. Marsh Farm, Twickenham. Bet. Twickenham and Worton.
IV. Bishop's Wood!; Hampstead Heath; *M. & G.* 471. Harrow Park, near Archery Ground ; *Melv.* 37.
VI. 'The Alders' Wood, Whetstone.
VII. Hornsey Wood; near Kentish Town ; *M. & G.* 471. [In a wood against the boarded river ; *Pet. Midd.*]
First record : *Petiver*, 1695.

LONICERA, *Linn.*
314. L. Periclymenum, *L.*   *Woodbine. Honeysuckle.*
Periclymenum, Ger. em. (Blackst.).

Cyb. Br. ii. 9 ; Comp. 200. Curt. F. L. f. 1.*
Hedges and woods ; very common ? Twining Shrub. June—September.
Frequent in all the districts except VII., for which we have no record.

First record ; *Blackstone*, 1737. The *form* (? *young state*) *with lobed leaves*, has been noticed—(I.) *P. fol quercinis*, Merr. In White-heath Wood, plentifully; *Blackst. Fasc.* 75.—(IV.) In Bishop's Wood, *M. & G.* 343. Highgate Woods ; *Irv. MSS.* and *Lond. Fl.* 158. Road from Harrow to Kenton, bet. turnpike and ry. bridge ; *Melv.* 38 (*v. s.*). —(V.) Green Lane, Greenford Road ; *Melv.* 38.—(VII.) Hornsey Wood, *M. & G.* 343.

L. Xylosteum, *L.* Cyb. Br. ii. 10; iii. 448. Syme E. B. iv. t. 641. III. Two shrubs in Harrow Grove; *Melv.* 38. VII. [In a hedge nearly opposite Lower Nursery, Haverstock Hill ; *Irv. MSS.* and *Lond. Fl.* 158.] No doubt planted in both localities.

RUBIACEÆ.

SHERARDIA, *Linn.*

315. S. arvensis, *L.*

Alisson Plinii (Turner). *Rubeola arvensis repens cærulea, C. B. Pin.* (Blackst.).

Cyb. Br. ii. 22. Syme E. B. iv. t. 663.

Fields and waste ground ; rather common. A. June—August.
I. Harefield ; *Blackst. Fasc.* 88. Ruislip ; *Melv.* 39.
II. Stanwell Moor. Tangley Park.
III. Drilling Ground, Hounslow. Grass lawn, Twickenham.
IV. Hampstead ; *Irv. MSS.* and *Cooper*, 102. Near Harrow Ry. Station ; *Farrar.*
V. A little from Syon ; *Turn. Names.* Among the corne beside Sion ; *Turn.* i. 39. Turnham Green ; near Kew Bridge Ry. Station ! ; *Newb.* Horsington Hill.
VI. About Highgate ; *Cooper*, 104. Trent Park, Southgate ; *Cole.*
VII. Waste ground opp. Veitch's Nursery ! ; *Newb.* Roadside, Prince's Gate, S. Kensington, 1867.

First record : *Turner*, 1548 ; also first as a British plant.

ASPERULA, *Linn.*

316. A. odorata, *L.* *Woodruff.*
Asperula, Ger. (Blackst.).
Cyb. Br. ii. 23. Syme E. B. iv. t. 660.

Woods and shady places ; rather rare. P. May, June.
I. Woods, Harefield ! ; *Blackst. Fasc.* 8. Ruislip ; *Farrar.*

* The only fault to be found with this drawing is, as was noticed by Sir J. E. Smith, that the stem is made to twist from right to left instead of from left to right.

IV. Bishop's Wood!; *Irv. MSS.* Stanmore. Scratch Wood, near Edgware. Hampstead; *Pet. H. B. Cat.* 30. Path by Julian Hill, Harrow; *Farrar.*

VI. Colney Hatch; *Herb. Hardw.* 1821; *Bennett* (*v. s.*). Forty Hill, Enfield.

VII. Cane (= Ken) Wood!, Pulteney; *Hill*, 74.

First record : *Petiver*, 1713.

[A. arvensis, *L. Asperula flore cæruleo* (Merrett). *A. cærulea* (Budd.). Cyb. Br. ii. 23. Syme E. B. iv. t. 662 (bis). IV. In the woods near Hamstead; *Merrett*, 11. I much question whether this be indigenous; *Budd. MSS.* and *Herb.* The first notice of the plant's occurrence in England. The specimen in *Herb. Budd.* vol. cxxi. fol. 3, was gathered in Yorkshire.]

GALIUM, *Linn.*

317. G. cruciatum, *With.* *Golden Mugwort. Crosswort.*
Cruciata, Dod. (Ger., Johns., Blackst.). *Valantia cruciata, L.* (Mart.). Cyb. Br. ii. 13. Syme E. B. iv. t. 647.

Open hedgebanks; rare. P. May—July.

I. Lane bet. Potter's Bar Station and North Mims Wood, 1856 ; *Phyt. N. S.* i. 407. (Probably the station given in *Fl. Herts.* 138.) Harefield ; *Blackst. Fasc.* 21.

II. Staines Moor Farm. Near Staines on the road to Hampton. Bet. Sunbury and Walton Bridge. Abundant in a dry ditch by the towing-path bet. Hampton Court and Kingston Bridge.

IV. Hampstead Heath; *Johns. Enum.* In a wood bet. the River Brent and Finchley ; *Irv. MSS.*

VII. [Hampsteede Churchyarde and a pasture adjoining thereto, by a mill ; *Ger.* 965, *Park. Theat.* 566, *and others.* It is lately lost in Hampstead Churchyard ; *Blackst. Spec.* 16. Refound there by Dr. Watson ; *Hill*, 513.]

First record : *Gerarde*, 1597.

318. G. Aparine, *L.* *Cleavers. Goose-grass.*
Aparine vulgaris, C. B. Pin. (Blackst.). Cyb. Br. ii. 20. Syme E. B. iv. t. 658.

Hedges and cultivated ground; very common. A. June—September. In all the districts.

VII. Abundant in all the outskirts of town, and even in London itself. A weed in most of the squares and on waste ground, but rarely growing to any size.

First record : *Blackstone*, 1737.

[**319. G. anglicum,** *Huds.*
Aparine minima (Ray).
Cyb. Br. ii. 18; iii. 449.   Syme E. B. iv. t. 657.
Old walls; very rare.   A.   June, July.
VII. Found at Hackney, on a wall, William Sherard; *Ray Syn.* i. 237 and *Pet. Midd.*

The only record of the plant in Middlesex, and the earliest notice of the species as British.   The figure given in *R. Syn.* iii (tab. ix. fig. 1), was probably drawn from a Kentish specimen.   The plant seems confined to the east side of England.]

320. G. Mollugo, *L.*   G. elatum, *Thuill.* (Syme E. B.).
Mollugo vulgatior, Park. (Blackst.).
Cyb. Br. ii. 16; Syme E. B. iv. t. 650.
Hedges; rather rare.   P.   July, August.
I. Harefield!; *Blackst. Spec.* 61.   Pinner.   Elstree.   Hillingdon!; *Warren.*
II. Tangley Park.
IV. Harrow district; very common; *Melv.* 39.   Hampstead; *Irv. MSS.* By the claypits, Harrow Weald.
V. Greenford; *Cooper*, 108.   Chiswick!; *Newb.*   By the Brent near Hanwell!; *Warren.*
VII. Lane near Kilburn, 1869; *Warren.*
First record: *Blackstone*, 1737.

321. G. verum, *L.*   *Yellow Bedstraw.*
G. luteum, Ger. em. (Blackst.).
Cyb. Br. ii. 13.   Syme E. B. iv. t. 648.
Dry fields, commons, roadsides, and hedgebanks; common.   P.   July, August.
I. Harefield; *Blackst. Fasc.* 31.   Harrow Weald Common.   South Mims.
II. Staines!; near Hampton Court!; *Newb.*   Teddington; abundant.
III. Wood End; *Melv.* 39.   Hounslow Heath, drilling ground, &c. abundant.   Twickenham.   Twickenham Park.   Near Isleworth.
IV. Road to Edgeware, 1815; *Herb. G. & R.*   Hampstead; *Irv. MSS.* Field near Harrow Station; hedges at Kenton; *Melv.* 39.   Near Finchley!; *Newb.*   Kingsbury; *Farrar.*
V. Apperton!; Ealing Common!; *Newb.*   Near Hanwell.   Horsington Hill.
VI. Edmonton.   Whetstone.
First record: *Blackstone*, 1737.   Abundant on the sandy parts of the county; less frequent on the clay.

322. G. saxatile, *L.*
Ray's Small Madder (Pet.).   *G. montanum, Huds.* (M. & G.).   ? *Aparine levis*, Park. (Merr.).
Cyb. Br. ii. 16.   Syme E. B. iv. t. 651.

RUBIACEÆ.

142

Heaths and roadsides; rather common.   P.   June—August.
I. Ruislip; *Melvill*, 39.   Harrow Weald Common.   Stanmore Heath.
II. Roadsides from Staines to Hampton; abundant.   Bushey Park.
III. Moist pits on Hounslow Heath; *Merrett*, 9.   Hounslow Heath; Drilling Ground, &c., common.
IV. Hampstead Heath!; *Pet. H. B. Cat.* 30.   Deacon's Hill, roadside.
VII. South Heath, Hampstead.   In the turf above the trench bet. Hyde Park and Kensington Gardens, a single tuft (1868), on the Hyde Park side!; *Warren.*

First record: *Merrett*, 1666?.   Not included in *Blackst. Fasc.*

323. G. uliginosum, *L.*
Aparine palustris minor parisiensis (Budd.).
Cyb. Br. ii. 15; iii. 448.   Syme E. B. iv. t. 655.

Ditches and wet places on commons; rare.   P.   June—August.
I. Near Harefield!; *Newb.*
III. Roxeth, W. M. H.; *Melv.* 39.
IV. Bogs on Hampstead Heath; *Herb. Budd., Huds.* i. 56, and many other authors.   1815; *Herb. G. & R.*
V. Bet. Acton and Turnham Green!; *Newb.*
VI. Finchley, 1827–30; *Varenne.*

First record: *Buddle*, about 1700.   Buddle's specimens are in *Herb. Budd.* vol. cxxi. fol. 2 and 10.

324. G. palustre, *L.*
Gallium album, Ger. em. (Blackst.).
Cyb. Br. ii. 14.   Syme E. B. iv. tt. 653, 654.
Sides of ditches and streams; common.   P.   June—August.

Var. a elongatum, Syme E. B. iv. t. 653.
I. Colne near Yewsley!; *Newb.*
II. Staines!; *Newb.* Bushey Park.
III. Roxeth; *Melv.* 39.   By the Cran in many places, as at Hospital Bridge.

Var. β genuinum, Syme.
I. Harefield; *Blackst. Fasc.* 31.   Stanmore Heath.   South Mims.
II. West Drayton; *Morris* (*v. s.*).   Staines!; *Newb.*
III. Hounslow Heath.   Isleworth.   Harrow; *Herb. Harr.*
IV. Deacon's Hill.   Near Sudbury.   Hampstead Heath; *Cooper,* 102.
V. Greenford; *Cooper,* 108.   Near Brentford; *Hemsley.*   Twyford!; Hanwell!; *Newb.*
VI. Edmonton.
VII. London Canal, Hackney; *Cherry* (*v. s.*).

Var. γ Witheringii. Syme E. B. iv. t. 654.
I. Harrow Weald Common.   Near South Mims.

II. Near Staines on road to Hampton. Staines Moor. Sunbury.

IV. Hampstead Heath; *Cooper*, 102.

V. Bet. Acton and Turnham Green!; *Newb.*

First record: *Blackstone*, 1737.

VALERIANACEÆ.

CENTRANTHUS, *Cand.*

C. ruber, *D.C.* Cyb. Br. i. 24; iii. 450. Syme E. B. iv. t. 664. III. Headstone Moat; *Melv.* 97. Probably cultivated there.

**[325. *C. Calcitrapa, *D.C.*

Valeriana, Calc. L. (With.).

Cyb. Br. ii. 25. Syme E. B. iv. t. 665.

Old walls; very rare. A. May.

VI. Walls belonging to the palace at Enfield, E. Forster; *Herb. Mus. Brit.* In 1843 E. Forster wrote (*Phyt.* i. 649), this grew ' 50 or 60 years ago . . . on the wall of a garden at Enfield . . . formerly that of Dr. Uvedale . . . I do not know the present state of the place.'

VII. Walls belonging to Chelsea Hospital, Mr. Dickson ; . . . Mr. Caley . . . reports it . . . completely naturalised ; *With.* ed. iv. i. 63 (1801), and *Hull*, ed. ii. i. 13 (1808). Chelsea, A. B. Lambert; *Herb. Sowerby*, and ' June 16th, 1815 ' ; *Herb. G. & R.*

First record: *Forster*, about 1790. These two extinct Middlesex localities, in each of which it seems to have been well naturalised, are much less generally known than that at Eltham, Kent, where it is still to be found. Dr. Uvedale, who lived in the old Palace from 1670 to 1722, probably introduced it at Enfield.]

VALERIANA, *Linn.*

326. V. officinalis, *L.* (Syme). V. sambucifolia, *Mikan.* (Bab.). *Wild Valerian.*

V. sylvestris major (Johns., Blackst.).

Cyb. Br. ii. 26; iii. 451 ; and Supp. 60, 90. Syme E. B. iv. t. 666. Curt. F. L. f. 6.

Sides of ditches and shaded streams; common. P. June—August.

I. Harefield !; *Blackst. Fasc.* 104. Ruislip ! ; *Melv.* 40.

II. About Staines. Bet. Hampton Court and Kingston Bridge. Side of Creek by Cross Deep House.

III. Near Hatton. River side bet. Twickenham and Richmond Bridge. Stream near Mother Ive's Bridge.

IV. Hampstead Heath ; *Johns. Enum.* Near Finchley !; *Newb.* Edgware. Turner's Wood.

V. Chiswick.

VI. Southgate; *Cole.*

VII. Bishop's Walk, Fulham; *Cooper,* 104. In great abundance in the I. of Dogs, [at Milbank and in Tothill Fields]; *M. & G.* 49. Millfield Lane. South Heath, Hampstead.

First record: *Johnson,* 1632. The variety *Mikanii* (Syme E. B. iv. 236), *V. officinalis, L.* (Bab.), has not been observed in the county.

327. V. dioica, *L.*

Valeriana minor, Ger. em. and *V. sylvestris vel palustris minor altera,* Ray Cat. (Blackst.).

Cyb. Br. ii. 25; iii. 450. Syme E. B. iv. t. 668.

Wet meadows; rather rare. P. April—June.

I. Harefield; abundantly!; *Blackst. Fasc.* 105. Ruislip, F. W. Farrar; *Hind.*

II. Bet. Hampton and Hampton Court, by the river!; *Newb.*

IV. Hampstead Heath; *Cooper,* 105; and *Irv. MSS.*

V. Bet. Acton and Turnham Green!; *Newb.*

VI. Near Trent Park, Southgate; *Cole.* Edmonton. 'The Alders' Wood, and meadows adjacent, Whetstone.

VII. Ken Wood, Hunter; *Park, Hampst.* Banks of Thames, Fulham; *Faulkner,* 22. Hackney Marsh, Warner; *Cooper,* 105.

First record: *Blackstone,* 1737.

· VALERIANELLA, *Moench.*

328. V. olitoria, *Moench.*   *Lamb's Lettuce.*

Latuca agnina, Ger. em. (Blackst.). *Fedia olitoria, Wahl.* (Melv.).

*. Cyb. Br. ii. 27. Syme E. B. iv. t. 669.

Hedgebanks, fields, and old walls; rather common. A. May—August.

I. Harefield; *Blackst. Fasc.* 48. Near Harefield Church, 1839, H. Kingsley; *Herb. Young.*

II. Near Staines on the Hampton Road. Bet. Hampton Court and Kingston Bridge!; *Newb.*

III. Roxeth; *Melv.* 40. Isleworth.

IV. Cornfields, Hampstead; *Irv. MSS. and Cooper,* 100. Kingsbury; *Farrar.*

V. Hanwell!; *Newb.* Walls about Turnham Green and Chiswick, abundant.

VI. [Near St. John's Wood Church, 1815; *Herb. G. & R.*] Various parts of West London Cemetery, Brompton, J. Britten; *Phyt. N. S.* vi. 592.

First record: *Blackstone,* 1737.

329. V. carinata, *Loisel.*

Cyb. Br. iii. 354. Syme E. B. iv. t. 670.

Open banks; very rare. A. May.

V. South-Western Ry. bank, at the Kew Bridge Station, abundant.   A
few specimens in several other neighbouring places; *Newb.*

First record : *Dr. T. Thompson* ?.   Not considered a native by Mr. H. C.
Watson.   It will probably be found in other parts of the county, but
is very liable to be passed by as *V. olitoria.*

330.   V. dentata, *Koch.*
Fedia dentata, Wahl. (Melv.).
Cyb. Br. ii. 28 ; iii. 451.   Syme E. B. iv. t. 672.
Cornfields; very rare.   A.   July—September.
I. Ruislip, W. M. Hind ; *Melv.* 40.
III. Near Hounslow Ry. Station! ; *Newb.*

First record: *Hind,* 1861.

DIPSACACEÆ.

DIPSACUS, *Linn.*

331.   D. sylvestris, *L.*   *Wild Teasel.*
D. sylv. major, C. B. P. (Blackst.).
Cyb. Br. ii. 29.   Syme E. B. iv. t. 674.

Hedges, waste places, and roadsides ; common.   B.   July—September.
I. Harefield ! ; *Blackst. Fasc.* 24.   Bet. Harefield and Ruislip.   Pinner.
II. Staines.   Towing-path bet. Hampton Court and Kingston Bridge,
abundant.
III. Hounslow Heath.   Hatton.   Twickenham.
IV. Harrow district. very common ; *Melv.* 40.   Mill Hill, A. A.; *Herb.*
Mus. Brit.   Hampstead ; *Irv. MSS.*   Edgware.
V. Near Brentford ; *Hemsley.*   Hanwell! ; *Warren.*   Ealing! ; near
Shepherd's Bush ! ; *Newb.*   Apperton.
VI. Near Waltham Cross ; *Cherry.*   Edmonton.
VII. Kilburn, 1815 ; *Herb. G. & R.*   Isle of Dogs ! ; *Young MSS.*
Finchley Road, 1845 ; *Morris* (*v. s.*).   Bet. Kentish Town and
Hampstead.

First record : *Blackstone,* 1737.

332.   D. pilosus, *L.*   *Shepherd's Rod.*
D. minor sive Virga pastoris, Ger. (Merr., Pet.).   *D. sylvestris capitulo
minore, &c., C. B. P.* (Blackst.).
Cyb. Br. ii. 30.   Syme E. B. iv. t. 676.
Woods, copses, and hedges ; rare.   B.   July, August.
I. About More Hall ; in a little wood near Sir George Cooke's house at
Harefield, plentifully ; *Blackst. Fasc.* 24.   Lane south of Rickmans-
worth Common Moor ; River Colne, near Stocker's End ; *Fl. Herts.*
143.
IV. By the roadside bet. Highgate and Finchley, just beyond the sign of

the 'Flower-de-Luce'; *Mart. Tourn.* i. 230. In a narrow shady lane bet. Golder's Green and the Brent; *Irv. MSS. and Lond. Fl.* 155. In a copse between Bishop's Wood and Finchley, 1865; *Grugeon (v. s.).*

VI. In Hornsey Churchyard; *M. & G.* 154. In a lane from Edmonton to the Hyde field, plentifully; *Forst. Midd. and B. G.* 400.

VII. [Beyond Fulham in the way to Hammersmith; *Merrett*, 33. By ye Bp. of London's at Fulham; *Newt. MSS.* About Fulham Moat; *Pet. Midd.* Bet. the Bishop's Palace and Fulham-field; *Mart. App. P C.* 66.]

First record : *Merrett*, 1666.

KNAUTIA, *Coult.*

333. K. arvensis, *Coult.* Scabiosa arvensis, *L.* (Syme E. B.). *Field Scabious.*

Scabiosa vulgaris pratensis (Park.). *S. prat. hirsuta quæ officinarum, C. B. P.* (Blackst.).

Cyb. Br. ii. 32. Syme E. B. iv. t. 679.

Cornfields, dry banks, and waste ground; rather rare. P. July—September.

I. Harefield ! ; *Blackst. Fasc.* 91. Pinner Chalkpits, sparingly.

II. Near West Drayton ! ; Hampton ! ; *Newb.* Stanwell Moor, abundant. Teddington. Tangley Park.

III. Twickenham Park.

IV. Stanmore, 1827–30; *Varenne.* Hampstead; *Irv. MSS.* Railway bank, Harrow; *Melv.* 41.

V. Greenford ; *Melv.* 41.

First record : *Blackstone*, 1737. Parkinson (*Theat.* 486) says: 'About London everywhere,' but he may have misunderstood the plant. Prefers a light sandy or chalky soil.

SCABIOSA, *Linn.*

334. S. succisa, *L. Devil's Bit.*

Morsus Diaboli, Lob. (Ger., Johns.). *S. radice succisa flore globoso, R. Syn.* (Blackst.).

Cyb. Br. ii. 31. Syme E. B. iv. t. 677.

Damp woods, heaths, and meadows ; common. P. August—October.

I. Harefield ! ; *Blackst. Fasc.* 91. Stanmore Heath.

II. Tangley Park.

III. Abundant about Harrow; *Melv.* 41. Hounslow Heath. Drilling Ground, very abundant, and frequent by roadsides near it. Whitton.

IV. Great store of it in Hampsteede Wood ; *Ger.* 587. Bishop's Wood, abundant. Hampstead Heath ! ; *Johns. Enum.* Railway bank, Harrow, *Melv.* 41. Stone Bridge ; *Farrar.*

V. Wormwood Scrubs; *Britten.*

VI. About Highgate, abundant; *Blackst. Spec.* 90. Winchmore Hill Wood.

VII. South Heath, Hampstead.

First record: *Gerarde,* 1597. *With pale flowers*; (III.) On Hounslow Drilling Ground.

335. * S. Columbaria, *L.*
Scabiosa minor campestris (Park.).

Cyb. Br. ii. 31. Syme E. B. iv. t. 678.

Fields and banks; very rare. P. July, August.

II. By the towing-path bet. Hampton Court and Kingston Bridge, nearer the former. In a field near Teddington Church, with *Campanula glomerata.*

First certain record: *the Authors,* 1867. Parkinson says: ' Dry fields about London' (*Theat.* 486); but the above are the only localities in which it has been observed in Middlesex. Probably not really native in either, but brought down by the river from chalky districts.

COMPOSITÆ.

EUPATORIUM, *Linn.*

336. E. cannabinum, *L.*   *Hemp Agrimony.*
Cyb. Br. ii. 94. Syme E. B. v. t. 785.

Sides of streams and ditches, and wet places; rather rare. P. July—September.

I. Harefield!; *Blackst. Fasc.* 27. Colne, near Yewsley!; *Newb.* Pinner, a single plant; *Hind.*

II. Staines and Stanwell Moors, abundant. By Queen's River, Hampton.

III. Thames bank bet. Twickenham and Richmond Bridge. By the Cran at Hospital Bridge. Duke's River, Isleworth.

IV. Kingsbury; *Varenne.* Hampstead; *Irv. MSS.*

V. Not far from Gunnersbury; *Newb.*

VI. Bury Street, Edmonton. 'The Alders' Wood, Whetstone.

First record: *Blackstone,* 1737.

PETASITES, *Gaertn.*

337. P. vulgaris, *Desf.*   *Butter-bur.*
Petasites, Ger. em. (Blackst.). *Tussilago Petasites, L.* (Curt.).

Cyb. Br. ii. 107; iii. 459. Curt. F. L. f. 2 (drawn from a Middlesex plant), and (*P. hybrida*) Hook. Curt. 4.

Sides of streams and ditches; rare. P. March, April.

I. By the river-side near Harefield Mill, abundantly!; *Blackst. Fasc.* 76.

IV. Near the lake, Bentley Priory, W. M. H.; *Melv.* 47.

V. Banks of Thames bet. Putney and Hammersmith; *Savage.*

VI. Banks of Lea, about Waltham Abbey; side of trench in Water

Lane, Edmonton; lane bet. Fortis Green and Hornsey, 1864, the
female plant (*P. hybrida*, L.); *Savage.*
VII. [About Chelsea Waterworks; *Blackst. Spec.* 71.] [Side of Thames
bet. Westminster Bridge and Chelsea; *Curt. F. L.*]

First record: *Blackstone*, 1737. Remarkably scarce in this county.

338. ***P. fragrans,*** *Presl.* Nardosmia fragrans, *Reich.* (Syme).
Winter Heliotrope.
Cyb. Br. ii. 109. Syme E. B. iv. t. 781.

Hedgebanks and shady places; rare. P. January—March.
II. West Drayton!; *Newb.*
VI. In a field near the Enfield foot-path, Bury Street, Edmonton.
VII. On the outside of a garden wall in Bayswater, and in a meadow;
Mag. Nat. Hist. viii. 389. On rubbish near 'The Warrington,'
Kilburn, 1868!; *Warren.*
Completely naturalised in . . . . Middlesex; *Syme E. B.* v. 117.

First record: 1835. Much cultivated in gardens, and easily establishing
itself.

TUSSILAGO, *Linn.*

339. T. Farfara, *L.*   *Coltsfoot.*
Tussilago, Ger. em. (Blackst.).
Cyb. Br. ii. 109. Curt. F. L. f. 2.

Roadsides, waste ground, and railway banks; very common. P. March,
April.
Generally distributed through the whole county.
VII. Abundant about London. Near Pancras, plentifully; *Blackst. MSS.*
West End Lane, Kilburn. Near Chelsea Hospital. By Buckingham
Palace. Downing Street. Thames Embankment. Gordon Square.
St. Paul's Churchyard. Seven Sisters Road. I. of Dogs, &c.

First record: *Blackstone*, 1737. Usually the first plant to appear on
railway banks, newly-made land, heaps of rubbish, and similar places.

ASTER, *Linn.*

340. A. Tripolium, *L.*
Cyb. Br. ii. 112. Syme E. B. v. t. 776.

Sides of ditches of brackish water; very rare. P. August, September.
VII. Three or four feeble plants by a ditch in the Isle of Dogs!, 1866;
Newb.
The only locality. Perhaps not uncommon in the I. of Dogs before that
district was built over and otherwise altered. With *Scirpus maritimus, Samolus Valerandi, Sclerochloa distans,* and a few other plants
of which but a few specimens have been gathered in late years, it
constitutes all that remains of the semi-maritime vegetation which

no doubt originally covered the marshy shores of the Thames, as far up as the influence of the tide extended.

A. Salignus, *Willd.* II. Side of Thames bet. Strawberry Hill and Teddington, 1867. An exotic garden species.†

ERIGERON, *Linn.*

E. acris, *L. Conyza cærulea acris, Ger.* (Merrett). Cyb. Br. ii. 110; iii. 459. Curt. F. L. f. 1. VII. (On Parson's Green, on the ditchbanks, plentifully; *Merrett,* 29.) A few plants on waste ground, west of Edgware Road Station of North London Railway, 1866 or 1867; *Warren.* Site of Exhibition of 1862, South Kensington. We have seen the plant nowhere native in the county; Merrett may have mistaken small plants of **341** *E. canadensis* for this at Parson's Green. The S. Kensington plant is a large form with a white pappus (*E. corymbosus,* Wallr. ?).

341. * **E. canadensis,** *L.*
Conyza annua acris alba Linariæ fol. Boccon (Ray, Budd., Pet.). *C. canadensis ann. alb. &c.* (R. Syn. iii.).
Cyb. Br. ii. 111. Syme E. B. v. t. 773.

Waste and cultivated ground; rather rare. A. July—September.
II. Hampton. In a lane near Hampton Court, F. Forster; *Herb. Mus. Brit.* and *B. G.* 410. Towing-path bet. Hampton Court and Kingston Bridge; *Bloxam.*
III. About Twickenham, abundant.
V. Railway bank at Kew Bridge Station, abundant. About Turnham Green.
VII. Frequently about London, but certainly not native, Dr. Tancred Robinson; *R. Syn.* i. 49, *Pet. Midd., Budd. MSS.,* &c. [Tothill Fields, 1830; *Herb. Hardw.*] Chelsea, common!; *Irv. H. B. P.* 525. Roads about Hackney Downs; *Grugeon* (*v. s.*). Sandy End, and elsewhere about Fulham. Parson's Green. Site of Exhibition, S. Kensington, in great quantity. Victoria Street, Westminster. Green Park. By Buckingham Palace. Thames Embankment.

First record: *Tancred Robinson,* 1690. A North American species. Thoroughly naturalised, and indeed one of the commonest weeds in and about London, especially in the dry sandy western suburbs, and extending up the Thames valley to Hampton. An American writer states, but without giving any reference or additional particulars, that this plant 'sprang up in Europe 200 years ago from a seed which dropped out of the stuffed skin of a bird.' Marsh, *Man and Nature,* 68.

† ? A. Amellus, *L. A. atticus,* and *A. Italorum,* upon Hampstead Heath; *Ger.* 393. These . . . are no other than two *Hieracia* ; *Johns. Ger.* 489.

BELLIS, *Linn.*

342. B. perennis, *L.*   *Daisy.*
Bellis sylvestris minor, C. B. P. (Blackst.).
Cyb. Br. ii. 124.   Syme E. B. v. t. 772.

Fields, meadows, grass-plots, and roadsides; very common.   P.   March—
September, but chiefly in May and June.
Throughout all the districts.
VII. Abundant in the parks and squares of London, and in waste ground
and gardens.

First record: *Blackstone,* 1737.

SOLIDAGO, *Linn.*

343. S. Virgaurea, *L.*   *Golden-rod.*
Virga-aurea (Ger., Blackst.).   *V.-a. vulgaris* (Park.).
Cyb. Br. ii. 112; iii. 460.   Syme E. B. v. t. 778.

Woods and heaths; rather rare.   P.   July—September.
I. Harefield; *Blackst. Fasc.* 111.   Piuner Wood; *Melv.* 47.
III. Drilling Ground, Hounslow.
IV. In Hampsteed Wood, plentifully, near the gate to Kentish Towne;
Ger. 349; *Park. Theat.* 543; *and many subsequent authors.*
Bishop's Wood.   Hampstead Heath!; *Blackst. Spec.* 105.
VI. Winchmore Hill Wood.
VII. Lower Heath, Hampstead.

First record: *Gerarde,* 1597; also the first notice as British.   A dwarf
form was distinguished by the old botanists.   Gerarde calls it *V.-a.
Arnoldi Villa novani,* and both he and Johnson (*Eric.*) mention it in
(IV.) Hampstead Wood.   It is probably *V.-a. vulgari humilior* (*R.
Syn.* iii. 176) on Hampstead Heath; *Huds.* i. 319; and nearly
answers to Smith's *var. Cambrica.*

INULA, *Linn.*

344. * I. Helenium, *L.*   *Elecampane.*
Helenium (Ger.).   *Enula campana* (Merr.).
Cyb. Br. ii. 121; iii. 462.   Syme E. B. v. t. 766.

Meadows; very rare.   P.   July—September.
I. In a close adjoining to Harefield Common; in a meadow called Gant-
lets near Breakspears; *Blackst. Fasc.* 25.
II. In an orchard bet. Colbrook and Ditton Ferry; *Ger.* 649.
IV. By Hamsted Heath; *Merr. MSS.*
VI. Bank of a pond bet. Chase Cottage and South Lodge, Enfield Chase;
Phyt. N. S. vi. 301.

First record: *Gerarde,* 1597.   Probably escaped from cultivation.

345. I. Conyza, *DC.* *Ploughman's Spikenard.*
Baccharis Monspeliensium, Ger. (Blackst.).
Cyb. Br. ii. 122. Syme E. B. v. t. 767.

Banks on a chalky soil ; very rare. P. July—September.

I. In the old chalk-pit near Harefield Mill, plentifully ; *Blackst. Fasc.* 9.
V. In a calcareous spot bet. Perivale and Horsington Hill ; *Lees.*

Not since recorded from Harefield, but probably still grows there.

PULICARIA, *Gaertn.*

346. P. vulgaris, *Gaertn.* Inula Pulicaria, *L.* (Syme and L. Cat.).
Conyza minima, sive Pulicaria (Lob., Blackst.). *Conyza minor, Trag.*
(Johns.).
Cyb. Br. ii. 123. Curt. F. L. f. 3.

Wet sandy commons and roadsides ; very rare. A. August, September.

I. Harefield ; *Blackst. Fasc.* 20.
II. Roadside bet. Hampton and Sunbury, H. C. W. ; *New B. G.* 100.
IV. [Hampstead Heath ; *Johns. Enum.*] [Golder's Green ; *Irv. MSS.* and
    Lond. Fl. 153. Not there in 1852, J. T. Syme ; *Phyt.* iv. 860.]
V. About Ealing towards the Harrow Road, plentifully, W. Pamplin ;
    New B. G. Supp. 588. On the common close to Apperton Bridge
    by the Brent ; *Lees.*

First (?) record: *Johnson,* 1632. Lobel (*Adv. Nov.* 145) first noticed this
    as British ; his locality is ' in Benard-Greyn ara et fossis, altero à
    Londino lapide.' This may be a Middlesex station, but we have not
    been able to identify Benard Greyn.

347. P. dysenterica, *Gaertn.* I. dysenterica, *L.* (Syme and L. Cat.).
Conyza media (Johns. and Blackst.).
Cyb. Br. ii. 123. Syme E. B. v. t. 770.

Ditchbanks, roadsides, &c. ; common. P. July—September.

I. Harefield ; *Blackst. Fasc.* 20. Pinner. Stanmore Heath. S. Mims.
II. Staines. Bet. Hampton Court and Kingston Bridge, abundant.
III. Twickenham.
IV. Common about Harrow; *Melv.* 48. Claypits, Harrow Weald. Hamp-
    stead ; *Irv. MSS.*
V. Ealing Common ! ; Twyford ! ; *Newb.* Apperton.
VI. Finchley ; *Herb. Hardw.* Edmonton.
VII. [In S. James his Park ; Tuthill Fields, &c.; *Johns. Ger.* 482.] I. of Dogs.
First record : *Johnson,* 1633. ' In Cheape-side the herbe-women call it
    Herbe Christopher, and sell it to Empericks ;' *Johns. Ger.* 483.

Galinsoga parviflora, *Cav.* Syme E. B. v. t. 765. V. In market-gardens
    in this district; *Newb.* VII. Parson's Green, 1862, J. Britten ; *Seem.*
    J. of Bot. i. 375. Plentiful on the Surrey side of the Thames about
    Kew and Richmond, whence it has migrated into our county.

BIDENS, *Linn.*

348. B. tripartita, *L.*

Eupatorium cannabinum fœmina, Lob. (Johns., Blackst.).

Cyb. Br. ii. 93; iii. 457. Curt. F. L. f. 4.

Side of streams, ditches, and ponds; common. A. August—October.

I. Harefield; *Blackst. Fasc.* 27. Bet. Yewsley and Iver Bridge!; *Newb.* Elstree Reservoir.

II. West Drayton; bet. Hampton and Hampton Court; *Newb.* Bet. Hampton Court and Kingston Bridge, abundant. Teddington.

III. Grove Pond; *Melv.* 46. By the Cran, Hounslow Heath. Twickenham.

IV. Greenhill; *Melv.* 46. Hampstead Heath; *Johns. Enum.* and *Irv. MSS.* Kingsbury; *Farrar.*

V. Near Greenford; *Melv.* 46. Bet. Acton and Ealing Common; *Newb.* Near Lampton.

VI. Southgate, F. Y. Brocas; *Phyt. N. S.* iii. 31. Edmonton.

VII. Marylebone Fields, 1815; *Herb. G. & R.* Hornsey Wood; *Herb. Hardw.* Upper Clapton; *Cherry.* Shepherd's Bush, abundant. Eel-brook Meadow, Parson's Green. South Heath, Hampstead. Hackney Wick. Side of canal, Camden Town.

First record: *Johnson,* 1632; also the first notice as British.

349. B. cernua, *L.*

Fœmina cannabina septentrionalium stellato et odore flore (Lob.). *Eupatorii cannabini fœminæ var. altera,* Ger. em. (Blackst.).

Cyb. Br. ii. 93. Curt. F. L. f. 3 (drawn from a Middlesex specimen).

With the former; common. A. August—October.

I. Harefield!; *Blackst. Fasc.* 28. Bet. Yewsley and Iver Bridge. Stanmore Heath. Uxbridge!; *Newb.*

II. West Drayton!; Bushey Park!; Hampton Court!; *Newb.*

III. Harrow; *Melv.* 46. Near Richmond Bridge. Behind Kneller Park. Behind Whitton Dean.

IV. Mill Hill, Mr. Children; *Herb. Mus. Brit.* North Heath, Hampstead. Near Brent Reservoir; *Farrar.*

V. Bet. Acton and Ealing!; Twyford!; *Newb.* Apperton.

VI. Southgate, F. Y. Brocas; *Phyt. N. S.* iii. 31. Edmonton.

VII. Everywhere about London; *Lob. Adv. Nov.* 227. Pond adjoining Hornsey Wood; *Curt. F. L.* Copenhagen Fields; *Herb. Hardw.* Bayswater; *Herb. G. & R.* Ditch bet. Hyde Park and Kensington Gardens.

First record: *Lobel,* 1570; also first notice as a British plant. A plant with the upper leaves quite undivided, but in other respects more like *B. tripartita,* was found on waste ground by the river near Twickenham Church: probably a hybrid.

ACHILLEA, *Linn.*

350. A. Ptarmica, *L.*   *Sneezewort.*

Ptarmica (Ger., Blackst.). *Ptarmica fol. long. serrato, fl. albo, J. B.*
(Blackst. Spec.).

Cyb. Br. ii. 132.   Syme E. B. v. t. 730.

Heaths and meadows; rather common. P. July, August.

I. Harefield!; *Blackst. Fasc.* 82. Pinner; *Melv.* 49. Stanmore Heath.
Elstree Reservoir.

III. Field behind Harrow Grove; *Melv.* 49. Drilling Ground, Hounslow.
Bet. Twickenham and Worton. About Whitton.

IV. Hampstead; *Pet. H. B. Cat.* North Heath. Pinner Drive; *Melv.* 49.
Bank of Midland Ry., Edgwarebury, abundant.

V. Twyford!; In a wet field bet. Acton and Turnham Green!; *Newb.*

VI. Near Highgate, plentifully; *Blackst. Spec.* 78. Hadley; *G. Johnson.*
Colney Hatch!; *Newb.*

VII. [In the three great fields adjoining to Kentish Town; *Ger.* 483.]
South Heath, Hampstead.

First record; *Gerarde,* 1597; also first notice as British.

351. A. Millefolium, *L.*   *Yarrow.*

Millefolium vulgare album, C. B. Pin. (Blackst.).

Cyb. Br. ii. 133.   Syme E. B. v. t. 727.

Meadows, waste ground, roadsides, &c.; very common. P. June—
September.

Throughout the county.

VII. Occurs in all the parks and in the turf of the squares, as in Lincoln's
Inn Fields and Hyde Park. South Heath, Hampstead. Isle of
Dogs. Hackney Wick, &c.

First record: *Blackstone,* 1737. *With dark rose-coloured flowers;* (I.)
On Harefield Common; (III.) in a meadow at Twickenham.

· ANTHEMIS, *Linn.*

352. *A. arvensis, *L.†*

Cyb. Br. ii. 130; iii. 462.   Syme E. B. v. t. 721.

Cultivated fields, roadsides, &c.; very rare. A. June—August.

II. Fields about Teddington; *Lond. Fl.* 155.

III. A single plant by the roadside near Whitton Park.

First certain record: *Irvine,* 1838.

353. A. Cotula, *L.*

Cyb. Br. ii. 131; iii. 463.   Curt. F. L. f. 5 (drawn from a specimen
gathered near London).

---

† *Cotula non fœtida, flore pleno,* (VII.) In S. James's field; *Merrett,* 30, is, perhaps, a
double-flowered variety of this species.

Fields, waste places, and roadsides; rather rare. A. July—September.

I. Stanmore Heath.

II. Near West Drayton!; *Newb.* Near Staines. Bet. Hampton Court and Kingston Bridge. Bet. Strawberry Hill and Teddington.

III. Waste ground by Twickenham Church.

IV. Harrow district, common; *Melv.* 49. Hampstead; *Irv. MSS.* Harrow Weald. Stanmore Marsh. Near Harrow Weald Park; *Farrar.*

VII. Marylebone Infirmary garden; *Varenne.* In the turf close to Kensington Palace, 1866; *Warren.*

First record: *Varenne,* 1827.

354. A. nobilis, *L.*   *Chamomile.*

Anthenus (Turn.). *Camemilla Romana* (Lob.). *Chamœmelum odoratissimum repens fl. simplici, J. B.* (Ray). **Chamœmelum, Ger.* (Merrett). *C. nobile seu odoratum, C. B.* (Pet.). *C. romanum, Ger. em.* (Blackst.). Cyb. Br. ii. 129; iii. 462. Syme E. B. v. t. 724.

Heaths, commons, and dry roadsides; common. P. June—September.

I. On Harefield Common, abundantly; *Blackst. Fasc.* 17. Hillingdon!; *Warren.* Bet. South Mims and Potter's Bar.

II. Towing-path, Hampton Court!; *Newb.* Feltham Green. Abundant on south side of high road bet. Harlington Corner and the 'Magpies.' Twickenham Common.

III. In mooste plenty of al in Hundsley (=Hounslow) Hethe, *Turn. Names, Merrett* 25, *R. Syn.* i. 57, &c. Drilling Ground, Hounslow. Near Hatton. Field near Hospital Bridge, abundant and very large.

IV. Golder's Green; *Irv. MSS.* Mill Hill; *Lond. Flora,* 154. North Heath, Hampstead.

V. In the wylde felde in Brantforde (=Brentford) Green; *Turn.* i. 47; *R. Syn.* i. 57, &c. Turnham Green; Acton Green; *Newb.* Ealing Common; *Lees.*

VI. Muswell Hill, plentifully, Dr. Watson; *Hill,* 435. Enfield Chase; *Burnet,* i. 38. Enfield Green. Hadley; *G. Johnson.*

VII. [In Tuthill Fields; *Pet. Midd.*] [Waste ground on this side Hide Park; *Merrett,* 25.] A patch of about four square yards in the gardens of Lincoln's Inn, 1869!; *Warren;* who has known it there for three years. South Heath, Hampstead, in several spots.

First notice: *Turner,* 1548; also first notice as British. Lobel's locality is 'in Benard Greyn'; *Lob. Adv.* 145. (See **346,** *Pulicaria vulgaris.*) Merrett records a variety as *Cham. nudum odoratum;* and Ray as *Cham. fl. nudo* (*R. Cat.* i. 67). (VII.) In Tuthill Fields; *Merrett,* 25.

* This may perhaps be *Matricaria Chamomilla,* L.; but the synonymy is greatly involved.

MATRICARIA, *Linn.*

355. ***M. Parthenium,** *L.* Chrysanthemum P., *Pers.* (Syme and
L. Cat.). *Feverfew.*
Parthenium, Ger. em. (Blackst.). *Pyrethrum P.* (Sm.).
Cyb. Br. ii. 127. Syme E. B. v. t. 715.

Hedgebanks and waste ground; rather common. P. July, August.
I. Harefield; *Blackst. Fasc.* 72. Moss Lane, Pinner!; *Melv.* 49. South
Mims.
II. Bet. West Drayton and Yewsley!; *Newb.* Bet. Sunbury and Walton
Bridge.
III. On walls at Twickenham and Isleworth.
IV. Hampstead!; *Irv. MSS.* Field behind King's Head, Harrow; Julian
Hill; *Melv.* 49.
V. Near Apperton!; *Newb.* About Perivale; *Lees.*
VI. Muswell Hill; *Herb. Hardw.* Garden weed at Edmonton.
VII. By Lea Canal; *Cherry.* South Heath, Hampstead. Shepherd's Bush.
Near Cremorne. Parson's Green. Site of Exhibition, South Ken-
sington (semi-double).

First record: *Blackstone*, 1737. Often the outcast of gardens.

356. **M. inodora,** *L.* Chrysanthemum inodorum, *L.* (Syme and L. Cat.).
Chamæmelum inodorum annuum humilius fol. obscure virentibus, H. Ox.†
(R. Syn. iii.). *Pyrethrum inodorum, Sm.* (Irv.).
Cyb. Br. ii. 127; iii. 462. Syme E. B. v. t. 717.

Waste ground and roadsides; rather common. A. July—September.
I. Harefield.
II. Bet. West Drayton and Yewsley!; bet. Sunbury and Hampton!;
Newb. Bet. Hampton Court and Kingston Bridge. Railway banks,
Fulwell Station.
III. Near Hounslow. Several places near Twickenham.
IV. Harrow district, very common; *Melv.* 49. Hampstead; *Irv. MSS.*
Burgess Hill.
V. Near Brentford; *Hemsley.* Acton!; *Newb.*
VII. Observed by Mr. Rand along the way to Chelsea; *Dill. in R. Syn.*
iii. 186. Site of Exhibition, South Kensington. Roadside, Ken-
sington Gore. Waste ground, Gordon Square. Thames Embank-
ment. Isle of Dogs.

First record: *Rand*, 1724. A small form of this is recorded as
Chamæmelum majus fol. tenuissimo, caule rubente, H. R. Monsp. Frequent about London; *Dill. in R. Syn.* iii. 186. (See *Herb. Pet.*
vol. cxlviii. fol. 136.)

† Petiver's specimens (*Herb. Pet.* vol. cli. fol. 116) with this name are *Anthemis arvensis*; but there is confusion between the two species in the old authors.

357. M. Chamomilla, *L.* Chrysanthemum Ch., *E. Mey* (Syme and L. Cat.). *Wild Chamomile.*

Chamæmelum sive Anthemis vulgatior, Lob. (Johns.).

Cyb. Br. ii. 128. Curt. F. L. f. 5 (from a London specimen).

Waste ground, roadsides, cornfields, and gardens; very common. A. June—August.

Abundant in all the districts.

VII. Kentish Town; *Herb. Hardw.* Parson's Green. Thames Embankment. Isle of Dogs. Hackney Wick.

First record: *Johnson*, 1632; also the first notice as British. By far the commonest of the Chamomiles in Middlesex; about Twickenham it is often an abundant cornfield weed.

CHRYSANTHEMUM, *Linn.*

358. C. Leucanthemum, *L.* *Ox-eye Daisy.*

Bellis sylvestris major, C. B. Pin. (Blackst.).

Cyb. Br. ii. 126. Syme E. B. v. t. 714.

Pastures, roadsides, railway banks, and waste ground; common. P. June—August.

I. Harefield!; *Blackst. Fasc.* 9. Near Yewsley!; *Newb.* Pinner. South Mims.

II. Near Hampton Court!; *Newb.* Bushey Park. Staines. Near Hayes Ry. Station.

III. Hounslow. Frequent about Twickenham.

IV. Hampstead. Stanmore. Edgware; *Varenne.*

V. Hanwell!; *Newb.* Near Lampton.

VI. Edmonton. Whetstone. Near Enfield; *Tucker.*

VII. Hyde Park, in the turf, 1868; Grosvenor Square, 1869; *Warren* (*v. s.*). Green Park, 1862. Site of Exhibition, South Kensington.

First record: *Blackstone*, 1737.

C. segetum, *L. Corn Marigold.* Cyb. Br. ii. 125; iii. 462. Curt. F. L. f. 1. I. Harefield; *Blackst. Fasc.* 18. A single plant on Harrow Weald Common. II. A single plant near Teddington, and in Teddington Park. IV. Hampstead; *Irv. MSS.* VI. Edmonton, in a gravel pit; *MS. note in Mart. Fl. Rust.* VII. [Bank of canal near St. John's Wood Chapel, 1815; *Herb. G. & R.*] Like many other cornfield plants, this is almost wanting in the county, our few cornfields being very free from weeds.

ARTEMISIA, *Linn.*

359. *A. Absinthium, *L.* *Wormwood.*

Absinthium vulgare, Ger. (Blackst.).

Cyb. Br. ii. 98. Syme E. B. v. t. 731.

Roadsides and waste ground; very rare. P. July, August.

I. About More Hall, plentifully, and elsewhere; *Blackst. Fasc.* 1.

IV. Roadside by the Swan Inn, bet. Hampstead and Hendon, E. H. Button; *Coop. Supp.* 11.

VII. Site of Exhibition of 1862, South Kensington, 1867.

First record: *Blackstone*, 1737. An obvious introduction at South Kensington.

360. A. vulgaris, *L.*   Mugwort.

A. vulg. major, C. B. Pin. (Blackst.).
Cyb. Br. ii. 98.   Syme E. B. v. t. 732.

Roadsides and waste ground; common. P.   July—September.
I. Harefield!; *Blackst. Fasc.* 7.   Woodridings, W. M. H.; *Melv.* 46. South Mims.
II. West Drayton!; *Newb.* Staines Moor. Hampton. Tangley Park. Bet. Hampton Court and Kingston Bridge. Teddington.
III. Teddington Park, abundant. Near Worton. Scarce about Twickenham.
IV. Greenhill; *Melv.* 46.   Hampstead; *Irv. MSS.* Near Harrow Weald Park; *Farrar.*
V. Near Ealing!; near Shepherd's Bush!; Acton; *Newb.* Near Lampton.
VI. Colney Hatch.   Edmonton.   Finchley Common; *Herb. Hardw.*
VII. Marylebone; *Varenne.* Upper Clapton; *Cherry.*

First record: *Blackstone*, 1737.

A. scoparia, *Waldst.* and *Kit.* VII. In great plenty on the site of the Exhibition of 1862, South Kensington, forming large bushes, 1865, 1866, and 1867.   A native of Austria, Turkey and Russia.

TANACETUM, *Linn.*

361. T. vulgare, *L.*   Chrysanthemum Tanacetum (*Syme and L. Cat.*). *Tansy.*

Tanacetum, Ger. em. (Blackst.).
Cyb. Br. ii. 96.   Syme E. B. v. t. 716.

Sides of rivers and streams, and in waste places; rather common. P. June—September.
I. In a meadow near Harefield Church; *Blackst. Fasc.* 97.   Denham Marsh near Uxbridge, 1839; *Herb. Young.*
II. Abundant in an islet in the Thames at Sunbury.   Extremely abundant by the towing-path bet. Hampton Court and Kingston Bridge. River-side bet. Teddington and Strawberry Hill.
III. By the river near Twickenham Church, and in a ditch in another place near.   Thames side near Isleworth.
IV. Waste places, Harrow; banks of the Brent; *Melv.* 46.
V. By the Thames, Chiswick. By the Brent near Hanwell!; *Warren.*
VI. Near Highgate Archway; *Irv. MSS.* and *Lond. Fl.* 146.   Hedge near Chase Cottage, Enfield; *Phyt. N. S.* vi. 303.   By Lea Canal bet.

Tottenham and Edmonton, 1867 ; *Cherry.* In the watery lane at
Edmonton ; *Forst. Midd.* Bury Street, Edmonton.

First record : *Blackstone,* 1737. Common by the Thames.

FILAGO, *Linn.*

362. F. germanica, *L.*   *Cudweed.*
Gnaphalium vulgare majus, C. B. Pin. (Blackst.).
Cyb. Br. ii. 105. Syme E. B. v. t. 736.

Dry fields and roadsides ; rather common. A. July, August.
  I. Harefield ; *Blackst. Fasc.* 33. Eastcot, W. M. H. ; *Melv.* 47. Road-
    side on south of Harrow Weald Common.
 II. Staines. Towing path bet. Hampton Court and Kingston Bridge, near
    the former. Fulwell. Near Teddington Ry. Station.
III. Drilling Ground, Hounslow, scarce.
 IV. Stanmore ; *Varenne.* Hampstead ; *Irv. MSS.*
  V. Hanwell ! ; *Newb.* Turnham Green.
 VI. Trent Park, near Southgate ; *Cole.*

First record : *Blackstone,* 1737.

363. F. spathulata, *Presl.*
Cyb. Br. ii. 106 ; iii. 459. Syme E. B. v. t. 738.

Dry banks and roadsides ; very rare. A. July—September.
 II. A few plants near Staines, on the Hampton road. Near Hampton
    Court, by the towing-path.
III. A few plants on waste ground by the river, near Twickenham Church.

First record : *H. C. Watson,* 1852. ('In Middlesex,' *Cyb. Br.* iii. 459.)

364. F. minima, *Fr.*
Least Cudweed (Pet.).
Cyb. Br. ii. 105 ; iii. 458. Syme E. B. v. t. 739.

Dry fields and heaths ; rare. A. June—September.
  I. Harrow Weald Common ! ; *Melv.* 47 ; in plenty.   Stanmore Heath,
    1827–30 ; *Varenne.*
 II. Field near Teddington Ry. Station.
III. Abundant on Hounslow Heath.
 IV. Hampstead ; *Pet. H. B. Cat.* and *Irv. MSS.*

First record ; *Petiver,* 1713.

F. gallica, *L.* should be looked for. It was gathered by Lightfoot 'near
  Iver, Bucks' (close to our boundary), and the specimen is preserved in
  the Banksian Herbarium in the British Museum. A note by Petiver
  (*Sloane MSS.* 3333, fol. 98) appears to imply that he found it near
  Enfield in 1696.

GNAPHALIUM, *Linn.*

365. G. uliginosum, *L.*

G. longifol. humile ramosum capit. nigris, Ray (Blackst.).

Cyb. Br. ii. 103. Syme E. B. v. t. 741.

Damp sandy ground, and roadsides where water has stood during winter; common. A. July—September.

I. Harefield; *Blackst. Fasc.* 33. Pinner; *Melv.* 47. Bet. Yewsley and Iver Bridge!; *Newb.* Stanmore Heath. South Mims. Ruislip; *Herb. Harr.*

II. Near Teddington Station.

III. Cricket-field, Harrow; *Melv.* 47. Near Hatton. Hounslow Heath. Twickenham.

IV. Hampstead; *Irv. MSS.* Near Brent Reservoir; *Farrar.* Harrow Weald.

V. Near Lampton. Horsington Hill.

VI. Edmonton.

VII. [Mary-le-bon; *Herb. Banks.*] Near Clapton; *Cherry.* South Heath, Hampstead. Back of Adelaide Road, N.W. Hackney Wick.

First record: *Blackstone*, 1737.

366. G. sylvaticum, *L.*

G. anglicum fol. longiore (Lob.). *G. anglicum* (Ger.). *G. folio longiore* (Johns.). *G. rectum* (Sm.).

Cyb. Br. ii. 101; iii. 453. Syme E. B. v. t. 743.

Heaths and woods; rare. P. August, September.

I. In a shady field near Battleswell, Harefield; *Blackst. Fasc.* 28. Harefield Common!; *Newb.* Harrow Weald Common. Stanmore Heath.

IV. [In the darke woods of Hampsteede; *Ger.* 518. Highgate Wood; *Johns. Eric.*] [Hampstead Heath, Mr. Watson; *Blackst. Spec.* 28.]

VI. Roadside bet. Highgate and Muswell Hill, Mr. Watson; *Blackst. Spec.* 28.

VII. [About Cane Wood, near Highgate; *Pet. MSS.*] On soil brought from elsewhere, near Paddington Cemetery, 1867! noticed for several years; *Warren.*

First record: *Lobel*, 1570. Lobel's locality (*Adv. Nov.* 202) is 'on this side of the Thames, in a dense wood three miles from London.' It is probably one of the Hampstead stations.

DORONICUM, *Linn.*

367. *D. Pardalianches,* *L.*

Cyb. Br. ii. 121. Syme E. B. v. t. 761.

IV. Fairly established in a little copse by the stream, near the road between Finchley and Hendon!, 1868; *Newb.*

An escape from cultivation.

SENECIO, *Linn.*

368. S. vulgaris, *L.*　　　　*Groundsel.*

Cyb. Br. ii. 113.　Syme E. B. v. t. 749.

Gardens and cultivated ground, waste places, &c.; very common. A.
Nearly all the year.

A common weed in all the districts, including London and its suburbs.
By ditch-sides about London, plentifully; *Brit. Phys.* 136.

First record: *Robert Turner,* 1664. Not included in *Blackst. Fasc.*;
doubtless passed over from its frequency.

[S. viscosus, *L.* Cyb. Br. ii. 114. Syme E. B. v. t. 752. VII. Copenhagen
Fields, near the Caledonian Hospital; *Lond. Flora,* 152. An accidental
waif.]

369. S. sylvaticus, *L.*

Cotton Groundsel (Pet.). *S. minor latiore folio seu montana, C. B. P.*
(R. Syn. iii.).

Cyb. Br. ii. 113; iii. 460.　Syme E. B. v. tt. 750, 751.

Heaths, dry roadsides, and waste places; common. A. July—September.

I. Harefield Common, sparingly. Harrow Weald Common, abundant.
Stanmore Heath. Near Uxbridge, towards Harefield; *Phyt. N. S.*
i. 62. Ruislip; *Melv.* 48.

II. Roadsides about Staines, Charlton, &c., to Hampton, abundant.

III. Hounslow Heath, the Drilling Ground, and roads adjacent, abundant.

IV. Hampstead!; *Pet. H. B. Cat.* North Heath and Child's Hill Lane.

V. Near Hanwell!; *Newb.*

VI. Near Hornsey; *Dill. in R. Syn.* iii. 178.

VII. [Marylebone Fields, 1818; *Herb. G. & R.*] [Primrose Hill, 1830;
Varenne.] Waste ground behind Adelaide Road.

First record: *Petiver,* 1713.

S. squalidus, *L.* Cyb. Br. ii. 115. Syme E. B. v. t. 753. II. In a newly
laid out road bet. Twickenham and Teddington, near Strawberry
Hill, 1867. A solitary plant.

370. S. erucifolius, *L.*

Groundsel Ragwort (Pet.). *Jacobæa Senecionis folio incano, perennis, R.
Syn.* (Blackst.).

Cyb. Br. ii. 115.　Curt. F. L. f. 5 (an admirable figure).

Roadsides, dry fields, &c.; rather common. P. July—September.

I. Harefield!; *Blackst. Fasc.* 45. Pinner. Stanmore Heath. South Mims.

II. Near Staines on road to Hampton.

III. Near Harrow.

IV. Hampstead; *Pet. H. B. Cat.* Abundant in the Harrow district; *Melv.* 48. Near Willesden; *Britten.* Mill Hill, Mr. Children; *Herb. Mus. Brit.* Bet. Penniwells and Woodcock Hill. Fields by Scratch Wood. Bet. Sudbury Station and Apperton.

V. Apperton!; *Newb.*

VI. Finchley Common, 1845; *Herb. Hardw.* Waste ground by Highgate Archway; *Lond. Flora,* 153.

First record: *Petiver,* 1713. Chiefly found on the clay and chalk, having much the range of **126,** *Hypericum hirsutum,* but somewhat more frequent than that species.

371. S. Jacobæa, *L.* *Ragwort. Seggrum.*

Jacobæa vulgaris, Ger. em. (Blackst.).

Cyb. Br. ii. 116. Syme E. B. v. t. 755.

Roadsides and waste places; rather common. P. July—September.

I. Harefield!; *Blackst. Fasc.* 46.

II. Staines Common. Road from Staines to Hampton. Bet. Sunbury and Walton Bridge. Harmondsworth. Abundant bet. Hampton Court and Kingston Bridge. Near Strawberry Hill.

III. Drilling Ground, Hounslow Heath, &c. Twickenham.

IV. North Heath, Hampstead.

V. Near Lampton.

VI. Roadside, Enfield Chase.

VII. Marylebone, 1817; *Herb. G. & R.* South Heath, Hampstead.

First record: *Blackstone,* 1737. Less common than *S. erucifolius.* Not included in the *Harrow Flora.*

372. S. aquaticus, *Huds.*

Jacobæa latifolia, Ger. em. (Blackst.).

Cyb. Br. ii. 116; iii. 461. Syme E. B. v. t. 756.

Sides of ponds, wet commons, and roadsides; common. P. July—September.

I. Harefield!; *Blackst. Fasc.* 45. Elstree.

II. Staines Moor, very abundant. Staines Common. Common by Walton Bridge. Bet. Hampton Court and Kingston Bridge. Teddington.

III. Hounslow Heath. Twickenham. Twickenham Park.

IV. Harrow district, common; *Melv.* 48. Near Finchley!; *Newb.* Stanmore. North Heath, Hampstead.

V. Occurs in this district; *Newb.*

VI. Edmonton. Colney Hatch.

VII. South Heath, Hampstead. Eel-brook Meadow. Bet. Lea Bridge and Bow; *Cherry* (*v. s.*).

First record; *Blackstone,* 1737.

CARLINA, *Linn.*

373. C. vulgaris, *L.*

, *C. sylvestris major, Ger. em.* (Blackst.).

Cyb. Br. ii. 87. Syme E. B. v. t. 698.

Dry banks ; very rare. B. July—September.

I. In the old chalk-pit at Harefield, plentifully ; *Blackst. Fasc.* 16.
   Ruislip Common, W. M. H.; *Melv.* 45.

VI. Finchley Common, 1845 ; *Herb. Hardw.*

First record: *Blackstone*, 1737. Has a strong attachment to chalk soils,
hence scarce in our county.

ARCTIUM, *Linn.*

374. A. majus, *Schk.*   *Great Burdock.*

Bardana major, Ger. (Blackst.).

Cyb. Br. Supp. 61. Syme E. B. v. t. 699.

Sides of streams, woods, and waste ground ; rare ? B. August.

I. Harefield ! ; *Blackst. Fasc.* 9.

II. Hampton ! ; *Newb.* Bet. Sunbury and Walton Bridge.

III. Harrow and Wood End ; *Hind.* Many places by the Cran near
    Twickenham.

V. By the Brent near Hanwell ! ; near Willesden Junction ! ; near Shepherd's Bush Ry. Station ! ; *Newb.*

First record: *Blackstone*, 1737. *Bardana rosea sive Lappa rosea, the
Rose Bur*, is a monstrous form of (probably) this species. VII. [On
the bankside bet. the Horseferry and the neat-houses near London ;
How, 14. About ye neat-houses in Chelsea, plentifully ; *Herbs. Budd.*
cxix. fol. 11, and *Pet.*] Bet. the second and third ponds in Cane
Wood ; *Herb. Pet.* cli. fol. 139.

375. A. minus, *Schk.* (Syme E. B.).   *Burdock.*

1. *A. nemorosum,* Lej.

A. intermedium (Bab. Man. v.). Cyb. Br. Supp. 61. Syme E. B. v. t. 701.

Woods ; very rare ?. B. July.

I. In a small copse north of Harefield towards Rickmansworth ! ; and in
   a copse above the chalk-pits, Harefield ! ; *Newb.*

We have seen no other specimens of this plant, which appears easily distinguishable from *intermedium* and *minus*.

2. *A. intermedium,* Lange.

*Lappa major capitulis minus tomentosis ; Bardana major lanuginosis
capitulis ; B. capitulis majoribus araneosis* (Pet.). *Lappa major capitulis majoribus parum tomentosis fere sphericis rufescentibus* (Budd.).
A. pubens (Bab. Man. v.). Cyb. Br. Supp. 61. Syme E. B. v. t. 700.

Roadsides, hedgebanks, and waste places; rather common?. B. July—September.

I. Harefield. On Burnt Oak Farm.
II. West Drayton! ; *Newb.*
III. Harrow ; *Herb. Harr.* Hanworth Road, near the Cran. Marsh Farm, Twickenham.
IV. Canon's Farm, Stanmore.
V. Apperton. Horsington Hill.
VII. [Above ye neat-houses in Chelsea ; *Budd. Herb., Pet. Herb., Pet. Midd.*] Hackney Wick.

Some of these stations may afford *A. minus.* Buddle (followed by Petiver), who studied this genus carefully, considered as distinct several plants, which from the specimens in his herbarium (vol. cxix. fol. 12), and in Petiver's (vol. cli. fol. 140, 141), seem referable to this sub-species, of which they may be varieties. Blackstone records under the name of *Bardana major altera,* Ger., a plant from Harefield, which is either this, or possibly *A. nemorosum,* Lej.

3. *A. minus,* Schk. (Bab.). *A. eu-minus* (Syme E. B.).
Lappa major ex omni parte vulgari minor, capitulis minime tomentosis (Budd.). *Bardana minor, Ger.* (Pet.). Cyb. Br. ii. 72 ('*A. Lappa*'). Curt. F. L. f. 4.

Roadsides and waste places ; rather common ?. B. July—September.
I. Harrow Weald, by the common. Harefield! ; *Newb.*
II. West Drayton! ; *Newb.* Hampton. Bet. Hampton Court and Kingston Bridge.
III. Roxeth ; *Hind.* Near Worton.
V. Lampton.
VI. Edmonton.
VII. Kentish Town, and elsewhere about London, plentifully ; *Herb. Budd.* Near the bridge at Newington; *Pet. Midd.* Victoria St., Westminster. Isle of Dogs.

First record : *Buddle,* 1710. Distribution not fully worked out.

SERRATULA, *Linn.*

376. S. tinctoria, *L.* *Saw-wort.*
Serratula (Ger., Blackst.).
Cyb. Br. ii. 74. Syme E. B. v. t. 704.

Woods and heaths ; rather rare. P. August, September.
I. Harefield ; *Blackst. Fasc.* 93. Elstree.
III. Roxeth ; Headstone, W. M. H.; *Melv.* 45. Drilling Ground, Hounslow.
IV. Hampstead Wood ; *Ger.* 577, and *Johns. Eric.* Bishop's Wood, abundant. Hampstead Heath! ; *Johns. Enum.* Kingsbury. Scratch Wood.

VII. [In great abundance in the wood adjoining to Islington; *Ger.* 577.]
[Marylebone Fields, 1815; *Herb. G. & R.*] Caen Wood; *Irv. MSS.*

First record: *Gerarde*, 1597. The *white-flowered* plant was noticed at
Hampstead by Johnson (*Eric.*) and Parkinson (*Theat.* 475); by Mr.
Irvine at Bollice Hill, Hendon; and by the Authors at Kingsbury.

CENTAUREA, *Linn.*

377. * C. Jacea, *L.*

Cyb. Br. ii. 88; iii. 456. Syme E. B. v. t. 705.

Meadows; very rare. P. August.

III. In considerable plenty in a grassy meadow close to the railway station
at Twickenham.

V. A stray plant or two at Acton, doubtless introduced; *Newb.*

First record: *Newbould*, about 1862. The Twickenham plant is
thoroughly established, and has the appearance of a native. It is,
however, most likely the remains of a garden, a supposition suggested
by the presence of horse-radish in the same field, and a garden rose
and periwinkles in a neighbouring hedge. Two forms grow together
here, the vars. α *genuina* and β *vulgaris* of Koch's *Synopsis Fl. Germ.*
(the former is *C. cuculligera*, Reich.). According to *Syme E. B. v.*
31, the Acton plant seems to belong to the continental form *C. amara*,
DC. (= *C. serotina*, Boreau).

378. C. nigra, *L.*   *Black Knapweed.*
Jacea nigra, Ger. em. (Blackst.).
Cyb. Br. ii. 89. Syme E. B. v. t. 706.

Fields, roadsides, and waste ground; very common. P. August, September.

In all the districts.

VII. Maiden Lane, St. Pancras, 1841; *Herb. Hardw.* Primrose Hill;
Morris (*v. s.*). Lower Heath, Hampstead. South Kensington.
Hackney Wick. Several patches in the turf of Lincoln's Inn
Fields, 1869.

First record: *Blackstone*, 1737. The '*rayed variety*' is recorded from
(III.) field behind the Grove, Harrow; *Melv.* 46. Another form with
long florets and small involucral scales, has been noticed sparingly
(VII.) near the Edgware Road Ry. Station (N. London), and elsewhere;
Newb. We have not seen the var. *decipiens*, Bab., in Middlesex.

379. C. Cyanus, *L.*   *Bluebottle.*
Cyanus vulgaris, Ger. em. (Blackst.).
Cyb. Br. ii. 89; iii. 457. Syme E. B. v. t. 709.

Waste ground and cornfields; rather rare. A. or B. July—September.
I. Harefield; *Blackst. Fasc.* 21. Near Pinner chalk-pits, W. M. H.;
Melv. 46.

II. Isolated plants by roadsides near Staines, Sunbury, Teddington, Strawberry Hill.

III. Waste ground by Twickenham Ry. Station.

IV. Stanmore; *Varenne.* Hampstead; *Irv. MSS.*

VI. By the Lea; *R. C. Powles (v. s.).* Starchness, Edmonton.

VII. Chelsea College, J. Britten; *Bot. Chron.* 58. South Heath, Hampstead. Cremorne. Kensington Gore, by the roadside.

First record: *Blackstone,* 1737.

380. C. Scabiosa, *L.*   *Great Knapweed. Matfellon.*

Jacea major, Ger. em. (Blackst.).

Cyb. Br. ii. 90. Syme E. B. v. t. 708.

Cornfields and borders of pastures; rare. P. July—September.

I. Harefield!; *Blackst. Fasc.* 45. Abundant in cornfields.

II. Stanwell Moor. Staines. Teddington. Tangley Park.

IV. Hampstead; *Irv. MSS.*

First record: *Blackstone,* 1737. Partial to dry soils, especially chalk.

[381. C. Calcitrapa, *L.*   *Star-Thistle.*

Carduus stellaris vulgaris (Park.). *C. stellatus, Ger.* (Merr., Pet., Blackst.

Cyb. Br. ii. 91; iii. 457. Syme E. B. v. t. 711.

Waste ground. A. or B. July—September.

I. Harefield; *Blackst. Fasc.* 15.

VII. In the fields about London in many places, as at Mile End Green; in Finsbury fields beyond the mills; *Park. Theat.* 990. Bet. London and Mile End, plentifully; *Merrett,* 21. In some barren fields near Whitechapel; *Pet. Midd.* Bethnal Green, Mr. Jones; *With.* ed. vii. iii. 961.

First record: *Parkinson,* 1640; last about 1830. May be refound.]

ONOPORDUM, *Linn.*

382. O. Acanthium, *L.*   *Cotton Thistle. White Thistle.*

Acantha leuke (Turner). *Acanthium* (Pet.). *Carduus tomentosus, Acanthium dict. vulgaris, R. Syn.* (Blackst.).

Cyb. Br. ii. 86; iii. 456. Curt. F. L. f. 6.

Waste places; rare. B. August, September.

I. Harefield; *Blackst. Fasc.* 14.

II. By the side of the Colne a little east of Colnbrook; *James Craig (v. s.).* Feltham Green, a single plant.

IV. Hampstead; *Irv. MSS.*

V. Sion; *Turn.* ii. 146.

VI. Bank of G. N. Ry. near Colney Hatch, one plant.

VII. [Most ditchbanks about town, very common; *Herb. Pet.*] [Gloucester Place, Kentish Town, 1844; *Herb. Hardw.*]

First record: *Turner,* 1562; also first as British. Often of garden origin.

CARDUUS, *Linn.*

383. C. nutans, *L.*

C. moschatus, Pet. (Blackst.).

Cyb. Br. ii. 74. Syme E. B. v. t. 683.

Fields and waste places; rather rare. B. June—August.

I. Harefield!; *Blackst. Fasc.* 15. Abundant by the canal. Bet. Yewsley and Iver Bridge!; *Newb.*

II. Laleham, 1815; *Herb. G. & R.* Staines Moor. Near Sipson. Bet.
Sunbury and Walton Bridge. Bet. Hampton Court and Kingston
Bridge. About Teddington in several places, abundant.

III. Hounslow Heath, abundant. Near Richmond Bridge, abundant.

IV. North Heath, Hampstead; *Lawson.*

VI. Edmonton.

First record: *Blackstone*, 1737.

384. C. crispus, *L.*

C. polyacanthos primus, Ger. em. (Blackst.). *C. polyacanthos* (Curt.).
C. acanthoides (Sm., With.).

Cyb. Br. ii. 75; iii. 454. Curt. F. L. f. 6 (drawn from a London specimen).

Hedgebanks and roadsides; rather rare. A. or B. June—August.

I. Harefield!; *Blackst. Fasc.* 15. Near Harrow Weald Common.

II. Bushey Park!; Hampton!; *Newb.* Towing-path bet. Hampton Court
and Kingston Bridge.

III. Hounslow Heath; *Cat. Lond.* 30.

VII. Very common in the environs of London; *Curt. F. L. and other
authors.* Kentish Town; *Irv. MSS.* Side of canal bet. Lea Bridge
and Bow; *Cherry (v. s.).* Isle of Dogs!; *Newb.*

First record: *Blackstone*, 1737. Curtis's plate well represents the usual
Middlesex form, which is the var. *litigiosus*, Gr. and Godr. (= *C. acanthoides* of *Koch's Synopsis*, but scarcely of *Bab. Man.*).

385. C. tenuiflorus, *Curt.*

C. vulgatiss. incanus, alato caule, confertis capit. parvis (Budd.).

Cyb. Br. ii. 76; iii. 454. Curt. F. L. f. 6 (drawn from a London plant).

Waste places; very rare. A. or B. June—September.

VII. Frequent about London; *Budd. MSS.* Growing in the very suburbs;
Curt. F. L. [Bethnal Green, L. W. Dillwyn; *B. G.* 410.] [Near
Canonbury House, Islington, 1841; *Herb. Hardw.*] Chelsea College;
Britten. Side of canal bet. Lea Bridge and Bow; *Cherry (v. s.).*
Green Lanes, Newington!; *Newb.* Side of Hackney Canal. Shep-
herd's Bush, abundant. North part of Holland Park, Kensington,
abundant.

First record: *Buddle*, about 1710. Seems confined to the metropolitan
district. Possibly an introduction.

386. C. lanceolatus, *L.*   *Spear Thistle.*

Cyb. Br. ii. 78.   Syme E. B. v. t. 686.

Waste places, fields, and roadsides; very common. A. or B.   July—
September.

In all the districts.

VII. Kensington Gardens, 1868 ; *Warren.* Green Park. Hyde Park.
Grosvenor Square. ˙Russell Square.   Victoria Street, Westminster,
and other places.

First record : *Merrett,* 1666.   Merrett (*Pin.* 21) noticed a variety, *C. lan-
ceatus, flore et capite minoribus* [on the ditch sides beyond St. James's].

387. C. arvensis, *Curt.*   *Common Creeping Thistle.*

C. ceanothos sive viarum repens (Park.).   *Cnicus arv., Hoffm.* (Melvill).

Cyb. Br. ii. 80.   Curt. F. L. f. 6.

Fields, gardens, and waste ground; very common. P. July—September.

In all the districts.

VII. Very common in waste places. On Kentish Town Green, abundantly ;
Park. Theat. 958. Kensington Gardens. Regent's Park. Green Park.
Gordon Square. Guildford Street. Hackney Wick. I. of Dogs, &c.

Var. β setosus.   Syme E. B. v. t. 694.

VII. Site of Exhibition of 1862 in some plenty. Plants with very nearly
the leaves of *setosus* were also noticed (III.) by the roadside bet.
Hounslow and Hanworth.

First record : *Parkinson,* 1640.

388. C. palustris, *L.*

Cnicus pal. Willd. (Melvill).

Cyb. Br. ii. 79 ; iii. 455.   Curt. F. L. f. 6.

Wet meadows, bogs and ditches; very common. A. or B. July—Sept.

In all the districts.

VII. South Heath, Hampstead.

First record : *Petiver,* 1713.

389. C. pratensis, *Huds.*

Cirsium anglicum and *C. montanum anglicum* (Johns., Blackst.).

Cyb. Br. ii. 82 ; iii. 455.   Syme E. B. v. t. 690.

Wet heaths and meadows; very rare. P. June—August.

I. [Stanmore Heath ; *Fl. Herts Supp.* 12.]   Ruislip ; *Hind.* Eastcott ;
Farrar.

III. [On Hounslow Heath near the end of Tuddington (= Teddington) town,
sparingly ; *Blackst. Spec.* 14.]

IV. [Hampstead Heath ; *Johns. Enum.*]

VII. [In a meadow on this side Highgate ; *Johns. Ger.* 1184.]

First record : *Johnson,* 1632 ; also the first notice as a British plant.
We have not noticed this species in Middlesex.

390. C. acaulis, *L.*

Cyb. Br. ii. 84; iii. 456. Syme E. B. v. tt. 692 and 692 *bis.*

Dry fields; rare. P. July—September.

I. Ruislip Common; *Melv.* 46. Field above the old chalkpit, Harefield, abundant.

II. On a bank by the Thames side opp. Thames Ditton Ferry.

III. Field by Hospital Bridge over the Cran, a few plants.

First record: *Hind*, 1861. The Harefield specimens grow amongst long grass, and have all a more or less evident stem.

(C. heterophyllus, *L.* Cyb. Br. ii. 85. Syme E. B. v. t. 691. *C. helenioides* (Mart.). *Cirsium Britannicum Clusii repens,* J. B.; (III.) on Hounslow Heath, Mr. Doody; *Dill. in Ray Syn.* iii. 193. Repeated by subsequent authors. As Mr. Watson (*Cyb. Br.* ii. 85) suggests, there can be little doubt that *C. pratensis* was mistaken for this.)

SILYBUM, *Gaert.*

391. S. Marianum, *Gaert.*

Carduus Mariæ, Ger. em. (Blackst.). *Carduus Marianus,* L. (Irv.).

Cyb. Br. ii. 77; iii. 454. Syme E. B. v. t. 681.

Waste ground, roadsides, and walls; rare. B. June—September.

I. Pinner, 1865; *Hind.* Harefield; *Blackst. Fasc.* 15.

III. Roxeth, not wild, W. M. H.; *Melv.* 45.

IV. Rubbish, Hampstead; *Irv. MSS.*

V. Several plants on walls at Turnham Green.

VII. [Maiden Lane, Islington; by the lodge, Tufnell Park, 1844; *Herb. Hardw.*] [Banks bet. London and Hackney, E. Forster; *Herb. Mus. Brit.*] Lincoln's Inn Fields, originally planted.

First record: *How*, 1650. A state of this with leaves without the white markings was distinguished by the older botanists, and seems to have been common about town. *Carduus Mariæ hirsutus non maculatus,* near London; *How*, 22. [In a ditch neer my Lord Southampton's house and about Islington; *Merrett*, 21.] [Shown me by Mr. George Horsnell near Clerkenwell, London; *Ray Cat.* i. 56.] [Near Clerkenwell, Mr. H. R. Bloss; *Herb. Budd.*] [Banks of New River; *Pet. Midd.*] [On a bank just beyond Shoreditch, towards Hackney, J. Sherard; *Ray Syn.* iii. 196.] [Bet. London Spaw and New River Head, Dr. Watson; *Hill*, 412.] [Marybone Park; *Mart. App. P. C.* 72.] [Brickfield near the 'Cat and Shoulder of Mutton,' Hackney; *Forst. Midd.*]

LAPSANA, *Linn.*

392. L. communis, *L.* *Nipplewort.*

Lampsana, Ger. em. (Blackst.).

Cyb. Br. ii. 70. Curt. F. L. f. 1.

Waste ground and hedgebanks; very common. A. June—August.
In all the districts.

VII. Chelsea. Parson's Green. Kentish Town. Thames Embankment.
Hackney Wick. Isle of Dogs.

First record: *Blackstone*, 1737.

[ARNOSERIS, *Gaert.*

393. A. pusilla, *Gaert.*

Hyoseris, Tab. (Ray). *Hieracium minimum Clusii* (Dill.). *Hieracium minus fol. sub rotundo, C. B. P.* (Blackst.). *Hyoseris minima, L*. (Huds.).

Cyb. Br. ii. 69; iii. 454. Syme E. B. v. t. 788.

Sandy fields; very rare. A. July—September.

II. In Hamton Court Park abundantly, and in the fields thereabouts,
Sam. Doody; *R. Syn.* ii. 344. In Teddington field, plentifully;
Blackst. Fasc. 42, and *Huds.* ii. 346.

First record: *Doody*, 1696; last, *Hudson*, 1778. This may be refound.]

CICHORIUM, *Linn.*

394. C. Intybus, *L.*   *Succory. Chicory.*
Cyb. Br. ii. 71. Curt. F. L. f. 4.

Roadsides, cornfields, and waste ground; rare. P. July—September.

I. Abundant by side of canal and in cornfields, Harefield.

II. Laleham; *Herb. G. & R.* About Teddington!, G. Francis; *Coop.
Supp.* 12. About Hampton!; *Newb.* Staines Moor. Bet. Sunbury
and Walton Bridge, abundant.

III. In a meadow near Richmond Bridge.

VII. Site of Exhibition of 1862, South Kensington.

First record: *Goodger* and *Rozea*, 1815. The cultivated variety (or
species) is probably the plant noticed in many places; certainly at
South Kensington.

HYPOCHÆRIS, *Linn.*

395. H. radicata, *L.*
Cyb. Br. ii. 41. Syme E. B. v. t. 790.

Fields, pastures, and waste land; very common. P. or B.? July—
September.

In all the districts, generally abundant.

VII. Hyde Park!; *Newb.* South Heath, Hampstead. Site of Exhibition,
South Kensington.

First record: *Irvine*, 1830.

396. T. hirta, *Dec.* Leontodon hirtus, *L.* (Syme E. B. and L. Cat.).
Hieracium pumilum sax. asp. præmorsa radice and *Dens Leonis minimus
asper* (Blackst.). *Leontodon hirtus, L.* (Curt.).
Cyb. Br. ii. 37. Curt. F. L. f. 6 (probably drawn from a Hampstead
specimen).

Heaths, commons, and fields; rather common. P. or B. June—
September.
   I. In the gardens at Breakspears, abundantly; *Blackst. Fasc.* 23. In
Harefield chalkpit; *Blackst. MSS.* Harrow Weald Common.
   II. Near Teddington.
   III. Drilling Ground, Hounslow!; *Newb.* Teddington Park. Near
Strawberry Hill.
   IV. Hampstead Heath!; *Curt. F. L.* Stanmore; *Varenne.* Very
common in the Harrow district; *Melv.* 42. Mill Hill; *Herb. Mus.
Brit.* Near Brent Reservoir; *Farrar.*
   V. Acton Green; *Newb.*
   VI. Enfield Chase pastures.
   VII. Near Cremorne, very large plants; *Fox (v.s.).* Side of canal,
Homerton; *Cherry (v.s.).* South Heath, Hampstead.

First record: *Blackstone,* ' 1736.'

APARGIA, *Schreb.*

397. A. hispida, *Willd.* Leontodon h., *L.* (Syme E. B. and L. Cat.).
Dens Leonis hirsutus leptocaulos, R. Syn. (Blackst.). *Leontodon h., L.*
(Curt.).
Cyb. Br. ii. 38; iii. 451. Curt. F. L. f. 5.

Meadows and pastures; rather common. P. June—September.
   I. Harefield!; *Blackst. Fasc.* 22. Pinner!, W. M. H.; *Melv.* 42.
Elstree.
   II. Staines. Hampton. Tangley Park. Teddington.
   III. Twickenham.
   IV. Hampstead; *Irv. MSS.* Stanmore; *Varenne.* Near Finchley!;
Newb.
   V. Apperton, W. M. H.; *Melv.* 42. Hanwell!; *Warren.* Horsington Hill.
   VI. Highgate; Finchley; *Herb. G. & R.* Edmonton. Enfield Chase.
   VII. Grosvenor Square, 1869!; *Warren.* Site of Exhibition, South
Kensington.

First record: *Blackstone,* 1737.

398. A. autumnalis, *Willd.* *Leontodon a., L.* (Syme E. B. and
L. Cat.).
Cyb. Br. ii. 39. Syme E. B. v. t. 794.

Meadows, heaths, and waste ground; rather common. P. August, September.

I. Near Harefield!; *Newb.*

II. Bushey Park!; *Newb.*

III. Harrow; *Melv.* 42. Drilling Ground, Hounslow!; *Newb.* Twickenham Park. Roadside near Whitton.

IV. Stanmore; *Varenne.* Hampstead Heath!; *Newb.*

V. Near Ealing!; *Newb.*

VI. Hadley!; *Warren.* Colney Hatch. Edmonton.

VII. Isle of Dogs!; *Newb.* South Heath, Hampstead. Turf of Lincoln's Inn Fields.

First record: *Petiver*, 1713. The variable leaves of this plant afforded opportunity to Petiver to distinguish several forms: *Hieracium præmorsum laciniatum, H. fol. acuto minus,* and *H. fol. obtuso minus;* all found about London. They are figured in *Pet. H. B. Cat.* t. xii.

TRAGOPOGON, *Linn.*

399. T. pratensis, *L.*   *Goat's-beard.*

Cyb. Br. ii. 33.

Meadows, waste ground, railway banks, &c.; rather common. B. May —July.

Var. a genuinus, Syme. *T. pratensis, L.* (Bab.). Syme E. B. v. t. 798.

II. Bet. Teddington and Strawberry Hill. River-side near the railway bridge, Twickenham.

V. Near Chiswick Ry. Station.

VI. Bank of G. N. Ry., Colney Hatch.

VII. [Banks of moat under Penitentiary, Millbank, 1836; *Herb. Young.*] Banks and side of line of G. N. Ry., about Copenhagen Fields Station, abundant.

Var. β. T. minor, Fries. *Bukkes bearde* (Turn.). *T. luteum* (Ger.). *T. lut. angustifolium* (Park.). *T. pratensis, L.* (Melv.). Syme E. B. v. t. 799.

I. Harefield!; *Blackst. Fasc.* 100.

III. Wood End; *Melv.* 41. Hounslow!; *Newb.*

IV. Hampstead; *Irv. MSS.* Football field, Harrow; *Melv.* 41. Railway bank; *Herb. Harr.* Midland Ry. bank near Edgware.

V. Greenford; *Melv.* 41. Brentford; *Hemsley.* Twyford!; Ealing!; near Chiswick!; *Newb.* Hanwell; *Warren.*

VI. Near Enfield; *Tucker.*

VII. About London; *Turn. Names.* Islington; *Ger.* 596. Pancras; *Newton MSS.* Paddington; *Budd. Herb.* and *Pet. H. B. Cat.* xv. 6. Brompton Cemetery; *Britten.* Side of London Canal; *Cherry* (*v. s.*). King's Road, Chelsea, abundant.

First record : *Turner*, 1548; also first as a British plant. *Var. β* is the common Middlesex plant, and when the variety was not stated, we referred the locality to it. Parkinson, Blackstone, and other old writers, distinguish the plants.

400. *T. porrifolius, *L.* **Salsify.**

T. purpuro-cæruleum Porri folio quod Artifi vulgo, C. B. P. (Blackst.).
Cyb. Br. ii. 34. Syme E. B. v. t. 801.

Meadows, waste places, and railway banks. B. June.

II. Several hundred plants on the embankment by the railway station at West Drayton ; *Cole.*

III. Several spots in neighbourhood of Greenhill and Roxeth, W. M. H. ; *Melv.* 42.

VI. In some meadows near Edmonton, Mr. Hurlock ; *Blackst. Spec.* 98. Gt. Northern Ry. bank (east side) at Barnet ; *Naylor.*

VII. Bromley near Bow ; *Forst. Midd.* Waste ground beyond World's End, bet. King's Road and Walham Green, 1830 ; *Pamplin.* Banks of a creek near Hammersmith, D. Cooper ; *Herb. Young.*

First record: *Hurlock*, 1746. The specimens we have seen are *var. a sativus* of *Syme E. B.* Probably often an outcast from gardens.

PICRIS, *Linn.*

401. P. hieracioides, *L.*

Hieracium asperum, Ger. em. (Blackst.).
Cyb. Br. ii. 35. Syme E. B. v. t. 796.

Dry fields and hedgebanks ; rare. B. or P. July, August.

I. In Harefield chalkpit ; *Blackst. Fasc.* 41. Near Harefield Common.

II. Plentiful in a field near Teddington.

IV. Hampstead Heath ; *Britten.*

First record : *Blackstone*, 1737.

HELMINTHIA, *Juss.*

402. H. echioides, *Gaert.* **Ox-tongue.**

Hieracium echioides capit. cardui benedicti, C. B. P. (Blackst.). *Picris e., L.* (Curt.).
Cyb. Br. ii. 35. Curt. F. L. f. 3.

Margins of fields, waste places, and roadsides ; rather common. A. or B. July—September.

I. Harefield, frequent ; *Blackst. Fasc.* 41. Wood Hall, Pinner.

III. Green Hill ; *Melv.* 42.

IV. Kenton ; *Melv.* 42. Edgware. Bet. Sudbury Station and Apperton.

V. Twyford! ; near Shepherd's Bush! ; *Newb.* Horsington Hill. Abundant on railway bank bet. Acton and Willesden Junction.

VI. Bet. Finchley Common and the Church ; *Herb. Hardw.* Edmonton.

VII. Kilburn Lane; *Herb. G. & R.* Bet. Kilburn and Kensal Green; *Herb. Hardw.* By creek near Brompton Cemetery; *Britten.*

First record: *Blackstone,* 1737.

LACTUCA, *Linn.*

[**403. L. saligna,** *L.*
Chondrilla viscosa humilis, C. B. and *Lactuca sylvestris minima* (Ray).
Cyb. Br. ii. 43. Syme E. B. v. t. 807.
Waste ground. A. or B. July—September.
VII. Tho. Willisel hath found it about Pancras Church near London; *R. Cat.* ii. 69 and *Pet. Midd.* By the path on the left hand in the close next this side Pancras, and by the roadside in the same close; *Newton MSS.* Banks of field bet. Pancras and Hampstead; *Forst. Midd.* Behind the Small-pox Hospital, Sir J. Banks; *E. Fl.* iii. 347.

First record: *Willisel,* 1677; not seen since 1820. *L. sylvestris angusto laciniato fol., Magnol.,* on the banks of some fields bet. Pancras and Hampstead; *Blackst. Spec.* 42, is probably the pinnatifid-leaved form of this species, and not *L. Scariola.*]

404. L. virosa, *L.*   *Wild Lettuce.*
L. agrestis odore opii, Lob. (Johns.). *L. sylvestris major od. opii, Ger. em.* (Merr., Blackst.).
Cyb. Br. ii. 41. Syme E. B. v. t. 805.
Hedgebanks and roadsides; rather rare. A. or B. August, September.
I. Harefield; *Blackst. Fasc.* 48. Eastcott; *Melv.* 43. Several places bet. Harefield and Ruislip. Bet. Wood Hall and Wood Ridings. Moss Lane, Pinner.
III. Hatch End.
IV. Abundant along Pinner and Greenford Roads; *Melv.* 43. Hampstead Heath; *Johns. Enum.* and *Irv. MSS.*
V. Acton, Ealing, Brentford, &c., Dr. Goodenough; *Fl. Brit.* 819. Road from Willesden Junction to the canal, abundant!; railway bank bet. Acton and Willesden; *Newb.* Bet. Perivale and Horsington Hill; *Lees.* Near the Asylum at Hanwell.
VII. About London, *Merrett,* 68; *Budd. MSS.* [Bet. Kinsinton (= Kensington) & Knights bridg by ye highway side; *Merr. MSS.*] [Marylebone fields; *Herb. G. & R.*] [World's End, near Stepney; banks of Thames, bet. Blackwall and Woolwich, Mr. Jones; *With.* iii. 677.] Gospel Oak Fields, Kentish Town; *Herb. Hardw.* [About Kilburn; in Maiden Lane, near Copenhagen House; *Burnett* i. 12.]

First record: *Johnson,* 1632.

405. L. Scariola, *L.*   *Prickly Lettuce.*
L. sylvestris prior, Trag. and *L. sylv. fol. dissectis* (Johns.). *L. sylv. laciniata* (Park.). *L. sylv. costa spinosa, C. B. P.* (Blackst.).
Cyb. Br. ii. 42; iii. 452. Syme E. B. v. t. 806.
Waste ground; rare. A. or B. August, September.

I. Harefield; *Blackst. Fasc.* 48.

IV. [Hampstead Heath; *Johns. Enum.*] Near Harrow; *Hemsley.*

VII. [Bet. London and Pancridge (= Pancras) Church, about the ditches and highway side; *Johns. Ger.* 310.] [On a high bank by the foot way going down Gray's Inn Lane unto Bradford Bridge; *Park. Theat.* 814.] [Banks near Pancras, E. Forster; *B. G.* 410.] [Bet. Hampstead and Kentish Town; *Irv. MSS.*] [Near Islington; *Macreight,* 142.] [World's End, near Stepney; *Mart. Mill. Dict.*]

First record: *Johnson,* 1632.

406. L. muralis, *DC.*

Ivy Lettuce (Pet.). *Sonchus lævis muralis,* Ger. em. (Blackst.). *Prenanthes mur.,* L. (Curt.).

Cyb. Br. ii. 43; iii. 452. Curt. F. L. f. 5 (drawn from a Hampstead specimen).

Old walls and hedgebanks; rather rare. A. or B. July—October.

I. In the lane from Harefield to Ruislip, abundantly; *Blackst. Fasc.* 95. Pinner Hill; Headstone and Payne's Lanes, Pinner, W. M. H.; *Melv.* 43. Wall in Harefield village. On Ruislip Church.

IV. Hampstead; *Pet. H. B. Cat.* Outside of the pales which terminate the terrace at the 'Spaniards,' Hampstead Heath; *Curt. F. L.* On a wall by the 'Spaniards,' and on the roof of the tap-room on the opposite side of the road, abundant.

VI. Near Hornsey; *Huds.* i. 296.

First record: *Petiver,* 1713.

LEONTODON, *Linn.*

407. L. Taraxacum, *L.* Taraxacum officinale, *Wiggers* (Syme E. B. and L. Cat.). *Dandelion.*

Dens Leonis, Ger. em. (Blackst.).

Cyb. Br. ii. 68. Syme E. B. t. 802.

Meadows, waste places and roadsides; very common. P. March—October. In all the districts.

VII. Common in the parks, squares, and gardens of the metropolis.

Var. β. T. erythrospermum DC. ? *Narrow-leaved Dandelion* (Pet.) Syme E. B. v. t. 803.

Old walls, dry waste places and roadsides. P. May—July. Common throughout the county.

VII. Islington Churchyard; *Pet. Bot. Lond.* Isle of Dogs!; *Newb.* Grosvenor Square!; *Warren.*

First record: *Petiver,* 1709.

SONCHUS, *Linn.*

408. S. oleraceus, *L.* *Smooth Sowthistle.*

S. lævis (Budd.).

Cyb. Br. ii. 46. Curt. F. L. f. 2.

Fields, gardens and waste places; very common. A. May—August.
In all the districts.

VII. In Gray's Inn garden; *Budd. MSS.* Kensington Gardens. Kensington Gore. Victoria Street, Westminster. Isle of Dogs, &c.

First record: *Buddle*, about 1710. *S. lævis laceratus*, ' Buddle's Jagged Sowthistle,' is a variety. Figured in *Pet. H. B. Cat.* xiv. 10.

409. S. asper, *Hoffm.* *Rough Sowthistle.*
Cyb. Br. ii. 45. Syme E. B. v. tt. 811, 812.

Cultivated and waste ground; very common. A. May—August.
In all the districts.

VII. Shepherd's Bush. Isle of Dogs.

First record: *Petiver*, 1713. *Sonchus asper dentatus, Pet. H. B. Cat.* xiv. 4, is a slight variety. *S. subrotundo fol. nost. læviss. spin. &c., Pluk.*, *Pet. H. B. Cat.* xiv. 1, is the round-leaved form in a young state.

410. S. arvensis, *L.*
S. arborescens, Ger. em. (Blackst.).
Cyb. Br. ii. 45. Curt. F. L. f. 4.

Cornfields, cultivated ground, and waste places; rather common. P. July—September.

I. Harefield; *Blackst. Spec.* 95. Near South Mims.
II. Stanwell Moor. Harmondsworth. Teddington. Hampton Court.
III. Twickenham.
IV. Kenton; Pinner Drive; *Melv.* 43. Harrow Weald Churchyard. Hampstead.
V. Apperton. Wyke Green.
VI. Edmonton.
VII. [Marylebone Fields; *Herb. G. & R.*] Near Upper Clapton; *Cherry* (*v. s.*). Isle of Dogs!; *Newb.*

First record: *Blackstone*, 1737.

[411. S. palustris, *L.*
S. arborescens alter, Ger. em. and *S. tricubitalis fol. cuspidato, Merr.* (Ray). *Sonchus asper arborescens, C. B.* (Budd.).
Cyb. Br. ii. 44; iii. 452. Curt. F. L. f. 5 (drawn from a Middlesex specimen).

River sides; very rare. P. July, August.

VII. About Blackwall; *R. Syn.* ii. 71. On the banks of the Thames in the Isle of Dogs; *Budd. MSS.* and *Pet. H. B. Cat.* Sparingly in the marshes about Blackwall and Poplar; *Curt. F. L.* In the Isle of Dogs, E. Forster; *B. G.* 410. Mr. Syme could not find it there in 1852 (see *Phyt.* iv. 860).

First record: *Ray*, 1696; last, 1805. The progress of London has destroyed
the localities for this fine species in Middlesex, though it still grows
further down the river on the Kentish shore, at Plumstead Butts.
(Two other localities are on record. IV. Hampstead Heath ; *Cooper*,
100. V. Among reeds in the claypits of the large brickfields at
Shepherd's Bush, 1835, Pamplin; *New B. G. Supp.*   *S. arvensis* was
in these cases probably mistaken for *S. palustris.*)]

CREPIS, *Linn.*

(C. fœtida, *L.*   *Hieracium castorei odore Monspeliensium*, *Ray* (Pet.).
Cyb. Br. ii. 66. Syme E. B. v. t. 815. VII. About Chelsea, Mr. Doody ;
Pt. Midd.   Possibly an accidental straggler, but probably an error of
Doody or Petiver, some other composite plant being intended.)

412. C. virens, *L.*
C. tectorum, L. (Curt.).
Cyb. Br. ii. 47.   Curt. F. L. f. 5.

Dry banks, roadsides, fields, &c.; common. A. or B. June—September.

I. Harefield, abundant. Near Yewsley! ; *Newb.* South Mims. Ruislip ;
Melv. 43.
II. Staines. Hampton.
III. Roxborough ; *Melv.* 43. Hounslow. Twickenham.
IV. Hampstead Heath ; *Irv. MSS.* Pinner Drive ; *Melv.* 43.
V. Hanwell! ; *Warren.* Kew Bridge Ry. Station. Chiswick.
VI. Wood Green, G. Munby ; *Nat.* 1867, 181. Edmonton.
VII. Churchyard of St. Clement Danes, Strand. Adelaide Road. Hackney
Wick. Isle of Dogs.

First record: *Petiver*, 1713. Curtis's figure represents the common large
form (*C. agrestis*, Kit.). The very variable leaves and amount of
luxuriance gave characters to several species of the old botanists.
Petiver figures (*H. B. Cat.* xii. 5 & 7) *Hieracium lactucæ folio* and
H. aphacoides acutum, both found about London.

C. biennis, *L.* Cyb. Br. ii. 48 ; iii. 452. Syme E. B. v. t. 819. V. A
single plant on waste ground near Chiswick Ry. Station. A casual
introduction.

HIERACIUM, *Linn.*
413. H. Pilosella, *L.*
Pilosella minima (Park.).   *P. repens, Ger. em.* (Blackst.).   *P. longifolia*
(Pet.).
Cyb. Br. ii. 51. Syme E. B. v. t. 822.

Heaths, dry banks, and pastures; common. P. May—August.

I. Harefield!; *Blackst. Fasc.* 76. Harrow Weald Common; *Melv.* 44. Stanmore Heath.

II. Towing-path bet. Hampton Court and Kingston Bridge. Hanworth Park.

III. Hounslow Heath. Drilling Ground, &c.

IV. Hampsteed Heath!; *Park. Theat.* 692. Railway Bank, Harrow; *Melv.* 44. Bet. Highgate and the 'Spaniards,' Rob. Walford; *Herb. Pet.*

V. Horsington Hill.

VI. Colney Hatch. Edmonton. Near Enfield; *Tucker.*

VII. Near Bayswater, 1817; *Herb. G. & R.*

First record: *Parkinson,* 1640.

H. aurantiacum, *L.* Cyb. Br. ii. 52. Syme E. B. v. t. 823. I. A single plant, in appearance perfectly wild, under furze on Stanmore Heath, 1866. III. Harrow; *Herb. Harr.* Outcast from cottage gardens; *Melv.* 44. Originally from gardens, no doubt.

414. H. murorum, *Fries.*

Pulmonaria gallorum Hieracii fl. (Johns.). *H. latifol. hirsut. fol. unico cauli insidente, Bob.*; *H. murorum fol. pilosissimo, C. B.* (Budd., Blackst.). *Pulm. gallica sive aurea* (Pet.).

Cyb. Br. ii. 55; iii. 453. Syme E. B. v. t. 846.

Walls and dry places; rare. P. May—August.

I. On a wall against Hillingdon Church; *Blackst. Fasc.* 42.

IV. Highgate Wood; *Johns. Eric.* On Hampstead Heath and woods adjacent, plentifully; *Budd. MSS.* and *Herb.* cxviii. fol. 16.

VI. At Dr. Uvedale's, Enfield; *Newton MSS.* Still grows on the wall of the old palace at Enfield where Uvedale lived.

First record: *Johnson,* 1829.

415. *H. maculatum,* *Sm.* (Syme E. B.).

Syme E. B. v. t. 849.

Walls; very rare. P. June—August.

III. On a high wall at Twickenham. First noticed in 1866.

Perhaps planted there. ·Mr. Syme considers the species 'very doubtfully native' in England.

416. H. vulgatum, *Fr.*

Pulmonaria gallica s. aurea latifol. and *P. g. angustifol.* (Merr.). *P. gallica fœmina* (Pet.). *H. murorum laciniatum minus pilosum, C. B.* (Budd.†).

Cyb. Br. ii. 57; iii. 453. Syme E. B. v. t. 850.

† Buddle says, 'Authors confound us much in this plant,' and indeed the old synonymy of this and 414 and 419 is much involved.

N

Heaths and hedgebanks; rather rare. P. July—September.

I. Harrow Weald Common!; Ruislip Wood; *Melv.* 44. Harefield Common. Stanmore Heath. Bentley Priory.

III. Road from the Drilling Ground to the Cemetery, plentiful. Hanworth Road; near the Cran.

IV. A great variety at Hampstead!; *Budd. MSS.* and *Herb.* cxviii., fol. 16. North Heath. Deacon's Hill.

VI. Wall in Southgate Lane, 1850 (*var. with spotted leaves*); *Herb. Hardw.* Colney Hatch. Enfield. Edmonton.

VII. Meadows on this side Hampstead; *Merrett*, 99. South Heath, Hampstead.

First record: *Merrett*, 1666.

417. H. tridentatum, *Fr.*

Cyb. Br. ii. 64. Syme E. B. v. t. 852.

Hedgebanks; very rare. P. July.

III. On a bank on south side of the road leading towards Whitton Park from Twickenham, just above the junction of the loop-line, 1867.

First record: *Buddle*, about 1705?. A specimen labelled *H. fruticosum latifol. glabrum, Park.* in *Budd. Herb.* cxviii., fol. 15, found at Hampstead, is perhaps this species.

418. H. umbellatum, *L.*

H. fruticosum angustiore fol. totum glabrum (Budd.). *H. frut. angustifol. majus* (Park., Pet.).

Cyb. Br. ii. 65; iii. 454.

Heaths and open woods; rare. P. August, September.

I. Woods near Pinner Lane; *Fl. Herts. Supp.* 13. Harefield Common, abundantly!; *Newb.*

III. Drilling Ground, Hounslow, and roads near it.

IV. Woods about Highgate and Hampstead; *Pet. Midd., R. Syn.* iii. 168, *Curt. F. L., Irv. MSS.*

VII. In the way to Hampstead Heath; *Park. Threat.* 802. Cane Wood, *Herb. Budd.*

First record: *Parkinson*, 1640. *Pulmonaria angustifol. glabra*, Hamstead Woods; *Pet. H. B. Cat.* xiii. 11, and *P. graminea*, about London; xiii. 12, are slight varieties.

419. H. boreale, *Fr.*

H. fruticosum latifol. hirsutum (Park., Ray, Pet., Blackst.).

Cyb. Br. ii. 64. Syme E. B. v. t. 854.

Heaths, hedgebanks, and woods; rather rare. P. August, September.

I. In the Old Park and other woods, Harefield; *Blackst. Fasc.* 42. Harrow Weald Common, W. M. H.; *Melv.* 44. Stanmore Heath.

III. Drilling Ground, Hounslow.

IV. Hampstead and Highgate; *R. Cat.* i. 165, and *Pet. Midd.* Bishop's

Wood, plentifully; *Herb. Budd.* North Heath, Hampstead. Scratch Wood near Edgware.

V. In Horsington Wood; *Lees.*

VI. Colney Hatch!; *Newb.* Winchmore Hill Wood.

VII. Cane Wood, plentifully; *Herb. Budd.* In the way to Hampstead Heath; *Park. Theat.* 802.

First record: *Parkinson*, 1640. Petiver's *Long hairy Hawklung*, Hamsted Woods; *Pet. H. B.* xiii. 8, seems a narrow-leaved variety. (See also *Budd. Herb.* cxviii. 15.)

[XANTHIUM, *Linn.*

420. *X. Strumarium, L.*

Bardana minor, Ger. (Blackst., Wilson).

Cyb. Br. ii. 135. Syme E. B. v. t. 860.

Waste places. A. August, September.

II. In the highway leading from Drayton to Iver, two miles from Colbrooke; *Ger.* 665. At Stanes (= Staines), Mr. Lawson; *Wilson*, 16.

VII. Near the footway in the field from Goswell Street to the New River, Islington; *Blackst. Spec.* 6. Mr. Woods has sought it here in vain; *B. G.* 411.

First record: *Gerarde*, 1597; last, *Blackstone*, 1746. Barely naturalised.]

CAMPANULACEÆ.

JASIONE, *Linn.*

421. *J. montana, L.*   *Sheep's-bit.*

Rapunculus Scabiosæ capit. cæruleo (Merr.).

Cyb. Br. ii. 143; iii. 465. Curt. F. L. f. 4.

Heaths and dry fields; rare. A or B. July, August.

III. Hounslow Heath, Drilling Ground, and other places near, abundant in 1866.

IV. Hampstead; *Pet. H. B. Cat., Curt. F. L.,* and *Irv. MSS.* Wood at Stanmore, 1815; *Herb. G. & R.*

VII. [In Hackney Common-field; *Forst. Midd.*]

First record: *Merrett*, 1666. *Fl. albo*; On Hamstead Heath; *Merr. MSS.*

(Phyteuma orbiculare, *L.* Cyb. Br. ii. 143; iii. 465. Syme E. B. vi. t. 864. In the hedges of a lane betwixt Kingsbury and Harrow; and in the meadows between Harrow and Pinner; *M. & G.* 331. Perhaps *Jasione* was intended.)

CAMPANULA, *Linn.*

422. *C. glomerata, L,*

Trachelium minus (Ger.).

Cyb. Br. ii. 140. Syme E. B. vi. t. 866.

Meadows; very rare. P. July—September.
II. Field at Laleham, 1815; 'grass mowed, dwarfish specimen;' *Herb. G. & R.* By the towing-path bet. Kingston Bridge and Hampton Court!; *Warren.* In a field near Teddington Church.

V. In Sion medow near unto Branford (=Brentford); *Ger.* 365.

First record: *Gerarde,* 1597. These localities are by the river, and the plant has probably in each case been imported from the chalk districts higher up its course.

423. ? C. latifolia, *L.*
Cyb. Br. ii. 137. Syme E. B. vi. t. 868.

Woods; very rare. P. June.
I. North Mims Wood, in a part flanking an old chalk-pit, very fine, plentifully, June 1856; *Phyt. N.S.* i. 407.
Perhaps this station is in Herts, and the same as Lord Cecil's 'chalkpit near Warren Gate, plentiful.' (See *Herts Flora,* 177.)

424. C. Trachelium, *L.*
Trachelium majus, Ger. em. (Blackst.).
Cyb. Br. ii. 139. Syme E. B. vi. t. 867.

Woods and thickets; very rare. P. July, August.
I. In the Old Park Wood, abundantly; *Blackst. Fasc.* 99. Near the mills, Harefield, 1844; *Herb. Young.* A little north of Harefield; *Newb.* In the old chalk-pits, 1868.
V. Borders of Horsenton Wood; *Lees.*

First record: *Blackstone,* 1737. Almost confined to the chalk.

[C. rapunculoides, *L.* Cyb. Br. ii. 138; iii. 464. Syme E. B. vi. t. 869. '*C. urticæ fol. oblongis minus asperis.* Now (1708) a troublesome weed in Darby's gardens at Hogsden (=Hoxton), being brought out of some woods in Oxfordshire with roots of yew trees;' *Budd. MSS.*]

425. C. rotundifolia, *L.*   *Harebell.*
Cyb. Br. ii. 135. Curt. F. L. f. 4.

Heaths and dry roadsides; common. P. July—September.
I. Harefield Common!; *Blackst. Fasc.* 13. Ruislip and Harrow Weald Commons, F. W. Longman; *Melv.* 49. South Mims.
II. Bushey Park.
III. Hounslow Heath, Drilling Ground, &c., abundant. About Twickenham, frequent.
IV. North Heath, Hampstead.
V. Side of ditch, Hanwell Road; *Cherry.*
VI. Bush Hill, Enfield. Hadley; *G. Johnson.*
VII. South Heath, Hampstead.

First record: *Blackstone,* 1737.

[**426. C. Rapunculus,** *L.* *Rampions.*
Cyb. Br. ii. 137 ; iii. 464. Syme E. B. vi. t. 872.
Fields ; very rare. P. July, August.
VI. Edge of Common field near Weir Hall, Edmonton ; near Baker Street,
Enfield, E. Forster, jun. ; in Enfield Churchyard, T. F. Forster, jun. ;
B. G. 401. Specimens from the first of these localities are in *Herb.*
Brit. Mus.
First record : *E. Forster,* 1805. It has not been since observed.]

SPECULARIA, *Heist.*

427. S. hybrida, *A. De C.* Campanula hyb., *L.* (Syme E. B. and
L. Cat.).
Speculum Veneris minus (Johns., Blackst.).
Cyb. Br. ii. 142 ; iii. 465. Syme E. B. vi. t. 874.
Cornfields ; very rare. A. June—September.
I. Cornfield adjoining to Harefield Chalkpit, plentifully ! ; *Blackst.*
Fasc. 96.
VII. [Divers times among the corn in Chelsey field ; *Johns. Ger.* 440.
Chelsea College ; *Irv. H. B. P.* 498.]
First record : *Johnson,* 1633. Now confined to the chalk.

ERICACEÆ.

CALLUNA, *Salisb.*

428. C. vulgaris, *Salisb.* *Ling. Heather.*
Erica vulgaris sive pumila (Ger., Blackst.). *Erica vulg. L.* (Curt.).
Cyb. Br. ii. 150. Curt. F. L. f. 5.
Heaths, roadsides, and woods ; common. Shrub. July—September.
I. Harefield Common ! ; *Blackst. Fasc.* 27. Ruislip Common ; *Melv.* 50.
Harrow Weald Common. Stanmore Heath.
II. Roadsides about Staines and Charlton, but not commonly.
III. Hounslow Heath. Drilling Ground. Near Hatton.
IV. Hampstead Heath ! ; *Ger.* 1199. Peterboro' Road, near Harrow ;
Melv. 50.
VI. Winchmore Hill Wood. Great Northern Ry. bank at Colney Hatch.
VII. South Heath, Hampstead.
First record : *Gerarde,* 1597. A very tomentose state is found at
Hounslow = *var. β incana,* Syme E. B. vi. 44.

ERICA, *Linn.*

429. E. Tetralix, *L.* *Cross-leaved Heath.*
E. maior fl. purp. (Ger., Blackst.). *E. pumila Belgarum Lobelii, Sco-*
paria nostras (Park.).

Cyb. Br. ii. 146.　Syme E. B. vi. t. 889.

Damp parts of heaths; rare.　Shrub.　June—August.

I. Harefield!; *Blackst. Fasc.* 26.　Harrow Weald Common!; *Melv.* 50. Stanmore Heath, abundant.

III. Drilling Ground, Hounslow, abundant.

IV. Hampstead Heath!; *Ger.* 1199, and *Park. Theat.* 1484.

First record: *Gerarde*, 1597.　*E. tenuifolia caliculata, Ger.* 1199, and *E. juniperifolia altera,* Lob., *Johns. Eric.,* Hampstead Heath; probably forms of *E. Tetralix.* Johnson says (*Ger. em.* 1385): 'There are not above three or four sorts of heath that I could ever observe to grow on Hampstead Heath,' whereas Gerarde records six.

430. E. cinerea, *L.*　　　*Fine-leaved Heath.*

Erica tenuifolia (Ger., Blackst.).　*E. pumila calyculata Unedonis fl., Lob.* (Johns.).

Cyb. Br. ii. 148; iii. 466.　Curt. F. L. f. 2.

Heaths and roadsides; rather rare.　Shrub.　June—September.

I. Harefield; *Blackst. Fasc.* 26.　Stanmore, 1827–30; *Varenne.* Harrow Weald Common, very sparingly.

III. Hounslow Heath, sparingly.　Drilling Ground, abundant.　By the road to the Drilling Ground, north of the Powder Mills.

IV. Hampstead Heath!; *Ger.* 1199, and *Johns. Eric.* Roadside at Deacon's Hill, a few plants.

VII. South Heath, Hampstead, very sparingly.

First record: *Gerarde*, 1597; also first as a British plant.　Less abundant than *E. Tetralix*, though found in a greater number of localities; both species are exceedingly rare in Hertfordshire.　*E. cruciata* and *E. pyramidalis* of Gerarde, on Hampstead Heath, may perhaps be slight varieties of this species.

VACCINIUM, *Linn.*

431. V. Myrtillus, *L.*　　　*Bilberry.*

Vaccinia nigra (Ger., Park.).　*Vitis Idæa angulosa, J. B.* (Budd.). *Vitis Idæa fol. oblong. crenat. fructu nigricante* (Blackst.).

Cyb. Br. ii. 156.　Syme E. B. vi. t. 879.

Heaths and woods; rather rare.　Small shrub.　April, May.

III. Isleworth; *Cole.*

IV. On Hampstead Heath!, and the woods adjoining; *Ger.* 1230 and *Budd. MSS.* Turner's Wood; *Herb. Hardw.* Near Hendon, 1823; *Herb. G. & R.* Bishop's Wood.

VI. Southgate; *Cole.* Finchley Wood; *Ger.* 1230.

VII. [St. John's Wood; *Park. Theat.* 1458.]　Cane (=Ken) Wood!; *Blackst. Spec.* 103.　Still abundant there.　South Heath, near the Vale of Health.

First record : *Gerarde*, 1597. The only station given for this plant in *Fl. Herts.*, Oxhey Wood, is almost on the Middlesex boundary.

AQUIFOLIACEÆ.

ILEX, *Linn.*

432. I. Aquifolium, *L.*   *Holly.*
Aquifolium, Ger. (Merrett, Blackst.).
Cyb. Br. ii. 164. Syme E. B. ii. t. 316.

Woods, heaths, and hedges ; rather rare. Tree. June.
I. Harefield! ; *Blackst. Fasc.* 3. Uxbridge ; *Cole.* South Mims.
II. Bushey Park ! ; *Newb.*
III. Near Hounslow ! ; *Newb.*
IV. Hampstead Heath! ; *M. & G.* 201. Stanmore ; *Varenne.* About Harrow, not unfrequent ; *Melv.* 50. Deacon's Hill.
VI. Hadley ! ; *Warren.* Whetstone. Winchmore Hill Wood.

First record : *Merrett*, 1666. With *variegated leaves*, 'neer Highgate' (*Merrett*, 3), and 'about Tottenham' (*M. & G.*). There are some fine old trees in the Old Park Woods at Harefield, and old stumps on Hampstead Heath.

OLEACEÆ.

LIGUSTRUM, *Linn.*

433. L. vulgare, *L.*   *Privet.*
Ligustrum (Ger., Johns., Blackst.).
Cyb. Br. ii. 164. Curt. F. L. f. 5.

Thickets and hedges ; rather common. Shrub. June, July.
I. Harefield ! ; *Blackst. Fasc.* 52. Pinner chalkpits. South Mims.
II. Hammonds, near Staines.
III. By the Cran, Hounslow Heath.
IV. Hampstead Heath ; *Johns. Eric.* Road from Finchley to Hendon ! ; *Newb.*
V. Near Hanwell ! ; *Warren.*
VI. Edmonton.
VII. Near Fulham ; *M. & G.* 15. Near Kilburn ; *Herb. G. & R.*

First record : *Gerarde*, 1597, who says it grows 'naturally in every wood and hedgerow of our London gardens' (p. 1208). Often planted.

FRAXINUS, *L.*

434. * F. excelsior, *L.*   *Ash.*
Fraxinus (Johns., Blackst.).
Cyb. Br. ii. 165. Syme E. B. vi. t. 902.

Hedges and woods; rather common. Tree. April.
I. Harefield; *Blackst. Fasc.* 29. South Mims.
II. Tangley Park. Young plants on the river bank bet. Kingston Bridge and Hampton Court, apparently spontaneous.
III. Hounslow Heath.
IV. Edgware. Hampstead Heath; *Johns. Enum.* Stanmore.
V. Near Hanwell!; *Warren.*. Bet. Acton and Ealing!; *Newb.*
VI. Whetstone. Edmonton.
First record: *Johnson*, 1632. Loudon mentions a fine tree at Mount Grove, Hampstead, 85 feet high, and with the trunk 3 feet 10 inches in diameter.

APOCYNACEÆ.

VINCA, *Linn.*

435. V. minor, *L.*   *Smaller Periwinkle.*
Clematis Daphnoides minor, seu Vinca Pervinca minor, Ger. (Merr., Pet., Doody).
Cyb. Br. ii. 166.   Curt. F. L. f. 3.

Woods and hedges; rather common. P. April—June.
I. In Little Grove Wood near Breakspears; *Blackst. Fasc.* 19. Hedgebank near Bacon's Hall, South Mims, abundant.
II. River bank bet. Richmond Bridge and Hampton Court!; *Newb.*
III. Abundant in Harrow Grove; *Melv.* 51.
IV. On the west of Hamsteed Heath, plentifully; *Merrett*, 27. Fieldway from Hampstead to Wilsdon (with variegated leaves); *M. & G.* 308. About Hendon; *Pamplin.* Wood near Ongate Farm; *Irv. MSS.* Harlesden Green; Willesden; *Cole.* Copse on the road between Finchley and Hendon!; *Newb.* Bentley Priory Woods, abundant; *Melv.* 51.
VI. Near Tottenham, in a footpath from the London road to the town; *M. & G.* 308. Abundant in almost every coppice about Enfield and Southgate; *Phyt. N. S.* vi. 301. Colney Hatch, common; *W. G. Smith.*
VII. [On the moatside at Jack Straw's Castle; *Pet. Midd.* and *Doody MSS.*] [Hedge of a field near Leeson Green; *Doody MSS.*] [In great abundance in Bellsize Lane, on the left-hand in going from Primrose Hill to Hampstead; *M. & G.* 308. Near Bellsize House; *Irv. MSS.*] [Road to Kilburn, 1817; *Herb. G. & R.*]

First record: *Merrett*, 1666. Seems truly wild in many places.

436. *V. major,* *L.*   *Greater Periwinkle.*
Clematis daphnoides major, Ger. em. (Blackst.). *Vinca pervinca, Park.* (How).
Cyb. Br. ii. 167.   Curt F. L. f. 4.

Hedges and meadows; rather common. P. June, July.
I. In a meadow near Harefield Church; *Blackst. Fasc.* 19. About a quarter of a mile from Uxbridge towards Harefield; *Cole.*
II. Hampton!; *Newb.* Hedge near Kingston Bridge, towards Hampton Court, abundant.
III. Hedge near Twickenham Ry. Station.
IV. Hendon; *Irv. MSS.* Station Road, Harrow; *Melv.* 51.
V. Horsington Lane; *Melv.* 51.
VI. Enfield; *Cat. Lond.* Lane at the foot of Muswell Hill not far from the 'Orange Tree' at Colney Hatch, 1841; *Herb. Hardw.* On Muswell Hill, W. Lloyd; *Cooper,* 104.
VII. [Neere Kingsland, going into Paradise Walk; *How,* 129.] [Bet. Bromley and Blackwall; *M. & G.* 312.] [Lane bet. Hampstead and Edgware Road, 1809; *Winch MSS.* [In Kentish Town Churchyard; *M. & G.* 312.] Green Lanes, Stoke Newington, 1869; *W. G. Smith.*

First record: *How,* 1650. In no place has this any claim to be considered a native; it is very readily established.

GENTIANACEÆ.

CHLORA, *Linn.*

437. C. perfoliata, *L.* *Yellow Centaury.*
Cyb. Br. ii. 176. Syme E. B. vi. t. 913.
Banks; very rare. A. or B. July—September.
I. Roadsides near Pinner Hill, R. Ord; *Melv.* 51. In and above the chalkpits, Harefield!, 1868; *Newb.*

First record: *Ord,* 1864. Confined to the chalk? Blackstone did not notice it nearer Harefield than Gerard's Cross, Bucks. Hudson, *Fl. Anglica* i. 146, separated this from *Gentiana,* and gave it the name of *Blackstonia* in honour of the author of the 'Fasciculus' and 'Specimen.' Linnæus afterwards adopted the long-prior appellation of Renealm's, which it now bears.

ERYTHRÆA, *Renealm.*

438. E. Centaurium, *Pers.* *Centaury.*
Centaurium minus, C. B. (Newton, Blackst.).
Cyb. Br. ii. 174. Syme E. B. vi. t. 909.
Sunny banks, heaths, &c.; rather rare. A. or B. July, August.
I. Harefield!; *Blackst. Fasc.* 17. Ruislip Wood; Pinner; *Melv.* 51. Pinner limepits abundant.
IV. In many places about Hampstead Heath; *Newton MSS.* Mill

Hill, 1837, Mr. Children; *Herb. Mus. Brit.* Stanmore, 1815; *Herb. G. & R.* Railway bank, Harrow; *Melv.* 51.

VI. About Highgate; *Blackst. MSS.* and *Irv. MSS.*

VII. West side of Hampstead Heath, near the Vale of Health, J. Bliss; *Park Hampst.* 29. By Paddington Canal, near Kensal Green Cemetery; *Britten.*

First record: *Newton*, about 1680. *With white flowers* at Ruislip Reservoir; *Hind.*

GENTIANA, *Linn.*

439. G. Amarella, *L.* G. eu-Amarella (*Syme E. B.*). *Fel-Wort.*
G. alpina verna (Johns.). *Gentianella fugax minor* (Ger.). *G. pratensis flore lanuginoso, C. B. P.* (Blackst.).*

Open pastures; very rare. A. August, September.

I. In the old chalkpit near Harefield Mill, plentifully; *Blackst. Fasc.* 32. A little north of Harefield, sparingly, 1868!; *Newb.*

II. On a heath by Colbrooke (=Colnbrook); *Ger.* 354 and *Johns. Ger.* 437.† This locality is perhaps in Bucks.

First record: *Gerarde*, 1597.

[**440. G. Pneumonanthe,** *L.* *Calathian Violet.*
Cyb. Br. ii. 170; iii. 468. Syme E. B. vi. t. 914.
Heaths; very rare. P. August, September.

(II. Heath by Colnbrook; *Cat. Lond.* Confounded with *G. Amarella?*)

III. On Hounslow Heath, sparingly, Rev. Dr. Goodenough; *Smith Fl. Brit.* i. 285, and *Dicks. H. S. Fasc.* 7, and *Herb. Mus. Brit.*

First record: *Dickson*, about 1795; last, about 1800.]

G. acaulis, *L.* is not a British plant. It is recorded as found 'near London, but certainly not native'; *Macreight,* 157.

VILLARSIA, *Vent.*

441. V. nymphæoides, *Vent.* Limnanthemum n., *Link.* (Syme E. B.).
Nymphæa lutea minor septentrionalium (Lob.). *N. lutea minor fl. fimbriato* (Blackst.). *Menyanthes nymph., L.* (Huds., &c.).
Cyb. Br. ii. 177; iii. 469. E. B. 217 (drawn from a Thames specimen), reproduced in Syme E. B. vi. t. 921.

Still water of rivers, and ponds; rare. P. July, August.

II. Near Walton Bridge!, Viscount Lewisham; *Smith Fl. Brit.* i. 226. Creeks of Thames near Sunbury, in vast plenty; and most creeks of the river in that district, Sir J. Banks; *B. G.* 401. Entrance to village of Sunbury, where the river is open to the road, J. E. Moxon; *Phyt.* i. 747. About Hampton; *Huds.* ii. 85. Bet. Hampton and

---

* Referred to *G. Campestris,* L. in *B. G.* 402.
† The figure represents a stunted specimen.

Hampton Court. In the Thames, just above Kingston Bridge, abundantly!; *Blackst. Spec.* 60 and *Winch MSS.*

III. Thames at Richmond, 1841 ; *Herb. Woodward.* Ornamental Water, Twickenham Park (? planted).

V. About Brentford ; *Robson* 72, and *Lees.* Ornamental Water in Sion Park, in great abundance, probably planted.

VII. [Pond in London Fields, Hackney, J. Woods ; *B. G.* 401.] Small pools near Hampstead, E. H. Button ; *Cooper Supp.* 11.

First record : *Lobel,* 1570 ; also first as a British plant. 'Juxta amœnissima Thamesis fluenta, udis scrobibus et lacustris pratensibus;' *Lob. Adv.* 258. Truly native to the Thames, but we have not noticed it high:r than a little above Walton Bridge, nor lower than Kingston.* In the stations away from the Thames it is introduced.

MENYANTHES, *Linn.*

442. M. trifoliata, *L.*　　　*Buckbean.*

Trifolium paludosum, Lob. (Johns.).　*T. palustre, C. B. P.* (Blackst.). Cyb. Br. ii. 178. Curt. F. L. f. 4.

Wet meadows and heaths ; rare. P. May—June.

I. On Harefield Moor in several places !; *Blackst. Fasc.* 103.

II. Marshes about Staines, in many of which it is the principal plant ; *Curt. F. L.*

IV. Hampstead Heath !; *Johns. Enum.* and *Blackst. Spec.* 100. Still abundant in the bog behind Jack Straw's Castle.

VII. Banks of Thames, Fulham ; *Faulkner,* 22.

First record ; *Johnson,* 1632.

POLEMONIACEÆ.

Polemonium cæruleum, L. Cyb. Br ii. 179 ; iii. 469. Syme E. B. vi. t. 922. I. Plantation at Pinner Hill ; *Hind.* IV. On Hampstead Heath ; *M. & G.* 304. A casual garden escape.

CONVOLVULACEÆ.

CONVOLVULUS, *Linn.*

443. C. arvensis, *L.*　　　*Field Convolvulus.*

C. minor arvensis, C. B. P. (Blackst.). Cyb. Br. ii. 179 ; iii. 469. Syme E. B. vi. t. 923.

Cultivated and waste ground ; common. P. June—September.

I. Harefield !; *Blackst. Fasc.* 20. Ruislip ; *Melv.* 53. Pinner chalkpits.

* From a remark in *Holiday Excursions of a Naturalist* (1867), p. 36, it would appear to have at one time been found in Chelsea Reach.

II. Hampton. Staines.

III. Twickenham.

IV. Station road and railway bank, Harrow ; *Melv.* 52.

V. Near Hanwell ! ; *Warren.* Chiswick.

VI. Edmonton.

VII. Upper Clapton, &c. ; *Cherry.* Isle of Dogs ! ; *Newb.* Where Bryan-
stone Square is now, 1815 ; *Herb. G. & R.* In the turf below the
summer-house in the centre of Grosvenor Square, 1869 ! ; *Warren.*

First record: *Merrett,* 1666. A *very narrow-leaved form* was distinguished by the older observers : *C. minimus spicæ foliis* [in Tuddington
field] ; *Merrett,* 29. *C. arvensis minimus* about London ; *Budd. MSS.*
M. & G. saw it *with a variegated leaf* at Chelsea (p. 281).

444. C. sepium, *L.*   *Bindweed. White Convolvulus.*

Smilax lævis, Matth. (Johns.).   *C. major albus, C. B. P.* (Blackst.).
Cyb. Br. ii. 180. Syme E. B. vi. t. 924.

Hedges and cultivated soil ; very common. P. July, August.

In all the districts : less common in the dry parts of II., III., and V.

VII. Near Fulham Gardens ; *Herb. G. & R.* Zoological Gardens. Shrubberies by Constitution Hill. Isle of Dogs.

First record: *Johnson,* 1632. The *pink-flowered variety* is not uncommon,
as at (III.) Twickenham in several places, by the river near Richmond
Bridge, and between Twickenham and Worton ; and (VI.) Near Park
Station, Edmonton.

CUSCUTA, *Linn.*

[**445. C. europæa,** *L.*   *Great Dodder.*

Cyb. Br. ii. 181. Syme E. B. vi. t. 927.

Parasitical on nettles and thistles ; very rare. A. August and September.

VII. Hyde Park, in the sunk ditch under the wall of Kensington Gardens,
on nettles and thistles, 1820 and 1821 ; *Bennett (v. s.).*

Not recorded since.]

446. C. Epithymum, *Murr.*

Cuscuta (Parkinson, Newt.).

Cyb. Br. ii. 182 ; iii. 470. Syme E. B. vi. t. 929.

Parasitical on *Ericæ, Calluna, Ulex nanus, Pteris,* and other heath
plants ; very rare. A. June—September.

III. On the Drilling Ground, Hounslow, abundant.

IV. Upon Hampstead Heath ; *Park. Theat.* 10. Sometimes about ye Fuz
(=Furze) ; *Newton MSS.*

First record: *Parkinson,* 1640.

447. * **C. Trifolii,** *Bab.*   *Clover Dodder. Hell-Weed.*

C. major, C. B. P. (Blackst.†)?.

Cyb. Br. ii. 183; iii. 470.   Syme E. B. vi. t. 929.

On Red Clover; very rare.   A.   July.

I. Harefield; *Blackst. Fasc.* 21.

V. Hanwell, 1847; *Herb. W. Mc Ivor* (*v. s.*).

First record: *Blackstone*, 1737?.

BORAGINACEÆ.

CYNOGLOSSUM, *Linn.*

448. C. officinale, *Linn.*   *Hound's-Tongue.*

C. majus vulgare, Ger. em. (Blackst.).

Cyb. Br. ii. 283; iii. 489.   Curt. F. L. f. 4.

Waste ground and roadsides; rare.   B.   June—August.

I. Harefield; *Blackst. Fasc.* 22.

II. Hampton!; *Newb.* Roadside near Walton Bridge. Under the trees in the avenue, Bushey Park. By the towing-path in several places bet. Hampton Court and Kingston Bridge.

III. Bet. Isleworth and Richmond, 1815; *Herb. G. & R.*

IV. Bet. Child's Hill and Hendon; *Irv. MSS.*

VII. Ken Wood, Hunter; *Park Hampst.* 28.

First record: *Blackstone*, 1737. Chiefly found in the S.W. portion of Middlesex.

[**449. C. montanum,** *Lam.*

C. m. virenti fol. minore fl., Tournf. (Budd.).   *C. sempervirens* (Pet.).

Cyb. Br. ii. 284; iii. 487.   Syme E. B. vii. t. 1119.

Waste ground; very rare.   B.   June, July.

VII. In a hedge facing the road on Stamford Hill, between Newington and Toddenham (=Tottenham); *Pet. Midd.* On Stroud Green, near the boarded river, near Islington; *Budd. MSS.* (Specimen in *Herb. Budd.* v. cxxi. fol. 5.)

First record: *Petiver*, 1695; also first as a British plant? Last: about 1705.]

BORAGO, *Linn.*

450. * **B. officinalis,** *L.*   *Borage.*

Cyb. Br. ii. 280; iii. 486.   Syme E. B. vii. t. 1114.

Rubbish and waste ground; rare.   A. or B.   July—September.

II. Bushey Park!; *Newb.*

---

† It is probable that Blackstone intended this species, from the remark he makes that it is ' the pest of growing crops;' but only one species of *Cuscuta* was recognised at that time.

III. About Twickenham in many places; near the sluice, a single plant; near the railway bridge over the Thames, and near the allotments, several plants; in some abundance on waste ground by Cole's Bridge.

IV. West End, near Hampstead; *M. & G.* 251. Waste heaps, Harrow; *Melv.* 53.

VII. Kentish Town, 1846; *Herb. Hardw.* About Chelsea; *Britten.* Lower Heath, Hampstead, a single plant.

First record: *Milne* and *Gordon*, 1793. Scarcely naturalised.

Anchusa sempervirens, *L.* Cyb. Br. ii. 281. Syme E. B. vii. t. 1113. Has been remarked growing spontaneously near London; *Mart. Mill. Dict.* Occasionally as a garden weed at Harrow; *Melv.* 53. Always of garden origin.

LYCOPSIS, *Linn.*

251. L. arvensis, *L.* Anchusa a., *Bieb.* (Syme E. B.) *Small Bugloss. Buglossa sylvestris minor* (Johns., Blackst.).
Cyb. Br. ii. 280. Syme E. B. vii. t. 1111.

Roadsides and hedgebanks; rare. A. or P. June—August.

I. Harefield, in fallow fields; *Blackst. Fasc.* 12.

II. Roadside near Teddington, a single plant.

III. By a path between gardens from Twickenham to Isleworth.

IV. Cornfields, Hampstead; *Irv. MSS.*

V. Sandy field near Chiswick!; *Newb.*

VII. [On the drie ditch-bankes about Pickadilla (=Piccadilly); *Johns. Ger.* 799.]

First record: *Johnson*, 1633; also first as a British plant.

SYMPHYTUM, *Linn.*

252. S. officinale, *L.* *Comfrey.*
Consolida major fl. albo, Ger. em. (Blackst.).
Cyb. Br. ii. 278. Curt. F. L. f. 4.

Sides of streams and ditches, and in damp places; very common. P. May—August.

In all the districts.

VII. Bet. Lea Bridge and Tottenham; *Cherry.* Regent's Park. Zoological Gardens. Hackney Wick Marshes, abundant.

Var β. S. patens, Sibth. Syme E. B. vii. t. 1116.

II. Near Staines; *Newb.*

III. River-side by the railway bridge, Twickenham.

IV. Near entrance to Bishop's Wood.

Less common than the type.

First record : *Blackstone,* 1737.   Often a persistent garden weed.   Not given in *Fl. Harrow.*

ECHIUM, *Linn.*

453. E. vulgare, *L.*   *Viper's Bugloss.*

Cyb. Br. ii. 286; iii. 488.   Syme E. B. vii. t. 1095.

Waste ground ; very rare.   A. or B.   July—September.

I. Harefield! ; *Blackst. Fasc.* 24.   West Drayton ; *Lightf. MSS.*

II. By the towing-path near Hampton Court, towards Kingston Bridge.

VII. [Haverstock Hill ; *Irv. MSS.*]   Chelsea College, 1859, J. Britten ; *Bot. Chron.* 59.

First record : *Blackstone,* 1737.   Prefers a chalk soil.

Pulmonaria officinalis, *L.*   *Lungwort.*   Cyb. Br. iii. 366, 487.   Syme E. B. vii. t. 1098.   III. One or two plants in the Grove, Harrow, probably not indigenous ; *Melv.* 52.

LITHOSPERMUM, *Linn.*

454. L. officinale, *L.*   *Gromill.*

L. majus, Ger. em. (Blackst.).

Cyb. Br. ii. 275 ; iii. 485.   Syme E. B. vii. t. 1101.

Dry fields and waste places ; very rare.   P.   May—July.

I. Harefield ; *Blackst. Fasc.* 54.

VI. Banks of the Lea navigation, Edmonton, 1865 ; *Grugeon* (*v. s.*).

VII. Isle of Dogs ; *Mart. App. P. C.* 65.

First record : *Blackstone,* 1737.

455. L. arvense, *L.*   *Corn Gromill.   False Alkanet.*

L. arvense rad. rubente, C. B. P. (Blackst.).

Cyb. Br. ii. 276.   Syme E. B. vii. t. 1102.

Cornfields and waste ground ; rare.   A.   May—July.

I. Harefield, abundant; *Blackst. Fasc.* 53.

III. On soil brought from elsewhere in Roxborough Park ; *Farrar.*

IV. Hampstead ; *Irv. MSS.*

V. Field by Acton Ry. Station! ; *Newb.*

VI. Near Enfield.

VII. [Banks of canal near St. John's Wood Chapel, 1815 ; *Herb. G. & R.*]   Waste ground opp. Veitch's Nursery, Chelsea ! ; *Newb.*

First record : *Blackstone,* 1737.

456. M. palustris, *With.*   *Forget-me-not,*

M. scorpioides palustris, Ger. em. (Blackst.).

Cyb. Br. ii. 269.   Syme E. B. vii. t. 1104.

Sides of streams, ditches, and wet places; very common.   P.   June—August.

In all the districts.

VII. Isle of Dogs; *Cooper*, 115.   Primrose Hill, 1834; *Herb. Mus. Brit.*
Lea Canal, bet. Tottenham and Edmonton; *Cherry.* Hackney
Marshes. Hornsey Wood. South Heath, Hampstead. Eel-brook
Meadow, Parson's Green.

Var. β.   M. strigulosa, Reich.

II. On Staines Common.

III. Bet. Richmond Bridge and Twickenham.

First record: *Blackstone*, 1737.

457. M. repens, *Don.*

Cyb. Br. ii. 271; iii. 484.   Syme E. B. vii. t. 1105.

Ditches and wet places; very rare.   P.   June—August.

I. Marshy ground by Ruislip Reservoir; *Melv.* 52.

IV. Ditches near Pinner Drive; *Melv.* 52.

The only records, 1864.   We have not met with this species in Middlesex, but the above localities are not improbable.

458. M. cæspitosa, *Schultz.*

Cyb. Br. ii. 271; iii. 485.   Syme E. B. vii. t. 1103.

Wet places, especially on heaths; rather rare.   B. or P.   June—August.

I. Stanmore Heath.   Elstree Reservoir.   Bet. South Mims and Potter's
Bar.

III. Hounslow Heath.   Harrow; *Herb. Harr.*

IV. Great bog on Hampstead Heath; *Irv. MSS.*

V. Greenford, &c.; *Melv.* 52.

VII. Hackney Marsh, E. F.; *Herb. Mus. Brit.*   Walham Green; *Irv. H.
B. P.* 468.   Reservoirs in Ken Wood Grounds.

First record: *Irvine*, about 1830.

(M. sylvatica, *Ehrh. M. scorpioides latifolia hirsuta, Merr. Pin.* (Blackst.).
Cyb. Br. ii. 273; iii. 485. Syme E. B. vii. t. 1107. I. In Gutter's Dean
Wood and the Old Park, sparingly; *Blackst. Fasc.* 62. V. Roadsides, Greenford, 1859; *Hind.* This was most likely the large wood
form (β *umbrosa*, Bab.) of *M. arvensis.* The Rev. Mr. Hind, however, has a strong belief that he found the true *sylvatica* (see *Phyt.
N. S. v.* 204), but the station is not a likely one. *M. sylvatica* occurs
in Essex.)

459. M. arvensis, *Hoffm.*

M. scorpioides arv. hirsuta, Ger. em. (Blackst.).

Cyb. Br. ii. 274. Syme É. B. vii. t. 1108.

Cornfields, meadows and waste ground, and woods ; common. A. May —July.

I. Harefield ! ; *Blackst. Fasc.* 62. Near Elstree. Reservoir. South Mims.

II. Staines.

III. Twickenham.

IV. About Harrow, abundant; *Melv.* 52. Brent Bridge, Hendon ; *Morris* (*v. s.*). Bet. Finchley and Hendon ! ; *Newb.* Hampstead, near Bishop's Wood.

V. Hanwell ! ; *Warren.* Ealing ! ; *Newb.* Near Brentford ; *Hemsley.* Horsington Hill.

VI. Edmonton. Enfield.

VII. [Marylebone Fields, 1817 ; *Herb. G. & R.*]

First record: *Blackstone*, 1737. The large-flowered form, *var. β umbrosa,* Bab., occurs at (III.) Hounslow Heath and (VI.) Whetstone.

460. M. collina, *Hoffm.*

Cyb. Br. ii. 274 ; iii. 485. Syme E. B. vii. t. 1109.

Heaths, dry banks and walls ; rare. A. or B. April—June.

I. Harefield Church, 1853 ; *Herb. Hardw.* Bank by roadside near Ruislip, abundant.

III. Hounslow Heath, abundant.

V. Greenford churchyard wall, very scarce; *Melv.* 53. Gravel-pit near Hanwell ; *Warren.* On a wall on north side of high road beyond Brentford.

First record : 1853.

461. M. versicolor, *Ehrh.*

M. scorpioides minor flosc. luteis. Park. (Blackst.).

Cyb. Br. ii. 275. Syme E. B. vii. t. 1110.

Hedgebanks, heaths, dry fields and roadsides ; common. A. or B. April— June.

I. Harefield ! ; *Blackst. Fasc.* 62. Roadside bet. Harefield and Ruislip.

II. Road from Staines to Hampton in several places.

III. Hounslow Heath Drilling Ground.

IV. Near the Brent ; *Melv.* 53. Stanmore Marsh. Hedgebank outside Bishop's Wood.

V. Northolt ; Apperton, abundant ; *Melv.* 53. Field near Chiswick ; *Fox.* Near Hanwell ! ; *Warren.* Acton ; *Tucker* (*v. s.*).

VI. Colney Hatch. Whetstone. Near Enfield ; *Tucker.*

VII. By the reservoirs in Ken Wood grounds.

First record : *Blackstone*, 1737.

o

SOLANACEÆ.

SOLANUM, *Linn.*

462. S. nigrum, *L.*　　　*Garden Nightshade.*
S. hortense, Ger. em. (Blackst.).
Cyb. Br. ii. 185.　Curt. F. L. f. 2.

Cultivated and waste ground; common.　A.　July—October.
I. Harefield!; *Blackst. Fasc.* 94.　Uxbridge Common!; *Newb.*
II. By the Thames under the wall of Hampton Court Gardens; *New. B.
G. Supp.* Sunbury.　Bet. Teddington and Bushey Park.
III. Hounslow Heath, a few plants.　Twickenham, common garden weed.
IV. About Harrow, not very common; *Melv.* 53.　North Heath, Hampstead.
V. Near Kew Bridge Ry. Station; *Hemsley.*　Under the wall of grounds of Chiswick House.
VI. Edmonton.
VII. Fulham; *Cole.* Kensington Gardens.　Kensington Gore.　Parson's Green.　Near Chelsea Hospital.　Sandy End.　South Heath, Hampstead.　Adelaide Road.　Victoria Street.　Hackney Wick.　Isle of Dogs.

First record: *Blackstone,* 1737.　A plant with deeply sinuate-dentate leaves, but the usual black berries, was found on the site of the Exhibition of 1862 at South Kensington.

463. S. Dulcamara, *L.*　　　*Nightshade.　Bitter-sweet.*
Solanum lignosum, Park. (Blackst.).
Cyb. Br. ii. 185.　Curt. F. L. f. 1.

Hedges and damp places; very common.　P.　June—August.
In all the districts.
VII. Maiden Lane, St. Pancras; *Herb. Hardw.*　South Heath, Hampstead.　Seven Sisters' Road.　Hackney Wick.　I. of Dogs.　Kensington Gardens, hedge near the Palace, 1868; *Warren (v. s.).*

First record: *Blackstone,* 1737.

ATROPA, *Linn.*

464. A. Belladonna, *L.*　　　*Deadly Nightshade.*
Solanum lethale (Ger., Mill., Blackst.).
Cyb. Br. ii. 186; iii. 470.　Curt. F. L. f. 5.

Waste places; very rare.　P.　June—August.
I. In a shady gravel-pit near the Old Park!; *Blackst. Fasc.* 95.　Gathered there in 1842 by Dr. Bromfield, and still grows in some plenty.

VI. [Without the gate of Highgate, near a pound or pinfold on the left hand; *Ger.* 269. By old gardens near Highgate; *Mart. Mill. Dict.*]

VII. [In a ditch at the end of Goswell Street in the road at Islington; *Mill. Bot. Off.* 416: whence it is now (1793) totally expelled; *M. & G.* 325.]

First record: *Gerarde*, 1597.

HYOSCYAMUS, *Linn.*

465. H. niger, *L.*   *Henbane.*
H. vulgaris vel niger, C. B. P. (Blackst.).
Cyb. Br. ii. 184.   Syme E. B. vi. t. 836.
Roadsides and waste ground; rather rare.   A. or B.   May—July.

I. Harefield; *Blackst. Fasc.* 43.   Bushey Heath, near Stanmore, T. Ralph; *Lond. Fl.* 139.   Cottage garden at Pinner Common, 1863–65; *Hind.*

II. Staines Common, 1841; *Herb. Young.*   About Hampton Court, G. Francis; *Coop. Supp.* 12.

V. Acton Green; *Newb.*

VII. [On Bow Common; *M. & G.* 287.]   [In the Vale of Health, Hampstead, Mr. Bliss; *Park Hampst.* 29.   We have seen it at Hampstead, but it has been for several years extirpated; *Burnett* i. 9.] [Near St. John's Wood Chapel, 1815; *Herb. G. & R.*]   Parson's Green, 1856; *Phyt. N. S.* ii. 168.   Chelsea College, 1860, Britten; *Bot. Chron.* 58.   Site of Exhibition of 1862, S. Kensington.

First record: *Blackstone*, 1737.   The variety *pallidus* (Syme vi. 106) occurred in 1865, on the site of the Exhibition at S. Kensington.

Lycium barbarum, *L.   Tea plant.*   Cyb. Br. ii. 187.   Syme E. B. vi. t. 833.   Has been noticed in a semi-wild state (I.) At Harefield!; *Newb.* and (IV.) Willesden; *Naylor.*

Nicandra physaloides, *Gaertn.*   Cyb. Br. ii. 187.   Site of Exhibition of 1862 at S. Kensington.   Of Peruvian origin, common in Europe.

Nicotiana rustica, *L.*   Occasionally about London on waste ground. Site of Exhibition of 1862, S. Kensington.   Victoria St., Westminster. Thames Embankment works.   [This is probably *Hyoscyamus luteus*, betwixt St. James's and Hide-park; *Merrett*, 64.]

DATURA, *Linn.*

466. *D. Stramonium, L.*   *Thorn-Apple.*
Cyb. Br. ii. 186.   Curt. F. L. f. 6.
Waste places; rather rare.   P.   July—October.

I. Rubbish at the Lodge; and Tweed Bank, Pinner; *Hind.*

III. Several places about Twickenham; near the railway station; Marsh Farm; Queen Street.

V. Chiswick, a single plant; *Fox.*

VII. Frequent about London; *Huds.* i. 78, and *many subsequent writers.* Walham Green, G. Francis; *Coop. Supp.* 12. Chelsea College, Britten; *Bot. Chròn.* 58. Notting Hill; *Cole.* Kilburn; *Warren.* Chelsea. Pimlico. Victoria St., Westminster. Garden at N.E. corner of Green Park. Site of Exhibition of 1862, S. Kensington. South Heath, Hampstead. Isle of Dogs.

First record: *Hudson,* 1762. The variety *D. Tatula,* L., occurred on the site of the Exhibition.

OROBANCHACEÆ.

OROBANCHE, *Linn.*

467. O. Rapum, *Thuill.*   *Broom Rape.*
Rapum Genistæ (Ger., Johns.).   *O. major,* L. (Curt., Irv.).
Cyb. Br. ii. 224; iii. 475.   Curt. F. L. f. 4.

Parasitical on broom, furze, &c.; very rare. P. May—July.
IV. On Hampstead Heath; *Ger.* 1132, *Johns. Enum., Curt. F. L.* Ibid. 1828; *Varenne.* Rising furzy ground by the Great Bog; *Irv. MSS.*

First record: *Gerarde,* 1597. Not seen since about 1830; perhaps extinct.

O. Hederæ, *Duby.* Cyb. Br. ii. 229. Syme E. B. vi. t. 1015. I. A single plant occurred in June and July, 1868, on the root of a scarlet *Pelargonium* in a pot, at Pinner; another is now (Jan. 1869) growing under the same circumstances; *Hind.* Imported from elsewhere.

468. O. minor, *Sutt.*   O. eu-minor (Syme E. B. and L. Cat.).
Cyb. Br. ii. 228; iii. 475.   Syme E. B. vi. t. 1016.

Parasitical on clovers; very rare. A. or P. June—August.
III. On *Trifolium arvense* at Twickenham; *Macreight,* 175.
The only record.

[LATHRÆA, *Linn.*

469. L. Squamaria, *L.*   *Toothwort.*
Dentaria major, Matthioli, Ger. em. (Blackst.).
Cyb. Br. ii. 232; iii. 475.   Syme E. B. vi. t. 1006.

Parasitical on roots of hazel, &c., in shady places; very rare. P. April.
I. In a shady lane near Harefield Town, leading from it to the river; *Blackst. Fasc.* 23 and *Spec.* 17.
It has not since been met with, but may still exist.]

SCROPHULARIACEÆ.

VERBASCUM, *Linn.*

470. V. Thapsus, *L.*   *Mullein. High-taper.*
Tapsus barbatus (Ger.).   *V. mas latifolium luteum, C. B. P.* (Blackst.).
Cyb. Br. ii. 187.   Syme E. B. vi. t. 937.

Waste ground, heaths and roadsides; rather common. B. June—September.
- I. Harefield!; *Blackst. Fasc.* 105. [Formerly common in Pinner Wood; *Melv.* 58.] A single plant at Pinner Lime Works.
- II. Many places on the road from Staines to Hampton. Tangley Park. By the river, Hampton Court, a single plant.
- III. Garden-wall by the racquet courts, and by Roxborough turnpike; *Melv.* 58. Twickenham Park, one plant. Hounslow Heath, abundant.
- IV. Hampstead; *Irv. MSS.* Stanmore.
- V. Near Chiswick Ry. Station, a few plants.
- VI. About Hiegate (=Highgate); *Ger.* 630. Edmonton, a single plant.
- VII. Parson's Green, A. I., 1856; *Phyt. N. S.* ii. 168. North End, Fulham, one plant. Site of Exhibition, S. Kensington.

First record: *Gerarde,* 1597.

V. Lychnitis, *L.* Cyb. Br. ii. 188. Syme E. B. vi. t. 939. A few plants on the site of the Exhibition of 1862 at S. Kensington!; *Naylor.*

471. V. nigrum, *L.*   *Dark Mullein.*
Cyb. Br. ii. 190. Syme E. B. vi. t. 940.

Waste ground and roadsides; rare. B. or P. July, August.
- I. Harefield!; *Blackst. Fasc.* 105.
- II. Bet. Hampton and Sunbury, by the roadside!; *New B. G.* 100. By the Thames, under the wall of Hampton Court Gardens!; *New B. G. Supp.*, and in other places by the towing-path towards Kingston Bridge.
- III. In a meadow by Richmond Bridge; *M. & G.* 299. Near the Ry. Station, Twickenham, one plant.
- IV. About Hampstead; *Mart. Mill. Dict.*
- V. At Strand on the Green; *Mart. Mill. Dict.*

First record: *Blackstone,* 1737.

472. *V. Blattaria, *L.*   *Moth Mullein.*
Blattaria fl. luteo (Doody). *Blattaria Plinii, Ger. em.* (Blackst.).
Cyb. Br. ii. 190. Syme E. B. vi. t. 942.

Waste ground; rare. A. or B. July—September.
- I. In a chalky field near Harefield Mill, but not plentifully, 1735; *Blackst. Fasc.* 11. In an orchard, Terrilands, Pinner, 1867; *Hind.*

II. Fields not far from Hampton Court; *Doody MSS.* In a piece of ground enclosed for a garden near Teddington Station, several plants.

IV. On Hampstead Heath, but in no abundance; *M. & G.* 301.

VII. South end of Hampstead Heath on rubbish; *Irv. MSS.* and *Lond. Flora,* 128. Site of Exhibition of 1862, South Kensington, one plant.

First record: *Doody,* about 1690. Scarcely naturalised.

V. virgatum, *With.* Cyb. Br. ii. 191. Syme E. B. vi. t. 941. Chelsea College; *Irv. H. B. P.* 460.

DIGITALIS, *Linn.*

473. D. purpurea, *L.*   *Foxglove.*
Cyb. Br. ii. 215. Syme E. B. vi. t. 952.

Woods, heaths, and waste ground; rather common. B. or P. June—August.

I. Harefield!; *Blackst. Fasc.* 23. Harrow Weald Common!; *Melv.* 56. Stanmore Heath.

II. Bushey Park!; *Newb.* Tangley Park, in some abundance.

IV. Hampstead Heath!; *Johns. Eric.* Harrow Weald; *Melv.* 56. Deacon's Hill. Scratch Wood.

VI. About Colney Hatch, W. Lloyd; *Cooper,* 107. Trent Park, Southgate; *Cole.* Hadley!; *Warren.* Near Forty Hill, Enfield. [Bet. Bury St. and Enfield, now destroyed.]

VII. Waste ground near Parson's Green, 1862.

First record: *Johnson,* 1629.

ANTIRRHINUM, *Linn.*

474. *A. majus, *L.*   *Snapdragon.*
Cyb. Br. ii. 216; iii. 473. Syme E. B. vi. t. 953.

On walls; rather common. P. June—September.

III. Kitchen garden wall, Grove, Harrow; *Melv.* 57. Walls at Twickenham and Isleworth, common.

IV. Hampstead, 1815; *Herb. G. & R.* Walls, Hendon, E. H. Button; *Coop. Supp.* 11. Near Stanmore Church, F. W. Longman; *Melv.* 57.

V. Wall of Chiswick House grounds, abundant; and other walls in the district.

VI. Wood Green; *Cole.* Bury Hall, Edmonton.

VII. Common about London; *Smith Fl. Brit.* ii. 661. Wall at Eel-brook meadow.

First record: *Smith,* 1800. Well naturalised on the old brick walls in III. and V.

475. A. Orontium, *L.*
Cyb. Br. ii. 216. Curt. F. L. f. 4.

Cultivated and waste ground; rare. A. July—September.

II. In a field near Strawberry Hill, plentiful. In some quantity in a newly-enclosed garden by Teddington Ry. Station. Cornfield, Tangley Park, a single plant.

III. Potato field near Twickenham Ry. Station, one plant.

VII. [Marylebone Infirmary garden ; *Varenne.*] Waste ground near Chelsea College, 1861, Britten; *Phyt. N. S.* vi. 349.

First record: *Varenne,* 1827–30. Partial to dry sandy soils.

LINARIA, *Mill.*

476. *** L. Cymbalaria,** *Mill. Ivy-leaved Toad-flax. Roving Jenny.*
L. hederaceo fol. glabro, seu Cymbalaria vulgaris, R. H. Inst. (Dill.).
Antirrhinum Cymbalaria, L. (Curt., Smith, &c.).
Cyb. Br. ii. 217. Curt. F. L. f. 1 (drawn from a Middlesex specimen).

Walls; common. A. June—September.
I. Eastcott; *Melv.* 57. Pinner.
II. On the water gallery at Hampton Court !, 1829; *Winch. MSS.* Sunbury.
III. Abundant about Isleworth, Twickenham, &c.
IV. Mill Hill, 1837, Mr. Children; *Herb. Mus. Brit.* Hampstead; *Irv. MSS.* Sudbury, W. M. H.; *Melv.* 57. Tombs in Stanmore Churchyard. Near Whetstone.
V. Ealing; *Lond. Fl.* 131. About Brentford.
VI. Highgate, Rev. S. Palmer; *Mag. Nat. Hist.* ii. 266. Upper Edmonton, abundant. Bet. Enfield and Winchmore Hill Wood; *Church.*
VII. Abundantly on the walls of Chelsea Garden, and in neighbouring places; *R. Syn.* iii. *282. Frequent about London; *Huds.* i. 271. Walls of the Thames; *Mart. App. P. C.* 65. [On the Temple wall ; † *Curt. F. L.*] [Sommerset House, 1802 ;] about Chelsea, 1809; *Winch. MSS.* On Battersea Bridge, Sowerby; *Herb. Mus. Brit.* Blackwall, 1836; *Herb. Young.* Kentish Town, 1841 ; *Herb. Hardw.* Ken Wood. Haverstock Hill. Eel-brook Meadow.

First record: *Dillenius,* 1724; also first as a British plant.‡ Dillenius (*loc. cit.*) considered that Chelsea Gardens was the point from which this plant, a native of South Europe, originated in England, or at all events about London ; and in this notion he was followed by Thos. Martyn, Curtis, and Smith. Dr. Bromfield, in *Phyt.* iii. 621, combats this view, holding *L. Cymbalaria* to have been known 'from an indefinitely remote period' in England, but to have been a comparative rarity till the general diffusion of a taste for gardening. He calls

† Curtis says, 'In all those parts near London that lay within reach of the Thames; seeds are carried by the flux and reflux of the tide up and down the river, and left at high-water mark in the crevices of old walls, where they take root and increase very fast.'
‡ Dr. Richardson was perhaps really the earliest observer : ' Everywhere in quarries at Darford, Yorkshire' ; *R. Syn.* iii. *282.

attention to its notice as a garden plant by Gerarde and Parkinson, and says that 'it seems as much at home here as in any country on the Continent.'

477. L. Elatine, *Mill.*

Elatine altera (Ger.). *E. fol. acuminato in basi auriculato fl. luteo, C. B. P.* (Blackst.). *Antirrhinum E., L.* (Curt.).

Cyb. Br. ii. 218; iii. 473. Curt. F. L. f. 1.

Cornfields and waste ground; rare. A. July—October.

I. Harefield!; *Blackst. Fasc.* 25. Pinner; *Melv.* 57.

II. Bet. Hampton and Sunbury, in garden ground!; *Newb.* Cornfields, Tangley Park.

IV. Stanmore; *Herb. G. & R.* Hampstead; *Varenne.*

V. [Middle of field next the churchyarde at Cheswicke; *Ger.* 501.]

VII. [Bet. Hammersmith and Turnham Green, 1826; *Herb. Hardw.*]

First record; *Gerarde,* 1597; also first as a British plant.

478. L. spuria, *Mill.*

Veronica fœmina Fuchsii (Ger.). *Elatine fol. subrotundo, C. B. P.* (Blackst.). *Antirrhinum s., L.* (Curt.).

Cyb. Br. ii. 217; iii. 473. Curt. F. L. f. 3.

Cornfields and waste ground; rare. A. July—October.

I. Harefield!; *Blackst. Fasc.* 25. Pinner; *Melv.* 57.

II. Bet. Hampton and Sunbury in garden ground!; *Newb.*

III. Field by Roxborough Turnpike; *Melv.* 57.

IV. Stanmore, 1817; *Herb. G. & R.* Garden weed, Aspen Lodge, Sudbury; Willesden Green; *Farrar.*

V. [Field next the churchyarde at Cheswicke; *Ger.* 501.] Greenford, W. M. H.; *Melv.* 57.

First record: *Gerarde,* 1597; also first as a British plant.

479. L. minor, *Mill.*

L. Antirrhinum dicta, R. Syn. iii. (Blackst.). *Antirrhinum m., L.* (Curt.).

Cyb. Br. ii. 222. Curt. F. L. f. 5.

Cornfields, gardens and waste ground; rare. A. July—October.

I. Harefield!, very common; *Blackst. Fasc.* 53. Roadside near the Hall, Pinner; *Hind.*

II. Towing-path bet. Hampton Court and Kingston Bridge; nearly opp. the chimney of the Chelsea Waterworks; *Bloxam.* Stanwell Moor.

III. Near Hounslow!; *Newb.*

First record: *Blackstone,* 1737.

L. purpurea, *Mill.* Cyb. Br. ii. 219. Syme E. B. vi. t. 960. I. Well established on an old garden wall at Pinner, and spontaneously in gardens near, W. M. H.; *Melvill,* 58. IV. Walls at Hampstead; *Irv. MSS.* and *Lond. Fl.* 130. Much cultivated on walls in gardens.

480. L. vulgaris, *Mill.* *Toad-Flax.*

L. lutea vulgaris, Ger. em. (Blackst.). *Antirrhinum Linaria, L.* (Curt.).

Cyb. Br. ii. 220 ; iii. 473. Syme E. B. vi. t. 962.

Hedges, roadsides, and waste ground ; common. P. June—September.

I. Harefield ! ; *Blackst. Fasc.* 53.

II. Roadsides about Charlton, Sunbury, and Staines. Hampton. Teddington Ry. Station. Towing-path bet. Hampton Court and Kingston Bridge.

III. Near Hounslow. Near Hatton. Abundant near Twickenham.

IV. Road to Edgware, 1815 ; *Herb. G. & R.* Hampstead ; *Irv. MSS.* Lane from Hendon to Finchley Road, 1848 ; *Morris (v. s.).* Wembley, W. M. H. ; *Melv.* 57. Near Willesden Junction, on railway banks. Kingsbury Churchyard ; *Farrar.*

V. Ealing ! ; *Newb.* Great Western Ry. bank, near Hanwell.

VI. Lea Canal bet. Tottenham and Enfield ; *Cherry.* Edmonton. Near Enfield ; *Tucker.*

VII. North-Western Ry. bank near Kensal Green. North London Ry. bank, Holloway. Site of Exhibition of 1862, S. Kensington.

First record : *Blackstone,* 1737. A monstrosity with all the flowers with two spurs was found in the hedge on left side of the road on the ascent of the hill to Highgate from the Spaniards, E. K. ; *Mag. Nat. Hist.* i. 379 ; an approach to the state called *Peloria* figured at Syme E. B. vi. t. 963, and Curt. F. L. f. 6.

SCROPHULARIA, *Linn.*

481. S. nodosa, *L.* *Figwort.*

S. major (Ger., Blackst.).

Cyb. Br. ii. 212. Syme E. B. vi. t. 949.

Sides of ditches and streams, woods, and damp places ; probably common. P. June—August.

I. Harefield ! ; *Blackst. Fasc.* 92. South Mims.

III. By the Cran near Isleworth, sparingly. Near Twickenham.

IV. Highgate Wood ; *Johns. Eric.* Common in Harrow district ; *Melv.* 56. Bet. Hendon and Finchley ! ; *Newb.* Bentley Priory. Bishop's Wood. Pinner Drive ; *Herb. Harr.*

V. Hanwell ! ; *Warren.* Apperton ! ; *Newb.* Brentford ; *Hemsley.* Greenford ; *Cooper,* 108. Heston, frequent.

VI. Hadley ! ; *Warren.* Edmonton.

VII. Wood bet. London and Harnesey (= Hornsey) ; *Ger.* 580. Kensal Green ; *Cole.* South Heath, Hampstead.

First record : *Gerarde,* 1597.

[482. S. Ehrharti, *Stev.**

Cyb. Br. ii. 212 ; iii. 472. E. B. S. 2875 (drawn from a Middlesex specimen), reproduced in Syme E. B. vi. t. 948.

* S. umbrosa, *Dumortier,* seems to be the prior name.

Sides of ditches ; very rare. P. July, August.

VII. Belsize Park, Mr. Sowerby, Aug. 24, 1841 ; *E. B. S.* Specimens were
exhibited at the meeting of the Linnæan Society on Dec. 6, 1841.
There are specimens collected at the same place, 'July 1841,
J. De C. Sowerby' in *Herb. Mus. Brit.*

It has not since been met with. This is the *S. aquatica* of Fries
and some other continental botanists. The Rev. C. A. Stevens's paper,
distinguishing it from the true *S. aquatica* of Linnæus, was read
Feb. 13, 1840, before the Edinburgh Botanical Society, and may be
also found in *Ann. Nat. Hist.* vol. i.]

483. S. aquatica, *L.* S. Balbisii, *Hornem.* (L. Cat.). *Water Betony.*
S. a. major, C. B. P. (Blackst.).
Cyb. Br. ii. 213. Curt. F. L. f. 5.

Sides of ditches and streams, and in wet places ; common. P. June—
September.

I. Harefield! ; *Blackst. Fasc.* 92. Ruislip Reservoir; Eastcot; Pinner
Melv. 56.
II. Staines. Hampton! ; Bet. Hampton Court and Kingston Bridge
Newb.
III. By the Cran at Hanworth Road and Twickenham. Duke's River
Isleworth, abundant. By the Brent ; *Farrar.*
IV. Hampstead ; *Irv. MSS.* Bet. Hendon and Finchley! ; *Newb.* Bishop's
Wood.
V. Greenford ; *Cooper,* 108. Horsington Hill. Near Ealing.
VI. Ponder's End ; *Cherry.* Edmonton. Whetstone.
VII. Lea, near Hackney. Isle of Dogs.

First record : *Blackstone,* 1737.

(S. vernalis, *L.* Cyb. Br. ii. 214 ; iii. 473. Syme E. B. vi. t. 951. Recorded for Middlesex, G. B. Wollaston ; *Phyt. N. S.* vi. 63. We have
not met with any record.)

LIMOSELLA, *Linn.*

484. L. aquatica, *L.*
Plantago aquatica minima Clusii, Park. (Merrett, Blackst.). *Plantaginella
palustris* (Pet., Doody).
Cyb. Br. ii. 222 ; iii. 475. Syme E. B. vi. t. 968.

Borders of ponds and muddy places ; rare. A. or P. July—September.
I. By the sides of the Warren Pond near Breakspears, plentifully ; in
the bogs on Harefield Common; *Blackst. Fasc.* 78. By the reservoir at Elstree, very scarce, *Fl. Herts.* 211. Wet ruts in the
cart-track leading to the keeper's house on the summit of the ridge,
North Mims, in profusion, June 1856 ; *Phyt. N. S.* i. 407, but probably in Herts.

III. On Hunslow-heath, a mile off that town in the way to Coln Brook; *Merrett*, 95. On Hounslow Heath towards Hampton, Doody ; *R. Syn.* ii. 344. Pond by roadside, near the Drilling Ground, Hounslow Heath, 1846 ; *Morris* (*v. s.*).

IV. Has been repeatedly but sparingly found in pools and ruts on the north side of the road bet. Hampstead and Caen Wood ; *Irv. MSS.*

V. [Pond on Acton Green, 1867 ! ; *Newb.*]

VI. Pond at Finchley, J. Woods ; *B. G.* 407.

VII. [In the back lane to Southgate, before the turning for Hornsey Wood ; *MS. note* (*Alchorne's*) *quoted in Phyt.* iii. 166. In a lane near the Devil's House, going to Hornsey ; *Pet. Midd.*]

First record : *Merrett*, 1666 ; also the first as a British plant. This seems very rare, but we have probably overlooked it.

MELAMPYRUM, *Linn.*

485. M. pratense, *L.*   *Cow Wheat.*
Crataeogonum rubrum et album (Ger.). *Crataeogonum vulgare* (Park.). *M. luteum latifolium, C. B. P.* (Blackst.).
Cyb. Br. ii. 209 ; iii. 472. Syme E. B. vi. t. 1003.

Woods and copses ; rather rare. A. June—August.

I. Harefield ; *Blackst. Fasc.* 59. Ruislip and Pinner Woods ; *Melv.* 56.

IV. Upon Hampstead Heath ! in all parts among the juniper and bilberrie bushes ; *Ger.* 85 and *Park. Theat.* 1326. Bishop's, Turner's, and Scratch Woods.

VI. Highgate, Rev. S. Palmer ; *Mag. Nat. Hist.* ii. 266. Wood near Muswell Hill ; *Church.* Winchmore Hill Wood.

VII. Hornsey Wood, 1815 ; *Herb. G. & R.*

First record : *Gerarde*, 1597 ; also first as a British plant. Quite absent from the open sandy districts II., III., and V.

PEDICULARIS, *Linn.*

486. P. palustris, *L.*   *Louscwort.*
P. palustris rubra elatior (R. Syn. iii., Blackst.).
Cyb. Br. ii. 211. Syme E. B. vi. t. 996.

Bogs and marshy places ; rare. A.? or B. May—September.

I. In the meadows near Harefield Moor! ; *Blackst. Fasc.* 73. Near Yewsley ! ; *Newb.*

II. Bet. Hampton and Hampton Court! ; *Newb.* Staines Common.

IV. North Heath, Hampstead (small stunted plants).

VII. [In the ditches bet. Bow and Blackwall, near to Blackwall, Mr. Newton ; *R. Syn.* iii. *284.]

First record : *Newton*, about 1690.

487. P. sylvatica, *L.* *Red Rattle.*

Alectorolophus, Tab. (Johns.). *P. pratensis purpurea, C. B. P.* (Blackst.).
Cyb. Br. ii. 221. Syme E. B. vi. t. 997.

Damp heaths and pastures; rare. A.? B. or P.? May—September.

I. Harefield!; *Blackst. Fasc.* 73. Ruislip Common; *Melv.* 56. Harrow
Weald Common, abundant. Stanmore Heath.

III. Drilling Ground, Hounslow.

IV. Hampstead Heath!; *Johns. Enum.*

VI. Hadley!; *Warren.*

VII. South Heath, Hampstead.

First record: *Johnson*, 1632.

RHINANTHUS, *Linn.*

488. R. Crista-galli, *L.* R. minor, *Ehrh.* (Syme E. B. and L. Cat.).
Yellow Rattle.

Pedicularis lutea pratensis, vel Crista-galli, C. B. P. (Blackst.).
Cyb. Br. ii. 207. Curt. F. L. f. 5.

· Meadows and pastures ; common. A. June.

I. Harefield!; *Blackst. Fasc.* 73. Harrow Weald Common. Elstree.
South Mims. Pinner; *Melv.* 56.

; II. Near Staines towards Hampton. Bet. Hampton Court and Kingston
Bridge.

III. Roxeth; *Melv.* 56.

IV. Football Field; Northwick Walk, Harrow; *Melv.* 56. Harlesden
Green; *Morris (v. s.).* Hampstead; *Lawson (v. s.).*

V. Near Brentford; *Hemsley.* Near Twyford!; *Newb.* Hanwell!;
Warren.

VI. Whetstone; *Cole.* Enfield.

VII. Lea Marshes, Essex; *Fl. Essex*, 224.

First record: *Blackstone*, 1737.

EUPHRASIA, *Linn.*

489. E. officinalis, *L.* *Eyebright.*

Cyb. Br. ii. 206. Syme E. B. vi. tt. 991, 992.

Heaths and commons; rare. A. June—September.

I. Harefield Common. Ruislip Common; *Melv.* 56. Harrow Weald
Common. Stanmore Heath. Pinner lime-pits.

III. Drilling Ground, Hounslow.

IV. On Hampstead Heath, plentifully; *Blackst. MSS.* and *Irv. MSS.*

V. Ealing Common!; *Newb.*

First record: *Blackstone*, about 1740. A small-flowered slender plant,
probably *E. nemorosa*, Soy.-Will., is the ordinary Middlesex form
(? *var. β gracilis,* Syme t. 992); but we have seen the type-form
at Harefield.

490. E. Odontites, *L.* Bartsia O., *Huds.* (Syme E. B. and L. Cat.).
Cratæogonon Euphrosine, Ger. em. (Blackst.).

Cyb. Br. ii. 206. Syme E. B. vi. t. 993.

Cornfields, roadsides and waste ground; rather common. A. July—
September.

I. Harefield!; *Blackst. Fasc.* 21. Wear's Pond, Elstree, 1825; *Herb.*
Hardw. Woodready; near Pinner Church; *Melv.* 55.
II. Staines Common, sparingly. Hampton.
III. Field by Hospital Bridge.
IV. Hampstead; *Pet. H. B. P.* and *Irv. MSS.* Mill Hill, 1837, Mr. Chil-
dren; *Herb. Mus. Brit.* Near Harrow; *Hemsley.* Stanmore.
V. Perivale; *Lees.* By the canal, Apperton.
VI. Enfield Chase.
VII. By the side of the Great North Road; *Burnett,* 105.

First record: *Petiver,* 1713. We believe the plant usually met with
is *E. serotina,* Lam. (*var. β,* Syme E. B.).

(Sibthorpia europæa, *L.* Cyb. Br. ii. 223. Syme E. B. vi. t. 969. Mr.
Mackay, of Totteridge, Herts, showed us growing specimens, which
he had been informed were collected among the grass on the lawn at
Ken Wood House. It is very improbable that it grows there, unless
cultivated.)

VERONICA, *Linn.*

491. V. scutellata, *L.*

Anagallis aquat. quarta, Lob. (Merrett). *V. aquat. angustifol. minor, R.*
Syn. iii. (Blackst.).

Cyb. Br. ii. 199. Curt. F. L. f. 5.

Sides of ponds and wet places on heaths; rather rare. P. June—
August.

I. On Harefield Common; *Blackst. Fasc.* 107. Ruislip Reservoir; Har-
row Weald Common; *Melv.* 54. Elstree Reservoir. Stanmore
Heath.
III. Hounslow Heath.
IV. Hampstead Heath; *Huds.* i. 5 and *Britten.* Stanmore Marsh.
VI. Finchley Common, 1841; *Herb. Hardw.* Hadley; *G. Johnson.*
VII. [Tuthil fields; *Merrett,* 7.]

First record: *Merrett,* 1666.

492. V. Anagallis, *L.*

V. aquat. longifol. media, R. Syn. iii. (Blackst.).

Cyb. Br. ii. 199. Curt. F. L. f. 5.

Sides of streams and ditches; common. B. or P. June—September.

I. Harefield!; *Blackst. Fasc.* 107. Near Ruislip, F. W. Farrar; *Hind.*
II. Bushey Park!; *Newb.* Stanwell Moor. Staines Common. By
Walton Bridge. Creek near Cross Deep House. Near Kingston
Bridge.

III. By the Thames, Twickenham.
IV. Hampstead Heath; *Varenne.* Near Kingsbury; *Farrar.*
V. Greenford; *Cooper,* 108. Near Chiswick; *Fox.* Boston fields, Brentford; *Cherry.*
VI. Near Waltham Cross; *Cherry.* Marsh side, Edmonton.
VII. [King's Road, Chelsea, 1827–30; *Varenne.*] Thames side, near Fulham. [Eel-brook Meadow, Parson's Green, 1862.] Hackney Marshes. Isle of Dogs.

First record: *Buddle,* about 1705. *Veronica aquat. minor fol. oblong.,* by the Thames, near Peterboro' House; *Budd. MSS.* and *Herb. Pet.* clii. f. 41, is a small form.

493. V. Beccabunga, *L.* *Brooklime.*
V. aquat. rotundifol., Beccabunga dicta, minor, R. *Syn.* iii. (Blackst.).
Cyb. Br. ii. 200. Curt. F. L. f. 2.

In and by ditches and streams; very common. P. May—August.
Through all the districts.
VII. Near Bayswater; Hampstead Lane, 1817; *Herb. G. & R.* Walham Green, 1845; *Morris (v. s.).* South Heath, Hampstead.

First record: *Blackstone,* 1737.

494. V. Chamædrys, *L.* *Germander Speedwell.*
Chamædrys sylvestris (Ger.). *V. Cham. sylv. dicta,* R. *Syn.* iii. (Blackst.).
Cyb. Br. ii. 201. Syme E. B. vi. t. 986.

Hedgebanks, pastures and woods; very common. P. May, June.
In all the districts.
VII. Many places about London; *Ger.* 531. Fulham Churchyard. Eel-brook Meadow. Hackney Wick.

First record: *Gerarde,* 1597.

495. V. montana, *L.*
Alyssum mont. Columnæ (Merrett). *V. chamædryoides, fol. pedic. oblong. insidentibus,* R. *Syn.* ii. (Blackst.).
Cyb. Br. ii. 201. Curt. F. L. f. 4.

Woods and shady hedgebanks; rather rare. P. May, June.
I. In the Grove Wood, near Breakspears; *Blackst. Fasc.* 106. Hedgebank near Harefield on the Ruislip Road. In the Old Park Wood, abundant.
III. Harrow Grove; *Melv.* 54.
IV. On Hampstead Heath; *Merrett,* 6, and *M. & G.* 26. Plentiful in Bishop's Wood; *Irv. MSS.* and *Lond. Fl.* 128. Abundant in Harrow Park; *Melv.* 54.
V. Brentford, Dr. M'Creight; *Herb. Young.*
VI. In ye narrow wood next Highgate on ye right hand, and in another wood nearer Hornsey; *Newton MSS.*

VII. [Bellsize Lane; *M. & G.* 26.] Lane near West End, 1869!; *Warren.*
First record: *Merrett*, 1666.

496. V. officinalis, *L.*   *Speedwell.*
V. mas vulgaris, and *V. vera et major* (Johns.). *V. mas supina et vulgatissima, C. B. P.* (Blackst.).
Cyb. Br. ii. 200. Curt. F. L. f. 3.

Heaths, commons and roadsides; rather common. P. June—August.
I. On Ruislip Common, abundantly; *Blackst. Fasc.* 107. Ruislip
    Wood; *Melv.* 54. Near Uxbridge, 1857, H. Kingsley; *Herb. Mus.*
    Brit. Harefield Common. Roadside at Pinner.
II. Abundant by the roadsides about Charlton, towards Staines and
    Hampton.
III. Hounslow Heath Drilling Ground.
IV. Hampstead Heath!; *Johns. Eric.* and *Enum.* Deacon's Hill.
V. Horsington Hill.
VI. Hadley!; *Warren.* Near Enfield; *Tucker.* Edmonton.
VII. Hammersmith, 1845; *Morris* (*v. s.*). South Heath, Hampstead.

First record: *Johnson,* 1629. (*V. recta minima,* Lob.; Hampstead; *Johns.*
Eric. Refers to *V. spicata,* L., but probably a form of *V. officinalis*
was intended.)

497. V. serpyllifolia, *L.*   V. eu-serpyllifolia (Syme E. B.).
Betonica Pauli (Turn.).   *Veronica pratensis* (Johns.). *V. prat. serpylli-folia, C. B. P.* (Blackst.).
Cyb. Br. ii. 195. Syme E. B. vi. t. 978.

Damp fields, hedgebanks and roadsides; very common. P. May—
August.
Throughout the districts.
VII. South Heath, Hampstead.

First record: *Turner,* 1548; also first as a British plant. His words
are, 'In a parke besyde London'; *Turn. Names:* and, 'In divers
woddes not far from Syon'; *Turn.*

V. peregrina, *L.* Syme E. B. vi. t. 977. Harrow and Pinner, from seed
of Belfast specimens, W. M. H.; *Melv.* 55.

498. V. arvensis, *L.*   *Wall Speedwell.*
V. flosc. sing. cauliculis adhærent., R. Syn. iii. (Blackst.).
Cyb. Br. ii. 194. Curt. F. L. f. 2.
Walls, dry places and roadsides; very common. A. May—July.
In all the districts.
VII. Near Vauxhall Bridge, 1817; *Herb. G. & R.* Hammersmith, 1845;

. *Morris* (*v. s.*). Haverstock Hill, N.W. Parson's Green. Kensington Gore.

First record : *Blackstone*, 1737.

499. V. agrestis, *L.*

V. fl. sing. in oblong. pedic. chamædryfolia, R. Syn. iii. (Blackst.).
Cyb. Br. ii. 203. Syme E. B. vi. t. 972.

Cultivated and waste ground, and roadsides ; rather common. A. April —September.

I. Harefield; *Blackst. Fasc.* 106. Pinner. Hillingdon ; *Warren.*
II. Ground by side of Queen's River, Hampton.
III. Isleworth ; *Cole.* Near Hounslow, a single plant.
IV. Abundant in the Harrow district; *Melv.* 55. Harrow Weald. North Heath, Hampstead.
V. Acton Green ; near Ealing ! ; *Newb.*
VI. Marked for this district, localities mislaid.
VII. [Marylebone Infirmary garden ; *Varenne.*] Isle of Dogs! ; opp. Veitch's Nursery, Fulham ! ; *Newb.* Kensington Gardens ; *Warren* (*v. s.*).

First record: *Blackstone*, 1737. A scarce plant in II., III., and V.

500. V. polita, *Fries.*

Cyb. Br. ii. 203. Syme E. B. vi. t. 971.

Cultivated and waste ground, walls, &c.; common. A. April—September.

I. Harefield ! ; *Newb.* Hillingdon! ; *Warren.* Pinner.
II. Stanwell Moor. Road from Staines to Hampton. About Hampton, abundant.
III. About Twickenham, abundant everywhere.
IV. Stanmore. Frequent in the Harrow district ; *Hind* ; *Herb. Harr.*
V. Common about Ealing!, Acton!, Turnham Green, &c. ; *Newb.* Chiswick. Horsington Hill. Northolt ; *Melv.* 55.
VII. Camden Town. Parson's Green. Site of Exhibition of 1862, S. Kensington.

First record : *Hind*, 1860. Probably universally distributed, but very easily overlooked.

501. * V. Buxbaumii, *Ten.*

Cyb. Br. ii. 204. Syme E. B. vi. t. 973.

Cornfields, gardens, and waste places ; rather rare ?. A. April—September.

I. Ruislip ; *Melv.* 55. Harefield! ; Uxbridge ; *Newb.*
II. West Drayton ! ; *Neub.* Stanwell Moor. Charlton. Teddington. Hanworth Park.

III. Roxeth; *Melv.* 55. Hounslow!; *Newb.* **Twickenham.** Twickenham Park.

V. Perivale, W. M. H.; *Melv.* 55. Bet. Acton and Turnham Green!; *Newb.* Apperton.

VII. Chelsea; *Britten.* West London Cemetery. South Kensington.

First record: *Hind*, 1860. Seems to have been introduced into England from the East about 1820.

502. V. hederifolia, *L.*

Alsine hederacea, Tab. (Johns.).

Cyb. Br. ii. 202. Curt. F. L. f. 2.

Fields and hedgebanks; rather common. A. April—June.

I. Ruislip; *Melv.* 55. Wood Hall, Pinner.

II. Staines. Teddington.

III. Near Hounslow. Twickenham.

IV. Hampstead Heath!; *Johns. Enum.*

V. Greenford; *Melv.* 55. Hanwell; *Warren.* Chiswick.

VI. Colney Hatch. Edmonton.

VII. Belsize Lane.

First record: *Johnson*, 1632.

LABIATÆ.

MENTHA,* *Linn.*

503. M. rotundifolia, *L.*

Menthastrum (Budd.). *M. sylv. rotundiore fol., C. B. P.* (Mart. and Blackst. Fasc.). *Menthastrum fol. rugoso rotund. spontaneum fl. spicato, odore gravi, J. B.* (Blackst. Spec.).

Cyb. Br. ii. 235; iii. 477. Syme E. B. vii. t. 1020.

Damp ground; very rare. P. July, August.

I. In Harefield Churchyard, abundantly; *Blackst. Fasc.* 60, *and subsequent authors.*

VI. [In Hornsey Churchyard; *Budd. MSS., Mart. Tourn.* ii. 116, and *Blackst. Spec.* 53.]

First record: *Buddle*, about 1710.

504. M. sylvestris, *L.* Horse Mint.

M. sylv. longiore fol. (Mart.). *Menthastrum spicat. fol. long. candicante* (Alchorne).

Cyb. Br. ii. 236; iii. 478. Syme E. B. vii. t. 1022.

Damp places; rare. P. July, August.

---

* In this genus we have followed the arrangement adopted by Mr. Syme, founded chiefly on Mr. J. G. Baker's paper in *Seem. J. of B.* iii. 233. We have relied on the latter botanist's determination there given of the numerous Mints in Buddle's herbarium.

IV. Hampstead; *Irv. MSS.*

VI. [In the churchyard at Hornsey on the north side; *MS. note (Alchorne's) quoted in Phyt.* iii. 166; and *Hill*, 292.] Foot of Muswell Hill, J. Woods, jun.; *B. G.* 406.

VII. [Side of a little brook at the bottom of a field near Hackney Church, plentifully; *Mart. Tourn.* ii. 117.]

Var. β nemorosa, Benth.　*M. villosa secunda* (Sole).　Sole, tab. 2.

II. In a home close of an inn at Hillington; *Sole*, 6.

First record: *J. Martyn*, 1732.

505.　*M. viridis, L.　　　*Spearmint.*
Cyb. Br. ii. 237; iii. 478.　Syme E. B. vii. t. 1023.

Wet places; very rare.　P.　July, August.

IV. Side of a stream at Harrow Weald, W. M. H.; *Melv.* 58.

V. Canal bank at Apperton; *Melv.* 58.

First record: *Hudson*, 1778.　By the banks of the Thames; *Huds.* ii. 250. An escape from gardens.

506.　*M. piperita, Huds.　　*Peppermint.*
Cyb. Br. ii. 238.　Syme E. B. vii. t. 1024.

Sides of streams and wet places; rare.　P.　July, August.

V. Side of canal, Greenford, probably not indigenous; *Melv.* 58.

VII. In the Green Lane to Southgate from Newington, before coming to Hornsey Wood; *MS. note (Alchorne's) quoted in Phyt.* iii. 166, and *Huds.* i. 222.　In a ditch by the roadside on Stamford Hill, on the right-hand side going down the hill, E. F.; *Herb. Mus. Brit.* and *B. G.* 406.

Var. β vulgaris, Sole.　*M. aquat. nigricans fervidi saporis* (Budd.).　Syme E. B. vii. t. 1025.

VII. By ye New River side towards Newington; *Budd. MSS.*

First record: *Buddle*, about 1710.

[507.　M. pubescens, *Willd.*
Sisymbrium hirsut. fol. angustiore et acutiore min. ramosum (Budd.).　Syme E. B. vii. t. 1026.

Sides of ponds and streams; very rare.　P.　July, August.

VII. First observed by Mr. Rand about some ponds near Marybone; *Budd. MSS.*

Var. β hircina, Syme　*M. aquat. fol. oblong. virid. glabro saporis fervidissimi* (Budd.).　Syme E. B. vii. t. 1027.

VII. By ye New River sides near Newington; *Budd. MSS.*

First record: *Rand*, about 1700.　Likely to occur elsewhere.]

508. M. hirsuta, *L.* M. aquatica, *L.* (Bab.). *Water Mint.*
M. aquat. s. sisymbrium, Ger. em., and *M. aquat. hirsuta s. Sisymb. hirsutum, Chabr.* (Blackst.).
Cyb. Br. ii. 239. Syme E. B. vii. t. 1030.
Sides of ditches and streams; very common. P. July—September.
In all the districts.
VII. Near Marybone abundantly, and bet. Newington and Hornsey; *Huds.*
i. 224. Marylebone Fields, 1815; *Herb. G. & R.* South Heath,
Hampstead.
First record: *Buddle,* about 1710. An almost glabrous-leaved form oc-
curred (II.) by the river-side bet. Hampton Court and Kingston Bridge.
Sisymbrium ramosissimum fl. in summ. ramulis in glob. conglom.,
near Stoke Newington; *Budd. MSS.,* is a much-branched form. (See
Herb. Budd. cxxi. f. 44.) A small form, *M. aquatica minor,* figured in
Sole, t. 10, was found in Hornsey Lane; *Sole,* 24.

509. M. sativa, *L.*
M. calaminthæ arvensis verticillatæ similis sed multo elatior (Budd.). *M.
verticillata, Riv.* (R. Syn. iii.). *M. hirsuta, var. ζ.* (Smith).
Cyb. Br. ii. 239; iii. 478. Syme E. B. vii. t. 1031.
Sides of streams; rather rare. P. July—September.
I. By the canal, north of Harefield!; *Newb.*
II. Canal, West Drayton!; *Newb.* Staines Moor. By the Thames at
Hampton; bet. Strawberry Hill and Teddington; bet. Hampton
Court and Kingston Bridge.
III. By the Thames bet. Twickenham and Richmond Bridge. Marble Hill.
IV. Stanmore, 1827–30; *Varenne.*
V. By the Brent, Apperton; *Lees.*
VII. By ye New River's side near Stoke Newington; *MSS. Budd.* In
Hackney River near the Ferry-house; *Dill. in R. Syn.* iii. 232. Ibid.,
E. Forster; *E. B.* 448.
First record: *Buddle,* about 1710. *M. verticillata odore fragrantissimo*;
by the New River's side near Stoke Newington; *Budd. MSS.;* a form
with narrow leaves. *M. aquatica exigua, Trag.,* observed by Mr.
Buddle, in company with Mr. Franc. Dale, by the side of the New
River, near the upper end of Stoke Newington; *R. Syn.* iii. 232, is
probably the same. (See *Sm. Fl. Brit.* ii. 618, where he refers both
plants to his *M. hirsuta var. θ.*) [*M. arvensis verticillata fol. rotund.
odore aromat.,* R. Syn. I think I have found this with Mr. Rand by
the Thames side neat houses, 1710; *Budd. MSS.*] There are two
specimens with this name in *Herb. Budd.* cxxi. f. 14, the other being
M. gentilis, L. var. γ.

510. M. rubra, *Sm.* M. pratensis, *Sole,* var. β (Bab.).
M. crispa verticillata, C. B. P. (Blackst.). *M. crispa verticillata fol.
rotundiore, J. B.* (Budd.).
Syme E. B. vii. 1033.

By ditches and streams; rare. P. July—September.

I. By a pondside just below Breakspears in the lane; *Blackst. Fasc.* 60.

IV. Ditch in a field near Harrow Ry. Station, through which the footway to Stanmore passes, 1859.

VI. By the road bet. Edmonton and Enfield, Mr. E. Forster; *Sm. Fl. Brit.* ii. 620.

VII. Near Stoke Newington; *Budd. MSS.* By the River Lea, near the Ferry House, Herb. Sherard; *Smith Fl. Brit.* ii. 620.

First record: *Buddle*, about 1710.

511. M. gracilis, *Sm.* M. pratensis, *Sole* (Bab.).
M. verticillata glabra odore M. sativæ, Sherard (Sm. Fl. Brit.).
Syme E. B. vii. t. 1034.

Wet places; very rare. P. July—September.

VII. At Stoke Newington, Sherard's Herbarium; *Sm. Fl. Brit.* ii. 622, and *Eng. Fl.* iii. 84.

* *Var. β Cardiaca.* Syme E. B. vii. t. 1035.

I. Roadside leading to Pinner, W. M. H.; *Melv.* 59.

VII. Banks of the Lea near Walthamstow, Middlesex, T. F. Forster; *Baker in Seem. J. of B.* iii. 247.

First record: *W. Sherard*, about 1720. A garden escape: it is much cultivated, e.g. about Twickenham.

[512. M. gentilis, *L.* M. sativa, *L. var. β* (Bab.).
Syme E. B. vii. t. 1037.

Ditches; very rare. P. June—September.

VII. In a ditch at Stroud's Green, near Hornsey, Month. Mag.; *With.* ed. 7, iii. 705.]

513. M. arvensis, *L.*
Cyb. Br. ii. 240. Syme E. B. vii. t. 1038.

Cornfields and waste ground, also by streams; rather rare. P. July—September.

I. Pinner; *Melv.* 59. Harefield.

II. West Drayton!; *Newb.* Staines Moor. Teddington. Bet. Hampton Court and Kingston Bridge. Bet. Strawberry Hill and Teddington.

III. Hounslow Heath; *Sole*, 30.

IV. Harrow; *Melv.* 59. Hampstead; *Irv. MSS.*

V. Near Twyford!; *Newb.*

First record: *Buddle*, about ·1710. *M. verticillata glabra fol. ex rotunditate acuminatis* of *Budd. Herb.* cxxi, f. 18, found by ye New River side near Stoke Newington, is referred by Mr. Baker to *var. ε Allionii,* Syme E. B. vii. t. 1040.

[514. M. Pulegium, *L.* *Penny-Royal. Pudding-Grass.*
Penny-ryall (Turn.). *Pulegium regale* (Lob.). *Pulegium regium* (Ger., Blackst.).

Cyb. Br. ii. 241 ; iii. 479. Syme E. B. vii. t. 1041.

Wet places on commons ; rare. P. July—September.

I. On Harefield Common, abundantly ; *Blackst. Fasc.* 82.

III. Beside Hundsley (= Hounslow) upon the heth beside a watery place ; *Turn.* ii.

VII. On the common neere London called Miles ende . . . whence poore women bring plentie to sell in London markets, and sundrie other commons neere London ; *Ger.* 546. Abounds round the edges of the ponds, South Heath, Hampstead ; *Irv. MSS.* and *Lond. Fl.* 133. Frequent there in 1850, but recently extinct ; *Syme.* Specimens collected by D. Cooper in *Herb. Young.*

First record : *Turner*, 1562 ; also first as a British plant. Last about 1855, or somewhat later.]

LYCOPUS, *Linn.*

515. L. europæus, *L.*   *Gipsy Wort.*

Marrubium aquaticum, Ger. em. (Blackst.).

Cyb. Br. ii. 235. Syme E. B. vii. t. 1019.

Sides of rivers, ditches, and ponds ; very common. P. July—September. Through all the districts.

VII. In most of the dirty ditches about London ; *Jenkinson*, 7. Millfield Lane, Highgate. Thames side, about Cremorne. Eel-brook Meadow. Isle of Dogs. Hackney Marshes.

First record : *Blackstone*, 1737.

[Elsholzia cristata, *Willd.* Chelsea and Parson's Green, 1856, plentiful ; in 1857 not noticed ; *Irv. H. B. P.* 437.]

SALVIA, *Linn.*

516. S. Verbenaca, *L.*   *Clary.*

Horminum sylvestre (Ger., Blackst.).

Cyb. Br. ii. 233 ; iii. 476. Curt. F. L. f. 6.

Dry fields, roadsides, and waste places ; rare. P. May—September.

I. Harefield, frequent ; *Blackst. Fasc.* 43.

II. Bet. Hampton and Hampton Court ! ; *Newb.* Bet. Sunbury and Walton Bridge. By the towing-path bet. Hampton Court and Kingston Bridge, abundant.

III. In a lane a quarter of a mile this side Richmond Bridge, 1815 ; *Herb. G. & R.*

V. Strand-on-the-Green.

VII. [In the fields of Holburne, neere unto Graies Inn ; at the ende of Chelsey next to London ; *Ger.* 628. Fields near Chelsea ; *M. & G.* 35.]

First record : *Gerarde*, 1597. Chiefly close by the Thames. Curtis says, ' Common about London, especially in churchyards.'

(S. pratensis, *L.* Cyb. Br. ii. 233 ; iii. 476. Syme E. B. vii. t. 1058. Hampstead Heath ; *Cooper*, 101. Undoubtedly an error.)

ORIGANUM, *Linn.*
517. O. vulgare, *L.*
O. sylvestre cunila Bubula Plinii, C. B. P. (Blackst.).
Cyb. Br. ii. 242 ; iii. 479. Curt. F. L. f. 5.

Dry banks and roadsides ; very rare. P. July—September.
I. Harefield, frequent ; *Blackst. Fasc.* 70. (Banks, Pinner ; *Hind. in Phyt. N. S.* v. 201. Mr. Hind now considers this to have been an error.)
Not since recorded. Extinct?.

THYMUS, *Linn.*
518. T. Serpyllum, *L.* *Wild Thyme.*
Serpyllum vulgare, Ger. em. (Blackst.).
Cyb. Br. ii. 242 ; iii. 478.

1. *T. eu-Serpyllum*, Syme. *T. Serpyllum, L.* (Bab.). Syme E. B. vii. t. 1043.
Dry banks ; very rare. P. June—August.
I. By the chalkpits, Harefield.
II. By the high road near Staines, towards Hampton.

2. *T. Chamædrys*, Fries. Syme E. B. vii. t. 1044.
Heaths and dry places ; rather rare. P. June—September.
I. Ruislip and Harrow Weald Commons ; Eastcott ; *Melv.* 59. Harefield Common. Stanmore Heath.
II. Towing-path bet. Hampton Court and Kingston Bridge. Teddington.
III. Drilling Ground and roadsides near it, Hounslow. Meadow near Richmond Bridge.
IV. Hampstead Heath ; *Varenne.*

First record : *Merrett*, 1666. A hairy form, *Serp. vulgare hirsutum*, was noticed by Blackstone on Oliver's Mount in Uxbridge Moor ; *Fasc.* 93. A white-flowered plant is recorded as found plentifully betwixt Tuddington and Hampton Court ; *Merrett* 112, and *R. Syn.* iii. 230.

CALAMINTHA, *Mönch.*
C. Nepeta, Clairv. Cyb. Br. ii. 244. Curt. F. L. f. 6. Likely to be found in the N.W. of the county if searched for. Blackstone's locality, by the roadside from Harefield to Chalfont St. Peters (*Fasc.* 13), whence we have seen specimens collected by Newb. in 1868, is just beyond the county boundary in Herts.

519. C. officinalis, *Mönch.* C. menthifolia, *Host.* (Syme E. B. and L. Cat.). *Calamint.*
C. vulgaris, Ger. em. (Blackst.).
Cyb. Br. ii. 244. Syme E. B. vii. t. 1050.

Dry hedgebanks and bushy places ; very rare. P. July—September.

I. In Harefield, street going to the river; *Blackst. Fasc.* 13. Old chalkpits, Harefield. Roadside north of South Mims.

First record : *Blackstone*, 1737. By the book-characters, our plant is *var. β Briggsii*, Syme E. B. vii. 35, and t. 1051.

520. C. Acinos, *Clairv.* *Wild Basil.*
Cyb. Br. ii. 243 ; iii. 479. Syme E. B. vii. 1048.
Dry places ; very rare. A. or B. July—October.
I. In the chalkpit, Harefield, a single specimen, 1868! ; *Newb.*
II. By the towing-path near Hampton Court, towards Kingston Bridge.

First record : *the Authors,* 1867. Blackstone's station (*Fasc.* 19) is in Bucks.

521. C. Clinopodium, *Benth.* *Great Wild Basil.*
Calamynte, first kind (Turn.). *Clinopodium majus, Park.* (Blackst.).
Cyb. Br. ii. 245. Syme E. B. vii. t. 1047.
Hedgebanks and bushy places ; rather rare. P. July—September.
I. Harefield, plentiful! ; *Blackst. Fasc.* 19. Harrow Weald Common ; *Melv.* 61.
II. Staines Moor. Teddington.
III. Bet. Twickenham and Whitton. By the Duke's River, near Chase Bridge.
IV. Stanmore, &c. ; *Varenne.* Woodcock Hill.
V. About Sion ; *Turn. Names.*

First record : *Turner,* 1548 ; also first as a British plant. Confined to the sands and gravels.

Melissa officinalis, *L.* *Balm.* Cyb. Br. ii. 246. Syme E. B. vii. t. 1053.
II. A large tuft by the roadside just outside the Teddington entrance of Bushey Park. VII. Bet. Little Chelsea and Parson's Green, 1856 ; *Phyt. N. S.* ii. 168 and *Irv. H. B. P.* 425.

SCUTELLARIA, *Linn.*

522. S. galericulata, *L.* *Skull-cap.*
Lysimachia galericulata cœruleo-purpurea (Lob.). *L. galeric.* (Johns., Blackst.).
Cyb. Br. ii. 269 ; iii. 484. Curt. F. L. f. 3.
Sides of streams and ponds ; common. P. July, August.
I. Harefield! ; *Blackst. Fasc.* 55. Ruislip Reservoir, W. M. H. ; *Melv.* 61. Elstree Reservoir.
II. Staines Common. Stanwell and Staines Moors. Hampton. Bet. Hampton and Hampton Court. By the river, Teddington. Bushey Park.
III. By the Cran at Hounslow, Hanworth Road, &c. Hatton. Whitton. Twickenham. Twickenham Park, abundant.

IV. Hampstead Heath; *Johns. Enum.* and *Cole.* Kingsbury; *Varenne.*
V. Copses, Turnham Green, Rev. S. Palmer; *Mag. Nat. Hist.* ii. 266. Canal at Greenford and Apperton!; *Melv.* 61. Canal bet. Brentford and Hanwell, abundant; *Hemsley.*
VI. New River, Enfield; *Forst. Midd.* New Cut, Edmonton, and by a pond in Bury Street.
VII. [By the ponds and water side in S. James his Parke; in Tuthill Fields; *Johns. Ger.* 479. In Rosamond's Pond in St. James's Park; bet. Newington and Islington; *Blackst. Spec.* 49. Banks of New River, plentifully; *Bot. Lond.* ii. New River at Islington; *Cooper,* 104.] Lea, bet. Tottenham and Clapton, G. Francis; *Coop. Supp.* 12. Tottenham; Lea Bridge; *Cherry.* Isle of Dogs, 1815; Kensington Gardens and Harrow Road, 1817; *Herb. G. & R.* Paddington Canal, near Kensal Green Cemetery, 1862, J. Britten; *Phyt. N. S.* vi. 348.

First record: *Lobel,* 1576 ('about London'); also first notice as a British plant.

523. S. minor, *L.*
Lysimachia minima galericulata (Lob.). *Gratiola latifolia* (Ger., Park.). *L. gal. minor* (Johns., Blackst.).
Cyb. Br. ii. 268. Curt. F. L. f. 4.

Wet parts of heaths and sides of streams; rather rare. P. June—September.

I. In the moist parts of Harefield Common!; *Blackst. Fasc.* 56. Harrow Weald Common!; *Melv.* 62. Abundant 1867.
III. Headstone; *Herb. Harr.*
IV. At the further end of Hampsteed Heath!; *Ger.* 466. *Johns. Eric., Park. Theat.* 221, and other authors. Still abundant. Sparingly among loose stones of Pinner Drive; *Melv.* 62.
V. (Copses, Turnham Green, Rev. S. Palmer; *Mag. Nat. Hist.* ii. 266.)
VI. Highgate; *Lob. Ill.* 66 and 68. Hadley; *G. Johnson.*
VII. Upon Hampstead Heath towards London, neere unto the head of the springs that were digged for water to be conveyed to London 1590 by Sir John Hart, Knight, Lord Mayor of the Citie of London!; *Ger.* 466. Now grows there (on the South Heath), sparingly. Banks of Lea, bet. Clapton and Tottenham, G. Francis; *Coop. Supp.* 12.

First record: *Gerarde,* 1597; also first as British.

PRUNELLA, *Linn.*

524. P. vulgaris, *L.* *Self-heal.*
Prunella, Ger. em. (Blackst.).
Cyb. Br. ii. 266. Syme E. B. vii. 1059.

Meadows, roadsides and waste ground; very common. P. July, August.
Throughout all the districts.

VII. Marylebone Fields, 1817; *Herb. G. & R.* Hackney Wick. S. Heath.

First record:' *Blackstone*, 1737?.* *With white flowers*, recorded from Harefield; *Blackst. Fasc.* 82, and South Heath, Hampstead; *Irv. MSS.*

NEPETA, *Linn.*

525. N. Cataria, *L.*   *Cat-Mint.*

Mentha Cataria vulgaris et major, C. B. P. (Blackst.).

Cyb. Br. ii. 265; iii. 483. Syme E. B. vii. t. 1054.

Hedges and bushy places; very rare. P. July—September.

I. In a meadow near Harefield Mill; about More Hall!; *Blackst. Fasc.* 60.

II. Bushy places by the towing-path bet. Kingston Bridge and Hampton Court.

V. Near Perivale Rectory; bet. Perivale and Apperton; *Lees.*

First record: *Blackstone*, 1737.

526. N. Glechoma, *Benth.*   *Ground Ivy.*

Hedera terrestris, Ger. em. (Blackst.). *Glechoma hederacea, L.* (Cyb. Br.).

Cyb. Br. ii. 264. Curt. F. L. f. 2.

Hedgebanks, meadows and damp ground; very common. P. April—June.

Through all the districts.

VII. Paddington; *Varenne.* Lane near Highgate. Isle of Dogs.

First record: *Blackstone*, 1737. *Vars. α* and *β* of Syme E. B. are about equally frequent; *var. γ, G. hirsuta,* W. & K., perhaps occurs about Harrow.

LAMIUM, *Linn.*

527. L. amplexicaule, *L.*   *Henbit.*

L. fol. caulem ambiente majus et minus, C. B. P. (Blackst.).

Cyb. Br. ii. 255; iii. 481. Curt. F. L. f. 2.

On and under walls, and as a weed in light soils; rather rare. A. May—August.

I. Harefield!; *Blackst. Fasc.* 49.

II. Staines. Tangley Park Fields.

III. Walls at Isleworth. Worton. Twickenham, a single specimen.

IV. Garden at the 'Welsh Harp,' Brent Reservoir; *Warren.*

V. Near Shepherd's Bush!; *Newb.* Near Chiswick!; *Warren.*

VI. Edmonton. Enfield.

First record: *Blackstone*, 1737.

528. L. incisum, *Willd.*

L. rubrum minus fol. profunde incisis (Ray). *L. purpureum β* (With.).

Cyb. Br. ii. 257. Syme E. B. vii. t. 1083.

---

* *Symphytum petræum:* about Syon; *Turn. Names.* If a wild plant was intended, it must have been this species.

Fields, roadsides, and waste ground; rare. A. April—September.
II. Teddington!; *Newb.* Near Strawberry Hill.
III. Roadside by the new cemetery near Hospital Bridge.
IV. Hampstead; *Irv. MSS.*
VII. [T. Willisel first shewed it me about St. James his fields at West-minster; *R. Cat.* i. 186.] It is found in some sallet gardens west of London; I have also met with it that way in ploughed fields; *Pet. Bot. Lond.* 284, and *Pet. Herb.* cli. f. 205.

First record: *Willisel*, about 1670 ; also first as a British plant.

529. L. purpureum, *L.*   *Red Dead-Nettle.*
L. rubrum, Ger. em. (Blackst.).
Cyb. Br. ii. 257. Curt. F. L. f. 1.

Fields, gardens, and waste ground; very common. A. March—October.
Throughout all the districts.
VII. Abundant in London. Hampstead. Brompton Cemetery. Eel-brook Meadow. Victoria Street. Isle of Dogs. Kensington Gardens.

First record: *Blackstone*, 1737. *With white flowers* at Edmonton.

530. L. album, *L.*   *White Dead-Nettle.*
Cyb. Br. ii. 253. Syme E. B. vii. t. 1086.

Hedgebanks and waste ground; very common. P. May—August.
Throughout all the districts.
VII. Hammersmith. Fulham. Victoria Street. St. Pancras Churchyard. Regent's Park. Hackney Wick. Isle of Dogs.

First record: *Blackstone*, 1737.

L. maculatum, *L.* Cyb. Br. ii. 254 ; iii. 481. Syme E. B. vii. t. 1085.
II. Site of an old garden, Fulwell, semi-wild. IV. Hedge at Kenton, two plants ; *Melv.* 60. VII. Formerly gathered about Bayswater, but suspected to be the outcast of some botanic garden; *Smith in E. B.* 2550, and *E. Fl.* iii. 90. Always the result of garden cultivation.

531. L. Galeobdolon, *Crantz.*   *Yellow Archangel.*
L. luteum (Ger., Johns., Merr., Blackst.). *Galeopsis Galeobdolon, L.* (Curt.).
Cyb. Br. ii. 253 ; iii. 481. Curt. F. L. f. 4.

Woods and hedges; rather rare. P. May, June.
I. Harefield!; *Blackst. Fasc.* 49. In the Old Park woods, abundant.
IV. Hampstead Heath ; *Johns. Eric.* Bishop's and Turner's Woods. Deacon's Hill. Kingsbury ; *Farrar.*
V. Horsington Lane ; near Greenford ; *Melv.* 60.
VI. On Muswell Hill; *Cooper*, 104. Betstile Lane, Colney Hatch ; *Morris* (*v.s.*). Abundant in the Alders Wood and hedges near Whetstone.
VII. Hedge left hand from the village of Hampsteed to the church, and in the wood thereby; *Ger.* 568. In a thicket near the lower end of Pond

LABIATÆ. 219

Street, Hampstead; *Blackst. Spec.* 43. [Stoke Newington; *Cat. Lond.* 16.] [In Hamsted Churchyard, and west side of Primrose Hill; *Merrett*, 69.] [Marylebone Fields, 1817; *Herb. G. & R.*] Still common in the hedges in the northern suburbs, Kentish Town, Hornsey Wood, Millfield Lane, Highgate, &c.

First record: *Gerarde*, 1597; also first as a British plant. A state with *pale cream-coloured flowers* was found at (I.) Harefield; and one *with variegated leaves* in (IV.) Turner's Wood, Hampstead. Does not occur in the sandy districts II. and III.

GALEOPSIS, *Linn.*

532. G. Ladanum, *L.* G. angustifolia, *Ehrh.* (Syme E. B. and L. Cat.). *Sideritis arvensis rubra, Park.* (Blackst.).
Cyb. Br. ii. 258; iii. 482. Syme E. B. vii. t. 1074.

Cornfields and waste places; very rare. A. August—October.
I. Harefield!; *Blackst. Fasc.* 93. Above the chalkpits!; *Newb.*
(IV. Hampstead Heath, Carter; *Cooper*, 101.)

First record: *Blackstone*, 1737. The Harefield plant seems Syme's var. β = *G. canescens*, Schultz.

533. G. Tetrahit, *L.* G. eu-Tetrahit (Syme E. B.). *Hemp-Nettle.*
Cannabis spuria (Ger.). *Lamium cannabino fol. vulgare, R. Syn.* iii. (Blackst.). *Purple Nettle Hemp* (Pet.).
Cyb. Br. ii. 259. Syme E. B. vii. t. 1078.

Borders of fields, hedgebanks, and waste land; rather common. A. June—September.
I. Harefield!; *Blackst. Fasc.* 49.
II. Bet. Staines and Hampton, common.
III. Roxeth; *Melv.* 60. Near Hounslow on the Staines Road. Near Hospital Bridge.
IV. About Hampstead Heath!; *Newton MSS., Pet. H. B. P.* t. 32, and others.
VI. Colney Hatch. Edmonton. Near Park Ry. Station.
VII. Side of Duckett's Canal, Hackney; *Cherry (v.s.).*

Var. β. G. bifida, Bönn. Syme E. B. vii. t. 1079.
I. Pinner; Ruislip Common; *Melv.* 60.
III. Tangley Park.
VII. Weed in garden of 70, Adelaide Road, N.W.; *Syme (v.s.).*

First record: *Gerarde*, 1597 ('about London;' p. 573); also first as a British plant. [*Lamium cannab. fl. alb. verticill. purpurascentibus*, Doody, is a white-flowered variety. Seen for many years near the neat-houses, and between them and Chelsea; *Doody in R. Syn.* ii. 342. (See *Budd. Herb.* cxxi. f. 21.)]

G. versicolor, *Curt.* Cyb. Br. ii. 260; iii. 482. Curt. F. L. f. 6. III. or
IV. Cornfields, Harrow, Hind; *Phyt. N. S.* v. 201 : we are informed
by Mr. Hind that only two plants were seen.

STACHYS, *Linn.*

534. S. Betonica, *Benth.*   *Betony.*
Betonica (Ger., Blackst.). *B. officinalis, L.* (Curt., Melv.).
Cyb. Br. ii. 261. Curt. F. L. f. 3.
Woods, hedges, and bushy heaths; common. P. July—September.
  I. Harefield ! ; *Blackst. Fasc.* 10. Ruislip Wood ; Harrow Weald
    Common ; *Melv.* 60. Stanmore Heath. Elstree.
  II. Stanwell Moor.
  III. Hounslow Heath. Drilling Ground.
  IV. Hampstead Heath. Bishop's and Turner's Woods. Scratch Wood.
    Deacon's Hill.
  V. Horsington Hill.
  VI. Hadley! ; *Warren.* Winchmore Hill Wood. Ry. bank, Colney Hatch.
  VII. South Heath, Hampstead.
  First record: *Gerarde,* 1597. [A *white-flowered plant* was noticed in
  (VII.) a wood by Hamsteede, near Master Wade's House; * *Ger.* 578.]

535. S. sylvatica, *L.*   *Hedge Nettle* or *Woundwort.*
Galeopsis genuina Dioscoridis (Park.).
Cyb. Br. ii. 262. Curt. F. L. f. 3.
Shady hedges and ditchbanks; very common. P. July, August.
Through all the districts ; less common in II. and III.
  VII. Kilburn ! ; *Warren.* Regent's Park. Site of Exhibition of 1862,
    South Kensington.

  Var. β. S. ambigua, Sm. S. sylvatici-palustris, *Wirtg.* (Syme E. B.).
  Syme E. B. vii. t. 1070.
  IV. On Golder's Green, bet. Hampstead and Hendon, 1840 ; *Herb. Hardw.*
  Exactly Smith's plant, which seems nearer to *S. sylvatica* than *S.
  palustris.*
  First record : *Parkinson,* 1640. Omitted in *Blackst. Fasc.*

536. S. palustris, *L.*   *All-heal.*
Sideritis anglica strumosa radice (Lob., Park., Blackst.).
Cyb. Br. ii. 261. Curt. F. L. f. 3.
Sides of streams and in damp places; common. P. July—September.
  I. Harefield, plentifully ! ; *Blackst. Fasc.* 93. Ruislip Reservoir ; *Melv.*
    61. Bet. Yewsley and Iver Bridge ! ; *Newb.*
  II. West Drayton ! ; *Newb.* Staines Moor. Hampton.

* Belsize House, then the residence of Sir William Wade, Lieutenant of the Tower
and Clerk of the Council, and afterwards occupied by Lord Wotton (v. **778**), has been
recently pulled down.

III. By the Cran on Hounslow Heath, and by Hospital Bridge. Twickenham. Twickenham Park.

IV. Moist places, Hampstead; *Irv. MSS.* By Brent Reservoir!; *Warren.*

V. Canal, Greenford; *Melv.* 61. Canal, Apperton.

VI. Trent Park, Southgate; *Cole.* Lea Canal bet. Edmonton and Tottenham; *Cherry.*

VII. Near Hackney; *Lob. Ill.* 111 and *Park. Theat.* 588. [In the fields
going to Chelsea and Kensington; *Park. Theat.* 588.] Near Paddington Cemetery!; *Warren.*

First record: *Lobel*, about 1600.

537. S. arvensis, *L.*
Sideritis humilis lato obtuso fol., Ger. em. (Blackst.).
Cyb. Br. ii. 264; iii. 483. Curt. F. L. f. 4.

Cornfields and waste ground; rather rare. A. or B. August—October.

I. Cornfields, Harefield; *Blackst. Fasc.* 93. Gardens at Pinner, W. M. H.;
Melv. 61.

II. West Drayton!; *Newb.* Tangley Park.

III. Waste ground near Twickenham Ry. Station.

IV. Hendon; *Irv. MSS.* Field in Bishop's Wood, Hampstead; abundant.

VI. Edmonton.

First record: *Blackstone*, 1737.

BALLOTA, *Linn.*

538. B. fœtida, *Lam.* B. nigra, *L., var. α* (Syme E. B. and L. Cat.).
Black Horehound.
Marrubium nigrum, Ger. em. (Blackst.).
Cyb. Br. ii. 251. Syme E. B. vii. t. 1065.

Hedgebanks and waste ground; very common. P. July—September.
In all the districts.

VII. Kentish Town. South Kensington. Shepherd's Bush. Chelsea.
Old St. Pancras Churchyard. Hackney Wick.

First record: *Merrett*, 1666. A *white-flowered state* is recorded from
nigh Chelsey; *Merr. MSS.*, and near Hammersmith, Mr. Woodward;
With. ii. 616. Another state, *B. fl. albo galea lutea*, often about
Fulham; *Herb. Pet.* cli. fol. 194.

B. ruderalis, *Svensk Bot.* Syme E.B. vii. t. 1066. Not yet certainly recorded
for Middlesex. Given as a native of the North Thames sub-province
in *Cyb. Br. Supp.* 65. In answer to enquiry, Mr. Watson writes:
'There is a specimen in my herbarium picked by Rev. Jas. Harris bet.
Hampton and Knighton which formerly I thought *B. ruderalis*, but lat-
terly only *B. fœtida* with more pointed calyx-teeth and white flowers:
in so far shading off towards (not exactly into) *B. rud.*' Such a form
was also brought from Harefield by Rev. W. W. Newbould.

MARRUBIUM, *Linn.*

539. *M. vulgare, *L.*   *White Horehound.*
M. album, (J. Mart., Blackst.).
Cyb. Br. ii. 266; iii. 484.   Syme E. B. vii. t. 1064.
Waste places; rare.   P.   July—September.
  I. On Uxbridge Moor, abundantly; *Blackst. Fasc.* 58.
  IV. On Hampstead Heath; *Huds.* i. 228.
  VI. In a pit on Finchley Common; *Mart. Tourn.* ii. 107.
  VII. At the foot of Highgate; *Mart. Tourn.* ii. 107.   Parson's Green;
     Britten.
  First record: *Martyn,* 1732.   Formerly much cultivated for medicinal
use.

TEUCRIUM, *Linn.*

540. T. Scorodonia, *L.*   *Wood Sage.*
Wood Sage (Pet.).   *Salvia agrestis sive Scorodonia, Ger. em.* (Blackst.).
Cyb. Br. ii. 247; iii. 479.   Curt. F. L. f. 6.
Heaths and hedgebanks; rather common.   P.   July—September.
  I. Harefield!; *Blackst. Fasc.* 89.   Ruislip Wood; Harrow Weald Com-
    mon!; *Melv.* 59.   Stanmore Heath.
  II. Tangley Park.
  III. Hounslow Heath.   Drilling Ground.   Near Hospital Bridge.
  IV. Hampstead; *Pet. H. B. P.* 34.   North Heath.   Deacon's Hill.
  V. Horsenton Wood; *Lees.*
  VI. Near Enfield; *Tucker.*   Hadley!; *Warren.*   Edmonton.   Colney
    Hatch.
  VII. [Marylebone Fields, 1815; *Herb. G. & R.*] South Heath, Hampstead.
  First record: *Petiver,* 1713.

AJUGA, *Linn.*

541. A. reptans, *L.*   *Bugle.*
Bugula, Ger. (Johns., Blackst.).
Cyb. Br. ii. 249.   Curt. F. L. f. 2.
Meadows and pastures; rather common.   P.   May, June.
  I. Harefield!; *Blackst. Fasc.* 12.   Pinner.
  II. Staines.
  IV. Hampstead!; *Johns. Eric.*   Stanmore.   Harrow.   Bet. Finchley and
    Hendon!; *Newb.*
  V. Near Brentford!; *Hemsley.*   Hanwell!; *Warren.*   Horsenton Wood;
    Lees.
  VI. Near Enfield; *Tucker.*   Colney Hatch.   Edmonton.
  VII. Belsize Park.   Millfield Lane, Highgate.   Kentish Town.   Highbury
    College; *Herb. Woodward.*
  First record: *Johnson,* 1629.

Plants *with white-flowers* have been seen in (III.) Harrow Park, W. F. Farrar; *Melv.* 59, and *with pink corollas* (IV.) By old Stanmore Church, W. F. Longman; *Melvill*, 59. (V.) Osterley Park; *Masters.* (VI.) Highgate, G. E. Dennes; *Mag. Nat. Hist.* viii. 390.

VERBENACEÆ.

VERBENA, *Linn.*

542.   V. officinalis, *L.*   *Vervain.*

V. communis, Ger. em. (Blackst.).

Cyb. Br. ii. 233; iii. 476.   Curt. F. L. f. 1.

Roadsides and waste land; rather common.   P.   June—August.

I. Harefield!; *Blackst. Fasc.* 105; by the paper mills, plentifully. Pinner Wood; Waxwell Lane; *Melv.* 62.   By Pinner Church.   Bet. Yewsley and Iver Bridge!; *Newb.*

II. Staines Moor.   Staines Common.   Bet. Staines and Hampton.   Bet. Hampton and Hampton Court!; *Newb.*   Hanworth Road, near the bridge over the artificial river.   By the towing-path bet. Hampton Court and Kingston Bridge, abundant.   Teddington.

III. Near the Church, Harrow; by the Grove gates; *Melv.* 62.   By the Duke's River, Isleworth.   About Twickenham, frequent.

IV. Bet. Finchley and Hampstead, 1815; *Herb. G. & R.*   Mill Hill, 1857, Mr. Children; *Herb. Mus. Brit.*   Hampstead; *Irv. MSS.*

V. Near Lampton in several places.   Bet. Perivale and Apperton; *Lees.*

VI. Bury Street, Edmonton.

VII. Site of the Exhibition, S. Kensington.

First record: *Blackstone*, 1737.

LENTIBULARIACEÆ.

UTRICULARIA, *Linn.*

[**543.   U. vulgaris,** *L.*

Cyb. Br. ii. 290.   Syme E. B. vii. t. 1125.

Ditches; very rare.   P.   June—August.

III. On Hounslow Heath towards Hampton; *Huds.* ii. 9.

VII. In the Green Lane leading from Newington to Southgate, not far from the sluice which opens the New River into the New Cut; *MS. note (Alchorne's) quoted in Phyt.* iii. 166.   In ditches near Hornsey; *Huds.* i. 8.

First record: *Alchorne*, about 1750; last, *Hudson*, 1778.]

[**544.   U. minor,** *L.*

Lentibularia minor (Dill., Blackst.).   *Millefolium palustre galericulatum minus fl. minore, R. Syn.* ii. (Doody).

Cyb. Br. ii. 291; iii. 488.   Syme E. B. vii. t. 1126.

Ditches and bogs; very rare. P. June—August.

I. In Uxbridge River, Mr. Hill; *Blackst. Spec.* 45.

III. On Hounslow Heath; *Doody MSS.* In the river on Hounslow Heath by Mr. Dandridge; *Dill. in R. Syn.* iii. 287; and *Cullum*, 6.

First record: *Doody*, about 1700; last, *Cullum*, 1774.]

PRIMULACEÆ.

HOTTONIA, *Linn.*

545. H. palustris, *L.*　　*Water Violet.*

Millefolium aquaticum, seu Viola aquatica caule nudo, C. B. P. (Blackst.).

Cyb. Br. ii. 296. Curt. F. L. f. 1.

Ditches; rather rare. P. May—July.

I. Harefield, frequent; *Blackst. Fasc.* 61.

II. Abundant in a ditch by roadside from Bucks to Staines; *Phyt. N. S.* iv. 263. Many places on Staines Moor.

IV. Lane leading from the 'Harp' to Kingsbury; *Lond. Fl.* 141. Lane bet. Edgware Road and Woodhouse, Hyde; *Irv. MSS.* Pool in a small plantation nearly opp. Woodford House, Kingsbury; *Farrar.*

VI. Ditch by side of Lea Navigation Canal, not far from Ponder's End Station; *Cherry* (*v. s.*).

VII. [About Chelsea; *M. & G.* 275.] [Banks of Thames, Fulham; *Faulkner*, 22.] [Tottenham Marshes; *Cat. Lond.* 16, and *Macreight*, 189.] [Abounds in most of our watery ditches near London; *Curt. F. L.*]

First record: *Blackstone*, 1757.

PRIMULA, *Linn.*

546. P. vulgaris, *Huds.*　　*Primrose.*

P. veris minor, Ger. em. (Blackst.). *Primula sylvarum* (Johns.).

Cyb. Br. ii. 291. Syme E. B. vii. t. 1129.

Woods, copses and hedgebanks; rather common. P. March—May.

I. Harefield!; *Blackst. Fasc.* 81. Abundant in Old Park Woods. Ruislip Wood.

III. Near Headstone Farm; very abundant. Harrow Grove; Roxeth; *Melv.* 62.

IV. Hampstead Heath; *Johns. Eric.* and *Herb. Rudge.* In Bishop's Wood, Hampstead, nearly eradicated. Bentley Priory.

V. Rather scarce in this neighbourhood; copse near Brentford; *Hemsley.*

VI. Hadley!; *Warren.* Winchmore Hill Wood. Copse near Warren Lodge, Edmonton, abundant.

VII. Kilburn Field, 1818; *Herb. G. & R.* [Primrose Hill is said to have derived its name from the former abundance of *P. vulgaris* there; v. *Park Hampst.* 258.]

First record: *Johnson*, 1629. Has become scarce round London, from being dug up and carried away for sale. Probably absent from the sandy S.W. parts. Plants *with red flowers* have been observed (VI.) at Colney Hatch; *W. G. Smith*, and at Bury Hall, Edmonton.

Hybrids between **546** and **547**. *P. officinali-vulgaris*, Syme E. B. vii. tt. 1132, 1133. *Great Cowslips. Oxlips. P. pratensis inodora lutea, Ger. em.* (Blackst.). *P. elatior*, and *P. caulescens* (Melvill). I. Pastures at Pinner, W. M. H.; *Melv.* 62. Pinner Hill. About Harefield, frequently; *Blackst. Fasc.* 81. IV. Occasionally in pastures about Harrow, W. M. H. One plant in Bentley Priory Woods with Primroses; *Melv.* 62. Stanmore; *Varenne.* The specimen in *Herb. Harr.* is nearer *P. vulgaris* than *P. veris ; = P. variabilis*, Goup. (Bab. Man.) ?. *P. elatior*, Jacq. does not occur.

547. P. veris, *Huds.* P. officinalis, *L.* (Syme E. B.). *Cowslip. Paigle.* *P. pratensis, Lob.* (Johns.). *P. veris major, Ger. em.* (Blackst.). Cyb. Br. ii. 293; iii. 490. Syme E. B. vii. t. 1130.

Meadows and banks; rather common. P. April—June.
I. Harefield!; *Blackst. Fasc.* 81. Ruislip. Bank of North-Western Ry. Elstree. Potter's Bar.
II. Staines.
III. Fields about Harrow; *Cooper*, 115. Bank of North-Western Ry.
IV. Abundant on railway bank!; Football Field, Harrow; *Melv.* 62. Hampstead Heath; *Johns. Enum.* and *Herb. Rudge.* Mill Hill, 1846; *Morris (v. s.).* Field bet. Bishop's Wood and Finchley; field near Golder's Green; *Davies.* Kingsbury; *Farrar.* Bentley Priory.
V. Northolt; *Cole.*
VI. Finchley Common, 1842; *Herb. Hardw.* Edmonton.
VII. Bet. Kentish Town and Hampstead; *Johns. Eric.*
First record: *Johnson*, 1629.

LYSIMACHIA, *Linn.*

548. L. vulgaris, *L.*   *Loosestrife.*
L. lutea, Ger. em. (Blackst.).
Cyb. Br. ii. 297; iii. 490. Curt. F. L. f. 5 (drawn from a Thames specimen).

Sides of streams, ditches, and ponds; rather rare. P. July—September.
I. Near Harefield!; *Blackst. Spec.* 50. In the meadows near Uxbridge Moor, very frequent; *Blackst. Fasc.* 56. Round Coppin, near Uxbridge, 1839, D. Cooper; *Herb. Young.*
II. Colne, near Yewsley!; Staines!; *Newb.* Thames at Laleham, 1815; *Herb. G. & R.* Bet. Sunbury and Hampton!; *Newb.* Bet. Hampton Court and Kingston Bridge!; *Bloxam.* Bet. Strawberry Hill and Teddington by the river, very abundant.
III. By the river-side near Twickenham!; *M. & G.* 264. Bet. Twickenham

Q

and Worton. Hounslow; *Cat. Lond.* 31. By the Cran in several places.

IV. Side of the River Brent; *Herb. Hardw.* and *Farrar.* Field near North End, Hampstead; *Burnett,* 105.

V. By the Brent, Perivale; *Lees.*

VII. Banks of Lea from Clapton to Tottenham, G. Francis; *Cooper Supp.* 12. Ditches, Hackney Wick.

First record: *Blackstone,* 1737. *L. lutea minor,* in Bow River; *Merr. MSS.,* a smaller form.

549. L. Nummularia, *L.*   *Creeping Jenny.*
Nummularia (Ger., Johns., Blackst.).
Cyb. Br. ii. 299. Curt. F. L. f. 3.
Damp meadows and ditchbanks; common. P. June, July.

'I. Harefield!; *Blackst. Fasc.* 65. Ruislip. Woodready. Stanmore Heath. Elstree. South Mims.

II. Staines. Bushey Park.

III. Hounslow Heath. Near Twickenham Ry. Station. Meadows near Richmond Bridge. Roxeth; *Melv.* 63.

IV. Highgate Wood; *Johns. Eric.* Hampstead Heath; *Morris* (v.s.). Pinner Drive; Football Field; *Melv.* 63. Harrow Weald.

V. Banks of Thames, Fulham; *Faulkner,* 22. Near Hanwell Church; *Cherry.* Horsington Hill.

VI. Hadley!; *Warren.* Whetstone. Edmonton.

VII. [Banks of Thames, right against the Queene's palace of Whitehall; *Ger.* 505.] Duval Lane, bet. Islington and Hornsey, 1816; *Bennett* (v. s.). Bet. Primrose Hill and Kilburn, 1831; *Herb. Hardw.* [Copenhagen Fields, 1841; *Herb. Hardw.*] By the canal near Kensal Green; *Warren.* South Heath, Hampstead. Millfield Lane. By the reservoirs in Ken Wood Grounds.

First record: *Gerarde,* 1597.

550. L. nemorum, *L.*   *Yellow Pimpernel.*
Anagallis lutea (Ger., Merr., Blackst.). *A. fl. luteo* (Johns.).
Cyb. Br. ii. 300; iii. 493. Curt. F. L. f. 5.
Damp woods; rather rare. P. May—August.

I. In Scarlet Spring, abundantly; *Blackst. Fasc.* 6. Pinner Wood; *Melv.* 63. In the Old Park Woods, Harefield.

IV. In Hampstead Wood; *Ger.* 497, *Johns. Eric., Merrett,* 7, *and other authors.* Bishop's Wood, abundant. Lane at Stanmore, 1815; *Herb. G. & R.* Wembley Park; *Cole.* Kingsbury; Stonebridge; *Farrar.*

VI. Road from Muswell Hill to Highgate; *M. & G.* 270. Hadley!; *Warren.* Winchmore Hill Wood. The Alders, Whetstone.

VII. Cain (=Ken) Wood; *Newton MSS.* and *Mart. App. P. C.* 68.

First record; *Gerarde,* 1597.

ANAGALLIS, *Linn.*

551. A. arvensis, *L.*   *Pimpernel.*

A. mas, Ger. em. (Blackst.).

Cyb. Br. ii. 301 ; iii. 493.   Syme E. B. vii. t. 1146.

Cultivated fields and waste ground ; rather common.   A.   June—Sept.

I. Harefield! ; *Blackst. Fasc.* 6.   Ruislip ; *Melv.* 63.

II. West Drayton! ; bet. Sunbury and Hampton! ; *Newb.*   Staines Moor.
Tangley Park.   Teddington.

III. Near Isleworth, 1815 ; *Herb. G. & R.*   Twickenham.

IV. Harrow Weald! ; *Melv.* 63.   Stanmore ; *Varenne.*   Hampstead ; *Irv.
MSS.*   Sudbury ; *Farrar.*

V. Near Brentford! ; *Newb.*   Chiswick.   Horsington Hill.

VI. Edmonton.

VII. Plentiful about Brompton and Kensington ; *Irv. H. B. P.* 411.   Garden
in High Street, Camden Town.

First record: *Blackstone,* 1737.   The plant with blue flowers (*A. cærulea,
Sm.*) has not been met with.

[552. A. tenella, *L.*

Nummularia minor fl. purpurascente, Ger. em. (Blackst.).

Cyb. Br. ii. 302.   Syme E. B. vii. t. 1148.

Wet places on heaths, &c. ; very rare.   P.   July—September.

I. On Harefield Moor, abundantly ; *Blackst. Fasc.* 65.

IV. On Hampstead Bog ; *Blackst. Spec.* 60.   Ibid., J. Bliss ; *Park Hampst.* 29.

VII. Ken Wood, Hunter ; *Park Hampst.* 29.

First record: *Blackstone,* 1737 ; last, *Hunter,* 1813.]

CENTUNCULUS, *Linn.*

553. C. minimus, *L.*

Centunculus, Cat. Giss. (Blackst.).

Cyb. Br. ii. 303 ; iii. 493.   Curt. F. L. f. 3 (drawn from a Hounslow
specimen).

Damp places on heaths ; very rare.   A.   July.

II. Near Hampton Court ; *Huds.* ii. 63.

III. Low marshy ground near the paper-mills on Hounslow Heath, Mr.
Watson ; *Blackst. Spec.* 13.   On Ashford Common in tolerable
plenty in moist depressed situations ; passing from Ashford to
Hounslow Heath in similar situations in greater plenty ; *Curt. F. L.*
Ibid. ; specimens in *Dicks. H. S.* fasc. 7.

First record: *W. Watson,* 1746 ; last, *Dickson.* about 1795.   Probably
still to be found in III., though not recently noticed.

SAMOLUS, *Linn.*

554. S. Valerandi, *L.*   *Brookweed.*

Cyb. Br. ii. 303.   Curt. F. L. f. 4.

Marsh ditches; very rare. P. July—September.

VII. In the Isle of Dogs!; *Mart. Tourn.* ii. 235. About Blackwall; *M. & G.* 329. Bet. Poplar and the Isle of Dogs, Mr. Jones; *With.* ii. 1, 221. Ditch close by the timber-dock, 1866.

First record: *J. Martyn*, 1732. Likely very soon to be extinct.

PLANTAGINACEÆ.

PLANTAGO, *Linn.*

555. P. Coronopus, *L.*　　*Buck's-horn.*
Cornu cervinum (Ger., Park.).　*P. fol. lacin. Coronopus dicta, R. Syn.* iii. (Blackst.).
Cyb. Br. ii. 311. Syme E. B. vii. t. 1168.

Dry heaths and roadsides; rather common. A.? or B. June—August.
I. On Uxbridge Common, plentifully!; *Blackst. Fasc.* 79. Harrow Weald Common; *Melv.* 64. Harefield!; *Newb.* Stanmore Heath. Hillingdon; *Warren.*
II. By Walton Bridge. Road bet. Staines and Hampton. Towing-path near Hampton Court. Hampton.
III. Drilling Ground. Hounslow Heath. Twickenham.
IV. Hampstead Heath, 1681; *Newton MSS., and other authors.* North Heath, abundant.
V. Ealing Common; *Lees.*
VII. [In Touthill Fields neere to Westminster; *Ger.* 347, and *Park. Theat.* 503.] [Hyde Park; *M. & G.* 189.]

First record: *Gerarde*, 1597.

556. P. lanceolata, *L.*　　*Ribwort. Rib-grass.*
P. quinquenervia, Ger. em. (Blackst.).
Cyb. Br. ii. 310; iii. 495. Syme E. B. iv. t. 1164.

Pastures, roadsides, &c.; very common. B. or P. May—August.
Throughout all the districts.
VII. Grosvenor Square; Hyde Park; Kensington Gardens; *Warren (v.s.).* Isle of Dogs. Hackney Wick.

＊Var. β. *P. Timbali,* Jord. Syme E. B. vii. t. 1165.
In corn and clover fields. August, September.
II. Tangley Park.
III. Near Whitton Lane.

First record: *Gerarde,* 1597. *P. quinquenervia rosea,* Ger., *the Rose Plantain,* is a monstrosity. It is recorded from (VII.) [a field in a village called Hoggesdon (=Hoxton), found by . . . Mr. James Cole;

Ger. 342.] Below Hamstead Church in the way to London ; *Merrett,* 95. Figured in *Pet. H. B. Cat.* iv. fig. 4.

557. P. media, *L.* *Lamb's Tongue.*

P. incana, Ger. em. (Blackst.).

Cyb. Br. ii. 309 ; iii. 495. Syme E. B. vii. t. 1165.

Pastures, roadsides, and waste ground ; rather common. P. June— August.

I. Abundant near Harefield ; *Blackst. Fasc.* 78. Pinner Wood, F. W. Longman ; *Melv.* 63. Bet. Yewsley and Iver Bridge ! ; *Newb.* Waxwell, Pinner, abundant.

II. Staines Churchyard ; *M. & G.* 185. Staines Common. Staines Moor. Bet. Staines and Hampton, frequent. Towing-path bet. Hampton Court and Kingston Bridge, abundant. Bet. Teddington and Strawberry Hill, abundant.

III. Twickenham, in several places.

IV. Hampstead ; *Irv. MSS.*

V. Near canal, Greenford ; *Melv.* 63.

VI. Railway near Edmonton.

VII. [On the grass plots in the British Museum gardens, in great plenty ; *Forst. Midd.*] Site of Exhibition of 1862, South Kensington ! ; *Naylor.*

First record : *Rand,* about 1720. *Pl. nostras latifol. minor incana trinervis,* Pluk. ; (III.) On Hounslow Heath, Rand ; *Dill. in R. Syn.* iii. 314, is a small form.

558. P. major, *L.* *Plantain. Way-bread.*

P. latifol. vulgaris, Park. (Blackst.).

Cyb. Br. ii. 309. Syme E. B. vii. t. 1162.

Roadsides, waste ground, &c. ; very common. P. June—September. Throughout all the districts.

VII. Kensington Gardens. Hyde Park. Green Park. Hackney Wick. St. Paul's Churchyard, &c.

Var. β. *P. intermedia,* Gilb.

I. Near Harefield ! ; *Newb.*

II. Hanworth Park.

III. Twickenham.

V. Acton ! ; *Newb.*

VII. Isle of Dogs.

Var. β is perhaps *P. latifolia minor annua,* Doody (*P. latifol. glabra minor,* C. B.), found on Hounslow Heath ; *Doody MSS.* and *R. Syn.* iii. 314 ; Petiver's *Dented Plantain, H. B. Cat.* t. iv. f. 2, may be a figure of it.

First record : *Doody,* about 1700. A small form with larger hairy leaves and slender spikes, *P. quinquenervia latifol. fol. sinuato hirsutiusculo,*

noticed about London by Rand ; *Budd. MSS.* A monstrosity with a
paniculate inflorescence is called the *Besom Plantain*, *P. latifol. spiralis*
(Pet.), *P. latifol. spica multiplici sparsa*, *C. B.* (Budd.), and *P. major
panicula sparsa*, *J. B.* (R. Syn. iii.). Doody found it in Enfield Marsh ;
Dill. in R. Syn. iii. 314 ; there are specimens in *Herb. Pet.* cli. f. 24 ;
and we have seen examples from Shepherd's Bush collected in 1848
by Dr. Morris. It is well figured in Petiver's *H. B. Cat.* iv. fig. 5. (See
Masters' *Vegetable Teratology*, pp. 108–111.)

[P. Psyllium, *L.* About Chelsea Waterworks, Miller's Gard. Dict. ;
Cullum, 59.]

LITTORELLA, *Linn.*

559. L. lacustris, *L.*

Plantago palustris gramineo fol. monanthos parisiense, Inst. R. H.
(Blackst.). *Ray's Long-threaded Toad-grass* (Pet.). *Plantago uniflora*
(Huds.).

Cyb. Br. ii. 312 ; iii. 496. Syme E. B. vii. t. 1169.

Wet places on heaths ; very rare. P. June, July.

I. In bogs on Harefield Common, abundantly ; *Blackst. Fasc.* 78.

III. Hounslow Heath ; *Pet. Conc. Gram.* 218. Near Whitton ; *Huds.* i. 53.
Particularly in the ditch on the south side of Whitton Gardens, Sir
Jos. Banks ; *B. G.* 412. Ashford Common ; *Curt. F. L.* f. 3 (under
Centunculus).

First record : *Petiver*, 1716 ; last, *Banks*, 1805. Perhaps extinct.

AMARANTHACEÆ.

AMARANTHUS, *Linn.*

A. Blitum, *L.* Cyb. Br. ii. 313. Syme E. B. vii. t. 1177. VII. Gathered
once at Stoke Newington by Mr. J. Woods, jun.; *B. G.* 412. Tottenham,
A. A.; *Herb. Mus. Brit.* Near Walham Green, in the footway to Hell-
brook Common, E. F.; *Herb. Mus. Brit.* and *Herb. Linn. Soc.* Parson's
Green, on rubbish heaps by the roadside ; *Pamplin.* Ibid., 1864; *Britten.*
Little Chelsea ; *Cole.* Very imperfectly naturalised.

560. *A. retroflexus, *L.*

Cyb. Br. ii. 313. Reich. Ic. Crit. v. 668.

Rubbish heaps and roadsides ; rare. A. August, September.

II. By the towing-path close by Hampton Court, a few plants.

III. Waste ground by the railway, Twickenham, two plants.

VII. A common weed in many gardens near London ; *Mart. Mill. Dict.*
About Chelsea ; *Irv. H. B. P.* 390. Camden Town. Hackney Wick,
a few plants.

First record : *T. Martyn*, about 1800. A hardier species than *A. Blitum*,
and apparently becoming completely naturalised.

CHENOPODIACEÆ.

CHENOPODIUM, *Linn.*

561. C. Vulvaria, *L.* C. olidum, *Curt.* (L. Cat.). *Stinking Orache.*
Atriplex olida, Ger. (Johns., Newt., Merr., Blackst.). *Vulvaria* (Brit.
Phys.).
Cyb. Br. ii. 314; iii. 496. Curt. F. L. f. 5 (drawn from a London
specimen).
Roadsides, under walls and palings, &c.; rather rare. A. August,
September.
I. Harefield; *Blackst. Fasc.* 8.
IV. Hampstead Heath; *Johns. Enum.*
V. Sandy field near Chiswick; *Fox.*
VII. About the mud walls in the fields about London, plentifully; *Brit.
Phys.* 14. Ibid.; *Merrett* 12, *Budd. MSS., Blackst. Spec.* 6, *Curt. F.
L.,* and *Burnett* iii. 176. By a dunghill on this side Hackney; [by
Sir Jno. Ouldcastle's, going from St. James's Park to Hyde Park];
Newton MSS. About Islington and Hampstead, J. Woods; *B. G.*
401. Bet. London and Hampstead in various places; in Hyde Park,
[in Pancras Churchyard]; *M. & G.* 448. [Under a wall opp. Belsize
House; *Lond. Fl.* 121.] Near ' Shepherd and Shepherdess,' Islington;
Cat. Lond. 14. [Under the wall of St. John's Wood Burial Ground,
1822; *Bennett (v. s.)*.] [King's Road, near St. Pancras Workhouse,
Camden Town, 1840; *Herb. Hardw.*] Kensington Gore, abundant on
the bank by the side of the road, and under the wall of Hyde Park.
First record: *Johnson,* 1632.

562. C. polyspermum, *L.*   *All-seed.*
Blitum polyspermum, C. B. P. (Blackst.).
Cyb. Br. ii. 314.
Gardens, waste grounds, and cornfields; common. A. July—September.
Var. a, genuinum (Syme), *obtusifolium* (L. Cat.). Syme E. B. viii. t. 1185.
I. Harefield; *Blackst. Fasc.* 11.
II. Staines Moor Farm.
III. Whitton.
IV. Hendon; *Irv. H. B. P.* 384, and *Lond. Fl.* 121. Hampstead, 1850,
Syme; *Phyt.* iv. 46.
V. Near Shepherd's Bush!; *Newb.* Perivale; *Lees.*
VI. Edmonton.
VII. Hackney, E. F.; *Herb. Mus. Brit.* Ditch bet. Pancras and Kentish
Town; *Doody MSS.* Near the Brecknock Arms, Camden Town,
1847; *Herb. Hardw.* Between Kentish Town and Highgate. Hack-
ney Wick. At the back of Adelaide Road, N.W. Site of Exhibition
of 1862, S. Kensington.

Var. β. C. acutifolium, Sm. Curt. F. L. f. 2.

I. Pinner Cemetery, W. M. H.; *Melv.* 64.

III. Hounslow Heath. Near Whitton Dean. Bet. Hounslow and Lampton.

IV. Bet. Sudbury Ry. Station and Apperton.

V. Apperton. Horsington Hill.

VII. On the Finchley Road; *Lond. Flora,* 121.   In a lane from Tottenham to West Green, E. F.; *Herb. Mus. Brit.* Bank of Paddington Canal by Harrow Road, 1818; *Bennett (v. s.).* Stoke Newington Common, 1865; *Grugeon (v. s.).* Eel-brook Meadow. Lower Heath, Hampstead. Craven Hill. Site of Exhibition of 1862, S. Kensington.

First record: *Blackstone,* 1737. The varieties seem about equally frequent. *Var. α* has been frequently mistaken for and labelled *Amaranthus Blitum.*

563. C. urbicum, *L.*

Cyb. Br. ii. 315; iii. 496.

Cultivated ground; very rare. A. July—September.

VII. Kensington, W. Pamplin; *New. B. G.* 101.

Var. β. C. intermedium, M. & K. *Broad Pointed Blite* (Pet.). Syme E. B. viii. t. 1194.

VII. About London; *Pet. H. B. Cat.* Near Hampstead, 1817; *Herb. G. & R.* Willshire's Garden, Kentish Town, 1846; Gloucester Place, Kentish Town, 1845; *Herb. Hardw.*

First record: *Petiver,* 1713. *Var. β* is figured in Petiver's *H. B. Cat.* viii. 8. Seems to have been at one time common about London.

564. C. album, *L.*

Cyb. Br. ii. 318; iii. 497.

Var. α, cornfields and waste ground. *Vars. β* and *γ,* gardens, waste ground, and roadsides. A. July—September.

Var. α. C. candicans, Lam. Curt. F. L. f. 2 (a large specimen).
Probably rather rare.

III. Twickenham.

IV. Cornfields, Hampstead; *Irv. MSS.*

V. Near Lampton. Horsington Hill.

VII. Site of Exhibition, S. Kensington.

Var. β. C. viride, L. Syme E. B. viii. t. 1189.
Probably common.

I. Harefield!; *Newb.*

III. Twickenham.

V. Near Apperton.

VII. Behind Adelaide Road; *Syme.* Opposite Veitch's Nursery, Chelsea!; *Newb.* Hackney Wick. Site of Exhibition, S. Kensington.

A form intermediate between vars. *β* and *γ* was observed at (II.) Hamp-

ton and Teddington, (III.) on a dunghill near Hospital Bridge, a single
large plant, and waste ground by the station, Twickenham.

Var. γ. C. paganum, Reich. Var. *virens* (L. Cat.). Syme E. B. viii. t. 1190.
Very common through all the districts.

VII. In many places. Parson's Green. Kensington Gore. St. James's
Park. Green Park. South Heath, Hampstead. Primrose Hill.
Hackney Wick. Isle of Dogs.

First record: *Buddle*, about 1710. *C. crasso et obtuso fol. oleæ*, bet. Stepney
and Hackney; *Dill. in R. Syn.* ii. 156, is a thick-leaved form of *C. album.* Others forms are *Atriplex sylvestris fol. paululum sinuato fructu
magno rariore spicata*, about London, very frequently; *Budd MSS.* vol.
vi., *Herb.* cxvii. f. 37: and *A. sylv. fol. breviore lato in rotunditatem
tendente, &c.*, about London, rarely, first noticed by Mr. Rand; *Budd.
MSS.* vi. Figured in *Pet. H. B. Cat.* viii. fig. 4.

C. opulifolium, Schrad. V. On waste heap by the Canal, Apperton, 1867.
VII. In several places near Parson's Green, shown us by Mr. Irvine,
1862. Likely to be found more commonly.

565. C. ficifolium, *Sm.*
*Atriplex sylv. fol. oblong. angusto, magno utrinque sinu donato fructu
multo racemoso* (Budd.). *Buddle's Fig Blite* (Pet.). *C. viride* (Curt.).
Cyb. Br. ii. 319; iii. 497. Curt. F. L. f. 2.
Gardens and waste places; common?. A. July—September.
I. Pinner Marsh; *Hind.* Bet. Yewsley and Iver Bridge!; *Newb.*
II. Bet. West Drayton and Yewsley!; *Newb.* In a potato field just above
Kingston Bridge.
III. Roxeth; *Melv.* 64. Hounslow.
IV. Hampstead, 1850, Syme; *Phyt.* iv. 46 & 861. Bet. Hendon and
Finchley!; *Newb.* Kenton; *Melv.* 64. Field in Bishop's Wood.
Burgess Hill.
V. Twyford!; Shepherd's Bush, abundant; *Newb.* About Apperton in
several places.
VII. Round London, not uncommon; *Budd. MSS.* vi. Portland Town, 1827
–30; *Varenne.* Fulham, Kensington, &c., Pamplin; *New B. G.* 101.
Primrose Hill, 1838, Mr. Borrer; *Varenne.* Walham Green,
G. Francis; *Coop. Supp.* 12. Bet. Kentish Town and Hampstead,
T. Lambert, 1839; *Herb. Hardw.* Very plentiful near Notting
Hill, 1852, Syme; *Phyt.* iv. 861; and *Herb. Mus. Brit.* Opp. Veitch's
Nursery, Chelsea!; Isle of Dogs!; *Newb.* Near Paddington
Cemetery; *Warren.* South Heath, Hampstead. Kensington Gore, one
plant. Site of Exhibition of 1862, South Kensington. Sandy End
Fulham. South Heath, Hampstead. Behind Adelaide Road, N.W.
First record: *Buddle*, about 1710; also first as a British plant. Speci-
mens in *Budd. Herb.* vol. cxvii. fol. 38. Well figured in *Pet. H. B. Cat.*
viii. fig. 4.

566. C. murale, *L.*

Blitum, Morisono Atriplex proc. fol. sinuat. &c dictum, R. Syn. iii. (Hill).
Cyb. Br. ii. 317; iii. 497. Curt. F. L. f. 6 (drawn from a London
specimen).

Roadsides and waste ground; rather rare. A. July—September.

II. Feltham, 1867.

III. In some plenty bet. Hounslow and Lampton. By palings of a wood-
yard opp. Post Office, Whitton.

V. [Near Acton, but scarce; *Newb.*]

VII. Frequent about London; *Hill,* 125. Plentifully on most of the great
roads leading from the metropolis; Edgware Road, abundant; *Curt.*
F. L. Regent's Park, 1817; *Herb. G. & R.* Kensington; *Pamplin*
(*v. s.*). Willshire's Garden, Kentish Town, 1846; back of Camden
Road Villas, 1847; *Herb. Hardw.* N.E. side of Kensington Gardens,
1868; Kilburn Park Estate!; *Warren.* Opposite Veitch's Nursery,
Chelsea, towards the river.

First record: *Hill,* 1760.

567. C. hybridum, *L.*

Cyb. Br. ii. 318; iii. 497. Curt. F. L. f. 4.

Cultivated ground; very rare. A. August, September.

III. A single plant in waste ground by the river near Twickenham Church,
in 1867.

VII. About London, Dr. Watson; *Hill,* 125. Not very scarce about Chelsea;
Irv. H. B. P. 386. Parson's Green; Chelsea College, 1864; *Britten.*

First record: *Petiver,* 1713. Petiver's figure, *H. B. Cat.* viii. fig. 7, is characteristic, and is marked as occurring in the London district.
Curtis's figure was drawn from a Battersea specimen, where the plant
still grows.

568. C. rubrum, *L.*   *Goosefoot.*

Atriplex sylvestris latifolia sive Pes anserinus * (Johns., Merr.). *Sharp*
Pointed Blite (Pet.).

Cyb. Br. ii. 316; iii. 496. Curt. F. L. f. 6.

Dungheaps and rich soil; common?. A. July—September.

II. Bet. W. Drayton and Yowsley!; *Newb.* Staines Moor Farm.

III. Harrow; *Melv.* 64.

IV. Hampstead Heath; *Johns. Eric.* Field bet. Sudbury Ry. Station and
Apperton.

V. Twyford!; *Newb.* Near Lampton. Wyke Green.

VI. Edmonton.

VII. About London; *Merrett,* 12, *Pet. H. B. Cat.* Bet. Kensal Green and
Notting Hill, 1862; *Britten.* [Among the new buildings near Bedford
Square, 1789, confirmed by Curtis himself to be *C. rubrum*; *Herb.*

. * Buddle, following Dale, who clearly distinguished the two species, makes this a synonym of 563, *C. intermedium.*

Linn. Soc.] Walham Green, G. Francis; *Coop. Supp.* 12. Eel-brook Meadow, Parson's Green. South Heath, Hampstead. Primrose Hill. College Street, Camden Town. Bet. Kentish Town and Highgate. Behind Adelaide Road, N.W. Thames Embankment opp. Somerset House, 1866. Hackney Wick, abundant. Isle of Dogs.

First record: *Johnson*, 1629. *Atriplex sylv. tota rubra*; about St. Giles in the Fields; *Merrett*, 13, is a common state. 'Mr. Leighton has sent me a variety with more triangular leaves, shorter spikes and larger seeds, grown in his garden from seed obtained near London;' *Bab. Man.* i. 250. This seems to be a plant of which we have a specimen labelled in Sir J. E. Smith's hand, '*Blitum sp. n.?* near London.' A dwarf state (var. *pseudo-botryoides*, Wats.) occurs at (IV.) near Sudbury, and in many other places.

569. C. glaucum, *L.*
Blitum procumbens fol. Botryoidis subtus incano, Rand (Budd.). *Atriplex Botrys fol. subtus glauco* (Pet.). *C. angustifol. laciniat. minus, Inst. R. H.* (Dill.).
Cyb. Br. ii. 320; iii. 497. Syme E. B. viii. t. 1198.
Waste places and roadsides; rather rare. A. August, September.
II. Staines Moor Farm, abundant, 1867.
III. Roxeth; *Melv.* 64.
IV. Kenton; *Melv.* 64. Beyond Kilburn and towards Edgware, 1825, Pamplin; *New B. G.* 101.
VII. Mr. Rand . . . first showed it me as pretty common about London; *Budd. Herb. and MSS.* [Plentifully just going into Tothill Fields, near the road next Westminster; *Pet. Herb. and Bot. Lond.* ii.] [By the roadside in various places bet. Little Chelsea and Hyde Park Corner, 1790–95, Mr. Haworth; *Pamplin.*] In Kates Lane bet. Newington Common and Clapton; by a dunghill near Kentish Town, E. F.; *Herb. Mus. Brit.* [Whitechapel Fields; *Forst. Midd.* Near the lead mills at Hoxton, Woods; *B. G.* 401.] Paddington, 1801, Dawson Turner; *Herb. Smith.* Walham Green, G. Francis; *Coop. Supp.* 12. [Abundant on waste ground behind Euston Square; *Irv. MSS.*] [Downshire Hill, Hampstead, 1852; *Herb. Hardw.*] In great abundance in 1867 at the back of the houses in Adelaide Road, N.W., where it was shown us by Mr. Boswell-Syme, who stated that there had been previously but two or three isolated plants there.

First record: *Rand*, about 1705; also first as a British plant. Petiver's figure of *Rand's Oak Blite, H. B. Cat.* viii. fig. 1, is a better one of the London plant than that in *E. B.*

570. C. Bonus-Henricus, *L.* *Allgood. Wild Spinach. Fat Hen. Good King Henry. English Mercury.*
Bonus Henricus, Ger. (Johns., Blackst.).
Cyb. Br. ii. 320. Curt. F. L. f. 3.

Roadsides and waste places; rather common. P. July—August.

I. Harefield, not uncommonly; *Blackst. Fasc.* 11.

II. Staines and Hillingdon Churchyards; *M. & G.* 446. Hampton Court Barracks; *Herb. G. & R.* By the river-side, Hampton Court. Under a large *Quercus Ilex*, bet. the cross avenue and the ponds, Bushey Park.

III. By garden wall, the Grove, Harrow; *Melv.* 65. By the river near Richmond Bridge. Bet. Twickenham Park and Isleworth.

IV. Stanmore; *Varenne.*

V. About Lampton.

VI. Bury Street, Edmonton. Enfield Green.

VII. [Bet. Kentish Town and Hampstead; *Johns. Eric.*]

First record: *Johnson*, 1629.

Beta vulgaris, *L.* VII. Site of Exhibition of 1862, South Kensington.

ATRIPLEX, *Linn.*

[**571. A. marina,** *L.* A. littoralis, *var. β.* (Syme E. B. and L. Cat.) *Narrow-leaved indented Orache* (Pet.).

Cyb. Br. ii. 328. Supp. 66, 98. Syme E. B. viii. t. 1201.

Ditches; very rare.

VII. Ditches between Old Street and Islington; differs little or nothing from that which grows in the salt marshes; *Pet. Bot. Lond.* 274. I have seen it between this and Islington about Mount Mill; *Pet. Herb.* cli. fol. 42. *Pet. H. B. Cat.* vii. 4, is a figure of *A. marina*, and is still marked as a London species. It still grows on the Kent and Essex shores of the Thames abundantly.]

572. A. patula, *Wahl.*

Var. a. A. angustifolia, Sm. *A. sylvestris angustifol., Ger.* (Blackst.). *A. sylv. Polygoni aut Helxcines fol., Lob.* (Johns.). Cyb. Br. ii. 325; iii. 499. Syme E. B. viii. t. 1202.

Cultivated and waste ground; probably very common. A. July—September.

I. Harefield; *Blackst. Fasc.* 8.

II. Staines. Bet. Hampton Court and Kingston Bridge.

III. Wood End; *Melv.* 65. Twickenham in several places. Hanworth Road.

IV. Hampstead; *Irv. MSS.*

V. Twyford!; *Newb.* Southall, frequent. Apperton.

VI. Edmonton.

VII. Bet. Kentish Town and Hampstead; *Johns. Eric.* Islington, Woods; *Herb. Linn. Soc.* South Heath, Hampstead.

Var. β serrata, Syme. *A. erecta, Huds.* (Bab. in part, and other authors). Cyb. Br. ii. 326. Supp. 66, 98.

Waste and cultivated land, especially cornfields; rather rare. A. August, September.

I. Wood Hall; Pinner; *Hind.* Harefield!; *Newb.*

II. Bet. Yewsley and West Drayton!; bet. Sunbury and Hampton!; *Newb.*

III. Wood End; *Hind.* Near Hounslow Ry. Station.

V. Northolt; *Hind.* Horsington Hill.

We have not seen a figure of this variety. Babington quotes *E. B.* 2223 with approval, but that evidently represents *var. γ.*

Var. γ. A. erecta, Huds. (Syme, Bab. in part). *A. hastata, var. γ* (L. Cat.). E. B. 2223,* reproduced in Syme E. B. viii. t. 1203.

Waste ground; apparently very rare. A. August, September.

III. On a piece of waste ground near the railway station, Twickenham, 1867. Pointed out to the Rev. W. W. Newbould, who recognised it as Hudson's *A. erecta,* only previously known by the specimen in the Smithian Herb., from which the E. B. figure was drawn (*v.* Syme E. B. viii. 30).

VII. Whitechapel Fields; *Mart. App. P. C.* 72. Perhaps this variety.

First record: *Johnson,* 1629.

573. A. hastata, *L.*

Cyb. Br. ii. 324; iii. 498.

1. *A. deltoidea,* Bab.

A. silv. sinuata, Lob.† (Johns.). Cyb. Br. Supp. 66, 98. E. B. Supp. 2860, reproduced in Syme E. B. viii. t. 1204 (drawn from a Middlesex specimen).

Waste ground and cultivated fields; very common. A. August, September.

In all the districts.

VII. Bet. Kentish Town and Hampstead; *Johns. Eric.* Portland Town, 1827–30; *Varenne.* Isle of Dogs!, 1838, E. F.; *Herb. Mus. Brit.* Near the railroad at Primrose Hill, 1839; *Babington in E. B. Supp.* Parson's Green. Chelsea. Kensington Gore. Site of Exhibition of 1862, South Kensington. Adelaide Road. Hackney Wick.

2. *A. hastata,* L. (Bab.). *A. Smithii* (Syme E. B.). *A. patula* (Smith). Syme E. B. viii. t. 1205.

Gardens, fields, and roadsides; probably common. A. July—September.

II. West Drayton!; *Newb.* Bet. Kingston Bridge and Hampton Court.

III. Harrow; *Hemsley.* Hounslow!; *Newb.* Near Twickenham Ry. Station.

IV. Harrow district, everywhere; *Melv.* 65. Hampstead; *Irv. MSS.*

V. Near Southall.

VII. Islington, J. Woods; *Herb. Linn. Soc.* Isle of Dogs!, E. F.; *Herb. Mus. Brit.* Near Lea Bridge; *Cherry* (*v. s.*). South Heath, Hampstead. Adelaide Road. South Kensington. Thames Embankment. Hackney Wick.

First record: *Johnson,* 1629; also first as a British plant?.

* Wrongly quoted as 259 in Syme E. B. viii. 29.

† Lobel's figure (*Icones,* 254) is a fair one of *A. deltoidea.*

(A. Babingtonii, *Woods*.   *A. maritima ad fol. basin auriculata procumbens et ne vix sinuata*, Pluk. *Alm.* (Dill.).   Cyb. Br. ii. 328 ; iii. 498.   Supp. 66, 98.   Syme E. B. viii. t. 1206.   III.   Roxeth, 1860 ; *Hind.*   VII. Dunghills about London, Doody ; *Dill. in R. Syn.* iii. 152.   Plentifully bet. Kilburn and Kensal Green, 1857 ; *Hind in Phyt. N. S.* v. 204.   A sea-coast species.   Smith (*E. Fl.* iv. 258) refers Plukenet's plant to his *A. patula, var.* γ, which (*fide* Syme) is, according to his herbarium, *A. Babingtonii*.   Mr. Hind considers his plants to be this species, but it is likely that both he and Doody may have mistaken forms of *A. hastata* for it.)

POLYGONACEÆ.

RUMEX, *Linn.*

[574. R. maritimus, *L.*
Lapathum fol. acuto, fl. aureo, C. B. (Doody).   *Bur Gold Dock* (Pet. H. B. Cat.).   *Lapathum aureum* (Pet. Herb. and Bot. Lond.).
Cyb. Br. ii. 347.   Syme E. B. viii. t. 1212.

Wet places ; very rare.
VII. Mr. Rand first showed me this beautiful dock growing plentifully in a moist place near Burlington House, and he has likewise found it behind Montague House ; *Doody MSS.*   Lately found about London ; *Pet. Bot. Lond.* ii.

First record : *Rand,* about 1700.   No recent record.   Marked as a London species in Petiver's characteristic figure, *H. B. Cat.* ii. 8 ; there are also specimens in his herbarium, vol. cli. fol. 9.   *Lapathum aureum angustifolium,* lately discovered by Mr. Isaac Rand about St. Giles ; *Pet. Bot. Lond.* ii.   The specimen in *Herb. Pet.* cli. fol. 9, is quite undeterminable.]

575. R. palustris, *Sm.*
Lapathum longo angustoque fol. Anthoxantho plurimum accedens, verticill. rar. caulem cingent. sem. majori (Pluk., Budd., Pet. Herb.).   *Lap. angustifol. anthoxantho, J. B. simile* (Doody).   *Gold Dock* (Pet. H. B. Cat.).   *R. maritimus, L.* (Curt.).
Cyb. Br. ii. 347 ; iii. 503.   Supp. 67, 99.   Curt. F. L. f. 3.

Damp waste ground, sides of ponds, &c.; rather rare.   B. or P.?   July —September.
III. Near Hounslow Ry. Station, one plant ; *Newb.*
IV. Hampstead!, 1850 and 1852, Syme ; *Phyt.* iv. 861.   Near Child's Hill Gate.   By the large pond on the North Heath.
V. Near Chiswick, J. Lindley ; *Brit. Ent.* 191.   Acton Green ; *Newb.*
VII. [Lately found in Tothill Fields by Mr. Isaac Rand ; *Pluk. Mant.* 112.] [By Lamb's Conduit ; betwixt the backs of Montague House and St. Giles's Road, and many places about London ; *Doody MSS.*]   About

London, frequent; *Budd. MSS.* and *Herb. Pet.* cli. 9. [By Annisy-clear,† in the ditch by the road before you go to Hoxton Square, J. Sherard; *R. Syn.* iii. 142.] Camden Town Fields, 1827–30; *Varenne.* Shepherd's Bush, 1849, W. Wing; *Herb. Mus. Brit.* [Notting Hill;] Isle of Dogs!, 1852, Syme; *Phyt.* iv. 861. Ibid., 1869; *Newb.* South Heath, Hampstead, 1865, only two or three plants; *Syme.* Green Lanes, Newington, 1867!; *Newb.* [Eel-brook Common, Parson's Green, 1862.] Bet. West India Dock basin and wall, abundant, 1867; *Cherry* (*v. s.*).

First record: *Rand,* about 1700; also the first as a British plant. Curtis's plate is less happy than usual, but is better than the new one in *Syme E. B.* Petiver's *H. B. Cat.* ii. 7,‡ is a good outline. Two forms seem to grow about London; a dark-green plant with spreading branches, and a taller plant with erect branches, and with foliage of a brighter green; but we have not been able to find any good technical characters to separate them.

576. R. conglomeratus, *Murr.*
Lapathum acutum, Ger. em. (Blackst.).
Cyb. Br. ii. 346; iii. 503. Supp. 67, 99. Syme E. B. viii. t. 1210 (bad figure). Pet. H. B. Cat. ii. 3, 4 (good outline drawings).
Damp roadsides, ditches, &c.; very common. P. June—August.
Throughout all the districts.
VII. Muddy banks of the Thames about Cremorne, abundant. Thames Embankment. South Heath, Hampstead Isle of Dogs.
First record: *Blackstone,* 1737. The veins of the leaves are frequently of a blood-red colour.

577. R. nemorosus, *Schrad.* (L. Cat.). R. sanguineus, *L.* (Bab. and Syme E. B.).
Cyb. Br. ii. 345.
Var. a. R. viridis, Sibth. Syme E. B. viii. t. 1211 (not good). Pet. H. B. Cat. ii. f. 6.
Shady hedges, roadsides, and waste ground; very common. P. June—August.
Throughout all the districts.
VII. Grosvenor Square, 1869!; *Warren.* Finchley Road, near New West End. South Heath, Hampstead.

* *Var. β. R. sanguineus,* L. (L. Cat.). *Lapathum sativum sanguineum* (Merrett). *L. fol. acuto rubente, C. B. P.* (Blackst.). *R. acutus* (Curt.).

† The spring of St. Agnes le Clere, or Dame Annis de Cleare, was used so lately as the beginning of this century as a bath for rheumatic and nervous complaints; Bath Street, Shoreditch, marks the site.
‡ Babington (*Man.* vi. 291) refers this figure to *R. limosus,* Thuill., because of the leaves being narrowed at the base. (See also *Bot. Gaz.* i. 296.) They vary in this respect, and specimens from Parson's Green and Hampstead are so far quite like Petiver's figure.

Curt. F. L. f. 3 * (except the dissected details, an excellent figure).

Woods and shady places; very rare ?. P. June—August.

I. In an orchard in the lane from Harefield to Ruislip; *Blackst. Fasc.* 50. Pinner Hill; *Hind.*

IV. In the woods about Hamsted; *Merrett*, 70.

VII. Among bushes at the bottom of Pond Street (Hampstead); *Hill*, 191.

If this has no other characters but the trivial ones of blood-red stems and leaf-veins, it grows in many places, notably in a brick-field on the South Heath, Hampstead. All authors, however, seem to consider it a garden escape, and not native (see *Ger. em.* 390, *Ray Syn.* ii. 56, *Syme E. B.* viii. 42). Linnæus gives only 'Habitat in Virginiâ' (*Sp. Plant.* 476); and Withering says (ed. 2, ii. 370), 'It migrated from Virginia to Hampstead.' It seems to have been formerly cultivated as a pot-herb, and according to Boreau (*Fl. du Centre de la France*, ii. p. 552) is still grown in France, under the name of '*Sang de Dragon.*'

First record: *Merrett*, 1666.

578. R. pulcher, *L.*   *Fiddle Dock.*

Lapathum pulchrum Bononicnse sinuatum, J. B. (Merr., Budd., Ray).

Cyb. Br. ii. 347; iii. 503.   Syme E. B. viii. t. 1214.

Waste ground and roadsides; rare. B. or P. July—September.

II. Roadside bet. Hampton and Hampton Court!, H. C. W.; *New B. G.* 110. Near Teddington!; *Newb.* Hampton. Sunbury, in several places.

III. Twickenham Park, near the ry. bridge. Orleans Road, Twickenham.

IV. Hampstead; *Irv. MSS.*

VII. [Common about London; *Merrett*, 69, and *Budd. MSS.*] [In St. James his fields by Westminster; *R. Cat.* i. 188.] [In the upper Moor-fields, and in the Teynter and Bunhill-fields; *Bot. Lond.* 271, and *Pet. Midd.*]

First record: *Merrett*, 1666.

579. R. obtusifolius, '*L.*'   *Common Dock.*

Cyb. Br. ii. 345. Syme E. B. viii. t. 1215 (probably drawn from a Middlesex specimen).

Roadsides, fields, and waste land; very common. P. June—August. Through all the districts.

VII. Common in London. Camden Terrace, Sowerby; *Herb. Mus. Brit.* Green Park. Hyde Park. Gordon Square. Hampstead. Thames Embankment. Lincoln's Inn. Hackney. Isle of Dogs, &c.

First record: *Irvine*, about 1830.

580. R. pratensis, *Mert. & Koch.*

Cyb. Br. ii. 344. E. B. Supp. 2757, reproduced in Syme E. B. viii. t. 1216 (drawn from a Middlesex specimen).

---

* In some copies this is coloured as *var. a*, or even as *R. conglomeratus*; the copy in the British Museum Reading-room is here referred to.

POLYGONACEÆ. 241

Roadsides and waste ground; rather common. P. June—August.

I. Harefield! ; *Newb.* Harrow Weald Common in several places.

II. Bet. West Drayton and Yewsley! ; *Newb.*

III. Road from Hounslow to Hanworth in several places.

IV. By the clay-pits, Harrow Weald. Waste ground by the roadside at Child's Hill turnpike gate.

V. Brickfields, Wyke Green.

VII. Camden Terrace, Camden Town, and other roads to the north of London, Sowerby ; *Herb. Mus. Brit.* and *E. B. Supp.* Bet. Hampstead and Camden Town, 1852, Syme; *Phyt.* iv. 861. On mud thrown out on the Lower Heath, Hampstead, 1865; *Syme.* Parson's Green; *Britten.* Abundant in a brickfield by the South Heath, 1866.

First record : *Sowerby,* about 1832. The figure in *Brit. Ent.* xvi. t. 191, called in the text *R. pratensis* and 'gathered at Chiswick by John Lindley, Esq.' is evidently *R. obtusifolius.*

581. R. crispus, *L.*　　*Curled Dock.*
Lapathum fol. acuto crispo, C. B. P. (Blackst.).
Cyb. Br. ii. 343. Syme E. B. viii. t. 1218.

Roadsides, pastures, and waste places ; very common. P. June—August.
Through all the districts.

VII. Frequent. [River wall of the Temple Gardens.] [Courtyard of Burlington House.] South Heath, Hampstead. Primrose Hill. Blackstock Lane, Highbury. Zoological Gardens. Parson's Green. Isle of Dogs.

First record : *Blackstone,* 1737.

582. R. Hydrolapathum, *Huds.*　　*Great Water Dock.*
Lapathum aquat. fol. cubitali, C. B. P. (Blackst.). *R. aquaticus, Sm.* (Winch, Irv.).

Sides of streams and ponds; common. P. July, August.

I. Harefield! ; *Blackst. Fasc.* 50. Colne, near Iver Bridge! ; *Newb.*

II. Ditches in Stanwell and Staines Moors, plentiful. Staines Common. Ponds in Bushey Park! ; *Winch MSS.* Hampton! ; *Newb.*

III. By the Cran, Hatton, Hounslow Heath, Twickenham. By the Duke's River, Isleworth. Twickenham Park.

IV. Ponds, Hampstead; *Irv. MSS.*

V. Canal; *Melv.* 66. Near Twyford! ; *Newb.* Canal at Brentford ; *Hemsley.* Banks of Thames, Fulham ; *Faulkner,* 22.

VI. Edmonton.

VII. By the reservoirs in Ken Wood grounds. Ditches, Hackney Wick.

First record : *Blackstone,* 1737.

583. R. Acetosa, *L.*　　*Sorrel.*
Acetosa pratensis, C. B. P. (Blackst.).
Cyb. Br. ii. 349. Syme E. B. viii. t. 1223.

Meadows and pastures; very common. P. May—July.

Throughout all the districts, but less common in II. and III.

VII. Forms a considerable proportion of the hay made in the northern outskirts of town, but few meadows are now left in this district. Tottenham; *Cherry.* Hammersmith; *Morris (v.s.).* Grosvenor Square;
Warren. Bet. Hornsey Wood and Hornsey. Highgate. Hampstead. Kilburn. Primrose Hill. Garden of Lincoln's Inn. Constitution Hill, Green Park.

First record: *Blackstone,* 1737.

584. R. Acetosella, *L.*   *Sheep's Sorrel.*

Acetosa arvensis lanceolata, C. B. P. (Blackst.).

Cyb. Br. ii. 349. Curt. F. L. f. 5.

Heaths, dry pastures, and roadsides; common. P. May—August.

I. Harefield!; *Blackst. Fasc.* 2. Harrow Weald! and Ruislip Commons; *Melv.* 67. Stanmore Heath. Near Elstree Reservoir. Near
Potter's Bar.

II. Staines. Near Sipson.

III. Hounslow Heath; Drilling Ground, &c. Allotments, Twickenham,
very luxuriant.

IV. North Heath, Hampstead. Green Hill, Pinner Drive. Peterboro'
Road; *Melv.* 67.

V. Near Ealing!; *Newb.* Hanwell; *Warren.* Horsington Hill. Kew
Bridge Ry. Station.

VI. Hadley!; *Warren.* Colney Hatch. Edmonton.

VII. Hyde Park, in ditch by Kensington Gardens, 1868; *Warren (v.s.).*
South Heath, Hampstead. Parson's Green. Gardens of Lincoln's
Inn, 1869. Grosvenor Square, 1869.

First record: *Blackstone,* 1737. Confined to poor sandy or gravelly
soils.

POLYGONUM, *Linn.*

585. P. Bistorta, *L.*   *Snake Weed. Bistort.*

Bistorta major, Ger. (Blackst.).

Cyb. Br. ii. 332. Curt. F. L. f. 1 (probably from a Hampstead specimen).

Moist meadows and woods; rather rare. P. June—September.

I. In the meadows near Uxbridge; *Blackst. Fasc.* 10 and *Lightf. MSS.*
Terrilands, Pinner; *Hind.* Field near Ruislip Reservoir.

III. On the inner slope of the moat, Headstone Farm; *Melv.* 66.

IV. Plentifully in a meadow by the side of Bishop's Wood, near Hampstead!; *Curt. F. L.* It is now almost eradicated from the meadow,
but grows on the bank at the border of the wood, and extends a
little way into the latter. Hampstead Common, 1809; *Winch MSS.*
Meadows bet. Mill Hill and Elstree; *Lond. Flora,* 123.

VI. Top of Muswell Hill, J. Woods; *B. G.* 404. Field at Muswell Hill,
1867; *Cherry (v.s.).* Near Warren Lodge, Edmonton.

VII. Ken Wood!; *Park Hampst.* 30.

First record: *Blackstone,* 1737.

586. P. amphibium, *L.*

Potamogeiton angustifol., Ger. (Merr.). *Persicaria salicis fol. perennis.*
Potamog. angustif. dicta, R. Syn. iii. (Blackst.).

Cyb. Br. ii. 333. Syme E. B. viii. t. 1242 (water form). Curt. F. L.
f. 4 (land form).

In ponds, streams, and marshes, also on rubbish and waste ground, and
by roadsides; very common. P. July—September.

Through all the districts.

VII. Bayswater Canal in Kensington Gardens, 1817; *Herb. G. & R.* Abundant in the Serpentine, 1868. Hackney Marsh!; *Mart. App. P. C.*
66. Round Pond, Kensington Gardens; *Pamplin.* At the back of
Adelaide Road; *Syme.* Zoological Gardens. Eel-brook Meadow.
Isle of Dogs.

First record: *Merrett,* 1666. Both states are figured in *Pet. H. B. Cat.*
iii. fig. 12. Merrett's *P. fol. salicinis,* in a pond in St. James's Park;
Merrett, 93, is probably this species.

587. P. lapathifolium, *L.*

P. Pensylvanicum, L. (Curt.) in part.

Cyb. Br. ii. 335. Curt. F. L. f. 1.

In cultivated ground; common. A. July—September.

I. Harefield!; *Newb.*
II. Bet. Sunbury and Hampton!; *Newb.* Staines. Hampton. Bet.
Hampton Court and Kingston Bridge.
III. Near Hounslow. Twickenham.
IV. Harrow Weald. Bet. Sudbury Ry. Station and Apperton. Waste
ground by Child's Hill Gate. Hampstead; *Irv. MSS.*
VI. Edmonton.
VII. Homerton; *Cherry* (v. s.). Near South Heath, Hampstead. Back of
Adelaide Road, N.W. Near Cremorne. Site of Exhibition of 1862,
S. Kensington. Mount Street. Victoria Street. Edgware Road.

First record: *Petiver,* 1713. About London; *H. B. Cat.* The form with
red perianths is probably *Persicaria major lapathifoliis, cal. floris purpureo,* pretty common about London; *Mart. Tourn.* ii. 166; mentioned
also by Curtis (*loc. cit.*). The site of the Exhibition of 1862, at South
Kensington, produced in 1867 a large crop of it.

588. P. nodosum, *Pers.* (Bab.). P. lapathifolium, *var. β nodosum*
(Syme E. B.); *var. β laxum* (L. Cat.).

P. laxum (E. B. Supp.).

Cyb. Br. Supp. 67, 98. E. B. Supp. 2822, reproduced in Syme E. B. viii.
t. 1240 (drawn from a Middlesex specimen).

Waste and cultivated ground, roadsides, &c.; very common. A. July—
September.

Through all the districts, but more abundant in those adjacent to the
metropolis.

VII. Near the railway station at Chalk Farm, 1837 (the more densely
spiked variety); *Babington in E. B. Supp.* Parson's Green; *Irv.
H. B. P.* 376. Eel-brook Meadow. Near St. Mark's College, Chel-
sea!; *Newb.* Gas Works, Sandy End. Victoria Street. Thames
Embankment. Behind Adelaide Road. Hackney Wick. Isle of
Dogs.

First record: *Doody*, about 1700. Very variable, but seems distinct from
lapathifolium in the sense of being easily recognisable. The stem is
frequently spotted and much swollen at the joints, when it is Rand's
Persicaria latifol. geniculat. caul. maculatis, frequent about London;
R. Syn. iii. 145, and the *P. Pensylvanicum var. caule maculato,* beauti-
fully figured in *Curt. F. L.* f. 1. Plants precisely like this drawing
occurred in a brickfield near the South Heath, Hampstead. A small ·
procumbent plant with the leaves white beneath, is called *Persicaria
macul. procumb. fol. subtus incanis*: about London; *Doody MSS., Budd.
MSS.* vi., and *Herb.* cxvii. f. 20. It is figured in *Pet. H. B. Cat.* iii.
figs. 9 and 10. We noticed it (V.) in cornfields on Horsington Hill, and
there is a specimen in *Herb. Harr.* from Greenford (= '*P. lapathifolium,*'
of *Melv.* 66).

589. P. Persicaria, *L.*

Persicaria maculosa, Ger. em. (Blackst.).

Cyb. Br. ii. 334. Curt. F. L. f. 1.

Damp ground, waste and cultivated; common. A. July—September.
I. Harefield!; *Blackst. Fasc.* 75.
II. River-bank, Hampton Court!; *Newb.* Bushey Park.
III. Abundant about Twickenham, where *P. nodosum* is rare. Hounslow.
IV. Hampstead; *Irv. MSS.* Harrow district, very common; *Melv.* 66.
V. Near Ealing!; near Shepherd's Bush!; *Newb.* Perivale; *Lees.*
VI. Edmonton.
VII. Eel-brook Meadow. Euston Square gardens, and gardens in the
Euston Road, 1867, as a weed. Hackney Wick. Isle of Dogs.

First record: *Petiver*, 1713. A variety with leaves hoary beneath is
figured in *Pet. H. B. Cat.* iii. fig. 8, and given as frequent about London.
It occurs (III) near Hounslow and at Fulham. Curtis's plate seems
to represent Syme's *var. β elatum,* Gren. and Godr., which is the
commoner variety about town.

590. P. mite, *Schrank.*

Cyb. Br. ii. 335; iii. 500. E. B. Supp. 2867, reproduced in Syme E. B.
viii. t. 1236 (drawn from a Middlesex specimen).

Wet places; rare. A. August—October.
II. Bushey Park. Bet. Hampton Court and Kingston Bridge by the
towing-path.
III. On soil brought from elsewhere, by the river near Twickenham Church,
abundant.

V. [Acton Green; *Newb.*]

VII. First detected by Prof. La Gasca of Madrid by the roadside at Chelsea, where . . . we have recently (1843) observed it; *Babington in E. B. Supp.* Roadside near Edgware Road Station, Kilburn ; *Warren.* Osier holt by the river, Sandy End, 1867.

First record : *La Gasca*, 1826 ; also first as a British plant.

591. P. Hydropiper, *L.*   *Water Pepper.*

Persicaria urens sive Hydropiper, C. B. P. (Blackst.).

Cyb. Br. ii. 336.   Curt. F. L. f. 1.

Sides of ditches and streams ; common. A. August—October.

I. Harefield ! ; *Blackst. Fasc.* 76. Near Potter's Bar.

II. Hampton. Bushey Park. Bet. Hampton Court and Kingston Bridge.

III. About Twickenham in many places.

IV. Hampstead ; *Irv. MSS.* Harrow district, very frequent; *Melv.* 66.

V. Perivale ; *Lees.* Sion Park. Heston.

VI. Bet. Tottenham and Edmonton. Edmonton.

VII. [Tothill Fields ; *Smith MSS.*] Hyde Park ! ; *Herb. G. & R.* Osier holt, Sandy End. South Heath, Hampstead. Kentish Town. Hackney Wick.

First record : *Blackstone*, 1737.

592. P. minus, *Huds.*

Persicaria angustifol. ex sing. genic. florens, and *Persicaria pusilla repens* (Pet.). *Pers. minor* (Budd.). *Pers. interrupta* (Gray). *P. Hydropiper,* var. β (Huds.).

Cyb. Br. ii. 337 ; iii. 501. Curt. F. L. f. 1 (drawn from a Middlesex specimen).

Marshy places ; rare. A. August, September.

II. Bushey Park, at the south corner of the first pond, 1867.

IV. Golder's Green, Hampstead ; *Irv. Lond. Fl.*

VI. Finchley Common, L. W. Dillwyn ; *B. G.* 404.

VII. [Ditchbanks in the meadows beyond Lord Peterborough's house at Westminster; *Pet. Mus.* i. p. 13. Tothill Fields ; *Pet. Midd., Blackst. Spec.* 70, *Huds.* ii. 171. Ibid., in the greatest abundance ; *Curt. F. L.*] Near London, A. H. Haworth ; *Gray,* ii. 732. Eelbrook Meadow, Parson's Green, 1864.*

First record : *Petiver*, 1695 ; also the first as a British species ? Figured in *Pet. H. B. Cat.* iii. fig. 6.

593. P. aviculare, *L.*   *Knot-grass.*

P. mas vulgare, Ger. em. (Blackst.).

Cyb. Br. ii. 337. Curt. F. L. f. 1. Syme E. B. viii. tt. 1229–1231.

---

* Rev. W. W. Newbould thinks Mr. Irvine showed it to him there so lately as 1867. We, however, failed to find it in that year, the meadow having become very dry after drainage.

By roadsides, in waste ground, cornfields, &c.; very common. A. June
—September.

Throughout all the districts.

VII. A common London plant. Chelsea. Green Park. Bedford Place.
Thames Embankment. Isle of Dogs.

First record: *Doody*, about 1700. Several well-marked forms are included
here, as was long ago noticed by Buddle and Petiver, the former of
whom distinguished four, and inclined to 'think them different species';
and the latter figured them as London plants in *H. B. Cat.* x. figs. 1–4.
Mr. Syme enumerates six sub-species, of which we have probably all
except *P. littorale*, Link. *P. agrestinum*, Jord., and *P. vulgatum*, Syme,
are certainly common enough. *P. arenastrum*, Bor., 'common near
London,' occurs at Harefield, and by the towing-path at Hampton
Court; and Mr. Newbould has seen it at Acton, &c. It is the *Thickset
Knotgrass* figured in *Pet. H. B. Cat.* x. fig. 2. *P. microspermum*, Jord.,
we have not noticed. *P. rurivagum*, Jord., is probably (see *Budd.
Herb.* cxvii. fol. 22) the *Narrow Knotgrass* of *Pet. H. B. Cat.* x. fig. 4,
and the *P. oblong. angustoque fol. C. B. P.*, found (VI.) amongst the corn
in Houndfield by Poundersend, plentifully; *Doody MSS.* and *Dill. in
R. Syn.* iii. 146. It grows (V.) in cornfields on Horsington Hill.

594. P. Convolvulus, *L.*   *Black Bindweed.*

Cyb. Br. ii. 340. Syme E. B. viii. t. 1227.

Cultivated and waste ground; probably very common. A. July—September.

I. Harefield!; *Newb.*

II. Hampton!; *Newb.*

III. Wood End; *Melv.* 66. Cornfields, Hounslow. Twickenham Park.

IV. Hampstead; *Irv. MSS.* Bet. Finchley and Hendon!; *Newb.*

V. Apperton.

VI. Edmonton.

VII. Eel-brook Meadow, Parson's Green. Green Park. Victoria Street.
Craven Hill. Site of Exhibition of 1862, South Kensington. Primrose Hill. Euston Road. Bedford Place. Isle of Dogs.

First record: *Irvine*, about 1830. No old authority. A form with the
perianth much winged (var. *pseudo-dumetorum?*) occurs at Harefield,
Twickenham, and Hackney Wick.

Fagopyrum esculentum, *Mönch.* Polygonum F., *L.* (Syme E. B.). *Buck-
wheat. Fagopyron* (Ger.). Cyb. Br. ii. 341. Syme E. B. viii. t. 1226.
About London; *Ger.* 83. Near Kingston Bridge among potatoes.
Rubbish heap near Twickenham Station. Cultivated as a food for
pheasants.

THYMELACEÆ.

DAPHNE, *Linn.*

595. * **D. Laureola,** *L.*   *Spurge Laurel.*
Lauriell or *Lowry* (Turn.†).   *Laureola, Ger. em.* (Blackst.).
Cyb. Br. ii. 351.   Syme E. B. viii. t. 1247.
Shady woods; rare.   Shrub.   February, March.
    I. In the little grove near Breakspears, plentifully; *Blackst. Fasc.* 51.
III. Harrow Park, one plant; *Melv.* 67.
  IV. Hampstead; *Irv. MSS.*
   V. At Sion; *Turn.* i. fol. 198.
VII. Kentish Town; *Mart. App. P. C.* 72.
    First record: *Turner,* 1568; also first as a British plant. In some of
the above stations probably planted for ornament.

    Aristolochia Clematitis, *L.* Cyb. Br. ii. 355. Syme E. B. viii. t. 1250.
By the river near Hampton Court, where the railings by the gardens
begin, coming from Ditton Ferry; *Watson.* River-side above Hammersmith Bridge, D. Cooper; *Herb. Young.* Escaped from gardens in
both cases.

    (Empetrum nigrum, *L.* Cyb. Br. ii. 355. Syme E. B. viii. t. 1251.
Ken Wood, Hunter; *Park Hampst.* 28. No doubt erroneous, as so
many of Hunter's statements are, unless intended as records of cultivated plants.)

EUPHORBIACEÆ.

Buxus sempervirens, *L. Box.* Cyb. Br. ii. 366. Syme E. B. viii. t. 1252.
Harefield, in woods, rarely; *Blackst. Fasc.* 12. Planted there.

EUPHORBIA, *Linn.*

596. E. Helioscopia, *L.*   *Wart-wort.*
Tithymalus helioscopius, Ger. em. (Blackst.).
Cyb. Br. ii. 356; iii. 501.   Syme E. B. viii. t. 1254.
Gardens, fields, and waste places; probably common.   A. June—October.
    I. Harefield! ; *Blackst. Fasc.* 99.
   II. Teddington! ; *Newb.* Staines Moor. Hampton.
III. Hounslow! ; *Newb.* Twickenham. Harrow; *Melv.* 68.
  IV. Hampstead; *Irv. MSS.*
   V. Lampton. Kew Bridge Ry. Station.
  VI. Hadley! ; *Warren.* Edmonton. Whetstone.
VII. Green Lanes, Newington! ; Isle of Dogs! ; *Newb.* Hackney Wick.
    First record: *Blackstone,* 1737.

† From his description; the figure is *D. Mezereum.*

597. * E. platyphylla, *Koch.*†

Tithymalus segetum longifolius, Cat. Cant. App. (Doody). *T. platyphyllos Fuchsii, Chabr.* (Blackst.). *E. platyphylla, L.* (Sm. Fl. Brit.). *E. stricta, var. β* (Sm. E. Fl.).

Cyb. Br. ii. 357. E. B. 333 (a starved form), reproduced in Syme E. B. viii. t. 1255.

Cornfields, roadsides, and waste ground; rare. A. June—September.

I. In the cornfields leading from Harefield Common to Battleswell, plentifully; *Blackst. Fasc.* 98. Near Harefield, Rev. Mr. Lightfoot; *Sm. Fl. Brit.* ii. 518. Gathered near Harefield in 1793; *E. Fl.* iv. 64. A few plants in a field a little south of Harefield Church!, 1868; *Newb.*

III. Amongst the corn near Twitnam (= Twickenham) Park, over against Richmond; *Doody MSS.* Roadside bet. Harrow and Eastcot, a single plant, W. M. H.; *Melv.* 68. Waste ground by the railway station, Twickenham!; *Syme.*

IV. Stanmore, 1827–30; *Varenne.*

VII. Site of Exhibition of 1862, South Kensington, 1867.

First record: *Doody,* about 1700. Sporadically; an introduced weed. Dillenius in *R. Syn.* iii. 312, erroneously quotes Doody's locality for *E. platyphylla* under *Tithymalus hibernicus* = *E. hiberna,* L., and in the *Hort. Elthamensis,* ii. 388, he says that Doody seems to have mistaken some other plant for it. But Doody evidently intended his MS. note to refer to the species opposite to which he wrote it, *Tith. segetum longifolius.* Dillenius has been quoted by Withering, Smith, and others.

(E. hiberna, *L.* Cyb. Br. ii. 359. Syme E. B. viii. t. 1257. Near Harefield; *Cat. Lond.* 30. No doubt *E. platyphylla* meant.)

598. E. amygdaloides, *L.*   *Wood Spurge.*

Characias (Turn.). *Tithymalus Characias amygdaloides* (Johns., Blackst.). Cyb. Br. ii. 365. Syme E. B. viii. t. 1260.

Woods and shady hedges; rather rare. P. April—June.

I. Harefield, abundantly!; *Blackst. Fasc.* 98. Pinner Wood; *Melv.* 69. Ruislip Wood.

IV. Highgate Wood; *Johns. Eric.* Stanmore; *Varenne.* Mill Hill, 1846; *Morris (v. s.).* Scratch Wood. Bishop's Wood.

V. In Sion Park; *Turn.* ii. Horsington Lane!; Greenford; *Melv.* 69.

VI. Enfield. Colney Hatch. Winchmore Hill Wood.

VII. Lane bet. Edgware Road and Hampstead, 1809; *Winch MSS.*

First record: *Turner,* 1562; also first as a British plant. Wanting in the sandy districts.

E. Esula, *L.* Cyb. Br. ii. 360; iii. 505. Syme E. B. viii. t. 1261. I. Elstree; *Herb. Rudge.* VII. North bank of North London Railway at Kilburn Station, 1862, W. P.; *Phyt. N. S.* vi. 349.

† Certainly *E. stricta* of the Linnæan Herbarium. The specimen is marked 'the Harefield plant,' in Sir J. E. Smith's handwriting.

599. E. Peplus, *L.*

Tithymalus parv. ann. fol. subrotund. non crenat., R. Syn. iii. (Blackst.).

Cyb. Br. ii. 363. Syme E. B. viii. t. 1265.

Gardens, fields and waste places; common. A. June—September.

I. Harefield!; *Blackst. Fasc.* 99.

II. Staines.

III. Harrow, abundant; *Melv.* 68. Twickenham.

IV. Hampstead; *Irv. MSS.*

V. Lampton.

VI. Edmonton.

VII. Victoria Street, Westminster. Thames Embankment. Isle of Dogs.

First record: *Blackstone*, 1737. Probably overlooked in many places.

600. E. exigua, *L.*

Tithymalus seu Esula exigua, C. B. P. (Blackst.).

Cyb. Br. ii. 363. Syme E. B. viii. t. 1266.

Cornfields, roadsides, and waste ground; rather rare. A. June—September.

I. Harefield!; *Blackst. Fasc.* 98 and *E. B.* 1336. Bet. Yewsley and Iver Bridge!; *Newb.* Pinner limepits.

II. Bet. Sunbury and Hampton!; *Newb.* Stanwell Moor.

III. Wood End; *Melv.* 68. Twickenham Park, abundant. On the railway at the level crossing, Twickenham.

IV. Hampstead; *Irv. MSS.* Stanmore; *Varenne.* By Brent Reservoir!; *Warren.* Harrow Weald.

V. Apperton.

First record: *Doody*, about 1700. A luxuriant but young plant is labelled *Tith. linariæ fol. sive Esula exigua elatior*; Mr. Doody first showed it me amongst ye corn near Thistleworth; *Herb. Budd.* cxxiii. fol. 30.

E. Lathyris, *L. Caper Spurge.* Cyb. Br. ii. 364; iii. 505. Syme E. B. viii. t. 1267. II. By the towing-path bet. Kingston Bridge and Hampton Court. III. Weed in gardens at Roxeth; Grove Garden, Harrow; *Melv.* 68. IV. Weed in gardens, Hampstead; *Irv. MSS.* Lawn of Aspen Lodge, Sudbury; *Farrar.* VII. Numerous plants on site of Exhibition of 1862, S. Kensington. A common plant in gardens, where it readily becomes a weed.

MERCURIALIS, *Linn.*

601. M. perennis, *L. Dog's Mercury.*

Cynocrambe (Ger., Johns., Blackst.).

Cyb. Br. ii. 367; iii. 506. Curt. F. L. f. 2.

Woods and shady hedges; common? March—June.

I. Harefield!; *Blackst. Fasc.* 22; in the Old Park Woods in vast quantity. Ruislip Wood. South Mims.

III. Roxeth; *Melv.* 68.

IV. In Hampsteede Wood!, and all the hedges thereabout!; *Ger.* 263 and *Johns. Eric.* Bet. Finchley and Hendon!; *Newb.* Mill Hill; *Morris* (*v. s.*).

V. Horsington Lane!; *Melv.* 68.

VI. Hadley!; *Warren.* Finchley, near the church; *Tucker.* Whetstone. Edmonton.

VII. [Primrose Hill, 1820; *Herb. G. & R.*] Fields bet. Kentish Town and Highgate, abundant. Hornsey Wood.

First record: *Gerarde,* 1597. Wanting in the S.W. parts.

602. M. annua, *L.*   *True Mercury. French Mercury.*
Right Mercury (Turn.).   *M. spicata seu fœmina Diosc.* (Budd.).   *M. mas et fœmina* (Pet.).

Cyb. Br. ii. 367.   Curt. F. L. f. 5 (drawn from London specimens).

Gardens, waste places, and on rubbish; rather common.   A.   June— October.

II. Hampton.   Near Strawberry Hill.

III. Abundant about Twickenham and Isleworth.

V. In great quantity on the railway bank at Kew Bridge Ry. Station.

VI. Edmonton.

VII. Beginneth now to be knowen in London, and in gentlemennis places not far from London; *Turn.* ii. 55.   About London, very frequent; *Budd. MSS.*   In most gardens in and about London; *Pet. Midd.* and *Bot. Lond.*   Near Paddington Church, 1757; *Hill,* 506. Humerton near Hackney; *Forst. Midd.*   Upper Clapton; *Cherry* (*v. s.*).   Spa Fields Chapel; Islington; Camden Town; *Lond. Fl.* 120.   Hammersmith, abundant.   Gardens at N.E. corner of Green Park.   Poplar.   Hackney Wick.

Var. β.   M. ambigua, L. fil.   Syme E. B. viii. t. 1270.

III. On waste ground near Twickenham Ry. Station.

VII. Parson's Green.   Site of Exhibition at S. Kensington.

This is merely the female plant with a few male flowers intermixed. It is worthy of remark that there are a very few male flowers drawn on the female plant in Curtis's figure.

First record: *Turner,* 1562; also first as a British plant.

CERATOPHYLLACEÆ.

CERATOPHYLLUM, *Linn.*

603. C. aquaticum (L. Cat. and Syme E. B.).   C. demersum, *L.* and C. submersum, *L.* (Bab. Man.).

Equisetum palustre ramosum et aquis immersum, &c. (Merr.).

Cyb. Br. i. 382; iii. 430; Comp. 174.   Syme E. B. viii. tt. 1276, 77.

Ponds and ditches; common. P. July, August.

I. Ruislip; *Melv.* 69. Harefield!; *Newb.* Eastcott.

II. Thames nr. Hampton Court, H. C. W.; *New B. G.* 99. Bet. Sunbury and Hampton!; *Newb.* Stanwell Moor. Staines Common. Hanworth Park.

III. Roxeth; Harrow Pond; *Melv.* 69. Headstone Moat; *Farrar.*

IV. Road fr. Finchley to Hendon!; *Newb.* Brook by Forty Farm; *Farrar.*

V. Canal near Brentford; *Hemsley.* Ealing Common!; not far from Kew Bridge Railway Station!; *Newb.*

VI. Edmonton.

VII. Betwixt Limehouse End and Blackwall; *Merrett*, 36. Near Goswell Street; Isle of Dogs!, E. F.; *Herb. Mus. Brit.* Lea Navigation Canal; *Cherry* (*v. s.*). Hackney Wick. Highgate Ponds; New River; Reservoir, Clifton Villas, Camden Square; *Jewitt.* [Ponds behind Park St., Camden Town (both species); *Irv. Lond. Fl.* 119.]

First record: *Merrett*, 1666. The few fruiting specimens we have seen are *C. demersum*, and this is probably the usual Middlesex plant. *C. submersum* is stated to have been found (VII.) in a piece of water at the top of Eel-brook Meadow, Parson's Green; *Irv. H. B. P.* 373. We have seen dried plants collected there by Dr. Morris in 1845, but without fruit. Mr. Watson says of *C. submersum*, 'Mr. Thos. Moore finds it in Middlesex,' *Cyb. Br.* iii. 430; and it is included in the Essex, but wanting in the Surrey and Herts Floras.

CALLITRICHACEÆ.

CALLITRICHE, *Linn.*

604. C. verna, *L.* C. vernalis, *Kütz.* (Syme E. B.). *Water Starwort.* Stellaria aquatica, *Park.* (Blackst.).

Cyb. Br. i. 379; Comp. 173. Syme E. B. viii. t. 1271.

Streams, ditches, and ponds; very common. A. or P. April—July. Throughout all the districts.

VII. Paddington, 1831; *Herb. Kew.* Hornsey Wood Ponds. Isle of Dogs.

First record: *Petiver*, before 1695. *Stellaria aquat. fol. long. tenuissimis*, R. Syn. ii. 280, *Long Water Starwort, Pet. H. B.Cat.* vi. 4, marked as a London species: from specimens similarly named in *Budd. Herb.* cxxii. fol. 52, it appears that the plant meant was a deep-water state of this species.

605. C. platycarpa, *Kütz.*
Alsine palustris serpyllifolia† (Johns.). *Stellaria pusilla pal. repens tetraspermos* (Pet.).

---

* Appears from the *Surrey Flora*, p. 88, to be *C. demersum*.

† This name is erroneously referred by Ray, *R. Cat.* i. 17, who could not have seen Johnson's figure, to his *Alsine palustris Portulacæ aquaticæ similis* (=*Montia fontana*, L.).

Cyb. Br. i. 380; iii. 429; Comp. 173.　Syme E. B. viii. t. 1272.

Ditches, ponds, and muddy places; rather rare.　A or P.　May—August.

I. Eastcott, in a deep pond.　Harrow Weald Common.　South Mims.

II. Bushey Park!; *Newb.*　Sunbury.　Common by Walton Bridge.

III. Hounslow Heath.　Near Hospital Bridge.　Whitton House grounds.
Duke's River near Isleworth.　Harrow, Hind; *Phyt. N. S.* iv. 112.

IV. Deacon's Hill.　North Heath, Hampstead.

VII. Bet. Kentish Town and Hampstead; *Johns. Ger.* 615.　[In a wood
near the boarded river; *Pet. Midd.*]

First record: *Johnson*, 1633, and figured by him in the plate affixed to
Enum. Hampst. fig. 5, under the name of *Serpyllifolia aquatica.*　The
first record as a British plant.

606.　C. pedunculata, *DC.*　C. hamulata, *Kütz., var. β* (Bab.).

Cyb. Br. i. 380; iii. 429; Comp. 173.　Syme E. B. viii. t. 1274.

On damp soil, borders of ponds, &c.; rather rare.　A or P.　May—
September.

I. In a feeder to Ruislip Reservoir; *Hind*, and *Herb. Harr.*　Bet. South
Mims and Potter's Bar.

II. Feltham Green.

III. Dockwell Lane, near Hatton.

IV. Hampstead Heath; *Newb.*

V. Bank of the river near Chiswick.

VI. Hadley!; *Warren.*

VII. Paddington Canal near Kensal Green!; *Warren.*

First record: *Newbould*, about 1860.

(C. autumnalis, *L.*　Cyb. Br. i. 380; iii. 429; Comp. 173.　Syme E. B.
viii. t. 1275.　Smith, *Fl. Brit.* i. 9, included under this name *C. hamulata*, Kütz. and *C. pedunculata*, Kütz., and he referred to it Petiver's
figure, *H. B. Cat.* vi. f. 4, already quoted under **603**, *C. verna* (q. v.).
Hence the locality, ' near London,' in *E. Fl.* i. 11, which has been copied
into many other books.　The true *C. autumnalis* is confined to a few
northern and western counties.)

URTICACEÆ.

PARIETARIA, *Linn.*

607.　P. diffusa, *Koch.*　P. officinalis, *L. var. α* (L. Cat.).

Parietaria, Ger. em. (Pet., Blackst.).

Cyb. Br. ii. 371.　Curt. F. L. f. 4 (drawn from a London specimen).

On and under old walls, hedgebanks, and rubbish; common.　P.　June
—September.

I. Harefield; *Blackst. Fasc.* 72.　Pinner Church wall, F. W. Longman;
Melv. 70.　Ruislip.

II. Hanworth Churchyard. Wall by the river of Hampton Court Gardens. Lane opp. Hampton Place, Twickenham.

III. By the river, Isleworth. Under palings of a flooded field by the Duke's River, Isleworth.

IV. On the tower of Edgware Church. Hampstead; *Irv. Lond. Fl.* 119.

V. Garden wall, Sion Park. Hedgebank near Turnham Green. Old tombs in Chiswick Churchyard. River front of Hammersmith Terrace.

VI. Highgate, Rev. S. Palmer; *Mag. Nat. Hist.* ii. 266, and *Irv. MSS.*

VII. [On the Charterhouse Cloisters, &c.; *Bot. Lond.* 274.] [Walls adjoining the Thames above and below Westminster Bridge; *Curt. F. L.*] Chelsea; Putney Bridge, 1817; *Herb. G. & R.* [Wall of fishpond at Canonbury House, 1818; *Bennett* (*v. s.*).]

First record: *Petiver*, 1709. Exhibits an attachment to the vicinity of the Thames.

URTICA, *Linn.*

[U. pilulifera, *L.* Cyb. Br. ii. 369. Syme E. B. viii. tt. 1280, 1281. Haverstock Hill, where the path leaves the road towards South End Hampstead; *Irv. MSS.* and *Lond. Flora*, 119.]

608. U. urens, *L.* *Small Nettle.*

U. racemifera minor annua, R. Syn. iii. (Blackst.).

Cyb. Br. ii. 368. Curt. F. L. f. 6.

Roadsides, waste grounds, and gardens; rather common. A. June—September.

I. Harefield; *Blackst. Fasc.* 112.

II. Teddington.

III. Harrow, frequent; *Melv.* 70. Hounslow Heath. Twickenham.

IV. Hampstead; *Irv. MSS.* Stanmore.

V. Hanwell!; *Warren.*

VI. Edmonton.

VII. Marylebone, 1827–30; *Varenne.* Isle of Dogs!; *Newb.* Primrose Hill.

First record: *Blackstone*, 1737.

609. U. dioica, *L.* *Nettle.*

U. racemifera major perennis, R. Syn. iii. (Blackst.).

Cyb. Br. ii. 369. Curt. F. L. f. 6.

Waste grounds, roadsides, &c.; very common. P. July—September. Throughout all the districts.

VII. Frequent in the suburbs. Primrose Hill. Isle of Dogs. Hackney Wick. Kentish Town.

First record: *Blackstone*, 1737.

Cannabis sativa, *L. Hemp.* Cyb. Br. ii. 372; iii. 506. Syme E. B. viii. t. 1283. III. By the river near Twickenham Church. V. By the

Canal, Apperton. About Acton, Turnham Green, Shepherd's Bush, &c.; *Newb.* Originates from the accidental distribution of the fruit ('hemp seed'), used as food for caged birds.

HUMULUS, *Linn.*

610. H. Lupulus, *L.*   *Hop.*
Lupulus mas et fœmina, C. B. P. (Blackst.).
Cyb. Br. ii. 372. Syme E. B. viii. t. 1284.

Hedges; rather common.   P.   July.

I. Harefield!; *Blackst. Fasc.* 54. Cowley, 1855; *Phyt. N. S.* i. 65. Pinner; *Melv.* 70.
II. Staines. Road from Staines to Hampton. Road from Sunbury to Walton Bridge. Hampton. Bet. Hampton Court and Kingston Bridge, plentiful. Teddington.
III. Roxeth; *Melv.* 70. Near Hatton. By the river, &c., Twickenham.
IV. Hendon, rare; *Lond. Fl.* 119. Stanmore; *Varenne.*
V. Apperton; *Melv.* 70. Near Shepherd's Bush!; *Newb.*
VI. Edmonton.
VII. Near Kensal Green, 1869; *Warren.*

First record: *Blackstone*, 1737.

ULMACEÆ.

ULMUS, *Linn.*

611. * U. suberosa, *Ehrh.*   *Elm.*
U. vulgatissima fol. lato scabro, Ger. em. (Blackst.).   *U. campestris* (Sm. Fl. Br.).   *U. camp.* and *U. sub.* (Sm. Eng. Fl.).
Cyb. Br. ii. 374; iii. 506. Syme E. B. viii. t. 1285.

Hedges and plantations; probably very common.   Tree.   March, April. In all the districts, including VII.; but probably always planted.

Var. β. U. glabra, Sm.
III. South side of Drilling Ground, Hounslow.
IV. Harrow Weald *Hind.* In Canons Park, some large trees.

First record: *Blackstone*, 1737. Middlesex is full of old and magnificent elms, for particulars of which reference must be made to *Loud. Arb. et Frut.* vol. iii. The largest are at Hampstead and in Kensington Gardens; but one at Twickenham was then (1838) 120 years old, and 90 feet high. Some of the elms in St. James's Park are older than this. Fine trees may be seen also at South Mims, Teddington, Heston, and Norwood. It is, however, a matter for regret that the elm is becoming more scarce year by year as old trees die off and no younger ones are found to supply their place. For an account of the celebrated hollow elm which formerly stood at Hampstead, see *Park Hampst.* pp. 33–40.

612. ***U. montana,** *L.*   *Wych Elm. Hertfordshire Elm. Wych Hazel.*
U. fol. latissimo scabro (Pet., Blackst.).
Cyb. Br. ii. 373 ; iii. 506.   Syme E. B. viii. t. 1287.
Hedges and plantations ; rather rare.   Tree.   March, April.
   I. Harefield, not frequent ; *Blackst. Fasc.* 112. Hedges at Pinner !
     Melv. 71. Bet. Ruislip and Pinner.
  II. Near Fulwell Station. Bet. Kingston Bridge and Hampton Court.
 III. Harrow Grove ; Harrow Park ; *Melv.* 71.
 IV. Hedges to the north of Stanmore, and about Harrow Weald, abundant.
 VI. Enfield Chase.
VII. At Hoxton neere London ; *Pet. Midd.* Hampstead. Regent's Park.
  First record : *Petiver*, 1695. This has a greater appearance of nativity
    than *U. suberosa*, yet it must be allowed to have been planted in most
    of its localities. It is very conspicuous in May, when the large leafy
    samaræ are nearly ripe, and give to the tree an appearance of developed
    foliage. Loudon mentions a fine tree at Muswell Hill, 85 ft. high, and
    with a trunk 3 ft. in diameter (*Arb. et Frut.* 1403) ; it is usually small.
    U. major, Sm., *Dutch Elm* (E. B. 2542), is probably to be referred to
    this species, as is done by Mr. Syme. In the neighbourhood of London,
    E. Forster ; *E. Fl.* ii. 22. Miller says that this tree was brought from
    Holland in King William's reign, and Loudon mentions (p. 1396) that
    the elms in the old part of Kensington Gardens, near the palace, many
    of which are more than 70 ft. high, are of this kind.

AMENTIFERÆ.

SALIX, *Linn.*†

[613. * **S. pentandra,** *L.*   *Bay-leaved Willow.*
S. fol. laureo, seu lato glabro odorato, Merr. (Blackst.).
Cyb. Br. ii. 387 ; iii. 508.   Syme E. B. viii. t. 1303.
Moist places ; very rare.   Tree.   May, June.
  I. Amongst the willows near Mr. Ashby's brick-kiln at Harefield ;
    Blackst. Fasc. 89.
VII. Chelsea, in the way to Fulham ; *Mart. App. P. C.* 65.
  First record : *Blackstone*, 1737 ; last, *T. Martyn*, 1763.]

614. **S. fragilis,** *L.*   *Crack Willow.*
Cyb. Br. ii. 388 ; iii. 508.   Syme E. B. viii. tt. 1306–7.
Sides of streams and ponds and damp situations ; common.   Tree.
   April, May.

---

† This genus contains many species which, though included in our Floras, have but
slight claims to be considered as native, or even really naturalised plants. This is the
case with many other trees and shrubs, and we cannot be sure that the localities here
given are natural ones.

I. Pinner lime-pits. Elstree Reservoir.

II. By the Thames, bet. Strawberry Hill and Twickenham.

III. Harrow and Roxeth; *Melv.* 71. By the Cran at Hounslow, Mother Ive's Bridge, Hospital Bridge, and Twickenham. By the Duke's River, Isleworth.

IV. Bishop's Wood.

V. Near Willesden Junction!; *Newb.* Brentford, by the ferry. Hanwell!; *Warren.*

VII. Several trees at Mill-bank, Westminster, and other parts of that neighbourhood; *Sm. E. B.* 1807 and *Herb. Smith.* [Marylebone Fields, 1818; *Herb. G. & R.*] Fields bet. Kentish Town and Highgate; *Lond. Fl.* 116. South Heath, Hampstead, planted. Osier-holt, Sandy End, Fulham. Hackney Wick.

At Brentford, 50 ft. high; *Loud. Arb. et Frut.* 1523.

Var. β. S. Russelliana, Sm. S. viridis, *Fries* (Syme E. B.). *Bedford Willow.*

Syme E. B. viii. t. 1308.

II. Bet. Sunbury and Hampton!; *Newb.*

III. Harrow, W. M. H.; *Melv.* 71. By the Cran at Mother Ive's Bridge.

V. Near Kew Bridge Ry. Station!; *Newb.* Perivale; *Lees.*

VII. Osier grounds at Stoke Newington, J. Woods; *B. G.* 412.

A departure from *S. fragilis* in the direction of *S. alba*, and apparently variable. At Sion Park 89 ft. high; *Loud. Arb.* 1523.

First record: *Woods*, 1805. Usually a planted tree.

615. S. alba, *L.* *White Willow.*

Cyb. Br. ii. 389. Syme E. B. viii. tt. 1309–10.

Banks of rivers and ponds, damp hedges, &c.; common. Tree. April, May.

I. Harefield!; *Newb.* Elstree Reservoir.

II. Near the Queen's River, Hampton.

III. Harrow; *Hind in Phyt. N. S.* v. 201, and *Herb. Harr.* By the Cran at Hounslow and Twickenham.

IV. North Heath, Hampstead.

V. Near Kew Bridge Ry. Station!; *Newb.* By the Brent; *Lees.*

VI. Bet. Whetstone and Barnet. Edmonton.

VII. Belsize Lane, Hampstead.

Var. β. S. vitellina, *L.* *Golden Willow.*

I. Pool in sandpits at Pinner!, W. M. H.; *Melv.* 71.

II. Hanworth Park, large trees.

III. Hatch End, near Pinner. Hedges on the Hounslow Road, Twickenham. Road between the New Cemetery and the Drilling Ground. A large tree by the river bet. Twickenham and Richmond Bridge.

VII. At Stoke Newington, J. Woods; *B. G.* 412.

First record: *Woods*, 1805.

616. S. undulata, *Ehrh.*

Cyb. Br. ii. 390. Syme E. B. viii. t. 1312.

Banks of rivers; rare. Shrub or Tree. April, May.

II. Bet. Hampton Court and Kingston Bridge, in several places by the river.

III. Near Richmond Bridge towards Twickenham!; *Baker.* Two bushes only. By the Duke's River near Worton.

VII. Osier-holt, Sandy End, Fulham.

First record: *Baker,* 1867. Specimens from North Surrey distributed by Mr. Watson come nearer to *S. triandra,* L. They were, however, named *S. undulata* by Dr. Andersson. (See *Fl. Surrey,* 214.

617. S. triandra, *L.*

S. fol. amygdalino aurito corticem abjiciens, R. Syn. (Pet.).

Cyb. Br. ii. 390. Curt. F. L. f. 6.* Syme E. B. viii. t. 1313.

Banks of streams and ponds; rather common. Shrub or Tree. April, May.

II. By the Queen's River, Hampton. River-side bet. Hampton Court and Kingston Bridge.

III. By the Thames bet. Twickenham and Richmond Bridge. Near Isleworth. By the Duke's River, Worton. Hedge bet. Twickenham and Worton.

IV. Hampstead; *Irv. MSS.*

V. By Brent near Apperton; *Melv.* 71.

VII. [Thames side bet. Westminster and Chelsea; *Pet. Midd.*] Banks of Lea at Hackney, E. F.; *Herb. Mus. Brit.*

Var. β. S. Hoffmanniana, Sm. Syme E. B. viii. t. 1314.

I. By Elstree Reservoir.

II. River-side bet. Hampton Court and Kingston Bridge.

V. By the Brent near Hanwell!; *Warren.*

Var. γ. S. amygdalina, L.

IV. Hampstead; *Irv. MSS.*

V. Canal near Brentford; *Hemsley.* By the Brent in several places, especially bet. Greenford and Perivale; *Lees.*

VII. King's Road bet. Chelsea and Fulham, A. B. Lambert; *Herb. Kew.* (named by Andersson). Osier ground near Homerton, E. F.; *Herb. Mus. Brit.*

First record: *Petiver,* 1695.

618. S. purpurea, *L.*

Cyb. Br. ii. 391.

Banks of streams and wet places; rather common. Shrub or Tree. March, April.

---

* Curtis's plate seems to represent a form intermediate between typical *triandra* and *Hoffmanniana.*

Var. a. S. Lambertiana, Sm. *Boyton Willow.* Syme E. B. viii. tt. 1317–18.

I. North of Harefield!; *Newb.*

II. Bet. Hampton Court and Kingston Bridge!; *Newb.* Staines, Mr. Lambert; *Sm. E. B.* 1359, and *Herb. Kew.* Hammonds near Staines.

III. River-side bet. Teddington and Strawberry Hill, abundant. Bet. Twickenham and Richmond Bridge!; *Baker.* By the Cran at Hospital and Mother Ive's Bridges.

IV. Hampstead; *Irv. MSS.*

VII. Near the Lea Bridge Mills, E. F.; *Herb. Mus. Brit.* Newington, J. Woods; *B. G.* 412.

Var. β. S. ramulosa, Borr. MSS. (Leefe Sal. Brit). *S. monandra*, L. (Curt.). *S. Helix* (Auct.). Curt. F. L. f. 6.

I. Side of pond bet. South Mims and Potter's Bar.

II. By Thames bet. Kingston and Hampton Court. By Mother Ive's Bridge.

III. By the Duke's River near Whitton.

IV. Hampstead; *Irv. MSS.*

VII. Near Temple Mills, E. F.; *Herb. Mus. Brit.*

Curtis's plate well represents the Middlesex plant. Smith's figure of *S. Helix* (*E. B.* 1343, reproduced in *Syme E. B.* viii. t. 1319) is made up of two different species, and has led Mr. Syme to refer the plant to *S. rubra.*

First record: *Lambert*, 1804.

619. S. rubra, *Huds.*   *Green Osier.*

S. minima fragilis fol. long. utrinque viridibus non serratis, R. Syn. App. (Pet.).

Cyb. Br. ii. 392; iii. 508.   Syme E. B. viii. t. 1320.

Sides of streams and wet places; rare. Shrub or Tree. April, May.

II. By the Queen's River, Hampton.   Thames side opp. Thames Ditton.

III. River-side bet. Twickenham and Richmond Bridge.

VII. [Among the willows on the Thames side bet. Westminster and Chelsea; *Pet. Midd.*] Osier grounds at Newington, J. Woods; *Herb. Mus. Brit.* and *B. G.* 412.

Var. β. S. Forbyana, Sm. Syme E. B. t. 1321.

II. By the Thames bet. Kingston and Hampton Court. Queen's River near Hampton.

First record: *Petiver*, 1695.

620. S. viminalis, *L.*   *Osier.*

? *S. vulgaris longis et angustis foliis* (Johns.).

Cyb. Br. ii. 392; iii. 508.   Syme E. B. viii. t. 1322.

Borders of ponds, wet meadows, &c.; rather common. Shrub or Tree. April.

I. Elstree Reservoir.

II. Hampton!; *Newb.* By the river bet. Strawberry Hill and Twickenham.

III. About Twickenham. Hounslow Heath. Duke's River, Isleworth.

IV. Harrow district, common; *Melv.* 72. Hampstead; *Johns. Enum.* Pond on the east border of Bishop's Wood.

V. Bet. Acton and Ealing!; not far from Kew Bridge Ry. Station!; *Newb.*

VI. Edmonton.

VII. Osier-holt, Sandy End, abundant.

First record: ? *Johnson*, 1632. *Salix caprea fol. longiore auriculato;* between Twitnam (= Twickenham) and Witton, and between Kingsland and Hackney; *Doody MSS.*, is perhaps this species.

621. S. stipularis, *Sm.*

Cyb. Br. ii. 392 ; iii. 508. Syme E. B. viii. t. 1323.

Osier holts and woods, very rare. Shrub. March.

IV. Hampstead ; *Irv. MSS.*

VII. Osier ground at Stoke Newington, J. Woods; *B.G.* 413.

First record : *Woods*, 1805. We have not seen this in the county.

622. S. Smithiana, *Willd.**

Cyb. Br. ii. 393 ; iii. 509. Syme E. B. viii. t. 1324.

Wet places ; very rare. Shrub or Tree. April—May.

I. Copse at Pinner Park, abundant; *Melv.* 72.

IV. Lowest part of North Heath, Hampstead.

VII. Side of marsh ditch, Hackney Wick, 1867.

First record : *Hind*, 1860.

623. S. cinerea, *L.*   *Sallow.*

Cyb. Br. ii. 395. Syme E. B. viii. t. 1327–29.

Hedges, woods, and damp places ; very common. Shrub or Tree. April. In all the districts.

VII. (*S. aquatica.*) In the lane from Devil's Lane to Hornsey Wood House, E. F. ; *Herb. Mus. Brit.* Kentish Town.

First record: *E. Forster*, about 1800. *S. aquatica*, Sm., is probably the usual Middlesex plant. *S. oleifolia*, Sm., and *S. cinerea*, Sm., however, doubtless occur. ' *S. Timmii* ; near London. 1800 '; *Herb. Smith*, is a form of *aquatica.* A monstrosity with the carpels represented by anthers is mentioned as found near London by Mr. Dillwyn and Mr. J. Woods in *Rees' Cyclop.* v. 31.

624. S. aurita, *L.*

Cyb. Br. ii. 395; iii. 509. Syme E. B. viii. t. 1330.

Moist woods and hedges ; rare. Shrub. April, May.

III. Hounslow Heath.

* *Var.* β, S. ferruginea, *Anders.* Banks of the Thames, Mr. G. Anderson ; *Hook. B. Fl.* iii. 428. But not in Middlesex : see *E. B. S.* 2665.

s 2

IV. Bishop's Wood, Hampstead, 1866. Hedge on road from Finchley to
Hendon !; *Newb.*

Probably occurs elsewhere, but has been overlooked.

625. S. caprea, *L.*   *Great Sallow.*
? *S. latioribus albidis fol.* (Johns.).
Cyb. Br. ii. 396. Syme E. B. viii. t. 1331.

Woods and hedges ; rather rare?   Tree or Shrub.   March, April.
  I. Ruislip ; *Melv.* 72. Harefield!; *Newb.* Pinner lime-pits. S. Mims.
 III. Harrow ; *Melv.* 72. Hounslow!; *Newb.* Near Whitton.
 IV. Hampstead Heath ; *Johns. Enum.* Bet. Finchley and Hendon!;
      Newb. Bishop's and Turner's Wood.
  V. Apperton ; *Melv.* 72. Twyford.
 VI. Hadley !; *Warren.* Whetstone. Edmonton.
VII. Ken Wood, Lambert; *Herb. Kew.* By second bridge from Edgware
      Road Station of North London Ry.; *Warren.*

First record: ? *Johnson*, 1632.

626. S. repens, *L.*   *Dwarf Willow.*
S. pumila linifolia incana, C. B. P. (Blackst.).   *S. humilis* (Ger.).   *S.
hum. repens Lob. bombifera* (Johns., Merr.).   *S. pumila latifolia* (Park.).
S. fusca, L. (Cyb. Br.).
Cyb. Br. ii. 399. Syme E. B. viii. tt. 1356–1362.

Heaths and commons in damp places; rather rare. Shrub. March,
April.
  I. Harefield Common!; *Blackst. Fasc.* 89. Harrow Weald Common;
      Melv. 72. Stanmore Heath.
 III. Waste ground, Whitton, G. Francis; *Coop. Supp.* 12. Whitton Park
      inclosure. Near Hatton. Drilling Ground, Hounslow.
 IV. Farther end of Hampstead Heath!; *Ger.* 1205, *Johns. Eric., Park.
      Theat.* 1394, *Merrett* 108, &c.
VII. South Heath, Hampstead.

First record: *Gerarde*, 1597; also first as a British species?. The six
or seven varieties included under this species have all been recorded
for Middlesex, but are scarcely distinguishable. The Harefield Common
plant seems *S. fusca*, L., and the Whitton specimens are probably to
be referred to the same. Edward Forster's specimens in *Herb. Mus.
Brit.*, from Hampstead Heath, are labelled *S. adscendens*, Sm.

(S. hastata, *L. S. malifolia*, Sm. Cyb. Br. ii. 398 ; iii. 509. E. B.
1617. Hollick Wood, Colney Hatch, J Woods ; *B. G.* 413. Not at all
likely to occur in England, either wild or cultivated. A willow labelled
' *Salix*, In a wood near Colney Hatch; E. F.' in *Herb. Mus. Brit.*,
may be the plant intended. It seems large *S. aquatica*.)

POPULUS, *Linn.*

627. P. alba, *L.*

River-sides and moist meadows; rare. Tree. March, April.

*1. *P. eu-alba,* Syme. *P. alba,* L. (Bab.). *White Poplar. Dutch Abele.*
Cyb. Br. ii. 382; iii. 507. Syme E. B. viii. t. 1299.

II. Staines Moor Farm. Hanworth Park. Sunbury, by the Thames, fine trees.

III. Harrow Grove; *Mclv.* 72.

IV. Hampstead; *Irv. MSS.*

VI. Finchley Common; Lane near the 'Orange Tree,' Colney Hatch; *Herb. Hardw.*

VII. [In a low meadow turning up a lane at the further end of Blackwall; *Ger.* 1302.] Kilburn, 1815; *Herb. G. & R.*

Introduced from Holland, probably not long before Gerarde's time. Loudon mentions specimens upwards of 100 feet high, on the banks of the Thames bet. Hampton Court and Chertsey. (*Arb. et Frut.* 1644.)

2. *P. canescens,* Sm. *Grey Poplar. English Abele.*
P. alb. similis fol. non incanis, &c (Doody). Cyb. Br. ii. 383; iii. 508. Syme E. B. viii. t. 1300.

I. Pinner Place, probably introduced, *Hind.*

II. By Hampton Court!; *Doody MSS.* Several bushes by the towing path bet. Hampton Court and Kingston Bridge.

III. Group of large trees by the river near Twickenham Church. By the Cran near Mother Ive's Bridge.

IV. Bishop's Wood. Road bet. Finchley and Hendon!; *Newb.*

VI. A clump of fine trees near Warren Lodge, Edmonton.

Probably native, but often planted.

First record: *Gerarde,* 1597; also first as British.

628. P. tremula, *L.* *Aspen.*
P. lybica, Matth. (Johns., Blackst.).
Cyb. Br. ii. 384. Syme E. B. viii. t. 1301.

Moist woods and heaths; rather rare. Tree. April.

I. Harefield!; *Blackst. Fasc.* 80. Harrow Weald Common. Near Potter's Bar.

III. Hounslow Drilling Ground!; *Newb.* By the Cran near Hounslow.

IV. Hampstead Wood; *Johns. Eric.* Bishop's and Turner's Woods, abundant. Edgware.

VII. Hornsey Wood.

First record: *Johnson,* 1629.

629. * P. nigra, *L.* *Black Poplar.*
Cyb. Br. ii. 384. Syme E. B. viii. t. 1302.

Near rivers and in damp places; rare. Tree. March.

I. Harefield; *Blackst. Fasc.* 80. Pinner Hall, &c.; probably planted; *Mclv.* 73. Wood Hall, Pinner.

III. A large tree by the Thames, near Richmond Bridge.

IV. Bishop's Wood, Hampstead.

VI. Edmonton.

First record: *Blackstone,* 1737. Much planted, especially in London, in the parks and squares, along with *P. monilifera, P. balsamifera, P. angulata,* &c., of North American origin.

[MYRICA, *Linn.*

630. M. Gale, *L.*   *Dutch* or *Bog Myrtle.*

Myrtus Brabantica seu Elæagnus Cordi (Ger., Merr.).

Cyb. Br. ii. 408. Syme E. B. viii. t. 1298.

Heaths; very rare. Shrub. May.

II. By Colbrooke; *Ger.* 1228.

III. On Hunslow Heath; *Merrett,* 82.

First record: *Gerarde,* 1597; last, *Merrett,* 1666. It is with some hesitation that we include this plant. It is not found in Herts, Essex, or Buckinghamshire, but its occurrence in abundance on Bagshot Heath, Surrey, much lessens the improbability of its former existence in the somewhat similar district of Hounslow Heath.]

BETULA, *Linn.*

631. B. alba, *L.*   *Birch.*

Cyb. Br. ii. 380.

Woods and heaths; rather rare. Tree. May.

1. *B. verrucosa,* Ehrh.

Syme E. B. viii. t. 1295.

II. Fulwell.

III. On and near the Drilling Ground, Hounslow. Whitton Park inclosure.

V. Near Hanwell!; *Warren.*

2. *B. glutinosa,* Fries. *B. pubescens,* Ehrh. (L. Cat.).

Betula, Lob. (Johns., Blackst.). Syme E. B. viii. t. 1296.

I. Harefield!; *Blackst. Fasc.* 10.

IV. Hampstead Heath; *Johns. Enum.* Bishop's Wood!; *Warren.*

First record: *Johnson,* 1632. An exotic species, often planted, is sometimes mistaken for the indigenous Birch.

ALNUS, *Tourn.*

632. A. glutinosa, *Gaertn.*   *Alder.*

Alnus, Ger. (Blackst.).

Cyb. Br. ii. 380; iii. 507. Syme E. B. viii. t. 1294.

Sides of streams, wet woods, &c.; very common. Tree. March—April. Throughout all the districts.

VII. Banks of Lea. South Heath, Hampstead.

First record : *Blackstone,* 1737. There are some picturesque old alders by the Cran on Hounslow Heath, and the swampy copse near Whetstone, called ' the Alders,' is mainly composed of this tree.

FAGUS, *Linn.*

633. F. sylvatica, *L.*

Cyb. Br. ii. 377 ; iii. 506. Syme E. B. viii. t. 1291.

Woods ; rare. Tree. May.

I. Harefield! ; *Blackst. Fasc.* 28. The trees in the Old Park Woods are very fine.

III. Harrow Grove, Harrow Park, &c. ; *Melv.* 73.

VI. Betstile Lane, Colney Hatch ; *Herb. Hardw.* Hadley! ; *Warren.* Winchmore Hill Wood.

VII. Ken Wood.

First record : *Blackstone,* 1737. No doubt a native at Harefield, and apparently so at Ken Wood.

Castanea vesca, *Lam. Edible Chestnut.* Cyb. Br. ii. 377. Syme E. B. viii. t. 1290. II. Hedges in Tangley Park. III. Harrow Grove, Harrow Park ; *Melv.* 73. VI. There are remains of old decayed chestnuts not far from London, particularly in Enfield Chase ; *Mart. Mill. Dict.* We have not seen the Enfield trees, and they probably no longer exist. The plant is not a native of England, though no doubt one of the earliest historic introductions. The oldest trees near London, says Loudon, are in Kensington Gardens ; they are mostly hollow, with a pollard-like head (*loc. cit.* p. 2000). On the question of the nativity of the chestnut in England, we may refer to a paper by the Hon. Daines Barrington, in vol. lix. p. 23 of the *Philosophical Transactions,* and to a series of letters in reply, from Dr. Ducarel, Mr. Thorpe, and Mr. Hasted, in vol. lxi. pp. 136–169.

QUERCUS, *Linn.*

634. Q. Robur, *L. Oak.*

1. *Q. pedunculata,* Willd.

Q. vulgaris, Ger. em. (Blackst.). Cyb. Br. ii. 375. Syme E. B. viii. t. 1288.

Woods, hedges, &c. ; very common. Tree. April, May.

In all the districts.

VII. Ken Wood.

Loudon (p. 1733) mentions a variety *heterophylla,* of which a single tree grows (IV.) at the London end of Mill Hill village. The largest oaks in Middlesex recorded by Loudon are, the Chandos Oak in the grounds of Michendon House, Southgate, chiefly remarkable for its immense head, which covers a space of ground of which the diameter is 118 ft. ; and one on Laleham Common with a trunk 23 ft. in girth. There are some fine trees in Bushey Park, and two magnificent ones in the enclosure of Twickenham Park.

2. *Q. sessiliflora,* Sm.

Cyb. Br. ii. 376. Syme E. B. viii. t. 1289.

Woods; very rare? Tree. April, May.

IV. One in the grounds of the Protestant Dissenters' School* at Mill Hill; *Loud. Arb. et Frut.* 1736.

VII. Most of the oaks in Ken† Wood are of this kind; *Loud. Suburban Gardener,* and *loc. cit.*; but *Q. pendunculata* is equally abundant there.

First record: *Blackstone,* 1737. The natural woods of the county probably consisted largely of oak as it is a prominent tree in the small portions of them which remain, not, however, usually growing to any large size.

Q. Cerris, *L. Turkey Oak.* Much planted as an ornamental and timber tree, and frequently collected as a native plant. Introduced about a century ago from South Europe.

CORYLUS, *Linn.*

635. C. Avellana, *L.* *Hazel.*

C. sylvestris (Johns., Blackst.).

Cyb. Br. ii. 378; iii. 507. Syme E. B. viii. t. 1292.

Woods and hedges; common. Shrub. February—April.

I. Harefield!; *Blackst. Fasc.* 21. South Mims. Pinner.

II. Staines.

III. Harrow, abundant; *Melv.*74. Twickenham. Worton. Hounslow Heath.

IV. Hampstead Heath; *Johns. Enum.* Bishop's Wood.

V. Horsington Hill.

VI. Edmonton.

VII. Certainly occurs in the northern suburbs.

First record: *Johnson,* 1632.

CARPINUS, *Linn.*

636. C. Betulus, *L.* *Hornbeam.*

Carpinus (Johns., Blackst.). *Betulus* (Merr.).

Cyb. Br. ii. 378; iii. 506. Syme E. B. viii. t. 1293.

Woods and hedges; common. Tree. April, May.

I. Harefield!; *Blackst. Fasc.* 10. Pinner Wood; *Melv.* 74. By Elstree reservoir, a single large tree. Near Pinner Ry. Station. Stanmore Heath. South Mims.

II. By Kingston Bridge. Hampton, probably planted.

III. Harrow, in plantations; *Melv.* 74. Near Twickenham in several places. Whitton.

IV. Hampstead Wood and Heath; *Johns. Eric.* and *Enum.* Stanmore. Scratch Wood, abundant. Bishop's and Turner's Wood. About Kingsbury; *Farrar.*

V. Twyford!; *Newb.* Horsington Hill.

* Formerly the residence of Peter Collinson. † *Ken=Kern,* or acorn.

VI. Enfield Chase. Finchley Common; *E. B.* 2032. Hadley!; *Warren.* Whetstone. Winchmore Hill Wood.

VII. On the west of Primrose Hill; *Merrett*, 15. Hedges bet. Regent's Park and Hampstead, 1819; *Bennett* (*v. s.*). In and near Hornsey Wood.

First record: *Johnson*, 1629. Formed a large, perhaps the chief part of the ancient forest on the clay north of London, of which Enfield Chase was the remains; it is still the chief tree in many parts of Epping Forest, Essex. It does not usually grow to any great height; but a fine tree is mentioned by Loudon (p. 2007) as growing in the grounds of Chiswick House.† The labyrinth in Hampton Court grounds is formed with hedges of Hornbeam.

CONIFERÆ.

TAXUS, *Linn.*

637. *T. baccata, L.*   *Yew.*

Cyb. Br. ii. 411. Syme E. B. viii. t. 1384.

In hedges and churchyards; rather rare. Tree. March, April.

I. Near Harefield!; *Newb.* Old trees by the roadside bet. Potter's Bar and South Mims.

II. Hanworth Churchyard and Harlington Churchyard, very large trees.

III. Harrow Grove, 'thoroughly naturalised;' *Melv.* 74.

IV. Near Hampstead, on Hendon Road.

VI. Winchmore Hill Wood. Edmonton Churchyard.

First record: *Hind*, 1860. Always planted?. For an account, illustrated with figures. of the Harlington Yew, which till 1790 was clipped into most fantastic shapes, see *Loud. Arb. et Frut.* 2077. In *Mart. Mill. Dict.* is a description of four beautiful Yew trees growing at Mill Hill.

[JUNIPERUS, *Linn.*

638. J. communis, L. J. eu-communis (Syme E. B.).   *Juniper. Juniperus, Ger. em.* (Blackst.). *J. vulgaris fruticosa* (Johns.).

Cyb. Br. ii. 410. Syme E. B. viii. t. 1382.

Heaths; very rare. Shrub. May.

I. On Harefield Common, abundantly; *Blackst. Fasc.* 48.

IV. Hampstead Heath; *Johns. Enum.*

VI. Finchley Common; *Coles,* 361, and *Mart. App. P. C.* 66.

First record: *Johnson*, 1632; last, *T. Martyn,* 1763.]

Pinus sylvestris, *L.* *Scotch Fir. Pinus* (Merr.). Cyb. Br. ii. 409. Syme E. B. viii. t. 1381. I. Spontaneously on Harrow Weald Common, the seeds from the plantations of Harrow Weald Parks; *Melv.* 74. Seedlings on Harefield Common!; *Newb.* III. Harrow Grove; *Melv.* 74. VII. In a wood on the left hand of Hamstead, called the Pine-walks; *Merr.* 94. Planted in many places besides those mentioned.

† Rev. W. W. Newbould was shown a very large Hornbeam in the grounds of Sutton Court School, Chiswick, which may be the one here alluded to.

MONOCOTYLEDONES.

TRILLIACEÆ.

PARIS, *Linn.*

639. P. quadrifolia, *L.*   *Herb True-love.   One-berry.*
Herba Paris (Merr., Blackst.).
Cyb. Br. ii. 470.   Brit. Ent. t. 138 (drawn from a Middlesex specimen).

Woods and copses; rare.   P.   May.
  I. In the Old Park!, Hanging Wood, and elsewhere near Harefield,
       plentifully; *Blackst. Fasc.* 41.   Abundant in a copse near Pinner
       Wood!, W. M. H.; *Melv.* 77.
  IV. In a wood near Hampstead Heath; *Merrett*, 61, and *Huds.* ii. 172.
       In great abundance in Grass Farm Wood opp. the eight-mile stone
       in the Mill Hill Road, 1766; *MS. note by Michael Collinson.*   In
       a wood near Hendon beyond Hampstead; *MS. note (Alchorne's)*
       quoted in Phyt. iii. 166.
  VI. The Alders Copse, near Whetstone, abundant.
  VII. Ken Wood, Hunter; *Park Hampst.* 30, and *Baxter*, vol. i.

  First record: *Merrett*, 1666.   Many of the Harefield plants are more
     than a yard high.

DIOSCOREACEÆ.

TAMUS, *Linn.*

640. T. communis, *L.*   *Black Bryony.*
Bryonia nigra, Ger. (Merr., Blackst.).
Cyb. Br. ii. 471.   Syme E. B. ix. t. 1508.

Hedges and woods; rather common.   P.   May, June.
  I. Harefield!; *Blackst. Fasc.* 11.   Pinner.   Elstree.   Potter's Bar.
       South Mims.
  II. Staines Moor.   Near Staines on road to Hampton.
  III. Marsh Farm, Twickenham.   By Whitton House Grounds.   Bet.
       Worton and Twickenham.
  IV. Harrow district, very common; *Melv.* 77.   Harrow Weald.   Stanmore.
       Deacon's Hill.   Bishop's Wood.
  VI. Hadley!; *Warren.*   Whetstone.   Colney Hatch.   Edmonton.   Enfield.
  VII. [In the pine walks going up to Hampstead; *Merrett*, 17.] [Marylebone
       Infirmary Gardens, 1815; *Herb. G. & R.*]

  First record: *Merrett*, 1666.

HYDROCHARIDACEÆ.

HYDROCHARIS, *Linn.*

641. H. Morsus-Ranæ, *L.*   *Frog-bit.*

Morsus Ranæ (Lob.). *Nymphæa alba minima, C. B. Pin.* (Blackst., Park.).
Cyb. Br. ii. 473 ; iii. 515. Curt. F. L. f. 3 (drawn from a London plant).
Ditches and stagnant water; rather common. P. July—September.

  I. Harefield; *Blackst. Fasc.* 65. Uxbridge, *Lightfoot MSS.* Colne near
     Iver Bridge! ; *Newb.*

 II. Staines Common; *Phyt. N. S.* iv. 263. Staines Moor ditches, abundant.
     Bet. Sunbury and Hampton!; *Newb.* Hampton Court, 1862 ; *W.*
     Bell (v. s.).

III. Thames at Twickenham, abundant, 1866; *Masters.* Marsh Farm,
     Twickenham. Pond on Hounslow Heath. Cran, by the bridge on
     Hanworth Road.

IV. Pond at Forty Green ; Wembley Park ; *Cole.*

 V. Canal near Brentford ; *Hemsley.* Pond near Acton, towards Ealing !;
     Newb.

VII. [By the Tower of London ; *Lob. Adv.* 258.] [Many ditches about
     London ; *Park. Theat.* 1253.] [Near Blackwall, 1822, J. Taylor;
     Herb. Young. Isle of Dogs, 1836; *Young MSS.* and *Lond. Fl.*
     109.] [Most abundant in the many stagnant ditches intersecting
     the fields and market-gardens at Ranelagh, the neat-houses, the
     willow-walk, and Tuthill fields, 1820-25 ; *Pamplin.*] Near Stok
     Newington, Winch; *New B. G.* 101.

  First record : *Lobel,* 1570; also first as a British plant. Seems to have
    nearly disappeared from the immediate vicinity of London, where it
    was formerly common.

ELODEA, *Rich.*

642. *E. canadensis, *Michaux.* Anacharis Alsinastrum (Bab.). *Water
Thyme.*

Cyb. Br. ii. 474; iii. 515. Syme E. B. ix. t. 1446.

Ditches, streams, ponds, &c.; very common. P. June—August.
Throughout all the districts.

VII. River Lea, bet. Lea Bridge and Upper Clapton, 1854, A. Evans;
     Phyt. N. S. i. 96. Ditches by the Thames near the Bishop's Palace,
     Fulham ; *Ibid.* Ditches and streams at Hackney, &c., abundant.
     Newington. Hornsey Wood. Hampstead Ponds. Serpentine,
     Kensington Gardens, where it flowers profusely.

  First record : *A. Evans,* 1854. Almost certainly introduced into Great
    Britain from North America about 1842 ; but it seems to have been
    noticed in Ireland as early as 1836 (see *Phyt.* v. 88). Is there a
    possibility of its being a native there ?

ORCHIDACEÆ.

ORCHIS, *Linn.*

643. O. Morio, *L.*   *Meadow Orchis.*

O. m. fœmina, C. B. P. (Blackst.).

Cyb. Br. ii. 422 ; iii. 511. Curt. F. L. f. 3.

Damp meadows; rather rare. P. May.

I. About Harefield ; *Blackst. Fasc.* 68. Pinner, abundant!; *Melv.* 76. Near Harefield Church ; *Cole.* On Ruislip Moor, plentiful.

III. Roxeth ; *Melv.* 76. Bet. Harrow and Pinner, 1862.

IV. Meadows opp. Swan Inn, bet. Hendon and Hampstead, E. H. Button; *Coop. Supp.* 11. Plentiful in a meadow near Hendon ; *Irv. MSS.* Near Mill Hill; *Michael Collinson MSS.* Near Highgate Wood; Stanmore, 1827–30 ; *Varenne.* ;

First record : *Blackstone*, 1737. Noticed *with white flowers* at (I.) Harefield, by Blackstone; and (IV.) Mill Hill, by Michael Collinson.

644. O. mascula, *L.*   *Early Purple Orchis.*

Cynosorchis Morio, Lob. (Johns.).   *O. morio mas fol. maculatis, C. B. P.* (Blackst.).

Cyb. Br. ii. 423. Curt. F. L. f. 2. Syme E. B. ix. t. 1455.

Meadows and woods; rare. P. April—May.

I. Harefield, frequent ; *Blackst. Fasc.* 67. Near Harefield Church ; *Cole.* Ganett Wood, 1855 ; *Phyt. N. S.* i. 62. Old Park Wood. Copse near Pinner Wood!, W. M. H.; *Melv.* 97.

IV. Hampstead Heath ; *Johns. Enum.* Side of wood behind 'Spaniards,' Hampstead, 1821 ; *Bennett (v. s.),* and *Irv. MSS.* In many places near Mill Hill ; *Michael Collinson MSS.*

First record : *Blackstone,* 1737. With *white flowers* at Mill Hill ; *Michael Collinson MSS.*

[645. O. purpurea, *Huds.*

O. magna latisfoliis galeâ fuscâ vel nigricante, Chabr. (Blackst.).

Cyb. Br. ii. 424 ; iii. 511. Curt. F. L. f. 6.

Chalk banks ; very rare. P. May, June.

I. In the chalkpit near the paper mill at Harefield ; *Blackst. Fasc.* 67. Since Blackstone's time it has been gathered frequently in . . . Middlesex ; *Bicheno in Linn. Trans.* xii. 30.

First record : *Blackstone,* 1737. Both Peter Collinson and his son Michael searched the chalkpit with great care, but could never discover this or *O. militaris,* L. after the most diligent investigation ; *Peter and Michael Collinson's MSS.* It is only certainly known now to occur in Kent.]

[**646. O. militaris,** '*L.*' *Jacq.*

O. latifol. hiante cucullo major, Inst. R. H. (Blackst.).

Cyb. Br. ii. 424. Syme E. B. ix. t. 1452.

Chalk banks; very rare. P. May, June.

I. In the chalkpit near the paper mill at Harefield, plentifully; *Blackst. Fasc.* 67.

Not found since. (See **645**, *O. purpurea.*) Bicheno (*Linn. Trans.* xii. 32) suspected Blackstone's plant to be his '*O. militaris*' = *O. Simia,* Lam. (*Syme E. B.* ix. t. 1453), only certainly known to grow in Berks and Oxford; but as the present species occurs in Bucks, and has also been found near Rickmansworth (*Fl. Herts.* 286) it is probable that it was the plant gathered in Middlesex.]

[**647. O. ustulata,** *L.*

O. sive Cynosorchis minor pannonica, Ger. (Blackst.).

Cyb. Br. ii. 424. Syme E. B. ix. t. 1450.

Chalk banks; very rare. P. May.

I. In Harefield chalkpit, sparingly; *Blackst. Fasc.* 69.

No other record. Michael Collinson says in his MSS. that he 'never found this sort;' but there is no reason to suppose Blackstone mistaken.]

648. O. maculata, *L.* *Spotted Hand Orchis.*

Palma Christi fœmina (Ger.). *Serapias fœmina pratensis, Lob.* and *S. candido fl. mont. maculatis fol., Lob.* (Johns.). *O. palmata pratensis maculata, C. B. P.* (Blackst.).

Cyb. Br. ii. 428; iii. 512. Syme E. B. ix. t. 1459.

Wet heaths, woods, and meadows; rather common. P. May, June.

I. Harefield chalkpit, plentifully; *Blackst. Fasc.* 68. Harrow Weald Common!; *Melv.* 76. Stanmore Heath, abundant.

III. Harrow Grove; *Melv.* 76. Drilling Ground, Hounslow.

IV. Hampstead Wood; *Ger.* 170. Wood beyond the 'Spaniards;' *Irv. MSS.* Hampstead Heath; *Johns. Enum.* Ibid. 1815; *Herb. G. & R.* Stanmore; *Varenne.* Harrow Park; *Melv.* 76. Bentley Priory Meadows. Scratch Wood.

V. Wood right hand of canal a mile beyond Brentford; *Hemsley.*

VI. The Alders Wood, Whetstone. Edmonton.

VII. [Margin of the pond in the Vale of Health; *Irv. MSS.*]

First record: *Gerarde,* 1597.

649. O. incarnata, *L.*

Palma Christi mas (Ger.). *O. palmata pratensis latifolia longis calcaribus, C. B. P.* (Blackst.). *O. latifolia* (Curt., Forst.).

Curt. F. L. f. 5.

Damp meadows; very rare. P. May, June.

I. Harefield, plentifully; *Blackst. Fasc.* 68.

IV. In Hampsted Wood; *Ger.* 170. Stanmore, 1827–30; *Varenne.**
VII. Marshes near Lea Bridge, E. F.; *Herb. Mus. Brit.*
First record; *Gerarde*, 1597; also first as a British plant.

[**650. O. pyramidalis,** *L.*
O. purpurea spica congesta pyramidali, R. Syn. ii. (Blackst.).
Cyb. Br. ii. 426; iii. 511. Syme E. B. ix. t. 1449.
Chalky banks; very rare. P. June.
I. In Harefield chalkpit, plentifully; *Blackst. Fasc.* 69. There in abundance; *Michael Collinson MSS.*
First record: *Blackstone*, 1737; last, *Michael Collinson*, about 1790.†
Likely to be refound.]

[GYMNADENIA, *R. Br.*

651. G. conopsea, *R. Br.*
O. palmata minor calcaribus oblongis, C. B. P. (Blackst.).
Cyb. Br. ii. 428. Syme E. B. ix. t. 1460.
Dry banks; very rare. P. June.
I. Harefield chalkpit; *Blackst. Fasc.* 69. In abundance there; *Michael Collinson MSS.*
First record: *Blackstone*, 1737; last, *Michael Collinson*, about 1790.]

HABENARIA, *R. Br.*

652. H. bifolia, *R. Br.* H. eu-bifolia (*Syme E. B.*).
O. alba bifolia minor (Budd., Doody, Alchorne). '*Smaller Butterfly Satyrion or Gnat-flower*' (M. Collinson).
Cyb. Br. Supp. 69, 101. Syme E. B. ix. t. 1664.‡
Heaths and woods; rare. P. June, July.
IV. On Hampstead Heath; *Budd. MSS.* and *MS. note* (*Alchorne's*) *quoted in Phyt.* iii. 169. In Bishop's Wood 'recently;' *Irv. MSS.*
VI. Near Enfield Chase by Dr. Uvedale; *Doody MSS.* In 1756 in great abundance for more than two miles amongst the bushes on Endfield Chase, bet. Southgate and the lodge now (1760) in the possession of Mr. Jalabert; *Michael Collinson MSS.* Highgate Wood; *Jewitt.*
VII. My father saw this in a wood bet. Hampstead and Highgate, now the property of Lord Mansfield, and since enclosed by him with pales; *Michael Collinson MSS.* In Cane Wood; *MS. note* (*Alchorne's*) *quoted in Phyt.* iii. 169.
First record: *Uvedale*, about 1700.

---

* Mr. Varenne's specimens are perhaps *O. latifolia,* L.
† The following remarks of Peter Collinson, quoted by Dillwyn in the *Hortus Collinsonianus,* p. 36, will help to account for the extinction of this and other orchids at Harefield. 'There is one Miles, a parson of Cowly, near Uxbridge, who is orchis mad, takes all up, leaves none to seed, so extirpates all wherever he comes, which is cruel, and deserves chastisement.'
‡ Erroneously lettered *H. chlorantha* on the plate.

653. H. chlorantha, *Bab.*

Orchis hermaphroditica (Ger., Blackst., M. Collinson). *O. bifolia.* (Curt.). Cyb. Br. 69, 101. Curt. F. L. f. 6.

Woods, &c.; rare. P. May, June.

I. Harefield, not frequent; *Blackst. Fasc.* 70.

IV. North end of Hampstead Heath; *Ger.* 166. In plenty about Mill Hill, 1757–66; *Michael Collinson MSS.*

VI. Fields adjoining to the pound or pinfolde without the gate at Highgate; *Ger.* 166. Highgate Wood, 1857; *Jewitt.*

First record: *Gerarde,* 1597. Perhaps extinct

OPHRYS, *Linn.*

[**654. O. apifera,** *Huds.*   *Bee Orchis.*

O. fucum referens major, &c., C. B. P. (Blackst.). Cyb. Br. ii. 433; iii. 513. Syme E. B. ix. t. 1467.

Chalky banks; very rare. P. May—July.

I. In Harefield chalkpit, sparingly; *Blackst. Fasc.* 70. On the bank of the chalkpit on the left-hand side, 1788, and in the arena of the same pit, 1790; *Michael Collinson MSS.*

First record: *Blackstone,* 1737; last, *M. Collinson,* 1790. It has not been noticed since,* but may perhaps be re-found.]

655. O. muscifera, *Huds.*

O. myodes minor, Park. (Blackst.). Cyb. Br. ii. 436. Syme E. B. ix. t. 1471.

Copses and bushy places on chalk; very rare. P. May—July.

I. Harefield chalkpit, particularly that side near the lane; *Blackst. Fasc.* 68. Several plants there July, 1790; *Michael Collinson MSS.* Chalkpit in Esquire Cook's park, Harefield, July, 1757; *Peter Collinson, quoted in Hortus Collinsonianus,* p. 36. In two copses on the road from Harefield to Uxbridge, near Moor Hall and West End, 1867–68; *Cole (v. s.).*

First record: *Blackstone,* 1737.

(Herminium Monorchis, *R. Br. Ophrys M.* (Mart.). Cyb. Br. ii. 432. Syme E. B. ix. t. 1466. VI. Enfield, near the town; *Mart. App. P. C.* 66, and *Mart. Mill. Dict.* No other record. The station in *Blackst. Fasc.* 69 is in Bucks. It seems probable that Martyn accidentally entered this for *Spiranthes autumnalis.*

SPIRANTHES, *Rich.*

656. S. autumnalis, *Rich.*   *Lady's Tresses.*

' *Satyrion, a certeyne kynde*' (Turn.). *Testiculus odoratus* and *Triorchis* (Ger.). *Orchis alba spiralis odorata* (Alchorne). *Ophrys spiralis* (Mart. Curt.).

Cyb. Br. ii. 413. Curt. F. L. f. 4 (drawn from a Middlesex specimen).

* See note to 650, *Orchis pyramidalis,* L.

Dry fields and banks ; rather rare. P. September.
I. Neighbourhood of Uxbridge ; *Lightf. MSS.* Pinner Hill, Mrs. Tooke ; *Melv.* 76.
III. In the field next Thistleworth as you go from Branford (=Brentford) to her Majesties house at Richmond ; *Ger.* 168.
IV. Plentifully in a field adjoining to our garden at Mill Hill ; in another field at Highwood Hill, abundantly ; *Michael Collinson MSS.*
V. Beside Syon ; *Turn. Names.* Hanwell Heath near Ealing, Dr. Goodenough ; *Curt. F. L.*
VI. At Enfield, on the Chase, in a dry ditch ; *Doody MSS.* On Enfield Chase, near the town ; *MS. note (Alchorne's) quoted in Phyt.* iii. 169 and *Mart. App. P. C.* 65.
VII. [In the field by Islington where there is a bouling place under a few old shrubby okes ; upon a common heath by a village called Stepney, Master John Coles ; *Ger.* 168.]
First record : *Turner,* 1548 ; also first as a British plant.

LISTERA, *R. Br.*

657. L. ovata, *R. Br.*　　　*Tway-blade.*
Ophris bifolia (Ger.). *Bifolium, Lob.* (Johns.). *Bifolium sylvestre vulgare, Park.,* (Blackst.). *Ophrys ovata* (Curt.).
Cyb. Br. ii. 416 ; iii. 510. Curt. F. L. f. 3.
Moist woods and copses ; rather rare. P. May, June.
I. In Whiteheath Wood and Scarlet Spring, and in meadows near the river, Harefield ; *Blackst. Fasc.* 10. Copse near Harefield towards Uxbridge ; *Cole.* Copse near Pinner Wood. Ruislip.
III. South End of Harrow Grove ; *Melv.* 75.
IV. Hampsteede Wood ; *Ger.* 326. Hampstead Heath ; *Johns. Enum.* Meadows opp. Swan Inn, near Hendon ; *Coop. Supp.* 11. Bentley Priory Woods ; *Melv.* 75. Bishop's Wood. Scratch Wood.
VI. In the fields by Highgate ; *Ger.* 326. The Alders Wood, Whetstone.
First record : *Gerarde,* 1597 ; also first as a British plant.

NEOTTIA, *Linn.*

658. N. Nidus-avis, *Rich.*　　　*Bird's-nest.*
Orchis abortiva fusca, C. B. P. (Blackst.).
Cyb. Br. ii. 414 ; iii. 510. Syme E. B. ix. f. 1478.
Parasitical on beech and other trees ; very rare. P.? May.
I. In Whiteheath Wood on Harefield Common, but very rarely, 1735 ; *Blackst. Fasc.* 67.
III. Harrow Grove, 1868 ; *Farrar.*
IV. On a common laurel in the grounds of Stanmore Cottage, about 1830 ; *Varenne.*
First record : *Blackstone,* 1737. Occurs in Oxhey Wood, Herts ; *Farrar.* Just beyond our bounds.

EPIPACTIS, *Rich.*

E. latifolia, *All.* Cyb. Br. Supp. 69, 101. Syme E. B. ix. t. 1480.
I. Neighbourhood of Uxbridge; *Lightf. MSS.* VI. Enfield; *Wollaston.*
A doubtful inhabitant of Middlesex, **659** being perhaps the species
intended in the above stations.

659. ? **E. media,** *Fr.**
' *E. latifolia* ' (Melvill).
Cyb. Br. Supp. 69, 101. Syme E. B. ix. t. 1480.
Shrubberies, perhaps parasitical; very rare. P. July, August.
IV. In some plenty on a raised mound formed about 19 years ago by the
soil removed in digging the foundations of the church, and planted
as a shrubbery bet. Harrow Weald Church and the Rectory, 1866.
First observed there by Rev. R. J. Knight (v. *Melv.* 76).
VI. In the shrubberies and plantations of Chase Cottage, near Enfield,
appearing spontaneously a few years before 1861 and rapidly increasing; *Phyt. N. S.* v. 380, and vi. 299.

First record: ' *M. A. W.*' 1861. The plants found at these two localities
are no doubt referable to the same species, but we have not seen specimens from Enfield. The situation of the plant in both localities is
such as to favour the view that it is parasitical, but its sudden appearance in both cases is not easily explicable. The figure quoted
does not well represent our plant, which is possibly Smith's *E. purpurata,* found in Bedfordshire and (*Fl. Herts.* 295) Herts.

660. E. palustris, *Sw.*
Cyb. Br. ii. 418. Syme E. B. ix. t. 1482.
I. Pinner Wood; *Miss Tooke.*
The only record. Requires confirmation. We have not seen specimens.

IRIDACEÆ.

IRIS, *Linn.*

661. I. pseud-Acorus, *L.*   *Water Flag.   Fleur-de-luce.*
I. palustris lutea, Ger. em. (Blackst.).
Cyb. Br. ii. 439.   Curt. F. L. f. 3 (seems to be *I. acoriformis,* Bor.).

Sides of ditches, ponds, and streams; common. P. June, July.
I. Harefield!; *Blackst. Fasc.* 46. Colne, near Yewsley!; *Newb.* Ruislip
Moor.
II. West Drayton; *Morris* (*v. s.*). Staines Moor. Staines Common. By
Thames to Hampton.

---

* *Helleborine*; in a wood 5 miles from London neere a bridge called Lock-bridge; *Ger.*
358. Middlesex ? Perhaps this species ?

III. By the Cran, Hounslow Heath.   Hatton, Hanworth Road.   Thames at Twickenham and Isleworth.
IV. Near Kenton; *Melv.* 76.   Stanmore; *Varenne.*   Bet. Finchley and Hendon!; *Newb.*
V. Canal at Greenford; *Melv.* 76.   At Hanwell; *Cole.*   Perivale; *Lees.*
VI. Copse near Warren Lodge, Edmonton.
VII. Ponds in Ken Wood Grounds.   Isle of Dogs.

First record: *Blackstone,* 1737. Mr. Syme says (*Seem. J. of B.* vi. 69) that *I. acoriformis,* Bor. (*Syme E. B.* ix. t. 1495) is the only one of the three forms into which Boreau divides this species which he has seen by the Thames, but that *I. Bastardi,* Bor., is the plant round Ken Wood ponds. The typical *I. pseud-Acorus,* however, probably occurs in the county.

662. I. fœtidissima, *L.*   *Stinking Gladdon or Gladwyn.*
Xyris, sive Spatula fœtida (Park., Pet., Mill.).
Cyb. Br. ii. 439.   Syme E. B. ix. t. 1494.
Hedges; very rare.   P.   June, July.
V. Pasture at Perivale, apparently native; *Lees.*
VI. Muswell Hill, in a hedge; *Mart. App. P. C.* 72.
VII. [Near to Kentish Town. . . . I do verily think it not natural in that place; *Park. Theat.* 258. In a hedge near Kentish Town; *Mart. Tourn.* ii. 45.] [On Jack Straw's Castle, and in a hedge near it; *Pet. Midd.* and *Mill. Bot. Off.* 421.] Near Hornsey; *Huds.* i. 14 and *subsequent writers.* Lord Mansfield's Park; *Cooper,* 102.

First record: *Parkinson,* 1640.

CROCUS, *Linn.*

663. *C. vernus, *Willd.*   *Spring Purple Crocus.*
Cyb. Br. ii. 442.   Syme E. B. ix. t. 1499.
Meadows; very rare.   P.   March, April.
VI. In meadows near the Church at Hornsey, in plenty, 1842, T. B. Flower; *Herb. Linn. Soc.* Southgate, F. W.; *Phyt. N. S.* vi. 349. Hornsey Meadows, 1852; *Herb. S. P. Woodward.* Abundant in the turf of a large meadow a little south of Colney Hatch, but in Hornsey parish. Prof. A. H. Church informs us that it grew in this last locality in the recollection of persons living in the neighbourhood, at the end of the last century [and that before houses were built it grew sparingly bet. Hornsey and Wood Green].

First record: before 1800. Is it certainly an alien?

AMARYLLIDACEÆ.

NARCISSUS, *Linn.*

664. ***N. biflorus,** *Curt.*   *Primrose Peerless.*

N. medio-luteus vulgaris, Park. (R. Syn. iii.). *N. medio-luteus, Ger. em.* (Blackst.).

Cyb. Br. ii. 444. Syme E. B. ix. tt. 1503.

Meadows; rare. P. May.

I. Near Harefield in several places; *Blackst. Fasc.* 64. In a meadow to the south of Ruislip Reservoir, in plenty.

IV. Waste ground by Mrs. Rotch's house at the bottom the Hill, Harrow, probably planted; *Melv.* 77.

VI. [Near Hornsey Church, J. Sherard; *R. Syn.* iii. 371. Mr. Dillwyn could not find it there; *B. G.* 403.]

First record: *J. Sherard*, 1724. Figured in *Pet. H. B. Cat.* lxvii. fig. 10.

N. poeticus, L. Cyb. Br. ii. 444. Syme E. B. ix. t. 1504. I. Meadow at Pinner Hill, likely escaped from a garden, Mrs. Tooke; *Melv.* 77. In a field bet. Ruislip Reservoir and the road to Harefield, 1866; *Griffith (v. s.).* Much cultivated.

665. **N. pseudo-Narcissus,** *L.*   *Daffodil.*

N. sylv. pallid. calice luteo, C. B. P. (Blackst.).

Cyb. Br. ii. 445; iii. 513. Syme E. B. ix. t. 1501.

Meadows; rather rare. P. April.

I. In the orchard at Breakspears, plentifully; *Blackst. Fasc.* 63. In a grove near Harefield Church, 1853; *Herb. Hardw.* In great plenty near the high road from Pinner to Rickmansworth, where the cross roads to Eastcott and Potter's Green are, 1867; *Cole.*

III. Roxeth, probably not indigenous, W. M. H.; *Melv.* 77.

IV. Mill Hill; *Salisbury in Trans. Hort. Soc.* i. 348. Ibid. 1840; *Herb. Hardw.* Field behind the 'Kings Head;' *Lond. Fl.* 109. Behind the 'Spaniards,' Hampstead, no doubt escaped or planted.

V. In a wood near Ealing, introduced; *Hemsley.*

VI. Hornsey; *Cat. Lond.* 14. Behind Bury House, Edmonton, garden escape.

VII. Stamford Hill; *Cat. Lond.* 16. Abundant in Ken Wood, but probably planted.

First record: *Clusius*, 1601, who says: 'In such abundance in the meadows close to London, that in that celebrated village of Ceapside the country women offer the flowers in profusion for sale in March, when all the taverns may be seen decked out with these blossoms;' *Rar. Plant. Hist.* 164. Lobel also (*Adv.* 51) says, that in February and

March the London flower market is full of it. Native in some parts of I. and IV., but usually a double-flowered plant (? *N. major*, Curt.), obviously from gardens.

[*N.* incomparabilis, *Miller.* Cyb. Br. ii. 446; iii. 513. Syme E. B. ix. t. 1502. *N. pallido-luteus anglicus alter calice magno brevi aureo*; I found this 1711 in some orchards and closes adjoining near Hornsey Church; *Budd. Herb.* cxxiv. fol. 30. We are indebted to Mr. Boswell-Syme for the determination of Buddle's specimen. The plant was figured in *Pet. H. B. Cat.* lxvii. p. 8 as the '*Hornsey Daffodil*,' and this figure has been thought to represent various species, e.g. *N. amplus*, Salisb., *N. pallidus*, St., *N. lætus*, &c. This old garden flower has been found semi-wild in many other places in England.]

'*N. serratus*.' In a field near the public-house at Mill Hill, E. F.; *Herb. Mus. Brit.* Several exotic *Narcissi* in a field behind the King's Head, Mill Hill; *Irv. Lond. Fl.* 304. The growth of these is explained by the residence in the village for seven years of R. A. Salisbury,* who cultivated such plants extensively. He occupied the house formerly Peter Collinson's, and afterwards a school. (See Salisbury's *Genera of Plants; a Fragment*: published in 1866.)

[LEUCOIUM, *Linn.*

666. L. æstivum, *L.*   *Summer Snowflake.*
Cyb. Br. ii. 448. Curt. F. L. f. 5.
Wet marshes by rivers; very rare. P. May.
VI. Hornsey, 1845; *Herb. Hardw.* From a garden?
VII. It has been found in the Isle of Dogs; *Curt. F. L.*

First record: *Curtis*, about 1790. Curtis's plate was drawn from specimens collected on the opposite shore, about half a mile below Greenwich, Kent, where it was still growing in 1837, but extinct in 1852 (see *Phyt.* iv. 862).]

ASPARAGACEÆ.

CONVALLARIA, *Linn.*

667. C. majalis, *L.*   *Lily of the Valley.   May Lily.*
Lilium convallium (Ger., Ray, &c.).
Cyb. Br. ii. 467; iii. 515. Curt. F. L. f. 5 (probably drawn from a Hampstead specimen).
Woods and bushy places; rather rare. P. May.
IV. [On Hampstead Heath in great abundance; *Ger.* 332, *Johns. Eric.*, *R. Cat.* i. 194, *Budd. MSS., &c.* In Curtis's and T. Martyn's time

* Born 1761, died 1827.

it had become scarce, 'since the trees have been destroyed,' *Mart.
Mill. Dict.* Very sparingly under the bushes near the bog, 1850–55;
Pamplin.] Ditchbank north of old target-bank; Turner's Wood,
Bliss; *Park Hampst.* 29. Bishop's Wood, 1864, a single root; for-
merly abundant.

(V. Norwood, Middlesex,† abundantly, Dr. Martyn; *Baxter*, i.)

VI. About Highgate; *Cooper*, 104. At Winchmore Hill, Dr. J. Mitchell;
Cooper, 119. Southgate, F. W.; *Phyt. N. S.* vi. 349. The Alders
Copse near Whetstone, 1863; we could not find it in 1867.

VII. Lord Mansfield's Wood, near the 'Spaniards;' *Curt. F. L.* and *Mart.
Mill. Dict.* Still abundant in the wooded parts of Ken Wood
Grounds.

First record: *Gerarde*, 1597; also first as a British species. Dug up for
gardens, and so nearly eradicated near London.

POLYGONATUM, *Tourn.*

(P. verticillatum, *All.* and P. officinale, *All.* are given for Ken Wood by
Hunter, but Bliss could not find them; *Park Hampst.* 29. Perhaps
they may have been in cultivation there.)

668. P. multiflorum, *All.* *Solomon's Seal.*
Cyb. Br. ii. 468. Syme E. B. ix. t. 1513.
Woods; very rare. P. May, June.

IV. A single plant found in 1864 by W. G. Smith in company with M. C.
Cooke and A. Grugeon in Bishop's Wood near the head of the bog
in which *Chrysosplenium* grows. In April 1866 we carefully searched,
in company with Mr. Smith, the spot indicated, but failed to find
the plant, which, however, may probably still exist there.

MAIANTHEMUM, *Wiggers.*

669. *M. bifolium, *DC.*
Convallaria bifolia, L. *Smilacina bif.*, Desf.
Cyb. Br. ii. 465; iii. 514. Syme E. B. ix. t. 1510.
Woods; very rare. P. May, June.

VII. A patch of about twenty square yards on an eminence under the shade
of a very large beech in the enclosure of Ken Wood Grounds near
its S.E. angle.

Known to have existed there for nearly ninety years, but first recorded by
Hunter, 1813, in *Park Hampst.* 29. Mr. Irvine collected it in 1829
(*Phyt. N. S.* iv. 233), shortly before which time another patch had been
recently destroyed. In 1835 Mr. E. Edwards found several patches
under the shade of fir trees (*Phyt.* i. 579), and since that time the spot
has been frequently visited by botanists. Possibly truly native. Lobel
suggests (*Adv.* 300) its nativity in England, Parkinson confirms it,

† Norwood, *Surrey,* probably intended.

Gerarde and How give special localities. It rarely produces fruit at Ken Wood, but Mr. G. Kay, one of the gardeners there, sent us, in August 1866, some nearly ripe berries.

RUSCUS, *Linn.*

670. R. aculeatus, *L.*   *Butcher's Broom.   Kneeholm.*
Ruscus sive Bruscus (Ger.).
Cyb. Br. ii. 464. Syme E. B. ix. t. 1516.
Woods; very rare. Shrub. March, April.
I.   In a little grove near Breakspears; *Blackst. Fasc.* 88.
IV. Hampstead Heath; *Ger.* 759. Formerly abundant there; *Loud. Arb. et Frut.* 2519. Bishop's Wood, in small quantity, 1861; not there in 1867.
VI. Near Finchley; *Varenne.*
VII. Lane near West End; *Burnett.*
First record: *Gerarde,* 1597. Perhaps extinct.

LILIACEÆ.

TULIPA, *Linn.*

671. * T. sylvestris, *L.*   *Wild Tulip.*
Cyb. Br. ii. 449. Syme E. B. ix. t. 1520.
Meadows; very rare. P. May, June.
I.   In a grove by Harefield Church, 1853; *Herb. Hardw.*
VI. Top of Muswell Hill, J. Woods; *B. G.* 403. Hundreds of plants there in 1855; *Phyt. N. S.* i. 391. Still growing there in 1860 or 1861; *Pamplin.* The Mus-well field in Rhode's fields; *Church.*
First record: *Woods,* 1805. Perhaps native. The plant very rarely flowers at Muswell Hill, the leaves being always cut with the hay. Prof. Church transplanted bulbs from the field into a garden, where flowers were produced.

FRITILLARIA, *Linn.*

672. F. Meleagris, *L.*   *Snake's-head.   Fritillary.   Guinea-hen flower.*
F. præcox purp. variegata, C. B. P. (Blackst.).
Cyb. Br. ii. 450. E. B. 622, reproduced in Syme E. B. ix. t. 1519 (drawn from a Middlesex specimen); but Curt. F. L. f. 3 is a better figure.
Meadows; rare. P. April, May.
I.   In Maud-fields near Rislip Common,† observed above forty years by Mr. Ashby of Breakspears; *Blackst. Fasc.* 29. Fields at Pinner, in

---

† In a letter to Richardson (*R. Corresp.* 354), Blackstone describes this locality as 'a meadow by a wood-side near Harefield,' 1736. There are no fields known as '*Maud-fields*' now at Ruislip, but *Ruislip Moor* answers to the above description.

one very abundant!; *Melv.* 78. This field is a little north of the church, and adjacent to Moss Lane. The plant occurs in three other meadows near, and also in an orchard; *Hind.*

VI. Near Enfield; *Huds.* ii. 144. Near Bury, Enfield; *With.* ii. 1, 346 Finchley, 1842, J. A. Hankey; *Proc. Linn. Soc.* i. 134; specimens in *Herb. Mus. Brit.* Field near the brook crossing Colney Hatch Lane near the Asylum, a single plant found; *Church.*

First record: *Blackstone*, 1736; also first as a British plant. Blackstone suspected this to be an outcast from gardens (see his MS. notes in *Johns. Merc. Bot.*), but there is no doubt it is a true native. Also grows abundantly in Lord Lytton's park, Totteridge, Herts.

Lilium Martagon, *L.* *Turk's Cap.* Cyb. Br. ii. 449. Syme E. B. ix. t. 1518. VI. Old pasture in Enfield Chase; *Phyt. N. S.* vi. 573. We know nothing of this locality. Rev. R. H. Webb records it from Totteridge Park, Herts, 1855, ascertained to have been there not less than one hundred years; *Phyt. N. S.* ii. 162.

Ornithogalum umbellatum, *L.* *Star of Bethlehem.* Cyb. Br. ii. 458. Syme E. B. ix. t. 1524. II. By the Thames side in the neighbourhood of Teddington Lock, E. K.; *Mag. Nat. Hist.* i. 83. To be looked for; we collected it in 1859 on the point of land (or island?) at Teddington Lock, Surrey.

O. pyrenaicum, *L.* Cyb. Br. ii. 457. Syme E. B. ix. t. 1525. II. Strawberry Hill, Mr. Woodward; *B. G.* 403. Perhaps cultivated there.

O. nutans, *L.* Cyb. Br. ii. 458. Syme E. B. ix. t. 1523. III. Twickenham, Mr. Chambers; *Herb. Mus. Brit.* A cultivated specimen?.

SCILLA, *Linn.*

673. S. autumnalis, *L.*

Hyacinthus autumnalis minor (Park., Johns., Merr.).

Cyb. Br. ii. 459. Curt. F. L. f. 6.

Sandy banks; very rare. P. August, September.

II. Sparingly in one spot on the sloping bank by the towing-path bet. Hampton Court and Ditton Ferry, 1868!; *Newb.*

VII. [Foot of a high bank by the Thames side at the hither side of Chelsey, before you come to the King's Barge-house; *Park. Parad.* 132, *Johns. Ger.* 111. Going to Hunslow Heath; *Merrett*, 64, probably the same locality.]

First record: *Parkinson*, 1629; also first as a British plant.

(S. verna, *Huds.* Syme E. B. ix. t. 1527. VII. Ken Wood, Hunter; *Park Hampst.* 29. An error; or cultivated?.)

ALLIUM, *Linn.*

674. A. vineale, *L.*

A. sylvestre (Ger., Blackst.). *A. sylv. cum flosc. bulb. intermixtis,* and *A. sylv. juncifol. min. nud. rarior. at majoribus* (Budd.).

Cyb. Br. ii. 454 ; iii. 514. Syme E. B. ix. t. 1534.

Fields and waste places; rare. P. June, July.

II. Lanes about Sunbury; *Winch MSS.* Bet. Kingston Bridge and Hampton Court! ; *Warren.*

III. Meadows at Roxeth, W. M. H. ; *Melv.* 78.

V. Among corn near Ealing, 1867 ! ; *Newb.*

VII. [In great plenty in a field called the Mantells on the back-side of Islington ; *Ger.* 142. About London, frequent ; in a pit in a field adjoining to Tyburn Road near Marylebone ; *Budd. MSS.*]

Var. β. A. compactum, Thuill. *A. camp. juncifol. capital. purpurasc. majus, C. B.* (Budd.).

III. Roxeth, with the type, W. M. H. ; *Melv.* 78.

VII. [In the pit by the Tyburn Road; *Budd. MSS.*]

First record : *Gerarde,* 1597.

675. A. oleraceum, *L.*

Cyb. Br. ii. 453. Syme E. B. ix. t. 1535.

Fields; very rare. P. July, August.

II. In a field very near Sunbury, bet. the Hampton Court Road and the Thames ; it looked wild enough there ; *H. C. Watson.*

The only record.

(*A. triquetrum, L.* Syme E. B. ix. t. 1539. 'Specimens were sent to the Botanical Society of London by the late Mr. J. Banker, of Devonport, with the locality "Isle of Dogs, May 1852";' *Syme E. B.* ix. 217. Some confusion may be presumed.)

676. A. ursinum, *L. Ramsons.*

A. sylv. latifolium, C. B. P. (Blackst.).

Cyb. Br. ii. 456. E. B. 122, reproduced in Syme E. B. ix. t. 1540 (drawn from a Middlesex specimen).

Damp woods and hedges ; rather common. P. May, June.

I. In a meadow near Gulch-well, Harefield ; *Blackst. Fasc.* 3. Near foot-crossing over railway, Pinner ; *Melv.* 78.

III. Bank on Pinner Road, Harrow ; *Melv.* 78.

IV. Hendon Place, near the church ; *Mart. App. P. C.* 66. Hendon ; *Lond. Fl.* 106. Sudbury ; Harrow Weald ; *Melv.* 78. Near Harrow Station, 1855 ; *Phyt. N. S.* i. 63. Cricklewood, beyond Kilburn, 1867 ; *Fox.* Golder's Green ; *Cole.* Renter's Lane ; *Davies.* Road from Finchley to Hendon! ; *Newb.* Copse in Bentley Priory, abundant.

VI. Near Finchley Common, 1793, J. Rayer ; *Smith MSS.* Highgate Wood. Alders copse and hedges near Whetstone, very abundant.

VII. Next field to Boobies Barn, also under the hedge of a lane to Hampstead; *Ger.* 142. [Ditch on ye back of Kentish Town; *Newton MSS.*] [Marylebone Fields, 1817; *Herb. G. & R.*] [By the ponds, Highgate; *Herb. Hardw.*] Field bet. Swiss Cottage and Hampstead, 1846; *Morris* (*v. s.*).

First record: *Gerarde*, 1597.

ENDYMION, *Dumort.*

677. E. nutans, *Dum.* Scilla nutans, *Sm.* (Syme). Hyacinthus non-scriptus, *L.* (L. Cat.). *Blue-bell.** *Wild Hyacinth.*
Hyacinthus (Turn.). *H. vulg. Ang. et Blg.* (Johns.). *H. anglicus, Ger. em.* (Blackst.).
Cyb. Br. ii. 460. Curt. F. L. f. 2.

Woods and shady places; rather common. P. April, May.

I. Harefield Common!; *Blackst. Fasc.* 43. Ruislip Wood. Pinner.
III. Harrow Grove, Roxeth, &c.; *Melv.* 78. Bet. Pinner and Harrow, abundant.
IV. Hampstead Heath!; *Johns. Eric.* Football Field, Harrow; *Melv.* 78. Bentley Priory. Bishop's and Turner's Woods.
V. Muche aboute Sion; *Turn. Names.*
VI. Hadley!; *Warren.* Whetstone. Winchmore Hill Wood.
VII. Ken Wood. Hornsey Wood.

First record: *Turner*, 1548; also first as a British plant. Noticed with *white flowers* in many places; with *rose-coloured flowers* (V.) in Osterley Park; *Masters.* Probably absent from the dry south-west part of Middlesex.

Muscari racemosum, *DC.* M. neglectum, *Guss.?* (Bab.). *Starch Hyacinth.* Cyb. Br. ii. 461; iii. 514. Syme E. B. ix. t. 1529. II. Hampton Court, perhaps escaped from the palace; *Coop. Supp.* 12.

COLCHICACEÆ.

COLCHICUM, *Linn.*

678. C. autumnale, *L.* *Meadow Saffron.*
C. commune (R. Syn. iii.).
Cyb. Br. ii. 471; iii. 515. Syme E. B. ix. tt. 1544–45.

Meadows; very rare. P. September.

VI. In Mr. Moor's meadow that comes down to the great fish-pond near his house at Southgate, Mr. J. Sherard; *R. Syn.* iii. 373. Forty Hill, Enfield Chase, W. Cullen; *Cooper*, 116.

First record: *J. Sherard*, 1724. We have not seen any specimens.

---

* Called *Harebell* by all British botanists, previous to Smith; but 425 usually now receives that name in England.

JUNCACEÆ.

JUNCUS, *Linn.*

679. J. effusus, *L.*   J. communis, *Meyer,* var. (Syme and L. Cat.).
  Soft Rush.
  J. lævis panicula sparsa major, *C. B. P.* (Blackst.).
  Cyb. Br. iii. 39.   Syme E. B. x. t. 1561.
  Damp waste ground, roadsides, and fields; common.   P.   July.
  I. Harefield!; *Blackst. Fasc.* 47.   Harrow Weald Common.   Stanmore
    Heath.   South Mims.
  II. Youveney, near Staines.   Fulwell.
  III. Twickenham Park inclosure.   Near Worton.
  IV. Abundant in the Harrow district; *Melv.* 79.   Golder's Green, 1846;
    Herb. Hardw.
  V. Bet. Acton and Ealing Station!; *Newb.*
  VI. Edmonton.
  VII. South Heath, Hampstead.
  First record: *Blackstone,* 1737.

680. J. conglomeratus, *L.*   J. communis, *Meyer,* var. (Syme and L. Cat.).
  Cyb. Br. iii. 38.   Syme E. B. x. t. 1560.
  Wet places, ditches, &c.; apparently rather rare.   P.   July.
  I. Harrow Weald Common!; *Melv.* 79.   Harefield!; *Newb.*   Stanmore
    Heath.
  III. Harrow; *Melv.* 79.
  IV. Hampstead Heath; *Cooper,* 103, and *Irv. MSS.*   Stanmore, 1837, D.
    Cooper; *Herb. Young.*
  VI. Edmonton.
  First record: *Irvine,* about 1830.   Probably overlooked, but no doubt
    considerably less frequent than **679,** *J. effusus,* L.

681. J. glaucus, *Sibth.*   *Hard Rush.*
  J. durus vulgaris, *R. Syn.* iii. (Blackst.).
  Cyb. Br. iii. 41.   Syme E. B. x. t. 1563.
  Wet places, banks of streams and ditches; very common.   P.   July, August.
  Through all the districts.
  VII. Eel-brook Meadow, Parson's Green.
  First record: *Blackstone,* 1737.

  J. diffusus, *Hoppe.*   J. communis, *Meyer,* var. (L. Cat.).   Cyb. Br. iii. 40.
    Syme E. B. x. t. 1562.   This should be looked for.   A locality is given
    (*Fl. Herts.* 308) near the Warren Gate, which is the extreme north
    point of Middlesex, and another (*Fl. Herts. Supp.* 19) in Oxhey Lane,
    just beyond our boundary.

682. J. acutiflorus, *Ehrh.*
Gramen junceum sylvatic., Tab. (Johns.). *J. fol. articulosis fl. umbellatis, Inst. R. H.* (Blackst.).
Cyb. Br. iii. 44. Syme E. B. x. t. 1567.
Sides of ponds and wet places, especially on heaths; very common. P. July. In all the districts.
VII. Hornsey Fields, 1815; *Herb. G. & R.* South Heath, Hampstead. Eel-brook Meadow.
First record: *Johnson,* 1632.

683. J. lamprocarpus, *Ehrh.*
Cyb. Br. iii. 44. Syme E. B. x. t. 1568.
Wet places; rather common. P. July, August.
I. Harrow Weald Common!; *Melv.* 79. Harefield!; *Newb.*
II. Bet. Hampton and Hampton Court!; *Newb.* Near Teddington Ry. Station.
III. Bet. Twickenham and Richmond Bridge. Meadow by the Duke's River at Worton, abundant,
IV. By the Brent near Willesden, 1842; Hampstead; *Herb. Hardw.* Pinner Drive; *Melv.* 79.
VI. Edmonton.
VII. South Heath, Hampstead.
First record: 1842.

684. J. supinus, *Mönch.*
Gram. junceum capsulis triang. minimum (Ray, Blackst.).
Cyb. Br. iii. 46. Syme E. B. x. t. 1570.
Boggy places, chiefly on heaths; rare. P. July, August.
I. Harefield Common!; *Blackst. Fasc.* 39. Harrow Weald Common; *Melv.* 79. Stanmore Heath.
IV. On Hampsted Heath!; *Ray Cat.* i. 150, *and subsequent writers.*
VII. South Heath, Hampstead. Eel-brook Meadow.
First record: *Ray,* 1670.

685. J. squarrosus, *L.* *Goose-corn.*
J. acutus cambro-britannicus, Park. (Ray). *Junco affinis panic. laxa seu longior. pedic. insid., Scheuz.* (Blackst.).
Cyb. Br. iii. 48. Syme E. B. x. t. 1571.
Wet places on heaths and commons; rare. P. June.
I. On Harefield Common, abundantly!; *Blackst. Fasc.* 47. Harrow Weald Common, abundant!; *Melv.* 79. Stanmore Heath.
III. Gravel-pits near Hounslow, sparingly.
IV. Moorish grounds about Hampsted Heath; *R. Cat.* i. 179. Ibid., 1846; *Herb. Hardw.*
First record: *Ray,* 1670.

686. J. compressus, *Jacq.*　J. bulbosus, *L. Subsp.* 2 (Syme E. B.).

Cyb. Br. iii. 47, 517.　Syme E. B. x. t. 1575.

Damp roadsides; rare.　P.　June.

IV. Stanmore; *Varenne.*　Golder's Green; *Irv. MSS.*

VII. Green Lanes, Newington, 1866!; Isle of Dogs, 1866!, very sparingly; *Newb.*

First record : *Varenne,* 1827.

687. J. bufonius, *L.*　　　*Toad-grass.*

Gramen alt. juncoides s. bufonis Flandricum, and *Gr. juncoides min. Anglo-Brit. Holosteo Matth. congener* (Lob.).　*G. holosteum min.* (Johns.). *Gr. junceum parvum s. Holosteum Math.* (Merr., Blackst.).

Cyb. Br. iii. 48.　Syme E. B. tt. 1572–73.

Moist places, roadsides, commons, &c.; very common.　P.　July, August. Throughout all the districts.

VII. [Lane by Tottenham Court towards Hampstead; *Johns. Ger.* 4.] South Heath, Hampstead. Eel-brook Meadow. Hackney Wick. Isle of Dogs.

First record: *Lobel,* about 1600; also first as a British plant.　Very variable in habit and general appearance.

LUZULA, *Cand.*

688. L. sylvatica, *Bich.*

Gramen nemorosum hirs. majus alt. præcox, tuberosa rad. (Lob.).　*Gr. nem. hirs. latifol. maximum* (Ray, Pet., Blackst.).　*Juncus sylvaticus, L.* (Curt.).

Cyb. Br. iii. 53.　Curt. F. L. f. 5 (drawn from a Hampstead specimen).

Woods; rather rare.　P.　April, May.

I. Field near Mr. Ashby's brick-kiln at Harefield; *Blackst. Fasc.* 37. Pinner Wood; *Melv.* 79.

IV. Plentifully in the ditch of a close adjoining to Hampstead Wood; *R. Cat.* i. 149, *and subsequent authors.*　Bishop's Wood!; *Curt. F. L.*

VI. Highgate; *Lob. Ill.* 39.　Near Highgate Archway; wood bet. Highgate and Muswell Hill, E. F.; *Herb. Mus. Brit.*　Winchmore Hill Wood. Bet. Whetstone and Colney Hatch.

VII. Cain Wood!; *Pet. Conc. Gr.* 227.　Hornsey Wood; *Blackst. Spec.* 31.

First record: *Lobel,* about 1600; also first as a British plant.

689. L. Forsteri, *DcC.*

Cyb. Br. iii. 53.　Syme E. B. x. t. 1547.

Woods; very rare.　P.

I. Woods, Pinner; *Hind in Phyt. N. S.* v. 201.

IV. Harrow Weald; *Melv.* 80.

First record: *Hind,* 1861.　We have not seen specimens.

690. L. pilosa, *Willd.*   *Wood Rush.*

Gr. nem. hirsut. minus angustifol. (Lob.). *Gr. nem. hirs. latifol. minus* *
(Merr.). *Gr. nem. hirsut. vulgare, R. Syn.* (Blackst.). *Juncus pilosus,*
L. (Curt.).

Cyb. Br. iii. 54. Curt. F. L. f. 5.

Woods and shady places; rather common ?. P. March—May.

I. Harefield!; *Blackst. Fasc.* 37. Pinner Wood; *Melv.* 80. Ruislip
Woods.

IV. Bishop's Wood. Turner's Wood. Scratch Wood. Stanmore; *Herb. Harr.*

V. Horsington Wood, W. M. H.; *Melv.* 80.

VI. About Highgate; *Lob. Ill.* 40. Edmonton.

VII. Below the pine-walk at Hampsted; *Merrett,* 54. Cain Wood; *Pet.*
Conc. Gr. 226.

First record: *Lobel,* about 1600.

691. L. campestris, *Willd.*

Gramen exile hirsut. cyperoides (Johns., Blackst.). *Gr. hirsutum exile,*
Ger. em. (Merr.). *Juncus campestris, L.* (Curt.).

Cyb. Br. iii. 55. Curt. F. L. f. 2.

Dry fields and pastures; rather common. P. April, May.

I. Harefield; *Blackst. Fasc.* 36. Pinner.

II. Near Strawberry Hill. About Staines. Near Charlton.

III. Hounslow Heath.

IV. Hampstead Heath!; *Johns. Enum.* Bishop's Wood. Harrow district,
abundant; *Melv.* 80.

VI. Finchley; Betstile Lane, Colney Hatch; *Herb. Hardw.* Edmonton.

VII. In Hide Park; *Merrett,* 53. Kensington; *Herb. Rudge.* Cain Wood;
Pet. Conc. Gr. 224. Fields bet. Highgate and Kentish Town.

First record: *Johnson,* 1632.

692. L. multiflora, *Lej.*

Gram. hirsut. elatius panic. juncea compacta, R. Syn. ii. (Blackst.). *Close-*
headed hairy Rush-grass (Pet.).

Cyb. Br. iii. 56. Syme E. B. x. t. 1550.

Heaths and commons; rather common. P. May, June.

I. Wood by roadside from Harefield Common to Busher Heath; *Blackst.*
Fasc. 37. Harefield Common. Harrow Weald Common, abundant.
Stanmore Heath.

III. Drilling Ground, Hounslow!; *Newb.* Near Hatton.

IV. Road bet. Finchley and Hendon!; *Newb.* North Heath, Hampstead.

V. Near Hanwell!; *Warren.*

VI. Hadley!; *Warren.* Great Northern Ry. bank, Colney Hatch.

VII. Cain Wood; *Pet. Conc. Gr.* 225, and *Herb. Young.*

First record: *Petiver,* 1716.

* Synonymy involved; perhaps refers to *L. sylvatica,* or possibly *L. Forsteri* confuses
the names.

ALISMACEÆ.

ALISMA, *Linn.*

693. A. Plantago, *L.*   *Water Plantain.*
Plantago aquat. major, Ger. em. (Blackst.).
Cyb. Br. ii. 475; iii. 516. Curt. F. L. f. 5.
Sides of ponds and ditches; very common. P. June—August.
Through all the districts.
VII. Dartmouth Park; *Jewitt.* Eel-brook Meadow, Parson's Green.

Var. β. *A. lanceolata,* With. Syme E. B. ix. t. 1438.
I. Bet. Yewsley and Iver Bridge!; *Newb.* Elstree Reservoir; Ruislip
   Reservoir; *Melv.* 80.
II. New Staines. Sunbury. By Queen's River, Hampton. By Thames
   bet. Hampton Court and Kingston Bridge.
V. Apperton; *Melv.* 80. Hanwell!; *Warren.*
VII. Hornsey Wood ponds.
First record: *Petiver,* 1715. *A. lanceolata,* With., is figured in *Pet. H. B.
Cat.* xliii. fig. 7, and marked as a London plant. It is perhaps only a
young state of the common Middlesex *A. Plantago.*

694. A. ranunculoides, *L.*
Plantago aquat. humilis, Ger. em. (Blackst.).
Cyb. Br. ii. 475. E. B. 326, reproduced in Syme E. B. ix. t. 1439 (drawn
   from a Middlesex specimen).
Marshy places; very rare. P. June—August.
I. On the banks of the upper pond near Mr. Ashby's brick-kiln, plenti-
   fully; *Blackst. Spec.* 78.
III. Near Hounslow, E. Forster; *B. G.* 404.
VI. [J. Rayer gathered it on Finchley Common 'August, 1795;' *Sm. E. B.*
   326. Frequent there, J. Woods; *B. G.* 404 and *Irv. Lond. Fl.*
   Specimen in *Herb. Hardw.*]
VII. [Eel-brook Common, abundantly, 1830; *Pamplin.*]
First record: *Blackstone,* 1737. Not collected since about 1845.

ACTINOCARPUS, *R. Br.*

695. A. Damasonium, *R. Br.*   *Thrum Wort.*
Plantago aquat. minor stellata (Johns., Blackst.). *P. aquat. min. muricata*
   (Park.). *Alisma Dam., L.* (Curt.).
Cyb. Br. ii. 477. Curt. F. L. f. 5.
Shallow pools on heaths and commons; rare. P.? June—August.
I. In a little bog on Harefield Common; *Blackst. Spec.* 75. Uxbridge,
   towards Denham; *Mart. App. P. C.* 74.

III. Mr. Goodyer found it on Hounslow Heath ; *Johns. Ger.* 418. Plentifully there, 1736 ; *Blackst. MSS.* Ibid. ; *Lambert's Herb. Kew.* Pond on Twickenham Common, 1842, Twining ; *Herb. Kew.*

IV. Golder's Green, Hampstead ; *Irv. MSS.* and *Lond. Fl.* 108.

V. [Acton Green, 1864 ; *Newb.*]

VI. [Finchley Common, J. Woods ; *B. G.* 404.]

VII. [Ditches left-hand of highway from Holloway to Highgate ; *Park. Theat.* 1245.]

First record : *Goodyer,** 1633 ; also the first as a British plant. Apparently less frequent now than formerly.

SAGITTARIA, *Linn.*

696. S. sagittifolia, *L.*   *Arrowhead.*

Pistana Magonis (Lob.). *Sagittaria major et minor* (Ger.). *Sagitta aquatica major et minor et S. angustifolia* (Blackst.). *Gramen bulbosum aquaticum, C. B. Prod.* (Merrett).

Cyb. Br. iii. 478. E. B. 84, reproduced in Syme E. B. ix. t. 1436 (drawn from a Middlesex specimen).

Sides of streams and ponds ; common. P. July, August.

I. Harefield ! ; *Blackst. Fasc.* 89.

II. Ponds in Bushey Park ! ; *New B. G.* 101. Near Hampton Court ; *Merrett,* 107. Queen's River, Hampton, abundant. Thames side bet. Hampton village and Court ; *New B. G. Supp.* Bet. Hampton Court and Kingston Bridge. Bet. Strawberry Hill and Teddington. Canal, W. Drayton ! ; *Newb.* Colne, Staines Moor.

III. Cran at Twickenham.

IV. Brent at Willesden ; *Herb. Hardw.* At Stonebridge ; Forty Farm ; *Farrar.*

V. Canal at Greenford ; *Melv.* 80. Apperton ; Wembley ; *Cole.* Brentford ; *Hemsley.*

VI. Lane from Hornsey to Tottenham Church ; *Coop. Supp.* 11. Ditch by the New Cut, Edmonton.

VII. [In the Tower ditch ; *Lob. Adv.* and *Ger.* 337.] Several places by the Thames banks near London ; *Merrett,* 49. [Ditches in Tuthill Fields ; *Blackst. Spec.* 86.] Isle of Dogs, 1792 ; *E. B.* Ibid. 1815 ; *Herb., G. & R.* Ibid. 1836 ; *Herb. Young.* Paddington Canal, 1862 ; *Britten.* Stream near the White House, Temple Mills, 1868 ; *Cherry.*

First record : *Lobel,* 1570, also first as a British plant. [A very narrow-leaved form, *Sagitta aquat. omnium minima,* was noticed by Plukenet plentifully before the Earl of Peterborough's house above the horse-ferry on Westminster side ; *R. Syn.* i. 242 and *Pet. Midd.* It is figured

---

* *John Goodyer,* of Maple-Durham, Hampshire, is frequently mentioned by Johnson, Parkinson, and How, as a zealous and successful botanist. Besides the present plant, he added several other interesting species to the British Flora, e. g. *Phyteuma orbiculare* and *Thesium.*

in *Pet. H. B. Cat.* xliii. t. 12.] The subaqueous grass-like leaves were mistaken by Doody for a species of *Potamogeton,* and called *P. gramineum majus fluviatile.* In the Thames and Hackney River; *Doody MSS.* Buddle (*MSS.* vi.) rightly referred the name to *Sagittaria,* and Petiver figured the leaves (*H. B. Cat.* xliii. 9) as the *Grass Arrowhead*; but Dillenius (*R. Syn.* iii. 150) again placed it among the *Potamogetons.*

BUTOMUS, *Linn.*

697. B. umbellatus, *L.*　　　*Flowering Rush. Water Gladiole.*

Juncus cyperoides floridus paludosus (Lob.).　*Gladiolus palustris Cordi, Ger. em.* (Blackst.).

Cyb. Br. ii. 478; iii. 516. Curt. F. L. f. 1 (drawn from a Middlesex specimen).

Sides of streams, ditches and ponds; common. P. June—September.

I. Harefield, frequent!; *Blackst. Fasc.* 33.

II. Bushey Park, 1802; *Winch. MSS.* Staines Common!, 1841; *Herb. Young.* Canal at W. Drayton!; *Newb.* Stanwell and Staines Moors.

III. Roxeth; *Melv.* 80. Hounslow Heath.

IV. River Brent near the stone bridge at Willesden; *Herb. Hardw.* Road bet. Finchley and Hendon!; *Newb.*

V. Canal near Greenford; *Melv.* 80. Twyford!; *Newb.* In the Brent profusely; *Lees.*

VI. New Cut, Edmonton.

VII. [By the Tower of London; *Lob. Adv.* 44.] Paddington Canal, near the 'Mitre,' 1815; [Bayswater Canal in Kensington Gardens, 1817; *Herb. G. & R.*] Marshes by Blackwall; *Curt. F. L.* and *Forst. Midd.* Very common in the Isle of Dogs; *Smith MSS.* Canal by Kensal Green Cemetery, 1862, Britten; *Phyt. N. S.* vi. 348, and *Cole.* Banks of Thames, Fulham; *Faulkner,* 22. Ken Wood, Hunter; *Park Hampst.* 30. By Ken Wood ponds, 1866; *Fox.* In the Lea at Hackney and ditches at Hackney Wick.

First record: *Lobel,* 1570; also first as a British plant.

TRIGLOCHIN, *Linn.*

698. T. palustre, *L.*

Gramen junceum spicatum, C. B. P. (Blackst.).

Cyb. Br. ii. 478. E. B. 366, reproduced in Syme E. B. ix. t. 1433 (drawn from a Middlesex specimen).

Wet places; rare. P. June—August.

I. Marshy grounds about Harefield, plentifully; *Blackst. Fasc.* 38. Bet. Yewsley and Iver Bridge!; *Newb.*

II. Bushey Park by the ponds, 1867.

V. Side of canal, Greenford; *Melv.* 81. Marshy spots by the Brent; *Lees.*

VI. New River near Hornsey; *Smith MSS.*

VII. [Eel-brook Meadow, Parson's Green; *Pamplin.*]

First record: *Blackstone*, 1737.

TYPHACEÆ.

TYPHA, *Linn.*

699. T. latifolia, *L.*   *Cat's-tail.   Reed-mace.*

T. palustris major, C. B. P. (Blackst.).   *T. major* (Curt.).

Cyb. Br. iii. 34.   Curt. F. L. f. 3.

Ponds and sides of streams; rather common.   P.   July, August.

I. Pool near Pinner Wood; Woodridings; *Melv.* 81.   Harefield; *Blackst. Fasc.* 104.

II. Near Teddington Ry. Station.

III. By the Cran near Hospital Bridge.

IV. Bet. Whitchurch and Stanmore.   Pond below the ' Spaniards,' Hampstead.

V. Canal at Greenford; *Melv.* 81.   By the Brent near Hanwell!; *Warren.* Shepherd's Bush!; not far from Kew Bridge Ry. Station!; *Newb.*

VII. Isle of Dogs; *Blackst. MSS.*   Blackwall; *Herb. Hardw.*   [Notting Hill, 1852, J. T. Syme; *Herb. Mus. Brit.*]

First record: *Blackstone,* 1737.

700. T. angustifolia, *L.*

T. media (Clus.).   *T. minor* (Curt.).

Cyb. Br. iii. 35.   Curt. F. L. f. 5.

Ponds, &c.; rare.   P.   July.

I. Side of Ruislip Reservoir; *Melv.* 81.

II. Shepperton; *Newb.*

III. Hounslow Heath; *Mart. App. P. C.* 71.

IV. Hampstead; *Irv. MSS.*   Stanmore; *Varenne.*

V. Canal bet. Brentford and Hanwell; *Hemsley.*

VI. Pond not far from Shepherd's Bush Ry. Station!; *Newb.*

VII. [Plentifully, 1581, in a pit by Tyburn churchyard, not far from the place where those who have been hung for crimes are buried, at the first milestone from London towards the west; *Clusius*, 215.]

First record: *Clusius*, 1581; also first as a British plant.

(T. minor, *Sm.* Cyb. Br. iii. 36.   E. B. 1457.   *T. palustris minor, C. B. P.,* minima, *Clus. Pan.*   Found by Mr. Dandridge on Hounslow Heath, where the *Sium alterum Olusatri facie* grows; *Doody MSS.*   An error on the part of Dandridge, a form of *T. angustifolia*, L. being the plant intended.*   The figure in *E. B.* was drawn from Geneva specimens.)

---

* Dillenius omits Clusius' name in his quotation of Doody MSS. at *R. Syn.* iii. 436. Parkinson's (*Theat.* 1204) *T. minima* seems small *angustifolia.*

U

SPARGANIUM, *Linn.*

701. S. ramosum, *Huds.*   *Bur-Reed.*

Cyb. Br. iii. 33. Curt. F. L. f. 5.

By ponds, ditches, and streams; common. P. July.

I. Harefield!; *Blackst. Fasc.* 96. Ruislip Reservoir; *Melv.* 81. Colne near Yewsley!; *Newb.*

II. Staines Common, 1860; *Phyt. N. S.* iv. 263. Towing-path bet. Kingston Bridge and Hampton Court; *Bloxam.* Staines Moor. Near the Queen's River, Hampton.

III. Harrow; Roxeth; *Melv.* 81. Cran at Hanworth Road and Twickenham.

IV. Stanmore.

V. Canal, Twyford!; Ealing Common!; *Newb.* Hanwell!; *Warren.*

VI. Canal bet. Edmonton and Tottenham; *Cherry.* Edmonton.

VII. Isle of Dogs, abundant!; *Curt. F. L.* Near Hornsey, 1815; *Herb. G. & R.* Stream near White House, Temple Mills; *Cherry.* Hackney Wick. Side of ponds in Ken Wood grounds.

First record: *Blackstone,* 1737.

702. S. simplex, *Huds.*

S. non ramosum, C. B. P. (Blackst.).

Cyb. Br. iii. 33. Curt. F. L. f. 5.

Sides of ponds and streams; rare. P. July, August.

I. Harefield!; *Blackst. Fasc.* 96. In a feeder to Ruislip Reservoir; *Hind.*

II. Bushey Park!; *Newb.*

. V. Canal bet. Brentford and Hanwell; *Hemsley.* Pools at Perivale; *Lees.*

VII. Lea bet. Clapton and Tottenham, G. Francis; *Coop. Supp.* 12. Sparingly in the old cut of the New River, S. Munby; *Naturalist* for 1867, p. 181.

First record: *P. Roberts,* about 1710. *Sparganium minimum,* Park., observed by Mr. P. Roberts about Bow, near London; *Herb. Pet.* cl. fol. 188; the specimen is *S. simplex,* Huds.

ARACEÆ.

ACORUS, *Linn.*

703. A. Calamus, *L.*   *Sweet Flag.*

A. verus sive Calamus officinarum, Park. (Pet., Blackst.).

Cyb. Br. iii. 30. E. B. 356, reproduced in Syme E. B. ix. t. 1391 (drawn from a Thames specimen). Burnett t. 32 (drawn from a Hampstead specimen).

Sides of streams and clear ponds; rather common. P. June—August.

I. In the ponds near Harefield Church; *Blackst. Fasc.* 2. Small pond in a field a little N. of Harefield Common!; *Newb.* Uxbridge; *Mart. App. P. C.* 71. On Hillingdon Common; *Sm. Fl. Brit.* i. 373.

II. Near Hampton Court, '1796,' A. B. Lambert; *E. B.* By the Thames bet. Hampton and Hampton Court!; *New. B. G. Supp.* Staines Common, abundant!; *Phyt. N. S.* iv. 263. Ponds on common by Walton Bridge. By the Thames bet. Hampton Court and Kingston Bridge and at Teddington. Near Strawberry Hill.

III. Hounslow Heath; *Huds.* i. 128. Ibid., 1820; *Bennett (v.s.).* Cran at Hospital Bridge. Duke's River in several places near Isleworth. Pond on Marsh Farm, Twickenham.

IV. Near Mill Hill; *Irv. Lond. Fl.* 87. Hampstead Heath; *Irv. H. B. P.* 279.

V. Greenford; *Cooper,* 108.

VII. About the moat of Fulham Palace, Doody; *Pet. Midd.* Reservoir in Earl of Mansfield's Park, Highgate!; [pond in Copenhagen Fields;] *Burnett,* 32. Marsh ditch, Hackney Wick.

First record: *Doody,* 1695. A decidedly common plant near the Thames in II. and III.

ARUM, *Linn.*

704. A. maculatum, *L.*   *Cuckoo-pint. Lords and Ladies.*
Arum (Johns.). *A. vulgare, Ger. em.* (Blackst.).
Cyb. Br. iii. 30. Curt. F. L. f. 2.

Under hedges and in shady places; very common. P. April, May.
In all the districts, but less frequent in II. and III.

VII. Bet. Kentish Town and Hampstead!; *Johns. Eric.* [Marylebone, 1817; *Herb. G. & R.*] Many places in the northern suburbs; Crouch End, Hornsey, abundant.

First record: *Johnson,* 1629.

LEMNACEÆ.

LEMNA, *Linn.*

705. L. trisulca, *L.*
Cyb. Br. iii. 29. Hook. Curt. vol. iv. (drawn from London specimens).

In ponds and ditches; rather common. P.? June.

I. Near Harefield!; *Newb.* Near Pinner Cemetery. Feeder to Elstree Reservoir.

II. Staines Moor, abundant. Staines Common. Common by Walton Bridge. Bushey Park, very abundant.

III. Grove Pond, Harrow; *Melv.* 82. Marsh Farm, Twickenham.

IV. Lane bet. Burroughs and Brent St., Hendon; *Irv. MSS.* Harrow Park Lake; *Melv.* 82. Golder's Green, Hampstead, 1853; *Herb. Hardw.*

V. Pond not far from Kew Ry. Station!; *Newb.*

VI. Near Whetstone.

VII. Plentifully in all the ditches in the neighbourhood of Tottenham; *W. G. Smith in Science Gossip,* vol. i.

First record; *Irvine*, about 1830. Usually entirely submerged, and forming a more or less complete subaqueous stratum beneath the floating species. J. de C. Sowerby says it is not rare ' in the vicinity of London' in flower; *Mag. Nat. Hist.* i. 290.

706. L. minor, *L.*   *Duckweed.*

Cyb. Br. iii. 27.   Syme E. B. ix. t. 1395 (bad figure, though the original drawing for E. B. 1045 is good).

Floating on stagnant water; very common. P.?   June, July.
Throughout all the districts.

VII. Common about town.   Hornsey Wood Ponds.   Hackney Wick.   Isle of Dogs.

First record: *Irvine*, about 1830. It is frequently found in flower. There are good figures of the flowers in *Hook. Curt.* vol. iv.

707. L. gibba, *L.*

Cyb. Br. iii. 27.   Syme E. B. ix. t. 1396.

Floating on ditches and ponds; rather common. P.?   June, July.

I. Eastcott; *Farrar.*   Near Potter's Bar Ry. Station, abundant.

II. Staines Moor.   Staines Common.   Ditch by side of Queen's River, Hampton, abundant.

III. Roxeth; *Hind. in Phyt. N. S.* iv. 116.   Cran on Hounslow Heath. Whitton, abundant.   About Twickenham in several places.   Isleworth.

IV. Silk Bridge near the Hyde, Hendon; *Irv. MSS.*   Harrow Weald.

V. Ponds on road from Harrow to Greenford; *Farrar.*

VII. Hackney Wick, abundant, 1867.

First record: *Irvine*, about 1830.   Much overlooked; it usually, though by no means always, excludes other species of *Lemna*, covering the whole surface of the water with a very dense sheet of vegetation. The flowers are well figured in *Hook. Curt.* vol. v.

708. L. polyrrhiza, *L.*

Cyb. Br. iii. 28.   Syme E. B. ix. t. 1397.

Floating on ponds and ditches; rather common. P.?   Flowers not yet seen in Britain.

I. Feeder to Elstree Reservoir.   Near South Mims.

II. Staines Moor.   Staines Common.   Bet. Hampton and Sunbury!; *Newb.*

III. Hounslow Heath. Isleworth, sparingly.   Ponds, Harrow; *Hind. in Phyt. N. S.* iv. 116.

IV. Hampstead Heath; *Herb. Hardw.*   Pond bet. Finchley and Hendon!; *Newb.* In the Brent; *Farrar.*

V. Ealing Common, abundant!; *Newb.*

VII. Hornsey; *Varenne.*   Bishop's Walk, Fulham; *Herb. Hardw.*   Serpentine; *Herb. Devon Institution, Exeter.*   In the Hampstead Ponds, abundant; *W. G. Smith in Science Gossip,* vol. i.

First record: *Varenne*, 1827. The least abundant in its localities, usually occurring sparingly along with *L. minor.*

WOLFFIA, *Horkel.*

709. W. arrhiza, *Wimm.*   Lemna arrhiza, *L.* (Bab. Man. and Syme E. B.).

Syme E. B. ix. t. 1398 (not at all characteristic).

Floating on clear water; very rare.   P.   Flowers not yet detected in England.

II. Piece of water which probably communicates with the Thames, but looks like a pond, near the railway bridge on Staines Common, June 14, 1866.   Still abundant there in 1868, but it has not spread into the other ponds near.

First record: *Trimen*, 1866; also first as a British plant.*   Probably occurs in other localities; but overlooked from its minuteness.   Since its discovery in Middlesex, it has been found at Walthamstow, Essex, by Mr. Moggridge; near Canterbury, Kent, by Mr. Gulliver; and in several spots in Surrey by Mr. Watson and Rev. W. W. Spicer. *Wolffia* is undoubtedly a good genus. (See Dr. Hegelmaier's monograph, *Die Lemnaceen* (1868), where twelve species are described.)

POTAMOGETONACEÆ.

POTAMOGETON, *Linn.*

710. P. natans, *L.*

P. majus vulgare, Matth. (Johns.).   *P. latifolium, Ger. em.* (Pet., Blackst.).

Cyb. Br. iii. 201.   Syme E. B. ix. t. 1399.

Ponds and ditches; common.   P.   June—August.

I. Harefield!; *Blackst. Fasc.* 80.   Uxbridge!; *Newb.* Elstree Reservoir. Near South Mims.

III. Near Hatton.

IV. Harrow district, abundant; *Melv.* 83.   Hampstead Heath; *Johns. Enum.* and *Irv. MSS.*   Harrow Weald.

V. Brent, near Ealing; *Herb. Devon Institution, Exeter.*   Near Willesden Junction!; *Warren.*   Perivale; *Lees.*

VI. Near Whetstone.   Ditch near Pickard's Lock, Edmonton.

---

* In *Seem. J. of Bot.* iv. (1866) p. 263, is a communication from Dr. J. E. Gray, stating that 'about fifty years ago,' some specimens were seen by him and Mr. J. J. Bennett, said to have been 'discovered in the neighbourhood of London,' by M. Gérard.   They were at the time considered to be 'a very young state of *Lemna minor,*' and as no specimens were preserved, the truth of this supposition can be now neither supported nor disproved.   Therefore, it is not possible to admit that the occurrence of the species in England was a *known fact* at that somewhat distant period, although there is no reason to suppose the plant a recent importation into this country.

VII. Ponds about London; *Herb. Pet.* Isle of Dogs, 1817; *Herb. G. & R.*
　　Lea Canal; *Cherry* (*v. s.*).

First record: *Petiver*, about 1710.

711.　P. polygonifolius, *Pourr.*
P. oblongus, Viv. (Melvill).

Cyb. Br. iii. 21.　Syme E. B. ix. t. 1400.

Ponds and ditches, on heaths; very rare.　P.　June—September.

I. Ditch at Harrow Weald Common!; *Melv.* 83.

IV. North Heath, Hampstead.

First record: *Hind,* 1861.　The plant at Harrow Weald Common grows
　　in rather deep water, and is quite like the figure quoted.　That at
　　Hampstead is the usual heath form (γ *ericetorum,* Syme).

(*P. fluitans, Roth.* A plant which may be this is in *Lamb. Herb.* at Kew;
　　it was gathered at Hounslow; *Bab. Man.* 362.　An error (see *Syme
　　E. B.* ix. 63).)

P. rufescens, Schrad. Cyb. Br. iii. 20.　Syme E. B. ix. t. 1402.　Should
　　be looked for.* It occurs in ditches by the Colne, bet. Rickmansworth
　　and Harefield Mill; *Fl. Herts.* 274, a station which if not within must
　　be very slightly beyond our boundaries.

712.　P. lucens, *L.*
P. aquis immersum fol. pellucido lato oblongo acuto (Budd., Pet.).

Cyb. Br. iii. 16.　Syme E. B. ix. t. 1408.

Rivers and streams; rather rare.　P.　June—August.

II. Thames above Teddington Lock.

III. On Hounslow Heath; *Budd. MSS.* vi.　Chase Bridge, Whitton, G.
　　Francis; *Coop. Supp.* 12.　Duke's River, near Isleworth.　Thames,
　　Twickenham.

V. Canal near Brentford; *Hemsley.*

VII. Many places in Thames bet. Fulham and Hampton Court; *Pet. Midd.*
　　[Ditches near Chelsea Waterworks, *M. & G.* 206.]　Lea Navigation!
　　and Hackney Canals!, abundant; *Cherry.*

Var. β. P. acuminatus, Schum.

VII. Floating in the Hackney Canal.

First record: *Petiver,* 1695.　[*P. fol. angust. pellucid. fere gramineo, vel
　　illi simillimum,* at Jack Straw's Castle; *Doody MSS.* and *R. Syn.* iii.
　　148, is perhaps a form of this.]

713.　P. perfoliatus, *L.*
Cyb. Br. iii. 15.　Syme E. B. ix. t. 1412.

Streams and ponds; rather common.　P.　June—August.

---

* The species of *Potamogeton,* together with many other water plants, have become
much less frequent, and are threatened with extinction in consequence of the extraordinary
increase of *Elodea.*

I. Harefield !; *Blackst. Fasc.* 89.

II. Grand Junction Canal, W. Drayton !; *Newb.* Queen's River, Hampton. Thames bet. Hampton and Kingston.

III. Thames at Twickenham, abundant. Colne, Staines Moor, &c., abundant. Cran at Hospital Bridge and Twickenham.

IV. Hampstead; *Irv. MSS.* North End; *M. & G.* 205.

V. Canal at Greenford and Apperton !; *Melv.* 83. Brent, near Ealing; *Herb. Devon Institution, Exeter.*

VII. New River Head, plentifully; *Pet. Midd.* Pond by Copenhagen House; *M. & G.* 205. Paddington Canal, 1818; *Bennett (v. s.).* Ibid., 1869; *Warren.* Lea and Hackney Canals, abundant; *Cherry (v. s.).* Thames, near Putney, 1817; *Herb. G. & R.* Bishop's Walk, Fulham; *Herb. Hardw.* Ornamental basins at head of Serpentine.

First record : *Petiver*, 1695.

714. P. crispus, *L.*

Tribulus aquat. minor quercûs fl., Ger. em.* (Johns., Blackst.).

Cyb. Br. iii. 15. Syme E. B. ix. t. 1413.

Ponds and streams ; rather rare ?. P. June—August.

I. Harefield; *Blackst. Fasc.* 100. Wax-well, Pinner. Pond by Church, South Mims.

III. Grove pond, Roxeth; *Melv.* 83. Hounslow ; *Herb. Kew.*

IV. Hampstead; *Irv. MSS.* Lake in Harrow Park; *Melv.* 83. Stanmore. Cricklewood !; *Warren.*

V. Canal, Greenford; *Melv.* 83. Near Willesden Junction !; *Warren.*

VII. New River, *M. & G.* 207. Marylebone Park; *Herb. Rudge.* Isle of Dogs, E. F.; *Herb. Mus. Brit.* About London ; *Johns. Ger.* 823. Paddington Canal; *Warren.* Hackney Canal. Pond beyond Primrose Hill. Ornamental basins, head of Serpentine.

First record : *Johnson*, 1633. *P. serratus*, Huds., is a state with the leaves not crisped (the young leaves are usually so); it is figured in flower in *Pet. H. B. Cat.* v. fig. 8. About West End, near Hampstead ; *M. & G.* 208. Canal, Greenford ; *Herb. Harr.*

715. P. zosterifolius, *Schum.* P. gramineus, *var.* (L. Cat.).

Cyb. Br. ii. 14. Syme E. B. ix. t. 1415.

Streams ; very rare. P. June.

V. In the Brent Canal by Brentford, A. Irvine ; *Herb. Syme.*

The only record. Was this merely *floating* in the water, or growing rooted in the soil ?

716. P. obtusifolius, *Koch.* P. gramineus, *var.* (L. Cat.).

? P. gramineum † (M. & G.).

Cyb. Br. iii. 12. Syme E. B. ix. t. 1417.

---

* The figure in *Ger. em.* is *P. crispus*, but the description is that of *P. densus.*

† By this name may have been intended either this or *P. zosterifolius*, or *P. acutifolius* not then differentiated.

Ponds and streams; very rare. P. June, July.

III. In the Thames at Twickenham; *M. & G.* 209.

VII. In the great circular pond opposite Kensington Palace ; *Herb. Hardw.* Isle of Dogs; *Irv. Lond. Fl.* 86.

First record : ? *Milne and Gordon*, 1793.

717. P. compressus, *L.* P. mucronatus, *Schrad.* (Syme E. B.).
Cyb. Br. iii. 12. Syme E. B. ix. t. 1418.
Streams and ditches ; very rare. P. June—August.

II. About Staines, abundantly; *Huds.* ii. 76.

IV. Streams at Harrow Weald ; *Melv.* 83.

VII. Not very uncommon about London; *Sm. E. B.* 418. Isle of Dogs, 1817 ; *Herb. G. & R.*

First record : *Hudson*, 1778.

718. P. pusillus, *L.*
P. pusill. gramineo fol. caule tereti (Pet.). *P. gramineum tenuifolium* (Budd.).
Cyb. Br. iii. 11. Syme E. B. ix. t. 1419.

Streams and ditches ; rather rare. P. June, July.

II. Ditches, Staines ; *Cooper*, 115.

III. In the river on Hounslow Heath with Mr. Doody, 1705 ; *Budd. MSS.* vi. and *Herb.* cxvii. fol. 29. Chase Bridge, Whitton, G. Francis ; *Coop. Supp.* 12.

IV. Pond by Pinner Drive ; *Melv.* 83.

V. Canal at Greenford ; *Melv.* 83.

VII. New River Head, plentifully; *Pet. Midd.* In the New River; [in the Green Park;] *M. & G.* 212. Isle of Dogs, 1817 ; *Herb. G. & R.* Reservoir at Clifton Villas, Camden Town ; in 1856 several cartloads were taken out ; *Jewitt.**

First record : *Buddle* and *Doody*, 1705.

719. P. pectinatus, *L.* (Syme E. B.).
Cyb. Br. iii. 10.
Streams and ponds ; rather rare. P. June, July.

1. *P. flabellatus*, Bab.
Syme E. B. ix. t. 1421.

II. Thames bet. Hampton and Kingston Bridges, 1867 ; *Bloxam.* Ibid., 1869 ; *Warren.*

III. Duke's River, near Chase Bridge.

VII. Paddington Canal, near Kensal Green, 1869 !; *Warren.*

2. *P. pectinatus*, L. (Bab.). *P. eu-pectinatus* (Syme).
E. B. 323, reproduced in Syme E. B. ix. t. 1422 (drawn from a Middlesex specimen).

* Mr. Orlando Jewitt, wood-engraver, died May 30, 1869, whilst these sheets were passing through the press.

I. Ruislip Reservoir, W. M. H.; *Melv.* 82. Canal north of Harefield!; *Newb.*

II. Grand Junction Canal, W. Drayton!; *Newb.* Staines Moor.

III. Harrow; *Herb. Harr.*

V. Canal at Apperton.

VII. Serpentine River, Hyde Park!; *Huds.* i. 62. Ibid. 1795; *Sm. E. B.* Lake in Regent's Park; River Lea, E. F.; *Herb. Mus. Brit.* Paddington Canal; *Irv. MSS.* Thames near Putney Bridge, 1817; *Herb. G. & R.* Round pond, Kensington Gardens; *Warren.*

First record: *Hudson,* 1762. Both subspecies are figured by Petiver, *H. B. Cat.* v. figs. 12, 13. The *P. zosteraceus* (*Bab. Man.* eds. i., ii.) of the 'Serpentine, Hyde Park, Dr. J. A. Power,' was a form of *pectinatus.* The locality was omitted in subsequent editions, when the name was changed to *P. flabellatus*; nor is it given in a list, by Mr. Babington, of localities for the latter plant in *Phyt.* iv. 1160.

720. P. densus, *L.*

Cyb. Br. iii. 9. Syme E. B. ix. t. 1414.

Ditches and streams; rather rare. P. June—August.

I. Colne near Iver Bridge!; *Newb.*

II. Colne, Staines Moor. Staines Common. Common by Walton Bridge. Bushey Park.

IV. Hampstead; *Irv. MSS.* Stanmore.

V. Canal at Greenford; *Melv.* 82.

VII. Canal towards Hampstead Road, 1815; *Herb. G. & R.* Marsh ditches, Hackney Wick. Isle of Dogs.

First record: *Goodger and Rozea,* 1815.

ZANNICHELLIA, *Linn.*

721. Z. palustris, *Linn.*

Potamogeito affinis graminifolia aquatica (Pet., Doody). *Aponogeton aquat. graminifol. stamin. sing.* (R. Syn. iii.).

Cyb. Br. iii. 23. Syme E. B. ix. t. 1425.

Ponds and ditches; rather rare. A. or P. June—August.

I. Several pools at Pinner, W. M. H.; *Melv.* 83. Pond by the church, South Mims.

II. Ditches, Staines; *Cooper,* 115.

IV. Pond below the 'Spaniards,' Hampstead.

VI. Pond by brick ground on Finchley side of Highgate Archway; *Davies.* Hadley!; *Warren.*

VII. Pond at upper end of Stroud Green; *Doody MSS.* Small pond on east side of Islington; *Pet. Midd.* By St. Pancras Church near London; *Dill.* in *R. Syn.* iii. 136. [Tothill fields, J. Sowerby; *Smith MSS.*] Carlton Road, Kilburn, 1869!; *Warren.* Ditches in Isle of Dogs, abundant.

First record: *Petiver*, 1695. *Z. pedicellata*, Fr. does not occur, but the fruit in the Isle of Dogs specimens has usually a short stalk, and the clusters of fruit are frequently found on a peduncle as long as the carpels, which, however, are not winged, but have on their dorsal margin from six to ten blunt prominences, which give to the profile a crenato-dentate appearance.

CYPERACEÆ.

[CYPERUS, *Linn.*

722. C. fuscus, *L.*

Cyb. Br. iii. 61. Hook. Curt. vol. iv.* E. B. S. 2626.† Brit. Ent. ix. t. 395 (all drawn from Middlesex specimens).

Marshy meadows; very rare. A. September.

VII. In some abundance on the sides of a ditch in a low marshy meadow in Little Chelsea!, A. H. Haworth; *Hook. Curt.* iv. This meadow is usually called Eel-brook Meadow. The plant has been probably seen there every year since its discovery up to 1865, though in 1836 it was said to be 'extinct' (*Hook.* in *Smith's Compendium*, 14), but has varied in abundance. In 1856, Mr. Irvine could find but a single plant; in 1862 we collected about a score of small specimens. The meadow was drained in 1864 or 1865, and its present dry state prevents damp-loving plants from growing there; the spot will soon be built over.

First record: *Haworth*,‡ 1819; also first as a British plant. Last, about 1865. Still grows on Shalford Common, Surrey.]

ELEOCHARIS, *R. Br.*

723. E. palustris, *R. Br.* Scirpus p., *L.* (L. Cat.). *Club Rush.*
Juncus aquat. minor capit. Equiseti, Ger. em. (Blackst.).
Cyb. Br. iii. 75. Reich. Ic. Germ. viii. 297.

Sides of ponds, wet meadows, &c.; common. P. June, July.

I. Harefield; *Blackst. Fasc.* 46. Harrow Weald Common; *Melv.* 83. Near Yewsley!; *Newb.* By Elstree Reservoir.

II. Bushey Park!; *Newb.* Staines Common. Sunbury.

III. Hounslow Heath.

---

* Not a characteristic specimen.
† Syme E. B. cannot be quoted further. Throughout the *Cyperaceæ*, the figures of vol. viii. of Reichenbach's *Icones Fl. Germ.* are referred to.
‡ Adrian Hardy Haworth, of Collingham, Yorkshire, was the author of several works on entomology, and also published *Observations on the Genus Mesembryanthemum* 1794; *Miscellanea Naturalia, &c.*, 1803; *Synopsis Plant. Succulent.*, 1812; and *Supplement*, 1819; *Narcissearum Revisio*, 1812 (a second edition, 1831); and *Saxifragearum Enumeratio*, 1821. He lived for some years at Little Chelsea near Eel-brook Meadow, and died there of cholera, Aug. 24, 1833. Gray (*Nat. Arr.* vol. ii. 730) called this plant after him *Cyperus Haworthii.*

IV. Hampstead; *Irv. MSS.*

V. By Canal, Apperton!; *Newb.* By the Brent; *Lees.*

VI. Edmonton.

VII. Kensington, 1817; *Herb. G. & R.* 1839; *Herb. Hardw.* Hackney Canal; *Cherry (v. s.).* Eel-brook Meadow. Isle of Dogs.

First record: *Buddle,* about 1705. [*A larger sort*; in ye River Thames near Peterborow House; *Budd. MSS.* and *Pet. Conc. Gr.* 208.]

? E. uniglumis, *Link.* Scripus u., *DeC.* Cyb. Br. iii. 76. Reich. Ic. Germ. viii. 296. I. Uxbridge Moor, E. F.; *Herb. Mus. Brit.* This is probably a small creeping form of *E. palustris.* What may be the same plant occurs (VII.) South Heath, Hampstead. Is *E. uniglumis* a coast plant only?

(E. multicaulis, *Sm.* Side of Ruislip Reservoir; *Melv.* 84. Probably an error, as a specimen in *Herb. Harr.* from the same locality is *E. palustris.*)

724. E. acicularis, *Sm.*' Scirpus a., *L.* (L. Cat.).
Juncellus omn. minimus cap. Equiseti (R. Syn. ii.).
Cyb. Br. iii. 79. Reich. Ic. Germ. viii. 294.

On the bottoms and margins of ponds; very rare. P. July, August.

I. Ruislip Reservoir, 1864, abundant; *Hind.* Plentiful in the reservoir at Elstree; *Herts. Fl.* 311.

III. On Hounslow Heath towards Hampton, S. Doody; *R. Syn.* ii. 344. Ibid., copiously; *Blackst. MSS.*

V. Heathy spot at Greenford near the Brent; *Lees.*

First record: *Doody,* 1696.

SCIRPUS, Linn.

725. S. maritimus, *L.*
Gramen cyp. palustre panic. sparsa, Park. (Ray). *Long Marsh Cyperus* (Pet.).
Cyb. Br. iii. 74. Curt. F. L. f. 4 (drawn from a Thames specimen). Reich. Ic. Germ. viii. 310, 311.

By the Thames; very rare. P. August, September.

VII. In the river of Thames; *R. Cat.* i. 147, *Pet. Conc. Gr.* 143. Thames where the water is not salt, and on the edges of the creeks running from it; *Curt. F. L.* Near Blackwall, 1815; *Herb. G. & R.* Isle of Dogs!; *Mart. App. P. C.* 65. Sparingly there in 1866. [Plentifully in a ditch bet. Ranelagh and the old Chelsea Waterworks near the neat-houses, 1820; *Pamplin.*]

First record: *Parkinson,* 1640. *Cyp. rotund. inodorus aquat. alter,* in the low marshes beyond Ratcliffe; *Park. Theat.* 1265, is 'a very distinct kind,' frequent by the Thames side near the town; *Budd. MSS.* v.; but there is no specimen so named in his herbarium. *Cyp. rotund. littor.*

inod. anglicus alter, Park, common in the Thames near London;
Budd. MSS. is a young slender specimen (*Budd. Herb.* liv. f. 52). These
are probably both referable to Smith's *var. β = S. tuberosus*, Desf. (see
Mart. Mill. Dict.), also found in Kent and Essex by the Rev. G. E.
Smith, and in Surrey (see *Seem. J. of Bot.* iii. 191).*

726. S. sylvaticus, *L.*

Cyp. gramineus s. miliaceus (Johns., Merr., Pet., Newt., Blackst.). *Pseudo-
cyperus gram. s. mil.* (Park.). *Gramen cyperoides miliaceum, C. B.*
(Merr.).

Cyb. Br. iii. 75. E. B. 919 (drawn from a Hampstead specimen). Reich.
Ic. Germ. viii. 313.

Wet places, especially in woods; rather rare. P. July, August.

I. On Uxbridge Moor, T. F. Forster; *B. G.* 399.

IV. Meadows bet. Hampstead Heath and Highgate; *Merr.* 33 and 52,
Pet. Midd. and *Conc. Gr.* 146. Hampstead Heath; *Johns. Enum.*
Turner's Wood; *Irv. MSS.* Bishop's Wood.

V. Canal at Northolt, W. M. H.; *Melv.* 84. Canal side, Apperton!;
Newb. Canal near Brentford; *Hemsley.* By the Brent, Hanwell!;
Warren. By the Brent, Perivale; *Lees.*

VI. Pond near Highgate; *Hill,* 33. Southgate, J. Woods; *B. G.* 399.
The Alders Wood, Whetstone. Near Park Station, Edmonton.

VII. [Bet. London and Kentish Town, in the bottom of a field; *Park. Theat.*
1173. A little beyond Pancras Church; *Newton MSS.* Road from
London to Highgate; *Blackst. Spec.* 16.] [Walcot Gardens, near the
Wilderness; *Hill,* 33.] [Meadow bet. Fortess Terrace, Kentish
Town, and Highgate, 1840; *Herb. Hardw.*]

First record: *Johnson,* 1632.

727. S. triqueter, *L.*

Juncus caule triangulari (Merr., Pet.). *J. acutus maritimus caule triangulo, C. B.* (R. Cat. and Syn. i.). *J. acutus marit. caule triquetro max.
molli procerior nostras, Pluk.* (R. Syn. ii.). *J. marit. caule triquetro†
molli* (Doody). *S. mucronatus, var. β* (Huds.).

Cyb. Br. iii. 73. E. B. 1694 (drawn from a Thames (Surrey) specimen).
Reich. Ic. Germ. viii. 305. Hook. Curt. vol. iv.

By the Thames within tidal influence; very rare. P. August, September.

VII. By the River Thames side, both above and below London; *R. Cat.* i.
179 and *Syn.* i. 201 and *Syn.* ii. 272. Isle of Dogs, copiously; in
the Thames by Limehouse; *Doody MSS.* At the horse-ferry at
Westminster, first shown me by Dr. Dale; *Merr.* 67. In the

* *Cyperus rotundus littoreus inodorus, J. B.*, by the Thames about London in many
places, Mr. Doody; *Ray Fasc.* 5. Plukenet (*Almag.* 127) thought this the same as the
ordinary plant, whilst Buddle (*MSS.* v.) says he never could 'learn there is any such
plant growing in England.' He quotes the figure in *Park. Theat.* 1264, 3, which represents apparently some *Carex*, certainly not a *Scirpus.*
† *Triangulari* in *R. Syn.* iii. 428, where Dillenius quotes Doody.

Thames bet. Peterborough House and the horse-ferry, Westminster; *Pet. Midd.* and *Conc. Gr.* 200. Millbank, E. F.; *Herb. Kew.* and *B. G.* 399. Thames, 1806; *Herb. Mus. Brit.* We think we have seen this bet. Hammersmith and Fulham; it is, or was, abundant on the opposite (Surrey) shore.

First record: *Dr. Dale,** 1666.

728. S. carinatus, *Sm.*

Juncus aquat. medius caule carinato (Doody). *Juncus aquaticus major carinatus, Doody* (Budd.).† *Doody's furrowed Bull-rush* (Pet.). *S. Duvalii, Hoppe* (Forster, Reich.).

Cyb. Br. iii. 70. E. B. 1983 (drawn from a Thames specimen). Reich. Ic. Germ. viii. 308.‡

By the Thames within tidal influence; very rare. P. August, September.

VII. In the Thames by Limehouse; *Doody MSS., Budd. MSS., Pet. Conc. Gr.* 199. Near Millbank, E. Forster; Thames near Westminster, F. Boott; *Herb. Mus. Brit.* Opposite Barnes, if not higher up the river; *Newb.* Almost certainly occurs about Hammersmith and Fulham; abundant on the Surrey side bet. Hammersmith Bridge and Wandsworth.

First record: *Doody*, about 1700.

[**729. S. Tabernæmontani,** *Gm.* S. lacustris, *var.* β, *glaucus* (L. Cat.). *Juncus s. Scirpus medius, C. B.*§ (Doody).

Cyb. Br. iii. 69. Reich. Ic. Germ. viii. 307.

By the Thames within tidal influence; very rare. P. August, September.

VII. In a pond of a breach a little beyond Limehouse; *Doody MSS.* and *R. Syn.* iii. 428.

The only record. This plant certainly comes very near to *S. carinatus*, Sm., and ought probably to be united with it, as the slight characters said to distinguish the plants do not hold good. A plant in *Herb. Mus. Brit.* labelled ' *Scirpus nova species, S. lacustris* γ, *Sm. Fl. Brit.* 52 ' was found by E. Forster in the Thames above Westminster Bridge. It is either this or *S. carinatus*, but differs from ordinary forms of both in the great size and length of its spikelets.]

730. S. lacustris, *L.*   *Bullrush.*

Juncus aquat. maximus, Ger. em. (Blackst.).

Cyb. Br. iii. 67. Reich. Ic. Germ. viii. 306.

---

* This cannot be *Samuel Dale* of Braintree, who was born in 1659.

† The specimens in *Budd. Herb.* cxxv. fol. 35, and liv. fol. 48, so labelled, have very small spikelets, and are more slender than usual.

‡ There is also a good figure in *Hook. Curt.* vol. iv., from specimens gathered by the Arun in Sussex by Mr. Borrer.

§ The specimens of *Juncus aquat. medius, C. B.* in *Budd. Herb.* cxxv. fol. 3, and liv. fol. 48, are almost certainly *S. Tabernæmontani.*

Streams and (rarely) ponds; rather rare?. P. June, July.
I. Harefield!; *Blackst. Fasc.* 46. Ruislip Reservoir; *Hind.*
II. Thames near Strawberry Hill.
III. River Cran on Hounslow Heath.
IV. Kingsbury, in the Brent.
V. Brent near Hanwell!; *Warren.* Canal at Apperton!; *Newb.* Near
  Ealing; *Herb. Devon Instit., Exeter.* Near Northolt; *Melv.* 84.
VII. River Lea at Temple Mills.

First record: *Blackstone*, 1737. The characteristic floating leaves were
very noticeable at Harefield and in the Cran.

[731. S. cæspitosus, *L.*
S. montanus capit. breviori, Inst. R. H. (Blackst.).
Cyb. Br. iii. 79. Reich. Ic. Germ. viii. 300.
Wet places; very rare. P.
I. On Harefield Moor, plentifully; *Blackst. Fasc.* 91.
No other record.]

732. S. fluitans, *L.*
Gramen junceum clavatum minimum, &c., R. Syn. (Blackst.). *Isolepis fl.*
  (Irv).
Cyb. Br. iii. 80. Reich. Ic. Germ. viii. 298.
Ponds on heaths; very rare. P. June, July.
I. In the bogs on Harefield Common; *Blackst. Fasc.* 37.
III. Hounslow Heath!; *Huds.* ii. 18, *E. B.* 216.
IV. Piece of open ground called Whitchurch Common, at the back of
  Canons Park, bet. Stanmore and Edgware, 1827–30; *Varenne.*
VI. [Finchley Common; *Irv. Lond. Fl.* 89.]
First record: *Blackstone*, 1737.

733. S. setaceus, *L.*
Juncellus omnium minimus chamæschænus, Lob. (Pet., Blackst.). *Isolepis*
  setacea, R. Br. (Coop.).
Cyb. Br. iii. 71. Reich. Ic. Germ. viii. 301.
Wet places on heaths and commons; rare. P.? July, August.
I. By the canal in Sir G. Cooke's garden; Harefield Moor; *Blackst.*
  Fasc. 47. Harefield Common.
IV. Woods about Highgate and Hampstead; *Pet. Midd.*
VI. Highgate; *Cooper*, 104.
VII. By the ponds on South Heath, Hampstead; *Irv. Lond. Fl.* 89. In
  several places on the heath, 1868.
First record: *Petiver*, 1695.

BLYSMUS, *Panz.*
734. B. compressus, *Panz.*
Cyb. Br. iii. 65. Reich. Ic. Germ. viii. 293.
I. Harefield, J. F. Young; *Herb. Young.*

Requires confirmation. The specimen is correctly named, and the plant not very unlikely to occur, being found in Essex, though not in Herts or Bucks, nor recently in Surrey. Still, transposition of labels or some other cause of error may have occurred. Liable to be passed over as *Carex disticha.*

ERIOPHORUM, *Linn.*

735. E. polystachion, *L.* E. angustifolium, *Roth.* (L. Cat.). *Cotton Grass.*

Gramen tomentarium (Ger., Blackst.). *Juncus bombycinus, Lob.* (Johns.). *Gram. juncoides lanatum* (Park.). *Gr. tomentosum pratense panic. sparsâ* (Merr.).

Cyb. Br. iii. 83. Curt. F. L. f. 4.

Bogs on heaths; rare. P. April, May.

I. Harefield; *Blackst. Fasc.* 39.

III. Piece of heathy ground by Dockwell Lane, near Hatton, 1867; a single specimen.

IV. Bog at further end of Hampstead!; *Ger.* 27, *Johns. Eric., Park. Theat.* 1272, *Merr.* 59. [Highgate Park; *Ger.* 27.] Still abundant in the bog behind Jack Straw's Castle.

First record: *Gerarde,* 1597.

CAREX, *Linn.*

736. C. pulicaris, *L.*

Gramen Cyperoides pulicare, Merr. (Blackst.).

Cyb. Br. iii. 89. Reich. Ic. Germ. viii. 195. Hook. Curt. vol. v.

Marshy places on heaths; very rare. P. June.

I. Harefield Moor, plentifully; *Blackst. Fasc.* 35. Harrow Weald Common in several places.

IV. Hampstead Heath; *Huds.* i. 347, *Irv. Lond. Fl.* 90.

First record: *Blackstone,* 1737.

737. C. disticha, *Huds.* C. intermedia, *Good.* (L. Cat.).

Cyb. Br. iii. 101. Reich. Ic. Germ. viii. 210.

Wet places; rare. P. May, June.

I. Harefield!; *Newb.* Ruislip Moor.

II. In a marshy meadow near Teddington, 1792, E. F.; *Herb. Mus. Brit.*

IV. Woods, Hampstead; *Irv. Lond. Fl.* 90.

V. Near the canal, Brentford; *Hemsley* (*v. s.*).

VI. Edmonton.

First record: *E. Forster,* 1792.

738. C. divisa, *Huds.*

Cyb. Br. iii. 103. Reich. Ic. Germ. viii. 205.

Meadows by the Thames; very rare. P. May, June.

VII. Isle of Dogs, plentifully, L. W. Dillwyn; *B. G.* 411, *Cooper* 115.

[Petiver says he observed either this or *C. disticha* in Hackney Marsh;

(v. *Sloane MSS.* 3333, fol. 97).] Common in the Thames marshes in Essex and Kent. Possibly still exists in the Isle of Dogs.

739. C. vulpina, *L.*

Gramen palustre echinatum, Lob. (Johns.).

Cyb. Br. iii. 106. Reich. Ic. Germ. viii. 217.

Ditches and wet places ; very common. P. June.

Throughout all the districts.

VII. Bet. Clapton and Tottenham, G. Francis ; *Coop. Supp.* 12. Near Lea Bridge ; *Cherry* (v. s.). Kentish Town, 1846 ; *Herb. Hardw.* Paddington Canal, 1869 ; *Warren.*

First record : *Johnson,* 1632 ; also first as a British plant.

740. C. muricata, *L.*

Cyb. Br. iii. 104. E. B. 1097 (drawn from a Middlesex specimen). Reich. Ic. Germ. viii. 215.

Hedgebanks and woods ; common. P. June.

I. Near Pinner Ry. Station. South Mims.

II. Staines Common. Road bet. Staines and Hampton.

III. Harrow ; *Melv.* 85. Whitton.

IV. Dry parts of woods about Hampstead and Highgate ; *Forst. Midd.* Bishop's Wood. Harrow Weald. Pinner Drive ; *Melv.* 85.

V. Twyford ! ; *Newb.*

VI. Edmonton.

VII. [Hyde Park, Mr. Groult ; *Smith MSS.* and *Herb. Kew.*] Towards Highgate, 1827–30 ; *Varenne.* Bishop's Walk, Fulham, 1848 ; *Herb. Hardw.*

First record : *E. Forster,* 1789.

741. C. divulsa, *Good.*

Gr. cyp. spic. minus, spica long. divulsa seu interrupta, R. Syn. (Blackst.).

Cyb. Br. iii. 105. Reich. Ic. Germ. viii. 220.

Ditches and meadows ; rather rare. P. June.

I. Harefield ; *Blackst. Fasc.* 36. Near Wood Hall, Pinner. Bet. Burnt Oak Farm and Harrow Weald Common.

II. Road bet. Staines and Hampton, sparingly.

III. Roxeth ; *Melv.* 84.

IV. Harrow Weald. Deacon's Hill. Bet. Hendon and Finchley ! ; *Newb.*

VI. Near south end of Great Northern Ry. tunnel bet. Wood Green and Colney Hatch, G. Munby ; *Nat.* for 1867, 181. Whetstone. Enfield.

First record : *Blackstone,* 1737.

742. C. paniculata, *L.*

Gr. cyp. palustre elatius, spic. long. laxa (R. Syn. ii.).

Cyb. Br. iii. 108. Reich. Ic. Germ. viii. 223.

Wet woods and thickets ; rare. P. June.

IV. In Hampstead Wood ; *Huds.* i. 347. Near Hampstead Heath ; *Doody*

MSS. and *Forst. Midd.* Lane bet. Bishop's Wood and Finchley;
Irv. MSS. Close to bridge over Brent at Stonebridge; *Farrar.*

V. Side of canal near Apperton, 1867!, sparingly; *Newb.*

VI. The Alders Copse, Whetstone, 1867, abundant.

VII. By the New Cut near Lea Bridge, probably accidental, E. F.; *Herb. Mus. Brit.*

First record: *Doody,* about 1700.

743. C. axillaris, *Good.*

Cyb. Br. 99. Reich. Ic. Germ. viii. 219.

Ditchbanks; very rare. P. July.

IV. Ditch on the east side of the road leading from Harrow Ry. Station to Harrow Weald, 1866.

VI. Behind the 'George and Vulture,' Tottenham; *Cat. Lond.* 16.

First record: *Cockfield,* 1813.

744. C. remota, *L.*

Gr. cyp. angustifol. spic. parv. sessil. in fol. alis, R. Syn. (Blackst.).

Cyb. Br. iii. 98. Reich. Ic. Germ. viii. 212.

Damp hedgebanks and woods; common. P. June.

I. Harefield; *Blackst. Fasc.* 36. South Mims, abundant.

II. Staines Moor. Bushey Park. Hanworth Park.

III. Harrow Park; Harrow Grove; *Melv.* 84. By Whitton House Grounds. Hatton. Hanworth Road. Near Twickenham.

IV. Stanmore; *Varenne.* Kenton Road; *Melv.* 84. Bet. Finchley and Hendon!; *Newb.* Woodcock Hill. Scratch Wood. Bishop's Wood.

V. Near Brentford; *Hemsley.* Bet. Acton and Ealing Common; *Young MSS.*

VI. Bury Street, Edmonton. Whetstone.

VII. Bet. Clapton and Tottenham, G. Francis; *Coop. Supp.* 12.

First record: *Blackstone,* 1737.

745. C. stellulata, *Good.*

Gr. cyp. spic. min. spica divulsa aculeata (Pet.).

Cyb. Br. iii. 94. Reich. Ic. Germ. viii. 214.

Marshy places on heaths; rather rare. P. June.

I. Harrow Weald Common, abundant!; *Melv.* 84. Stanmore Heath. South Mims.

III. Hounslow Heath; *Herb. Rudge.*

IV. Woods at Hampstead and Highgate; *Pet. Midd.* North Heath, in the bog. Stanmore, 1837, D. Cooper; *Herb. Young.*

VII. South Heath, Hampstead.

First record: *Petiver,* 1695.

746. C. ovalis, *Good.*

Gr. cyp. spic. e pluribus spic. brev. moll. comp. (Pet.).

Cyb. Br. iii. 95. E. B. 306 (drawn from a Middlesex specimen). Reich. Ic. Germ. viii. 211.

Heaths, fields, and meadows; very common. P. June, July.

In many places in all the districts; less common in V. and VI.

VII. [In a wood near the boarded river, plentifully; *Pet. Midd.*] [Kensington Gardens; *E. B.*] South Heath, Hampstead, 1866.

First record: *Petiver*, 1695.

747. C. acuta, *L.*

Gram. cyp. majus angustifol., Park. (Blackst.). *C. gracilis* (Curt.).

Cyb. Br. iii. 115. Curt. F. L. f. 4. Reich. Ic. Germ. viii. 231-2.

Sides of streams and ditches; rather rare. P. May.

I. Harefield; *Blackst. Fasc.* 35.

II. River-bank bet. Hampton Court and Kingston Bridge!; *Newb.* Staines Moor.

III. Creek of the Duke's River, Isleworth. Twickenham Park inclosure.

V. Brentford; *Hemsley.* Canal, Apperton!; *Newb.* By the Brent near Hanwell!; *Warren.*

VII. Lea bet. Tottenham and Clapton, G. Francis; *Coop. Supp.* 12.

First record: *Blackstone*, 1737.

748. C. vulgaris, *Fries.*

C. cæspitosa, Sm. (Irv.).

Cyb. Br. iii. 110. Reich. Ic. Germ. viii. 216-8.

Damp heaths and meadows; rather rare. P. May—July.

I. Harefield!; *Newb.* Harrow Weald Common. Stanmore Heath.

III. Whitton Park inclosure.

IV. Harrow district, not uncommon; *Melv.* 85. North Heath, Hampstead!; *Irv. MSS.*

VI. Edmonton, in several places.

VII. South Heath, Hampstead.

First record: *Irvine*, about 1830. A very tall form, looking like *C. acuta*, in a pond on the west side of Harrow Weald Common.

749. C. pallescens, *L.*

Gram. cyp. polystach. flavicans spic. brev. &c. (Pet.).

Cyb. Br. iii. 118. Hook. Curt. v. Reich. Ic. Germ. viii. 251.

Woods; rare. P. May, June.

I. Woods by Pinner Lane; *Fl. Herts.* 318 (? Middlesex).

IV. Near the lake, Bentley Priory, W. M. H.; *Melv.* 85. Bishop's Wood, Hampstead.

VI. Highgate Wood, 1839; *Herb. Hardw.*

VII. [Wood against the boarded river; *Pet. Midd.*]

First record: *Petiver*, 1695.

750. C. panicea, *L.*

Cyb. Br. iii. 124. Reich. Ic. Germ. viii. 245.

Damp heaths, commons, and meadows; rather rare. P. May, June.

I. Harefield Common. Harrow Weald Common, abundant. Stanmore Heath.

III. Hounslow Heath; *Herb. Rudge.* Common at Harrow; *Hind in Phyt. N. S.* v. 202. Drilling Ground, Hounslow.

IV. North Heath, Hampstead.

VI. Whetstone. Edmonton.

First record: *Rudge*, about 1810.

751. C. strigosa, *Huds.*

Cyb. Br. iii. 130. Reich. Ic. Germ. viii. 242.

Hedgebanks; very rare. P. June.

I. Moss Lane, Pinner!, W. M. H.; *Melv.* 85. In some quantity on the ditchbanks on both sides of the lane, 1866.

First record: *Hind*, 1864.

752. C. pendula, *Huds.*

Gr. cyp. spic. pend. longior. et angustiore, C. B. (Merr., Pet.).

Cyb. Br. iii. 132. Curt. F. L. f. 3 (drawn from a Middlesex specimen). Reich. Ic. Germ. viii. 243.

Damp woods and hedges; rather common. P. May.

I. Pinner!; *Melv.* 85. Elstree.

III. Harrow Grove; *Melv.* 85.

IV. Woods at Hampstead and Highgate; *Pet. Midd., Huds.* i. 353. Bentley Priory Woods; *Melv.* 85. Bet. Finchley and Hendon!; *Newb.* Bishop's and Turner's Woods. Scratch Wood. Hedges near Harrow Weald Church.

V. Apperton; *Melv.* 85.

VI. Alders Copse, Whetstone. Colney Hatch. Near Maiden Bridge, Enfield, T. F. Forster; *B. G.* 411.

VII. [On the west side of Primrose Hill; *Merrett,* 52. Bet. Marylebon and Kilbourn; *Huds.* ii. 411. Marylebone Fields, 1827–30; *Varenne.*] [Field bet. Pancras and Kentish Town; *M. & G.* 81.] Bet. Stoke Newington and Hornsey, J. Woods; *B. G.* 411. By Lea bet. Clapton and Tottenham, G. Francis; *Coop. Supp.* 12. [Isle of Dogs; *Coop.* 115.] Bet. Kentish Town and South End, Hampstead, 1841; *Herb. Hardw.* Hornsey Wood; *Herb. Newb.* Millfield Lane, Highgate, 1867.

First record: *Merrett*, 1666.

753. C. præcox, *Jacq.*

Gram. spic. fol. veton. caryoph., Lob. (Johns.).

Cyb. Br. iii. 134. Hook. Curt. iv. Reich. Ic. Germ. viii. 261 (very large plant).

Dry places on heaths, &c.; rare. P. May.

I. Ruislip Common; *Melv.* 86. Harefield Cricket-ground. Woodready, Pinner.

IV. Hampstead Heath; *Johns. Enum.* and *Irv. MSS.*

V. Greenford; *Melv.* 86.

First record: *Johnson*, 1632; also first as a British plant.   Probably
. elsewhere overlooked.*

754.   C. pilulifera, *L.*

*Gr. cyp. fol. moll. tenuibus, spic. brev. coacervatis tribus quatuorve in
summ. caulis* (R. Hist.).   *Gr. cyp. spic. brev. congest. fol. molli* (Doody).
Gr. cyp. tenuifol. spic. ad summ. caul. sessil. globul. æmulis, Pluk.
(Blackst.).

Cyb. Br. iii. 135.   Reich. Ic. Germ. viii. 260 (a small specimen). ____

Heaths and roadsides; rare.   P.   May, June.

I. Harefield Common; *Blackst. Fasc.* 36.   Harrow Weald Common,
W. M. H.; *Melv.* 86.   Stanmore Heath, sparingly.

II. Near Staines on the Hampton Road, a single specimen.

III. Drilling Ground, Hounslow!; *Warren.*   Heston gravel-pits, a few
plants.

IV. Hampstead Heath!, Doody; *Ray Fasc.* 10, *R. Hist.* ii. 1911, *R.
Syn.* i. 196, &c.

First record: *Doody*, about 1688; also first as a British plant.   A specimen in *Budd. Herb.* liv. fol. 55, labelled '*varietas major*, near Hampstead,' is an ordinary plant of this to which a large and long fruiting
stalk of *C. præcox* has been accidentally added.   Is this the ' *var. β*,'
near Hampstead Heath; *Forst. Midd.?*

755.   C. glauca, *Scop.*

Gr. cyp. fol. caryoph. spic. oblong. e pedic. longior. pendulis, R. Syn.
(Blackst.).   *C. recurva*, Huds. (Irv.).

Cyb. Br. iii. 133.   Reich. Ic. Germ. viii. 269.

Dry banks, heaths, &c.; rather common.   P.   April, May.

I. Harefield!; *Blackst. Fasc.* 35.   Stanmore Heath; *Varenne.*   Harrow
Weald Common.   About Pinner.   Elstree.

III. Harrow, W. M. H.; *Melv.* 85.

IV. Side of pond near Finchley Church, 1842; *Herb. Hardw.*   Bet.
Finchley and Hendon!; *Newb.*   Hampstead Heath; *Irv. MSS.*
Bishop's Wood.

V. Near Brentford; *Hemsley.*   Near Twyford!; *Newb.*

VI. Hadley!; *Warren.*   Edmonton.   Near Colney Hatch.

First record: *Blackstone*, 1737.

756.   C. flava, *L.*

Cyb. Br. iii. 116.   Reich. Ic. Germ. viii. 263.

---

* C. ericetorum, Poll. (E. B. S. 2971) should be looked for.   One of the origina
drawings for E. B. 1099 is it, and the want of any locality seems to indicate that
Sowerby gathered the specimen near London.   Smith caused the glumes to be altered t
suit *C. præcox*.   Sowerby's specimens, however, in Herb. Mus. Brit., are ordinary *præcox*
from Shirley Common

Heaths and commons, in wet places; rare. P. May, June.
I. Harefield Common. Harrow Weald Common!; *Melv.* 85. Stanmore
   Heath, abundant.
IV. North Heath, Hampstead!; *Irv. MSS.*
VII. [Hyde Park, 1817; *Herb. G. & R.*] South Heath, Hampstead.
   First record: *Goodger & Rozea*, 1817. The Hampstead Heath plant
   has been considered *C. Œderi*, Ehrh. (*C. flava, var. β*, L. Cat.), Reich.
   Ic. Germ. viii. 262; and that plant is said (see *Fl. Herts.* 319) to be
   the commoner of the two in Herts, but small *flava* has been pro
   bably the plant meant.

   (C. fulva, *Good.* Cyb. Br. iii. 118. Reich. Ic. Germ. viii. 252. I. Field
   near Harrow Weald Common; *Melv.* 85. Probably an error of name,
   as ' *C. fulva* ' from Harrow in *Herb. Harr.* is *C. panicea.*)

   (C. distans, *L.* Cyb. Br. iii. 119. Reich. Ic. Germ. viii. 253. *Gr.
   cyp. polystachion spic. parv. longiss. distantibus*; (IV.) On Hampstead Heath, Mr. Doody; *R. Fasc.* 10. VII. Banks of Lea bet.
   Clapton and Tottenham, G. Francis; *Coop. Supp.* 12. It has been
   again recently reported from Hampstead, but in this case, as probably
   in those above, *C. binervis* was the plant intended.)

757. C. binervis, *Sm.*
Cyb. Br. iii. 122. Reich. Ic. Germ. viii. 255.
Heaths; rare. P. May, June.
I. Ruislip Common, sparingly; Harrow Weald Common!; *Melv.* 85.
   Stanmore Heath.
IV. Hampstead Heath; *Irv. MSS.*
First record: *Irvine*, about 1830.

758. C. lævigata, *Sm.*
Cyb. Br. iii. 123. Reich. Ic. Germ. viii. 254.
Wet places; very rare. P. June.
I. Wet places on Stanmore Heath by the side of what was called the
   Banqueting House Ground, 1827–30; *Varenne.*
We have not seen this, but Mr. Varenne has no doubt of the species.

759. C. sylvatica, *Huds.*
Gram. cyp. sylv. tenuius spicatum, Park. (Pet.).
Cyb. Br. iii. 132. Reich. Ic. Germ. viii. 242.
Woods; rather rare. P. May, June.
I. Old Park Wood, Harefield. Ruislip Wood. South Mims.
IV. Harrow Park; Bentley Priory Woods; *Melv.* 85. Bet. Finchley and
   Hendon!; *Newb.* Bishop's Wood. Scratch Wood.
V. Bet. Acton and Ealing, 1835; *Herb. Young.*
VI. Betstile Lane, Colney Hatch, 1841; *Herb. Hardw.* Bet. Wood Green

and Colney Hatch, G. Munby; *Nat.* for 1867, 181. Alders Wood, Whetstone. Winchmore Hill Wood.

VII. [Wood against the boarded river; *Pet. Midd.*]

First record : *Petiver*, 1695.

760. C. pseudo-Cyperus, *L.*

Gram. cyp. spic. pend. brev. C. B. (Pet.).

Cyb. Br. iii. 132. Reich. Ic. Germ. viii. 275.

Borders of ponds and ditches ; rather rare. P. May, June.

II. Near Longford, E. Forster; *B. G.* 411. Youveney, near Staines, plentiful.

III. Hounslow Heath, Rev. Dr. Goodenough; *Sm. Fl. Brit.* iii. 986, and *Herb. Rudge.* Marsh Farm, Twickenham.

IV. Stanmore, in a grove opposite Drummond's House; *Herb. G. & R.* Finchley Road ; *Newb.*

VI. Bet. Highgate and Muswell Hill ; *Irv. Lond. Fl.* 92. Near Bacon's Farm, Bounds Green, 1848 ; *Herb. Hardw.* Bounds Green, about 200 yards west of the high road, G. Munby; *Nat. for* 1867, 180. Marsh side, Edmonton.

VII. [In a ditch bet. the boarded river and Islington Road; *Pet. Midd.*] [About Stoke Newington, common, J. Woods ; Near Hackney, E. Forster; *B. G.* 411.] [Maiden Lane, near Highgate Cemetery, 1842, E. Edwards; *Phyt.* i. 428.]

First record : *Petiver*, 1695.

761. C. hirta, *L.*

Gram. cyp. nortvegicum ima fol. basi tantillum lanuginosum (Lob.). *Gram. cyp. polystachion lanug. R. Syn.* (Blackst.).

Cyb. Br. iii. 139. Reich. Ic. Germ. viii. 257.

Damp ground, especially on heaths ; rather common. P. May.

I. Harefield ; *Blackst. Fasc.* 35. Stanmore Heath. Bet. South Mims and Potter's Bar.

II. Near Staines on Hampton Road. Near Sunbury.

III. Near Hatton. Hounslow Heath. Twickenham Park. Harrow ; *Herb. Harr.*

IV. Hampstead Heath ; *Irv. MSS.* Kenton, W. M. H.; *Melv.* 86.

VI. Near Highgate ; *Lob. Ill.* 51. Hornsey ; *Varenne.* Edmonton.

VII. South Heath, Hampstead. Isle of Dogs.

First record : *Lobel*, about 1600; also first as a British plant.

762. C. vesicaria, *L.*

Cyb. Br. iii. 141. Reich. Ic. Germ. viii. 276.

Sides of streams and ponds ; rare. P. May.

I. Elstree Reservoir; *Fl. Herts.* 322. Boggy ground near Ruislip Reservoir, W. M. H.; *Melv.* 86.

IV. Side of the Brent near Willesden, 1842 ; *Herb. Hardw.*

V. By the Brent near Hanwell! ; *Hemsley* and *Warren.*

VII. [By the waterworks at Pimlico, and in other wet places about London ;
     Sm. Fl. Brit. 1006, and *E. B.* 779.]

First record : *J. E. Smith*, 1800.

763. C. paludosa, *Good.*

Gram. cyp. minus angustifol., *Park.* (Pet.).   *C. acuta* (Curt.).

Cyb. Br. iii. 142.   Curt. F. L. f. 4.   Reich. Ic. Germ. viii. 266.

Ditches and sides of streams ; rather rare.   P.   May.

I.   Ruislip Moor.

III.   By Baber Bridge, Hounslow.

IV.   Bet. Stanmore and Harrow Weald, 1817; *Herb. G. & R.*   Stream in
      Bishop's Wood, 1866.

V.   Bet. Ealing and Acton, 1835 ; *Herb. Young.*   Bet. Acton and Turn-
     ham Green ! ; *Newb.*   By Canal, Greenford, abundant ; *Melv.* 85.

VI.   Copse near Whetstone.   Edmonton.

VII.   [In the ditches against the ' King's Arms' at Whitehall, Mr. Rand ;
       Pet. Conc. Gr. 159, and *R. Syn.* iii. 418.]   [By the Lea bet. Clapton
       and Tottenham, G. Francis ; *Coop. Supp.* 12.]

First record : *Rand*, 1716.

764. C. riparia, *Curt.*

Gram. cyp. latifol. spica rufa sive caule triangulo, C. B. P. (Blackst.).

Cyb. Br. iii. 143.   Curt. F. L. f. 4.   Reich. Ic. Germ. viii. 268.

Sides of streams, ditches, and ponds ; rather common.   P.   May.

I.   Harefield ! ; *Blackst. Fasc.* 35.

II.   Staines Common.   By the Thames at Sunbury.   Walton Bridge, and
      bet. Strawberry Hill and Teddington.

III.   By the Cran on Hounslow Heath, at Hanworth Road and at Hospital
       Bridge.   Duke's River at Isleworth.

IV.   Hampstead Heath ; *Cooper*, 103, and *Irv. MSS.*

V.   Canal, Greenford ; *Melv.* 85.   Not far from Kew Bridge Ry. Station ! ;
     Newb.

VII.   [Marylebone Fields, 1817 ; *Herb. G. & R.*]   Isle of Dogs.

First record : *Blackstone*, 1737.   *Gram. cyp. species fol. angust. et durior.
pan. parv. nigra congesta* ; in the Thames by Limehouse ; *Doody MSS.*,
is either this or some allied species in an undeveloped state.

GRAMINEÆ.†

DIGITARIA, *Scop.*

765.   * D. sanguinalis, *Scop.*

Gr. dactylon latiore fol. C. B. (Pet.).   *Panicum s.*, L. (Curt.).

Cyb. Br. iii. 148.   Curt. F. L. f. 4.

† A complete set of good figures of the British Grasses is still a desideratum. We
have quoted, when Curtis (whose plates are all that can be wished) has not figured the
species, Lowe's *British Grasses*, in which the plates, though deficient in accuracy, are yet
fairly characteristic of habit and general appearance.

Rich waste ground ; very rare. A. August.

VII. [Thames bank about the neat-houses † ; *Pet. Midd.*] Chelsea Gardens, spontaneous, 1849 ; *Thos. Moore* (*v. s.*). Ibid., 1852 ; *Herb. Mus. Brit.*

First record : *Petiver*, 1695. Scarcely naturalised. Extinct ?

ECHINOCHLOA, *Pal. de Beauv.*

766. * E. Crus-galli, *Beauv.* Panicum C.-g., *L.* (L. Cat.).
Gr. paniceum sylv. spica divisa, C. B. (Merr., Budd., Pet., Dill.).
Cyb. Br. iii. 148. Curt. F. L. f. 4.

Waste and cultivated ground ; rather rare. A. July—September.

II. By the towing-path, Hampton Court, a single plant, 1867.

V. Waste heaps by the canal, Apperton.

VI. Near Barnet ; *Herb. Dev. Instit. Exeter.* Amongst ' seeds,' Starchness, Edmonton, 1858, abundant.

VII. [In a lane by the neat-house gardens ; *Merrett,* 56, and *R. Syn.* iii. 394.] Physic garden, Chelsea ; *Budd. MSS.* About Fulham moat ; *Pet. Midd.* and *Gram. Conc.* 43. Parson's Green. Site of Exhibition, South Kensington. Victoria Street, Westminster.

First record : *Merrett,* 1666 ; also first as a British plant. The Middlesex plant usually has the upper glumes awned, as figured in *Knapp,* t. 11.

SETARIA, *Pal. de Beauv.*

767. * S. viridis, *Beauv.*
Gr. Panici effigie spica simplici, Ger. (Merr.). *Panicum v., L.* (Curt. Mart.).
Cyb. Br. ii. 149. Curt. F. L. f. 4.

Waste and cultivated ground ; rather rare. A. July—September.

II. In a field betwixt Tuddington and Hampton Court ; *Merrett,* 56. By the towing-path, Hampton Court, a few plants.

III. Twickenham, near the railway, a few plants, and by the Thames, two specimens.

IV. Dungheap in a field bet. Sudbury Ry. Station and Apperton, 1867.

VII. [Near the neat-houses, Chelsea ; *Mart. App. P. C.* 65.] Market garden grounds bet. Earl's Court and Kensington, 1829 ; *Pamplin.* Near Down's Park Road, Clapton, 1865 ; *Grugeon* (*v. s.*). West India Docks, bet. basin and wall, 1867 ; *Cherry* (*v. s.*). A weed in Chelsea Gardens ; *T. Moore.* Site of Exhibition of 1862, South Kensington. Kensington Gore, a single plant. Camden Town, near the canal. Back of Adelaide Road, N.W.

First record : *Merrett,* 1666 ; also first as British. An ancient introduction. Considered native by many botanists. but undoubtedly of foreign origin in several (probably in all) of the above localities.

---

† The neat-houses, so frequently mentioned, stood on that part of the Thames bank along which Grosvenor Road now passes, between Vauxhall and Chelsea Bridges. The south part of modern Pimlico was built on the neat-house gardens.

S. verticillata, *Beauv. Gr. paniceum, Tab.* (Budd.). *Gr. pan. spica aspera, C. B.* (Ray). Cyb. Br. iii. 150. Curt. F. L. f. 4. VII. [Beyond the neat-houses towards Chelsea, Newton; *R. Hist.* ii. 1263, *Syn.* i. 181, *Pet. Midd.*, &c.] In the Bishop of London's garden at Fulham, and in ye physic garden at Chelsea; *Budd. MSS.* Still a weed in Chelsea Gardens; *T. Moore.*

768. * S. glauca, *Beauv.*
Cyb. Br. iii. 150. Reich. Ic. Germ. i. 47.
Waste and cultivated ground; rare. A. July—September.
II. By the towing-path, Hampton Court, a few plants, 1867.
III. By the Thames near Twickenham Church, a single specimen.
V. Waste heap by canal, Apperton, a few plants.
VII. A single plant at Parson's Green.
First record: *the Authors*, 1867. Likely to become naturalised. Often abundant at Battersea, on the Surrey bank of the Thames.

S. italica, *Beauv. Millet.* V. Waste heaps by the canal, Apperton, a few plants. VII. Sparingly behind Adelaide Road, N.W.; *Syme.*

Panicum miliaceum, *L.* V. Waste heaps by the canal, Apperton, 1867. VII. Site of Exhibition of 1862, S. Kensington.

P. capillare, *L.* II. A single plant by the towing-path, Hampton Court, 1867. A North American species.

(Spartina stricta, *Roth. Dactylis cynosuroides* (M. & G.). Cyb. Br. iii. 144. Lowe, 67. VII. One plant on a wall near Whitehall Stairs; *M. & G.* 83. Some other grass mistaken for it.)

PHALARIS, *Linn.*
769. * P. Canariensis, *L.* *Canary Grass.*
Cyb. Br. iii. 151. Lowe t. 7.
Waste ground, roadsides, &c.; common. A. June—August.
II. Bet. Yewsley and W. Drayton!; *Newb.*
III. Several places about Twickenham.
IV. Northwick Walk, Harrow; *Melv.* 87. Brickfield at Burgess Hill.
V. Greenford; *Melv.* 87. Chiswick; *Fox.* By the canal, Brentford; *Hemsley.* By the canal, Apperton.
VI. Hornsey Lane, 1815; *Herb. G. & R.* Highgate, 1835, G. E. Dennes; *Coop. Supp.* 11.
VII. Bet. Camden Town and the Regent's Park; *Irv. MSS.* Kentish Town; *Herb. Hardw.* Primrose Hill, 1845; *Morris* (v. s.). Notting Hill; *Cole.* By Duckett's Canal; *Cherry* (v. s.). Hampstead. Near Chelsea Hospital. Thames Embankment. Isle of Dogs.
First record: *Goodger and Rozea*, 1815. Spread by the sweepings of bird-cages, and by bird-catchers.

770. P. arundinacea, *L.*

Calamagrostis aquat. Anglo-Brit. acerosa gluma (Lob.). *Gr. arundina-ceum acerosa gluma* (How.). *Chaffy Reed* (Pet.). *Arundo colorata* (G. & R.).

Cyb. Br. iii. 151. Lowe t. 7.

Borders of streams and ponds; very common. P. June—August.

In all the districts; especially frequent by the Thames, and less noticed in IV. and VI.

VII. Not far from London, 'via qua itur Ratteam*;' *Lob. Ill.* 45. By the Thames banks; *How* 50, *Pet. Conc. Gr.* 70. Isle of Dogs, &c.; *Herb. G. & R.* Serpentine, 1813; *Herb. Dev. Inst. Exeter.* Hackney Wick.

First record: *Lobel*, about 1600; also first as a British plant.

ANTHOXANTHUM, *Linn.*

771. A. odoratum, *L.* *Sweet Vernal Grass.*

Cyb. Br. iii. 152. Lowe t. 1.

Meadows, heaths, and woods; very common. P. May, June.

Throughout all the districts.

VII. [Marylebone Fields, 1817; *Herb. G. & R.*] Meadows to the north of London; as about Hornsey Wood, Kilburn, Kentish Town, &c., abundant.

First record: *Goodger and Rozea*, 1817. One of the staple grasses of the rich meadows of the county. Both the forms mentioned in *Bab. Man.* 402 are found.

PHLEUM, *Linn.*

772. P. pratense, *L.* *Timothy Grass. Meadow Catstail.*

Gramen typhinum iii. (Johns.). *Gr. typhoides asperum primum* and *G. t. a. alterum* (Blackst.).

Cyb. Br. iii. 154. Lowe t. 8.

Meadows and roadsides, old cornfields; very common. P. June.

Throughout all the districts.

VII. [Plentifully about London; Chelsey Field; *Johns. Ger.* 12.] Fulham. Isle of Dogs. Common in the meadows north of London.

First record: *Johnson*, 1633; also first as a British plant. A valuable grass. Very variable in size and habit.

? P. asperum, *Jacq.* Cyb. Br. iii. 156. Lowe t. 9. There is a specimen in *Herb. Mus. Brit.* labelled '*Phalaris aspera,* Willd.; near Elstree, Herb. S. R.' This, like many other Elstree plants in Rudge's collection, would appear, however, to have come from a garden.

* Possibly Ratcliff Highway, described by Stow as 'a large highway with fair elm trees on both the sides.'

ALOPECURUS, *Linn.*

773. A. pratensis, *L.*   *Meadow Foxtail.*

Cyb. Br. iii. 159. Lowe t. 3.

Meadows and waste places; very common. P. May, June.
Throughout all the districts.

VII. Grosvenor Square, 1869!; *Warren.* Isle of Dogs!; *Newb.* Meadows
to the north of London. Site of Exhibition of 1862, S. Kensington.

First record; *Irvine,* 1830.

774. A. geniculatus, *L.*

Cyb. Br. iii. 160. Lowe t. 6.

Margins of ponds and wet places; common. P. June—August.

I. Harrow Weald Common. Stanmore Heath. Elstree Reservoir.
South Mims.

II. Bushey Park.

III. Roxeth; *Melv.* 87. Near Hatton.

IV. Kenton; *Melv.* 87. North Heath, Hampstead. By Turner's Wood.

V. Near Brentford; *Hemsley.* Near Shepherd's Bush!; Acton!; Twyford!; *Newb.*

VI. Edmonton.

VII. Hyde Park; *M. & G.* 76. Paddington, 1817; *Herb. G. & R.* Copen-
hagen Fields; *Herb. Hardw.* Homerton; *Cherry (v. s.).* Isle of
Dogs!; *Newb.* Kensal Green; *Warren.* Hackney Wick. South
Heath, Hampstead.

First record: *Milne and Gordon,* 1793.

775. A. fulvus, *Sm.*

Cyb. Br. iii. 161. Lowe t. 5.

Wet margins of ponds; rare?. P. June—September.

I. Margin of Elstree Reservoir, abundantly; *Fl. Herts.* 324. Edge of
Ruislip Reservoir, W. M. H.; *Melv.* 86.

VI. Finchley, J. Woods; *Herb. Mus. Brit.*

VII. Margin of the lowest pond in Ken Wood grounds.

First record: *J. Woods,* about 1805. Probably passed over as *A. geniculatus.*

776. A. agrestis, *L.*   *Mousetail Grass. Black Bent.*

Gr. typhoides spic. angustiore, C. B. and *Gr. myosuroides minus spic.
brevior. aristis recurvis,* Dale (Blackst.). *Gr. myos. majus spic. long.
aristis rectis* (R. Syn.). *A. myosuroides* (Huds., Curt.).

Cyb. Br. iii. 163. Curt. F. L. f. 2.

Cornfields, waste ground, and roadsides; common. P. May—September.

I. Harefield!; *Blackst. Fasc.* 40, *Spec.* 32.

II. Bet. Hampton and Sunbury!; *Newb.* Stanwell Moor. Sunbury.

III. Twickenham. Hounslow ; *Hemsley.*
IV. Stanmore ; *Melv.* 86. Hampstead; *Irv. MSS.* Nr. Harrow Ry. Station.
V. Greenford; *Melv.* 86. Near Shepherd's Bush!; Twyford!; *Newb.* Horsington Hill. Chiswick.
VI. Enfield. Edmonton.
VII. Near Paddington, Buddle ; *Dill. in R. Syn.* iii. 397. Camden Town ; *Herb. Hardw.* Bow ; *Cherry (v. s.).* Chelsea!; *Newb.* Isle of Dogs. Hackney Wick. Kentish Town. South Heath, Hampstead. Victoria Street. Thames Embankment. St. George's Road, S.W. ; Carlton Road, Kilburn ; *Warren.*
First record : *Buddle,* about 1710.

NARDUS, *Linn.*

777. N. stricta, *L.* *Mat Grass.*
? *Gr. sparteum capillaceo fol. minimum* (Johns.). *Gr. sparteum juncifolium, C. B. P.* (Blackst.).
Cyb. Br. iii. 245. Hook. Curt. v. Lowe t. 2.
Heaths ; rather rare. P. June—August.
I. Harefield Common, abundantly!; *Blackst. Fasc.* 39. Harrow Weald Common, abundant!; *Melv.* 87. Stanmore Heath.
III. Hounslow Heath, Mr. Clements ; *Hill,* 35. Drilling Ground, Hounslow.
IV. Hampstead Heath; *Johns. Enum.* and *Johns. Ger.* 30. Ibid. 1850; *Morris (v.s.).*
V. On Hanwell Heath near Boston Lane ; *Hill,* 35.
VII. South Heath, Hampstead, rather sparingly, 1866.
First record : ? *Johnson,* 1632 ; also first as a British plant.

(Stipa pennata, *L.* Cyb. Br. iii. 183. Lowe t. 13. Ken Wood, Hunter ; *Park Hampst.* 28. Cultivated in the garden there ?)

MILIUM, *Linn.*

778. M. effusum, *L.* *Millet Grass.*
Gram. miliaceum, Ger. em. (Pet., Blackst.).
Cyb. Br. iii. 167. Curt. F. L. f. 4.
Woods ; rather rare. P. May.
I. Harefield!; *Blackst. Fasc.* 38.
III. Thickets at Wood End, W. M. H.; *Melv.* 88.
IV. Bishop's Wood; *Warren.* Turner's Wood.
VI. Colney Hatch; *Herb. Hardw.* Copse near Whetstone. Winchmore Hill Wood.
VII. [In Lord Wotton's grove at Bellsize, Hampstead *; *Pet. Herb.* cl. f. 129.] Ken Wood!; *Macreight,* 260.
First record : *Petiver,* about 1683 ?.

* Charles, Lord Wotton, lived at Belsize from 1673 to his death in 1683.

779. P. communis, *L.*   Arundo Phragmites, *L.* (L. Cat.).   *Reed.*
Arundo vallatoria, Ger. (Blackst.).
Cyb. Br. iii. 173.   Lowe t. 15.

Borders of streams, ditches, and ponds ; rather rare.   P.   August.

I. Harefield ! ; *Blackst. Fasc.* 8.   Near Woodridings, Pinner ; *Melv.* 89.
II. Stanwell Moor.   Staines.
IV. Hampstead ; *Irv. MSS.*
V. Bet. Acton and Turnham Green ! ; Ealing Common ! ; Near Shepherd's
Bush and Kew Bridge Ry. Station ! ; *Newb.*
VII. Lea Canal bet. Tottenham and Edmonton ; *Cherry (v. s.).*   Hackney
Wick.   Isle of Dogs.

First record : *Blackstone*, 1737.

CALAMAGROSTIS, *Adans.*

C. lanceolata, *Roth.*   Arundo Cal., *L.* (L. Cat.).   Cyb. Br. iii. 173.   Lowe
t. 15.   Is likely to be found.   Occurs in Newlands Wood, bet. Harefield
and Rickmansworth ; *Fl. Herts.* 328, a locality part of which is perhaps in Middlesex.

780. C. Epigeios, *Roth.*   Arundo Ep. *L.* (L. Cat.).
C. sylvæ Di. Joannis (Park., Merr.).   *Gr. arund. panic. molli spadicea
majus* (Pet.).   *Arundo Calamagrostis* (Huds., M. & G.).
Cyb. Br. iii. 174.   Lowe t. 15.

Woods and shady places ; very rare.   P.   August.

VII. [In St. John's Wood ; *Park.* 1181, *Merrett,* 48.   Wood against the
boarded river ; *Pet. Midd.* 338.]   [Bet. Newington and Hornsey ;
Huds. i. 43.]   [Field bet. Kentish Town and the 'Adam and Eve,'
Pancras ; *M. & G.* 127.]   [Hedge bet. Kilburn and Primrose Hill,
near the railway ; *Irv. MSS.*]   Near Lea Bridge, below Clapton,
Pamplin ; *Lond. Fl.* 94.

First record : *Parkinson*, 1640 ; also first as a British plant.   Perhaps
extinct.

APERA, *Adans.*

781. A. Spica-venti, *Beauv.*   *Corn Bent.*   *Windlestraws.*
Gr. agrorum, Lob. (Ray.).   *Gr. agr. ventispica, Park.** (Blackst.).   *Agros-
tis Spica-venti, L.* (B. G.).
Cyb. Br. iii. 167.   Lowe t. 18.

Cornfields, cultivated and waste ground ; rather common   A.   July.

I. Harefield, frequent ; *Blackst. Fasc.* 34.   Near Uxbridge ; *Lightf. MSS.*
II. Teddington ; *Lond. Fl.* 94.   West Drayton, 1848 ; *Morris (v. s.).*
Near Hanworth, very abundant.

---

* Buddle (*Herb.* liv. fol. 36) makes this a synonym of **787**, *Aira cæspitosa* ; but Parkinson's figure (*Theat.* 1158. 3) seems more like the present species, to which also Ray
refers it.

III. Twickenham, Borrer; *B. G.* 400. Roxeth, W. M. H.; *Melv.* 88. Hounslow. Near Hospital Bridge. Whitton. About Twickenham and Isleworth.

V. Wyke Green. Waste heaps by canal, Apperton.

VI. Edmonton, J. Woods; bet. Tottenham High Cross and the Mills, Dillwyn; *B. G.* 400. About Edmonton, abundant.

VII. About London, Willisel; *R. Cat.* i. 137. [In Pancras Churchyard; *M. & G.* 79.] Bet. Tottenham and the Lea; *Lond. Fl.* 94. Opp. Veitch's Nursery, Chelsea!; *Newb.* Waste ground, King's Road, Chelsea.

First record: *Willisel*, 1670. A troublesome and abundant weed in II. and III.; perhaps originally introduced. Johnson says (*Ger. em.* 6) that this is the grass wherewith they in London usually adorn their chimnies, that is, set up in the fire-place during the summer (v. *Park. Theat.* 1159).

AGROSTIS, *Linn.*

782. A. canina, *L.*

A. vinealis (With.).

Cyb. Br. iii. 170. Lowe t. 16.

Damp places on heaths and meadows; rather rare. P. July, August.

I. Harrow Weald Common!; *Melv.* 88. Bank of Ruislip Reservoir, *Hind.* Stanmore Heath. Elstree Reservoir; *Fl. Herts.* 326.

III. On some parts of Hounslow Heath, abundant, Dr. Goodenough; *With.* iii. 2127.

IV. Hampstead Heath, plentifully; *M. & G.* 80, *Irv. MSS.*

V. Ealing Common!; *Newb.*

VII. [Hyde Park; *M. & G.* 80. Ibid. 1833; *Varenne.*] [In Pancras and Hackney Churchyards; *M. & G.* 80.] [Regent's Park; *Herb. Hardw.*] Bet. London and Hampstead; *Herb. Lambert.*

First record: *Goodenough*, 1796.

783. A. vulgaris, *With.*

A. capillaris, L. (M. & G.).

Cyb. Br. iii. 170. Lowe t. 18.

Dry fields, pastures, and roadsides; common?. P. July.

I. Harrow Weald Common. Stanmore Heath. Elstree.

III. Whitton Park inclosure. About Twickenham.

IV. Bishop's Wood; *Cooper*, 103. Harrow district, common everywhere; *Melv.* 88.

VI. Finchley; *Herb. Hardw.* Edmonton.

VII. In the Green Park; bet. Pancras and Kentish Town; *M. & G.* 81. Hyde Park, 1833; *Varenne.* Isle of Dogs!; *Newb.* Kensington Gardens. South Heath, Hampstead. Grosvenor Square Gardens; *Warren.*

Var. β. A. pumila, Lightf. *M. polymorpha var. γ, Huds.* (M. & G.).

I. Harrow Weald Common.

III. Bet. Twickenham and Hounslow Heath ; *Herb. Lambert.*

IV. Bet. North End and Hampstead ; *M. & G.* 82.

First record : *Buddle,* about 1710. An abnormal or monstrous state was found by Buddle in Bishop's Wood ; his specimens (*Budd. Herb.* cxxv. fol. 13) are labelled *Gr. panic. fere arundinacea glumis oblongis.* It is *Gr. miliaceum sylv. glumis oblongis* of *Pet. Gr. Conc.* 121 and *Dill. in R. Syn.* iii. 404 ; Hudson (ii. 32) who found it also in Hornsey Wood, referred it to his *A. polymorpha, var.* η ; Linnæus placed it at first (*Sp. Plant.* ii. 91) under *A. arundinacea,* and afterwards (ii. 1665) *A. sylvatica* ; and Smith (*E. Fl.* i. 93) makes it *var* γ of *A. alba.*

784. A. alba, *L.   Marsh Bent. Creeping Bent. Fiorin.*
Cyb. Br. iii. 171. Lowe t. 17.

Damp meadows, sides of ponds, &c.; rather common. P. July.

II. Bet. Sunbury and Hampton ! ; *Newb.*

III. Twickenham.

IV. Hampstead Heath ! ; *M. & G.* 82. Harrow district, very common ; *Melv.* 88. Bet. Whitchurch and Stanmore.

V. Canal, Brentford ; *Hemsley.*

VI. Edmonton.

VII. Pound Lane, Clapton ; Hackney Marsh, E. F. ; *Herb. Mus. Brit.* Kilburn ; *Herb. Hardw.* South Heath, Hampstead. West London Cemetery.

First record : *Milne and Gordon,* 1793. Includes *A. stolonifera,* L.

Polypogon monspeliensis, *Desf.* Cyb. Br. iii. 166. Lowe t. 45 (bad). III. In a field by the turnpike on the road bet. Pinner and Harrow ; *Farrar.* Introduced with grain.

HOLCUS, *Linn.*

785. H. lanatus, *L.   Woolly Soft Grass.*
Cyb. Br. iii. 189. Curt. F. L. f. 4.

Meadows and roadsides ; rather rare ?. P. June, July.

I. Harefield ! ; *Newb.* Potter's Bar. South Mims.

III. Twickenham.

IV. Hampstead ; *Irv. MSS.* Harrow district ; abundant ; *Melv.* 89.

VI. Wood Green ; *Munby in Gard. Chron.* for 1868, 499. Edmonton.

VII. South Heath, Hampstead.

First record : *Irvine,* 1830. Other localities probably overlooked.

786. H. mollis, *L.   Creeping Soft Grass.*
Cyb. Br. iii. 189. Curt. F. L. f. 5.

Dry places, heaths, and woods ; rather common. P. July.

I. Near Uxbridge ; *Lightf. MSS.* Harefield ! ; *Newb.* Harrow Weald Common. Stanmore Heath. South Mims.

II. Hanworth Road near the bridge over the New River.

III. Roxeth; *Melv.* 89. Drilling Ground, Hounslow. About Twickenham.

IV. Hampstead Heath; *Cooper*, 102. Bishop's Wood!; *Newb.*

V. Chiswick!; *Newb.*

VI. Hornsey Lane, 1815; *Herb. G. & R.* Wood Green, Munby; *Gard. Chron. for* 1868, 499.

VII. [Primrose Hill, 1845; *Morris (v. s.).*]

First record: *Goodger and Rozea*, 1815. Though apparently more generally distributed, this does not give the impression of being so frequent a plant as *H. lanatus.*

AIRA, *Linn.*

787. A. cæspitosa, *L.* *Tufted Hair Grass.*

Cyb. Br. iii. 177. Lowe t. 20.

Fields, woods, &c.; very common. P. June, July.

In all the districts.

VII. Kensington Gardens, 1850; *Morris (v. s.).* Side of London Canal, Hackney; *Cherry (v. s.).* South Heath, Hampstead.

First record: *Irvine*, 1830.

788. A. flexuosa, *L.*

Cyb. Br. iii. 179. Lowe t. 22.

Heaths and woods; rare. P. June, July.

I. Harrow Weald Common, abundant!; *Melv.* 88. Stanmore Heath. Elstree.

IV. Bishop's Wood, Hampstead; *Newb.*

VII. South Heath, Hampstead!; *Irv. MSS.*

First record: *Irvine*, 1830.

789. A. caryophyllea, *L.*

Cyb. Br. iii. 180. Curt. F. L. f. 6.

Dry sandy places; very rare. A. June, July.

II. Near Teddington Ry. Station in plenty.

IV. Hampstead Heath; *Irv. Lond. Fl.* 96.

VII. [Hyde Park; *Dicks. H. S.*]

First record: *Dickson*, about 1795.

790. A. præcox, *L.*

Gr. præcox parvum spic. lax. canescente, Ray (Blackst.).

Cyb. Br. iii. 181. Curt. F. L. f. 3.

Heaths, dry roadsides, &c.; rather rare. A. May, June.

I. Harefield Common!; *Blackst. Fasc.* 38. Harrow Weald Common.

II. Near Charlton, sparingly.

III. Hounslow Heath. Drilling Ground and roads near.

IV. Hampstead Heath!; *Herb. Hardw.*

V. Near Hanwell!; *Warren.*

VII. Hyde Park, 1816; *Herb. Dev. Inst., Exeter.* South Heath, Hampstead.

First record : *Blackstone,* 1737.

TRISETUM, *Linn.*

791. T. flavescens, *Beauv.* Avena flav., *L.* (L. Cat.). *Yellow Oat Grass.*
Cyb. Br. iii. 187. Curt. F. L. f. 3.

Meadows, heaths, and roadsides ; rather common. P. June.

I. Harrow Weald Common. Stanmore Heath. Elstree. Near Potter's Bar.

II. Near Staines on road to Hampton. Shepperton!; *Newb.*

III. Harrow ; *Melv.* 92. Near Whitton. Twickenham.

IV. Hampstead ; *Irv. MSS.*

V. Horsington Hill.

VI. Wood Green, Munby; *Gard. Chron. for* 1868, 499. Edmonton.

VII. Hyde Park, 1817; *Herb. G. & R.* Kilburn, 1847, T. Moore; *Herb. Mus. Brit.* Ibid., 1869; *Warren.* Kensington Gardens, 1850 ; *Morris (v. s.).* Site of Exhibition of 1862, S. Kensington.

First record : *Goodger and Rozea,* 1817.

AVENA, *Linn.*

792. A. fatua, *L.* *Wild Oat. Haver.*
Festuca utriculis lanugine flavescentibus, C. B. (Newt.).
Cyb. Br. iii. 183. Lowe t. 59.

Cornfields and waste ground ; rather rare. A. July.

II. Bet. Hampton and Hampton Court !; *Newb.*

III. Twickenham Park.

V. Waste heap by canal, Kensington ; *Melv.* 92. Perivale ; *Lees.*

VII. [Road to Kentish Town; *Newt. MSS.*] Near Homerton, E. F.; *Herb. Mus. Brit.*

First record: *Newton,* about 1680. Probably overlooked.

A. sativa, *L.* *Oat.* Is frequently met with as a straggler from cultivation. The *A. nuda* of *Lob. Adv.* 9, found near the Thames, is a cultivated race, called also *A. nuda* by Linnæus.

792 *bis*. A. pratensis, *L.*
Cyb. Br. iii. 185. Lowe t. 60.

Fields ; very rare. P. June.

III. Meadow near the Thames by Richmond Bridge!; *Newb.*

The only record, 1869.

[793. A. pubescens, *L.*
Gr. avenaceum glabrum panic. purpuro-argentea splendente (Ray).
Cyb. Br. iii. 187. Lowe t. 60.

III. In the pastures about the Earl Cardigan's house at Twittenham ; *R. Hist.* ii. 1910, *Fasc.* 9, and subsequent works. Figured in *R. Syn.* iii. t. **xxi.** fig. 2.

The only record, 1688 ; also first as a British plant. Brought down by the river from some chalk district ?.]

ARRHENATHERUM, *Pal. de Beauv.*

794. A. avenaceum, *Beauv.*   *Tall Oat Grass.*

Gr. caninum nodosum (Ger.). *Gr. caninum avenacea panic. non nodosum* (Merr.). *Gr. avenac. pan. acerosa sem. papposo* (Dill.). *Avena elatior* (Curt.).

Cyb. Br. iii. 188.   Curt. F. L. f. 3.

Hedges, banks, and borders of fields ; very common. P. June, and again in September.

In all the districts.

VII. Frequent in Middlesex ; *Merr.* 50.   About London ; *R. Syn.* iii. 406. [Fields next S. James's wall as ye go to Chelsey ; fields bet. the Tower Hill and Radcliffe ; *Ger.* 32.]   Tottenham ; London Canal ; *Cherry* (*v. s.*). Isle of Dogs! ; *Newb.*   Hackney Wick.   Green Park, by Constitution Hill, 1865.

First record : *Gerarde,* 1597.   We cannot state the relative abundance of the ' bulbous ' and ' non-bulbous ' varieties.   The former seems to be a weed of cultivation.

TRIODIA, *R. Br.*

795. T. decumbens, *Beauv.*

Gr. avenaceum parv. procumbens, panic. non aristatis, R. Syn. (Blackst.).
Cyb. Br. iii. 190.   Lowe t. 41.

Heaths and commons ; rather rare. P. May, June.

I. Harefield Common! ; near Battleswell ; *Blackst. Fasc.* 34.   Harrow Weald Common, abundant ; *Melv.* 90.   Stanmore Heath.

II. Bet. Bushey Park and Teddington! ; *Newb.*

III. Drilling Ground, Hounslow.

IV. Hampstead Heath, 1850 ; *Morris* (*v. s.*).

VII. South Heath, Hampstead, abundant.

First record : *Blackstone,* 1737.

KOEHLERIA, *Pers.*

796. K. cristata, *Pers.*

Cyb. Br. iii. 191.   Lowe t. 29.

Dry places ; very rare. P. June.

II. By the roadside not far from Staines on the Hampton Road, 1866.

III. Meadow near Richmond Bridge! ; *Newb.*

First record : *the Authors,* 1866.   Possibly overlooked elsewhere.

MELICA, *Linn.*

797. M. uniflora, *Retz.*   *Wood Melic Grass.*

Gr. avenaceum rariori grano nemorense Danicum (Merr.). *Gr. aven nemorense glum. rar. ex fusco xerampelinis* (Dill., Blackst.).
Cyb. Br. iii. 191.   Curt. F. L. f. 5.

Woods; rather rare. P. May, June.
I. In the Old Park Wood!; lane near the chalkpit; *Blackst. Fasc.* 34.
IV. Hampstead Wood!; *Merrett*, 49, and *R Syn.* iii. 403. Harrow Park
near the ruins; *Melv.* 89. Bishop's and Turner's Woods. Scratch
Wood.
VI. Colney Hatch; *Herb. Hardw.* Hadley!; *Warren.* Near Whetstone.
Winchmore Hill Wood.
VII. Hornsey Wood near Highgate; *Blackst. Spec.* 29. Ken Wood.
First record: *Merrett*, 1666.

(M. nutans, *L.* Copse near Hornsey; *M. & G.* 86. No doubt *M. uniflora*
was intended.)

MOLINIA, *Schrank.*

798. M. cærulea, *Mönch.*
Gr. *pratense scrot. panic. long. purpurascente, R. Syn.* (Blackst.). *Aira
cærulea* (M. & G.). *Melica cærulea* (Curt.).
Cyb. Br. iii. 193. Curt. F. L. f. 5.
Wet heaths and meadows; rare. P. July—September.
I. Moist meadows near Harefield Moor; *Blackst. Fasc.* 39. Harefield
Common. Harrow Weald Common; *Melv.* 89.
III. Drilling Ground, Hounslow.
IV. In the bog on North Heath, Hampstead.
VI. [In Hornsey Churchyard; *M. & G.* 89.]
VII. [In a field near Pancras; *M. & G.* 89.]
First record: *Blackstone*, 1737.

POA, *Linn.*

799. P. annua, *L.*   *Annual Meadow Grass.*
Cyb. Br. iii. 201. Curt. F. L. f. 1.
Fields, meadows, gardens, waste ground, gravel paths, walls, &c.; very
common. A. All the year; in perfection in May.
Throughout all the districts.
VII. Forms by far the largest part of the turf in the London parks, squares
and gardens, being one of the few grasses capable of enduring the
adverse influences of atmosphere, soil and drought.* Appears spon-
taneously in every piece of waste garden ground in town, e. g. St.
Paul's Churchyard, and is probably the commonest plant we have.
No old record. First: *Irvine*, about 1830.

(P. bulbosa, *L.* Cyb. Br. iii. 202. Lowe t. 39. IV. On Hampstead Heath;
M. & G. 103, *Cooper* 102. VII. On a wall in Fulham; *M. & G.* 103.
Probably the bulbous state of *P. pratensis,* was intended.)

---

* When fresh grass-seed is sown to renew the turf in London, very few species ever
flower, and *Poa annua* with *Lolium perenne* and *Dactylis* are almost the only ones able to
flourish.

800. P. nemoralis, *L.*

Cyb. Br. iii. 206. Lowe t. 40.

Woods and shady places; rather common. P. June, July.

I. Harefield!; *Newb.* Moss Lane, Pinner. Road from Elstree to Penniwells. South Mims, abundant.

III. Roxeth; Headstone; *Melv.* 90.

IV. Harrow Park; *Melv.* 90. Scratch Wood.

V. On walls near Ealing!; *Newb.* Sion Park.

VI. Colney Hatch!; *Newb.* Highgate; *Irv. MSS.*

VII. Bet. Kentish Town and Highgate, 1847; *Herb. Hardw.* Kensington Gardens, 1850; *Morris (v. s.).* Kilburn, 1869!; *Warren.*

First record: *Irvine,* 1830.

801. P. trivialis, *L.*   *Rough-stalked Meadow Grass.*

P. setacea (Huds.). *Gr. prat. panic. majus angustiore fol.* (Pet.).

Cyb. Br. iii. 204. Curt. F. L. f. 2.

Meadows and pastures and waste ground; probably very common. P. June

Through all the districts; less common in the dry II. and III.

VII. Beyond the half-way house to Hampstead; *Herb. Pet.* Marylebone; *Huds.* i. 34. Kentish Town; *Herb. Hardw.* Bet. Lea Bridge and Bow; Upper Clapton; *Cherry (v. s.).* Kilburn; *Warren.* South Heath, Hampstead. Isle of Dogs.

First record: *Petiver,* about 1715. [*P. palustris*; at Whitehall Stairs; in Tothill Fields; *M. & G.* 102; a form of *trivialis.*]

802. P. pratensis, *L.*   *Smooth Meadow Grass.*

Cyb. Br. iii. 203. Curt. F. L. f. 2.

Meadows, roadsides, and walls; probably very common. P. June. Through all the districts.

VII. Near Marylebon; *Huds.* i. 33. Haverstock Hill; *Herb. Hardw.* Paddington; *Varenne.* Near Temple Mills, Hackney; *Cherry (v. s.).* Hyde Park and Kensington Gardens, sparingly; Kilburn!; *Warren.*

Var. β, subcærulea.

III. Hounslow.

IV. Hampstead Heath.

V. Ealing!; *Newb.*

First record: *Hudson,* 1762.

803. P. compressa, *L.*

Cyb. Br. iii. 205. Lowe t. 37.

Walls and dry banks; rather rare. P. June—August.

II. Teddington.

III. Isleworth, abundant.

IV. Hampstead; *Irv. MSS.*

V. Turnham Green; *Newb.* Osterley Park wall.

VII. Near Marylebone; *Huds.* i. 33. West India Docks, bet. the basin and wall; *Cherry (v. s.).* Well Walk, Hampstead. Kensington Gore, 1867.

First record: *Hudson*, 1762.

GLYCERIA, *R. Br.*

804. G. aquatica, *Sm.*

Poa aquat. L. (Curt.).

Cyb. Br. iii. 195. Curt. F. L. f. 5.

Borders of slow streams, ditches and ponds; common. P. July—September.

I. Canal, Harefield.

II. Thames above Hampton Court!; *New B. G.* 103. Bet. Hampton Court and Kingston Bridge. Staines Moor. Staines Common. Queen's River, Hampton.

III. Thames bet. Twickenham and Richmond Bridge. Twickenham Park. By Cran at Hospital Bridge.

IV. Hampstead; *Irv. MSS.*

V. Greenford; *Melv.* 89. Twyford; near Kew Bridge and Shepherd's Bush Ry. Stations; *Newb.* Hanwell!; *Warren.* By Thames, Strand on the Green.

VI. Whetstone. Edmonton.

VII. [Serpentine, 1813; *Herb. Dev. Inst. Exeter.*] By Thames at Chelsea: *Morris (v. s.).* Thames near Fulham; *New B. G.* 103. London Canal; *Cherry.* [Walham Green, abundant, 1862.] South Heath, Hampstead. Isle of Dogs.

First record: 1813.

805. G. fluitans, *R. Br.* Flote Grass. Manna Grass.

Gr. aquaticum panicula longissima (How). *Poa fluitans, Scop.* (Melv.). *Festuca fl., L.* (M. & G.).

Cyb. Br. iii. 196. E. B. S. 2975.

Borders of ponds, slow streams, &c.; very common?. P. June, July. Throughout all the districts.

VII. About London; *How,* 50. [Abundantly in Tower ditch; about Chelsea Waterworks; *M. & G.* 113.] West End, Hampstead; *Herb. Hardw.* South Heath, Hampstead, sparingly.

First record: *How,* 1650.

806. G. plicata, *Fr.* G. fluitans, *var. β* (L. Cat.).

Cyb. Br. iii. 197, 519. E. B. 1520.

Sides of ponds and wet places; common?. P. June—September.

I. North of Harefield!; *Newb.*

VII. Field near Hampstead, Mr. Moore; *Phyt.* ii. 500. South Heath; *Newb.* Eel-brook Meadow, abundant, 1862. Isle of Dogs, 1866.

These perhaps should all be referred to *var. β.*

Var. β. G. pedicellata (Towns.). *G. fluitans, var.* γ (L. Cat.). *Festuca fluitans*; Curt. F. L. f. 1.

I. Eastcott.

III. Near Hounslow Ry. Station!; *Newb.*

IV. North Heath, Hampstead.

V. Seen in this district; *Newb.*

VI. Pond near Whetstone.

VII. South Heath, Hampstead, in plenty. Ditches at Hackney Wick. Much commoner than the type.

First record: *T. Moore*, 1845; also first as a British plant. Distribution not worked out.

SCLEROCHLOA, *Pal. de Beauv.*

807. S. distans, *Bab.*

Poa retroflexa (Curt.).

Cyb. Br. iii. 198. Curt. F. L. f. 6 (drawn from a Middlesex specimen).

Waste places; very rare. P. June—September.

VII. [Right-hand side of the road leading up the hill to Hampstead in tolerable plenty; *Curt. F. L.*] Hampstead Heath; *Cooper*, 102. [Isle of Dogs, a single plant, 1866!; *Newb.*] On the new soil of the Thames Embankment opp. Somerset House, 1866.

First record: *Curtis*, 1786.

808. S. rigida, *Link.*

Poa rigida, L. (Curt., Melv.).

Cyb. Br. iii. 200. Curt. F. L. f. 2 (drawn from a London specimen).

Walls; rather rare. A. May, June.

II. Hampton Court walls. Stone facing of the river-bank near Hampton Court.

III. Harrow Grove; Headstone Farm; *Melv.* 90. Near Harrow Church, very fine; *Hemsley.* Isleworth, abundant.

V. Wall of Chiswick House grounds.

VII. Most walls about London; *Curt. F. L.* Chelsea; *Morris* (v. s.). Parson's Green.

First record: *Curtis*, about 1780.

BRIZA, *L.*

B. minor, *L.* Cyb. Br. iii. 212. Lowe t. 43. III. Isleworth, in waste ground; a single large plant, 1866. Perhaps introduced with garden plants; it does not readily spread.

809. B. media, *L.* *Quaking Grass.*

Gr. tremulum s. phalaris pratensis (Johns.). *Gr. tremulum majus, C. B. P.* (Blackst.).

Cyb. Br. iii. 211. Lowe t. 42.

Meadows; rare. P. May, June.

I. Harefield, very common!; *Blackst. Fasc.* 39.

III. Roxeth; *Melv.* 90.

IV. Hampstead Heath; *Johns. Enum., Irv. MSS.* Near Stanmore, 1817; *Herb. G. & R.* Kenton; Northwick Walk and football field, Harrow; *Melv.* 90.

First record: *Johnson*, 1632.

CATABROSA, *Pal. de Beauv.*

810. C. aquatica, *Presl.*

Gramen dulce (Lob.). *Gr. miliaceum fluitans suavis saporis* (Merr). *Gr. paniculatum aquat. minus* and *Gr. miliaceum aquaticum* (Ray). *Aria aquat., L.* (Curt. &c.).

Cyb. Br. iii. 194. Curt. F. L. f. 1.

Ditches, borders of ponds and wet ground; rather common. P. May — August.

II. Colne below Iver, 1847; *Morris (v.s.).* Stanwell Moor. Near Strawberry Hill.

III. Isleworth.

IV. Hampstead; *Irv. MSS.*

V. Brentford.

VI. Hornsey; *Cat. Lond.* 13. Near Whetstone.

VII. About London near the Thames; *Lob. Ill.* 10. [Knightsbridge near London; *Merrett*, 53.] [Bank of New River behind Islington, and many muddy places about London; *Pet. Midd.*] [Milbank; by Chelsea Waterworks; on the sides of the canal in St. James's Park; *M. & G.* 90.] Lea bet. Clapton and Tottenham, G. Francis; *Coop. Supp.* 12.

First record: *Lobel*, about 1600; also first as a British plant.

CYNOSURUS, *Linn.*

811. C. cristatus, *L.*   *Crested Dog's Tail.*

Gr. cristatum, C. B. (Lob., Blackst.).

Cyb. Br. iii. 213. Lowe t. 19.

Pasture fields; very common. P. July, August.

Through all the districts; the chief grass in many pastures about Twickenham.

VII. James Cargill, D. C., showed me this about London; *Lob. Ill.* 31. [Marylebone Fields, 1817; *Herb. G. & R.*]

First record: *Cargill*, about 1600. The *Gr. cristatum Britannicum* found about Hackney; *Lob. Ill.* 30, is the same (see *Ray Hist.* ii. 1269).

DACTYLIS, *Linn.*

812. D. glomerata, *L.*   *Cocksfoot.*

Cyb. Br. iii. 216. Lowe t. 41.

Meadows, pastures, hedges, &c.; very common. P. June—August.
Throughout all the districts.

VII. Kensington Gardens, 1845 ; *Morris (v. s.).* Grosvenor Square ; Hyde
Park, 1868; *Warren (v. s.).* Green Park. Kilburn. Lincoln's Inn
Fields.

First record: *Doody*, about 1700. *Gram. asp. simile sed lævius*, observed
by Mr. Doody pretty common about London ; *Pet. Gr. Conc.* 63, is a
more glabrous form.

FESTUCA, *Linn.*

813. F. sciuroides, *Roth.* F. Myurus, *subsp.* (L. Cat.).

F. bromoides, (Sm.)

Cyb. Br. iii. 217. Lowe t. 50.

Heaths, walls, roadsides, and waste ground ; very common. A ?. June.
Through all the districts ?, no record in VI.

VII. [Fields near Marylebone ; *Herb. Rudge.*] [Kensington Gardens,
Winch MSS. ; *New B. G.* 103.] South Heath, Hampstead.

First record : *Dillenius*, about 1730. On Hampstead Heath ; *Herb. Dill.*
at Oxford, fide *Newb.*

814. F. Myurus, *L.*   *Capon's-tail Grass.*

Gr. murorum spica longissima (Merr.). *F. pseudo-Myurus, Koch.* (Melv.).

Cyb. Br. iii. 218. E. B. 1412.

Walls, rarely on dry ground ; rather rare. A.? June.

II. On Walton Bridge!; *Newb.*

III. The Grove, Harrow ; Headstone Farm ; *Melv.* 91. Near the church,
Harrow. Near the sluice, Twickenham, on rubbish. Brick walls
about Isleworth, abundant.

IV. Wall opp. Jack Straw's Castle, Hampstead ; *Irv. MSS.* Stanmore.

V. Brentford, in plenty.

VI. Edmonton.

VII. [On the walls by Pickadyllee (= Piccadilly); *Merrett*, 55.] Path
from Dalston to Pigwell or Pitwell, E. F. ; *Herb. Mus. Brit.* Near
Parson's Green.

First record : *Merrett*, 1666.

815. F. ovina, *L.*   *Sheep's Fescue.*

Cyb. Br. iii. 219. Lowe t. 50.

Dry fields, heaths, and roadsides ; common ?. P. June.

I. Bet. Harefield and Ruislip. Harrow Weald Common. Stanmore
Heath. South Mims.

II. Roadsides about Charlton.

III. Hounslow Heath ; *Winch MSS.* Drilling Ground, abundant.

IV. North Heath, Hampstead.

V. Northolt; *Melv.* 91.

VI. Bet. Edmonton and Ponder's End ; *Cherry (v. s.).*

VII. Hyde Park, J. Gray ; *Herb. Dev. Inst. Exeter.* Garden of Lincoln's Inn.

First record : *Winch,* about 1810. *F. tenuifolia,* Sibth., *F. heterophylla,* Auct., &c., are slight varieties.

816. F. duriuscula, *L.* F. ovina, *var.* γ (Bab.*)?. *Hard Fescue.*
Cyb. Br. iii. 219. E. B. 470.

Heaths, dry roadsides, hedgebanks, &c. ; common. P. June.

I. Harrow Weald Common. Potter's Bar.

II. Abundant about Charlton, and from Staines to Hampton.

III. Near Isleworth. Roxeth ; *Herb. Harr.*

IV. Hampstead ; *Irv. MSS.*

V. Near Shepherd's Bush Ry. Station! ; *Newb.*

VI. Near Edmonton ; Muswell Hill ; *Cherry (v. s.).* Wood Green, Munby ; *Gard. Chron.* for 1868, 499.

VII. Kentish Town ; *Herb. Hardw.* Kensington Gardens, 1850 ; *Morris(v.s.).* Near King's Road, Chelsea. West London Cemetery, Brompton.

First record : *Irvine,* 1830.

817. F. gigantea, *Vill.* Bromus g., *L.* (L. Cat.).
Gr. avenac. glabrum panic. e spic. rar. strigosis comp. arist. tenuissimis (Ray).
Cyb. Br. iii. 225. Curt. F. L. f. 5 (drawn from a Thames specimen).

Wet places, ditches, and woods ; common. P. July, August.

I. Harefield. Elstree, by the reservoir. Bet. Elstree and Penniwells.

II. Staines Moor. Bet. Staines and Kingston. River-side bet. Hampton Court and Kingston Bridge.

III. Harrow Park; *Melv.* 91. Twickenham. Near Whitton.

IV. About Hampstead, plentifully ; *M. & G.* 119. Near the Brent ; *Irv. MSS.* Harrow Park; *Melv.* 91. Deacon's Hill.

V. Sion Park.

VI. Muswell Hill ; *Cherry (v. s.).*

VII. Fulham, near the Bishop's Palace, Doody ; *R. Hist.* ii. 1909. *Hill* 56. Thames above Putney ; *Herb. Mus. Brit.* Sandy End, Fulham.

First record : *Doody,* 1688 ; also first as a British plant.

818. F. elatior, *L.* F. arundinacea, *var.* β (Bab.). *Tall Fescue.*
Gr. arund. aquat. panic. avenacea (Ray). *F. arundinacea, Schreb.* (Melv.).
Cyb. Br. iii. 222, 519. Curt. F. L. f. 6 (drawn from a Thames-side specimen).

Wet meadows, sides of streams and ditches ; rather rare. P. June— August.

I. Elstree Reservoir. Potter's Bar.

---

* Perhaps our plant is the *F. rubra* of Bab. Man.

III. By the Cran near Whitton. By the Thames, Twickenham.

IV. Bet. Ken Wood and Finchley; *Irv. MSS.* By a stream in the football field, Harrow; *Melv.* 91.

V. By canal near Apperton!; bet. Acton and Turnham Green!; *Newb.*

VII. [By the Thames bet. London and Chelsea, Doody; *R. Hist.* ii. 1909.] [Woods by the boarded river; *Hill*, 37.] By the river, Chelsea Hospital, 1861.

First record: *Doody*, 1688; also first as a British plant.

819. F. pratensis, *Huds.*   *Meadow Fescue.*

Cyb. Br. iii. 223.   Curt. F. L. f. 6.

Rich meadows and pastures; very common. P.   June, July.

Throughout all the districts, but less common in II. and III.

VII. Bet. Lea Bridge and Bow; *Cherry (v.s.).* Near Hornsey. South Heath, Hampstead. Isle of Dogs.

Var. β. F. loliacea, Huds.   *Festuca fluitans, var. β* (M. & G.).   Curt. F. L. f. 6.

I. Potter's Bar.

III. About Harrow, not uncommon, W. M. H.; *Melv.* 91. Twickenham, near the church.

V. Apperton!; *Newb.*

VII. [In the privy garden, Whitehall, a single plant; *M. & G.* 113.] Common about Stoke Newington, J. Woods; *B. G.* 400. About London; *Curt. F. L., and other authors.* About Kentish Town; *Irv. MSS.*

First record: *Curtis*, about 1790.

BROMUS, *Linn.*

820. B. erectus, *Huds.*

Cyb. Br. iii. 228.   Lowe t. 51.

Dry pastures; very rare. P.   June, July.

I. Neighbourhood of Uxbridge; *Lightf. MSS.* Field above the chalk-pit, Harefield, abundant.

First record: *Lightfoot*, about 1780.   Confined to the chalk.

821. B. asper, *L.*

B. hirsutus (Curt.).

Cyb. Br. iii. 225.   Curt. F. L. f. 2 (drawn from a Middlesex specimen).

Moist hedges; rather common. A., B., or P.   July, August.

I. Harefield, abundant. About South Mims, abundant.

II. Staines Moor.

III. Harrow, very common; *Melv.* 92. Near Whitton. By Baber Bridge. Hounslow. Near Twickenham, very abundant.

IV. About Hampstead; *Curt. F. L.* Near Harrow Weald Church. Bet. Sudbury Station and Apperton.

V. Near Kew Bridge Ry. Station!; *Newb.* Norwood. Horsington Hill.
VII. Upper Clapton; near Hornsey; *Cherry* (*v. s.*). South Heath, Hampstead, in the hedge by the brickfield.

First record: *Curtis*, about 1780.

822. B. sterilis, *L.*

Festuca avenacea sterilis elatior, R. Syn. (Blackst.).

Cyb. Br. iii. 226. Curt. F. L. f. 1.

Dry waste ground, under hedges, on walls, &c.; probably very common.
A. May, June.

Through all the districts.

VII. Holloway, 1840; Primrose Hill; bet. Swiss Cottage and Hampstead,
1847; *Herb. Hardw.* Kensington Gore. Isle of Dogs. Kilburn!;
Warren.

First record: *Blackstone*, 1737.

B. diandrus, *Curt.* About London; *Huds.* ii. 50, *Cyb. Br.* iii. 227, &c.
The locality meant, we presume, is in Surrey, near Battersea Church,
whence the specimen figured in *Curt. F. L.* f. 6 was obtained.

B. tectorum, *L.* Several plants on the road near Cricklewood, brought
with grain to the Shoot-up Hill Mill, 1869!; *Warren.* A common
European species.

SERRAFALCUS, *Parl.*

823. *S. secalinus, *Bab.* Bromus s., *L.* (L. Cat.).

Gr. bromoides latiore panicula, Park. (Merr.).

Cyb. Br. iii. 229. Lowe t. 54.

Cornfields and waste places; rare. A. June—August.

IV. Cornfield near Ken Wood; *Irv. MSS.*
V. Seen in this district; *Newb.*
VI. Cornfield, Finchley Common, 1841; *Herb. Hardw.*
VII. [In Pancras Churchyard, in great abundance; *M. & G.* 116.] Woods
below Hamsted; *Merr.* 49.

Var. β. Bromus velutinus, Sm. E. B. 1884.

I. Cornfield weed in fields above the canal, Harefield.
V. Field bet. Kew Ry. Station and Brentford in small quantity!; *Newb.*
VII. Bet. Kentish Town and Hampstead; *Irv. MSS.*

First record: *Merrett*, 1666.

824. S. racemosus, *Parl.*

Bromus arvensis (Knapp).

Cyb. Br. iii. 230. E. B. 1079.

Meadows and roadsides; common?. A. or B. May, June.

I. Near Potter's Bar Ry. Station.
II. Shepperton!; *Newb.* Meadows by the Thames, Sunbury, in plenty.

IV. About Hampstead; *Irv. MSS.* By Brent Reservoir!; Cricklewood!; *Warren.* Harrow Weald.

V. Near Twyford!; near Ealing!; near Acton!; *Newb.* Hanwell!; *Warren.* Near Chiswick.

VI. About Whetstone.

VII. Lamb Lane, Hackney; marshes near Lea Bridge, E. F.; *Herb. Mus. Brit.* Kensal Green Cemetery!; *Newb.* About the West India Docks at Limehouse, plentifully; *Knapp,* i. 82. Kilburn!; *Warren.*

Var. β. S. commutatus, Bab. E. B. 920.

III. Roxeth; *Hind.* Near Whitton.

IV. Harrow district, frequent; *Melv.* 92.

V. Near Twyford!; amongst vetches near Brentford!; *Newb.*

VII. In Clapton Field; banks of Lea above Lea Bridge; *Herb. Mus. Brit.*

First record: *E. Forster,* about 1800.

825. S. mollis, *Parl.* Bromus m., *L.* (L. Cat.). *Soft Brome Grass. Lop Grass.*

Cyb. Br. iii. 231. Curt. F. L. f. 1.

Meadows and roadsides; very common. A. or B. May, June.

Throughout all the districts.

VII. Abundant in the meadows in the northern suburbs. Hyde Park. Isle of Dogs.

First record: *Irvine,* 1830. Forms a very considerable part of the hay made in this county. The glabrous or sub-glabrous form (*B. racemosus,* Parn. not Parl.) frequently occurs, e.g. at Hounslow, near Twyford and Shepherd's Bush, at Hornsey, &c.

S. arvensis, *Godr.* Bromus a., *L.* (L. Cat.). Cyb. Br. iii. 232. Lowe t. 56.

II. Unused road over the railway bridge, Teddington, abundant.

BRACHYPODIUM, *Pal. de Beauv.*

826. B. sylvaticum, *R. & S.*

Festuca sylvatica, Huds. (Irv.).

Cyb. Br. iii. 233. Lowe t. 66.

Hedgebanks and woods; rather common. P. July, August.

I. Bet. Yewsley and Iver Bridge!; *Newb.* Harefield. South Mims. Near Pinner Wood.

II. Staines. Sunbury. Teddington.

III. Near Whitton. About Twickenham, very common.

IV. Edgware. Kenton, &c.; *Melv.* 93. Scratch Wood. Lane from Hendon to Finchley Road, 1845; *Morris (v. s.).* Turner's Wood.

V. Bet. Acton and Ealing!; *Newb.* Lampton. Sion Park.

VI. Woods beyond Highgate; *Irv. MSS.* Enfield.

First record: *Irvine,* 1830.

(B. pinnatum, *Beauv.* Cyb. Br. iii. 233. Lowe t. 66. *Gramen spica Brizæ majus,* Highgate Wood; *Johns. Eric. Festuca pinnata,* bet. Kentish Town and Hampstead; *Irv. Lond. Fl.* 99. *B. sylvaticum* was probably mistaken for it.)

TRITICUM, *Linn.*

827. T. caninum, *Huds.*

Cyb. Br. iii. 234. Lowe t. 65.

Hedgebanks; very rare. P. June, July.

I. Near Harefield, on road to Rickmansworth. South Mims, near the county boundary.

VII. [Marylebone Fields, 1817; *Herb. G. & R.*]

First record: *Goodger and Rozea,* 1817.

828. T. repens, *L.*   *Couch Grass.*

Gr. caninum (Ger.).

Cyb. Br. iii. 235. Lowe t. 65.

Hedges, roadsides, &c.; very common. P. June, July.

In all the districts.

VII. [Fields next to S. James' wall; fields bet. Tower Hill and Radcliffe; *Ger.* 22.] Gospel Oak Fields, Kentish Town, 1846; *Herb. Hardw.* Lea Bridge, &c.; *Cherry (v.s.).* Kilburn. Isle of Dogs.

First record: *Gerarde,* 1597.

HORDEUM, *Linn.*

H. sylvaticum, *Huds. Elymus europæus,* Sm. Cyb. Br. iii. 241. Lowe t. 62. Probably occurs, as it is recorded from a wood near Stocker's Farm, bet. Rickmansworth and Harefield; *Fl. Herts.* 342.

829. H. pratense, *Huds.*   *Meadow Barley.*

Cyb. Br. iii. 242. Lowe t. 63.

Meadows; common. P. June, July.

I. Near Harrow Weald Common, very abundant. By Elstree Reservoir. Potter's Bar.

II. Staines!; Shepperton!; *Newb.* Hampton.

III. Whitton. Near Hounslow. Twickenham.

IV. Hampstead; *Morris (v.s.).* Harrow district, very common; *Melv.* 92. Harrow Weald.

V. Acton Green!; *Newb.*

VI. About Hornsey, abundant.

VII. [Camden Town, 1847; *Herb. Hardw.*] Hackney Marshes; *Cherry.* Isle of Dogs.

First record: *Morris,* 1845. A common grass in meadows on a strong clay soil.

830. H. murinum, *L.*   *Way Bennet.   Wall Barley.   Wild Rye.*
Cyb. Br. iii. 243.   Curt. F. L. f. 5.

Foot of walls, on rubbish, waysides, &c.; very common.   A.   May—
September.

Through all the districts.

VII. Especially frequent in the suburbs.   Path by the Serpentine, near
Albert Gate, 1868; *Warren.*   Chelsea.   Kilburn.   Hackney Wick.
Isle of Dogs.   Camden Town.   Hackney Churchyard.   Green Park.

First record: *Irvine*, 1830.

LOLIUM, *Linn.*

831. L. perenne, *L.*   *Perennial Darnel.   Rye Grass.   Ray Grass.*
Cyb. Br. iii. 238.   Lowe t. 67.

Meadows, roadsides, waste ground, &c.; very common.   P.   June—August.
Throughout all the districts.

VII. After *Poa annua*, the commonest grass in London.   Abounds in all
the parks and in most pieces of waste ground, e.g. Leicester Square.

First record: *Irvine*, 1830.   Varies with awned paleæ, and with the spikes
branched or abbreviated.   Agriculturists also distinguish numerous races.

832.   * L. italicum, *A. Braun.*   *Italian Rye Grass.*
Cyb. Br. iii. 238.   Lowe t. 67.

Meadows and fields and waste ground; rather common.   B. or P.   June.
I. Harefield!; bet. Yewsley and Iver Bridge!; *Newb.*
II. Hampton!; *Newb.*   Stanwell Moor.   Bet. Hampton Court and
Kingston Bridge.
III. By Headstone Lane, W. M. H.; *Melv.* 93.   Near Hounslow.   Near
    Worton.   Twickenham.
V. Bet. Acton and Turnham Green!; *Newb.*
VII. Isle of Dogs!; Chelsea!; *Newb.*

First record: *Hind*, 1864.   A continental grass, introduced in 1831,
extensively sown for agricultural purposes, and now quite established
as a weed.

833.   L. temulentum, *L.*   *Darnel.*
Gr. loliaceum (Johns.).
Cyb. Br. iii. 239.   Lowe t. 68.

Cultivated and waste ground; rare.   A.   July, August.
III. Near Twickenham Church, 1867.
IV. Hampstead Heath; *Johns. Enum.*   Kenton Road; *Melv.* 93.   Corn-
field near Ken Wood; *Irv. MSS.*
V. Apperton, W. M. H.; *Melv.* 93.
VII. Gloucester Terrace, Camden Town, 1847; *Herb. Hardw.*   Waste land
bet. Chelsea Hospital and the river, 1861.

Var. β.   L. arvense, With.   E. B. 1125.
VII. On Hackney Downs, E. F.; *Herb. Mus. Brit.*

First record: *Johnson*, 1632.   Usually introduced with foreign seed,
improved farming having almost extirpated it in British cornfields.

CRYPTOGAMIA.

EQUISETACEÆ.

EQUISETUM, *Linn.*

834. E. arvense, *L.*   *Horsetail.*
Equiset. arv. longioribus setis, C. B. P. (Blackst.).
Cyb. Br. iii. 304.   Curt. F. L. f. 4.

Damp fields, cultivated and waste ground; very common.   P.   April.
In all the districts.
VII. Isle of Dogs!; *Newb.*   Millfield Lane, Highgate.   Sandy End, Fulham.

First record: *Blackstone*, 1737.

835. E. maximum, *Lam.*
E. fluviatile, Sm. (Irv.).   *E. Telmateia, Ehrh.* (Melv.).
Cyb. Br. iii. 300.   Newman, 67.

Moist hedges and woods, sides of ponds, &c.; rare.   P.   April.
II. Near Teddington.
IV. Hampstead Heath, 1823; *Bennett* (*v.s.*).   Turner's Wood, abundant!;
     Irv. MSS.   Bentley Priory, Harrow Weald, W. M. H.; *Melv.* 96.
     Near Edgwarebury.
VI. The Alders Copse, Whetstone, abundant.
VII. Banks of Reservoir near Ken Wood.   Near New West End.

First record: *J. J. Bennett*, 1823.

836. E. sylvaticum, *L.*
E. omnium minus tenuifolium (Lob., Park., Ray).
Cyb. Br. iii. 305.   Newman, 59.

Damp shady places on heaths and in woods; rather rare.   P.   April,
May.
I. Harefield; *Blackst. Fasc.* 26.   Harrow Weald Common, abundant!;
     Melv. 97.
IV. In the meadows by Hampstead Heath; *Merrett* 35, *R. Cat.* i. 100, &c.
     North Heath.   Bishop's and Turner's Woods, and meadows near
     them, abundant.
VI. Highgate; *Lob. Ill.* 143, *Park. Theat.* 1201.   The Alders near Whet-
     stone, 1867; *J. S. Mill.*
VII. Field bet. Kentish Town and Kenwood near the ponds, E. F.; *Herb.
     Mus. Brit.*   Fields near Kenwood, J. Rayer; *B. G.* 413.

First record: *Lobel*, about 1600.

837. E. limosum, *L.*
E. *nudum, sive junceum, Ger.* (Merr.). *E. nudum lævius nostras, R. Syn.*
(Blackst.).
Cyb. Br. iii. 306. Newman, 51.
Marshy places, ponds, and ditches; common. P. May, June.
I. In the Warren pond at Breakspears; *Blackst. Fasc.* 26. Ruislip Re-
servoir; Ruislip Common; *Melv.* 96. Near Yewsley!; *Newb.*
Stanmore Heath.
II. Meadows near Sunbury.
III. Near Hatton.
IV. Pits on Hampsted Heath; *Merrett,* 35. North Heath, abundant.
VI. Whetstone. Edmonton.
VII. Hammersmith, near Sir Nicholas Crisp's brick-pit; *Merrett,* 35. Isle
of Dogs, Newman; *Phyt.* i. 691. Sides of ponds, Ken Wood grounds.
First record: *Merrett,* 1666; also first as a British plant. The form
with long branches, *E. fluviatile,* L. is *E. fol. nudum ramosum,* C. B. P.;
Breakspears; *Blackst. Fasc.* 26; also common at Hampstead.

838. E. palustre, *L.*
Cyb. Br. iii. 306. Newman 43.
Wet places, sides of streams, &c.; rather rare. P. July, August.
I. Near Harefield!; near Yewsley!; *Newb.* Ruislip Moor.
II. Staines. Near Teddington. Hampton. Bet. Hampton and Hampton
Court.
IV. Bog on North Heath, Hampstead!; *Irv. MSS.*
V. Canal side near Greenford; *Melv.* 97.
VII. South Heath, Hampstead.
First record: *Buddle,* about 1705. An unbranched form, *E. nudum
ramosum* (III.), on a bog on Hounslow Heath, where nothing as yet
but this had sprung up after the digging of peat; *Budd. MSS.* vi. and
Budd. Herb. cxvii. fol. 11. This we suppose is *E. læve pene nudum* of
Pet. Gr. Conc. 238 (see also *Dill.* in *R. Syn.* iii. 131).

(E. hyemale, *L.* Cyb. Br. iii. 307. Newman, 17. III. Hounslow
Heath; *Mart. App. P. C.* 71. Ibid., E. Forster; *B. G.* 413. Specimens of *E. limosum* and *E. palustre* have been probably taken for this.)

FILICES.*
POLYPODIUM, *Linn.*
839. P. vulgare, *L.*	*Polypody.*
Polypodium, Ger. em. (Blackst.).
Cyb. Br. iii. 252. Moore tt. 1–7. Curt. F. L. f. 1.

* The Ferns are naturally poorly represented in Middlesex, and in consequence of
being marketable, have become of late years very scarce in the vicinity of London; some
have been quite eradicated. We quote the nature-printed plates in the octavo edition of
Mr. Thomas Moore's *British Ferns,* where are figures and descriptions of very many
'varieties.'

Hedgebanks, old walls and trees; rather common. P. May—October.

I. About Harefield!; *Blackst. Fasc.* 80. Pinner, W. M. H.; *Melv.* 94.

II. Near Staines on Hampton Road. Tangley Park.

III. Wall on Teddington Road, Twickenham.

IV. Woods, Hampstead; *Irv. MSS.* Lane from Hendon to Finchley Road, 1845; *Morris* (*v. s.*).

VI. Bet. Edmonton and Winchmore Hill; *Mart. App. P. C.* 65. Highgate; *Cooper*, 104. Betstile Lane, 1842; *Herb. Hardw.* Bury St., Edmonton. Enfield. Winchmore Hill Wood.

VII. Chelsea (state with long pinnæ); *Herb. Dill.* Angler's Lane, Kentish Town, E. H. Button; *Coop. Supp.* 11.

First record: *Blackstone*, 1737. A variety with serrated pinnæ; lane bet. Norwood and Hanwell, 1851, S. O. Gray; *Herb. Mus. Brit.*

(P. Phegopteris, *L.* Cyb. Br. iii. 254. Moore t. 8. V. Norwood, three miles from Brentford, Mr. J. Beevis; *Francis*, 17: but in the fifth edition it reads, 'Norwood, Surrey, and near Brentford.' Not a likely locality.)

LASTREA, *Presl.*

840. L. Oreopteris, *Presl.*

Cyb. Br. iii. 265. *L. montana*, Moore tt. 30, 31.

Woods and heaths; very rare. P. June, July.

I. Harrow Weald Common, B. F. Westcott; *Melv.* 94. By the hedge or fence separating the farm buildings and yard of the Priory from Stanmore Heath, 1827–30; *Varenne.*

IV. On the edge of Hampstead Heath, J. Rayer; *B. G.* 413. Ibid., 1818; *Bennett* (*v. s.*). In Bishop's Wood, plentiful; *Irv. Lond. Fl.* 80.

First record: *Rayer*, 1805; last, *Newman*, 1855. Extinct?

841. L. Filix-mas, *Presl.* *Male Fern.*

Filix mas, Ger. (Blackst.). *Aspidium F.* (Coop.).

Cyb. Br. iii. 267. Moore tt. 32–38.

Woods, heaths, hedgebanks, &c.; common? P. June—August.

I. Harefield!; *Blackst. Fasc.* 29. Pinner Wood; *Melv.* 95. Harrow Weald Common. Stanmore Heath. By Elstree Reservoir. Potter's Bar. South Mims.

II. About Teddington, G. Francis; *Coop. Supp.* 12. Tangley Park. Hampton.

III. Whitton!; G. Francis; *Coop. Supp.* 12. Harrow Grove; *Melv.* 95. Heston. Bet. Worton and Twickenham, scarce.

IV. Harrow Park; *Melv.* 95. Bet. Mill Hill Station and Old Finchley Church; *J. B. George.* Bishop's and Turner's Woods.

VI. Whetstone. Edmonton. Enfield.

VII. Wall of kitchen garden, Ken Wood (a very small form).

First record: *Doody*, about 1705. A var. *pinnulis non serratis* was observed by Doody in Cane Wood; *R. Syn.* iii. 120. A remarkable form (*erosa?*) occurred (I.) in Moss Lane, Pinner.

842. L. spinulosa, *Presl.*

Cyb. Br. iii. 268. *L. cristata, var. β.* Moore t. 42.

Woods and hedges; very rare. P. June—September.

IV. On Hampstead Heath, and hedgebanks towards Hendon; *Pamplin.*

VI. Winchmore Hill Wood, 1859; *Church.*

I possess specimens from Middlesex; *Wats. Cyb. Br.* iii. 268. Middlesex; *Moore,* i. 220.

First record: *Pamplin,* about 1830. Abundant in Oxhey Woods; *Melv.* 95: just beyond our border in Herts.

843. L. dilatata, *Presl.*

Filix-mas ramosa pinn. dentatis, Ger. em. (Pet., Blackst.). *Filix tenuissime sectâ ex monte Ballon* (Budd.). *Polypodium cristatum* (Huds.).

Cyb. Br. iii. 270. Moore tt. 43–50.

Woods and heaths; rather rare. P. June—September.

I. In a bog near More Hall; Battleswell, Harefield; *Backst. Spec.* 22. Pinner Wood; Harrow Weald Common; *Melv.* 95. Stanmore Heath; *Herts. Fl. Supp.* 20.

III. Whitton Park, abundant. Outside Kneller Park Grounds. Hanworth Road near Hounslow.

IV. Woods near Hampstead!; *Budd. MSS., Pet. Midd.,* &c. Hampstead Heath!; *Huds.* i. 390. Stanmore; *Cole.* Bishop's Wood.

VI. Winchmore Hill Wood, 1859; *Church.* The Alders, near Whetstone, abundant.

VII. [By pond in the Vale of Health, Hampstead; *Irv. MSS.*] Ken Wood grounds.

First record: *Buddle,* about 1705. Mr. Moore records a *glandular* form from Hampstead (i. p. 248), and also finds there his vars. *pumila* (p. 212), *obtusa* (p. 236), and *tanacetifolia* (p. 242).

POLYSTICHUM, *Roth.*

844. P. aculeatum, *Roth.*

Filix-mas non ramosa pinn. latis auriculat. spinosis. (Pet., Blackst.). *Polypodium ac., L.* (Irv.).

Cyb. Br. iii. 261, 262. Moore tt. 16, 17.

Hedgebanks and woods; rather rare. P. June—September.

I. In the Old Park Wood at Harefield, plentifully; *Blackst. Fasc.* 29. Banks at Pinner near the Station Road; *Melv.* 94. Bet. Harefield and Ruislip. Moss Lane, Pinner. South Mims.

IV. Woods about Hampstead and Highgate; *Pet. Midd., Newton MSS.* About Hendon; *Irv. Lond. Fl.* 80. Renter's Lane, Hendon; *Davies.* Road bet. Finchley and Hendon, 1868!; *Newb.* Stanmore; *Varenne.* Bentley Priory.

V. Osterley Park; Lampton Lane; Sion Lane, J. Beevis; *Francis,* 28.

VI. Hornsey Lane; *Hill,* 535. Bet. Colney Hatch and Whetstone, 1845; *Herb. Hardw.*

VII. Ken Wood, Mr. Hunter ; *Park Hampst.* 30.

First record : *Newton,* about 1690. The form *lobatum (Asp. lobatum,* Sm.) is recorded from (IV.) Lanes bet. Hendon and Finchley ; *Pamplin.* (VI.) Lane from Bound's Green towards the New River near Colney Hatch, 1848 ; *Herb. Hardw.* And (V.) Norwood, S. F. Gray ; *Moore,* i. 132.

845. P. angulare, *Newm.*

Cyb. Br. iii. 263. Moore tt. 18–28.

Hedgebanks ; very rare. P. June—September.

I. Lane from Pinner to Harrow Weald ; *Melv.* 94.

V. Lane bet. Hanwell and Norwood, S. O. Gray ; *Herb. Mus. Brit.* Brentford ; *Moore* i. 140.

First record : *S. O. Gray,* 1851. Mr. Moore (i. 148) mentions the *var. biserratum* at (V.) Osterley Park ; Lampton Lane ; S. F. Gray.

ATHYRIUM, *Roth.*

846. A. Filix-fœmina, *Roth.* Lady Fern.

Filix-mas non ramosa pinn. angustis raris prof. dentatis (Pet.). *Lastrea F.* (Irv.).

Cyb. Br. iii. 273. Moore tt. 52–66.

Woods and shady hedgebanks ; very rare. P. June—August.

I. Inclosure near Harrow Weald Common, T. Wilkinson ; *Melv.* 95.

IV. Woods about Hampstead and Highgate ; *Pet. Midd.* Stanmore ; *Varenne.* Hampstead Heath ; *Irv. MSS.*

VI. Winchmore Hill Wood, 1859 ; *Church.* Middlesex ; *Moore,* ii. 17.

First record : *Petiver,* 1695. We have not met with this species. Extinct ?

ASPLENIUM, *Linn.*

847. A. Adiantum-nigrum, *L.* Black Spleenwort.

Adiantum nigrum vulgare, Park. (Blackst.).

Cyb. Br. iii. 280. Moore tt. 70–72.

Hedgebanks and old walls ; rare. P. June—September.

I. Lane from Harefield to Rickmansworth, abundantly ; *Blackst. Fasc.* 2.

II. On Teddington Church, G. Francis ; *Coop. Supp.* 12.

III. Headstone Lane, Mrs. G. W. Stuart ; *Melv.* 95.

IV. Hampstead, Hunter ; *Park Hampst.* 30. Bet. Burroughs and Brent Street, Hendon ; bet. Hendon and Finchley ; *Irv. MSS.* Hedgebank, Harrow Weald, T. Wilkinson ; *Melv.* 95.

VI. Near the Church, Finchley ; *Mart. App. P. C.* 66.

First record : *Blackstone,* 1737.

848. A. Trichomanes, *L.* Spleenwort.

Trichomanes mas, Ger. em. (Blackst.).

Cyb. Br. iii. 277. Moore tt. 75–76 bis.

Old walls; rare. P. May—September.

I. Orchard at Harefield Place, plentifully; about Breakspears; *Blackst. Fasc.* 100. For about twenty yards of garden wall, Hillingdon Place, 1868!; *Warren.*

II. River wall of Hampton Court Gardens; *Davies.*

IV. Hampstead, Wheeler; *Park Hampst.* 30. Under the wall or paling by the Pines, near the 'Spaniards'; *Irv. Lond. Fl.* 81. Sudbury, T. Wilkinson; *Hind.*

VII. [On the walls of Chelsea College next the Thames, Mr. Nicholls; *Blackst. Spec.* 99.]

First record: *Blackstone*, 1737. Perhaps not native in all its stations.

A. viride, *Huds.* Cyb. Br. iii. 276. Moore t. 77. VI. Wall near Arno's Grove, Southgate, O. E. Walker; probably an escape from cultivation; *Moore* ii. 115.

849. A. Ruta-muraria, *L.*   *Wall Rue.*
Adiantum album, Tab. (Pet.).   *Adiant. album sive Salvia Vitæ, Ger. em.* (Blackst.).
Cyb. Br. iii. 281.   Moore tt. 78, 79.

Old brickwork of walls and buildings; common. P. May—September.

I. On Pinner Church!; *Blackst. Fasc.* 2, *Melv.* 95. Tomb in Pinner Churchyard.

II. On Teddington Church, G. Francis; *Coop. Supp.* 12. River wall of Hampton Court Gardens!; *Newb.*

III. On a wall by high road beyond Brentford.

IV. On Willesden old Church, 1820-24; *Pamplin.* Whitchurch!; *Varenne.* About Hampstead; *Irv. MSS.* Sudbury, T. Wilkinson; *Hind.*

V. Near Hanwell; *Blackst. Spec.* 1. Old wall near Greenford Church; *Melv.* 95. Old brick tombs in Chiswick Churchyard, 1863.

VI. On Finchley Church; *Blackst. Spec.* 1, and *Irv. MSS.* Enfield. Bet. Highgate and the Archway Road; *Church.*

VII. [On an old stone conduit bet. Islington and Jack Straw's Castle; *Pet. Midd.*] Fulham Churchyard; *Blackst. Spec.* 1. Ibid. 1820; *Bennett* (*v. s.*). Kitchen garden wall, Ken Wood. East Heath Street, Hampstead. [Brickwork of a kitchen area in Bloomsbury Street, W. C., 1866; planted?]

First record: *Petiver*, 1695.

SCOLOPENDRIUM, *Sm.*

850. S. vulgare, *Symons.*   *Hart's Tongue.*
Phyllitis, Ger. (Blackst.).   *Asplenium Scol.* (Mart.).
Cyb. Br. iii. 283, 520.   Moore tt. 82-92 bis.

Shady places, ditches, and old walls; rare? P. August, September.

I. In the place orchard at Harefield; *Blackst. Spec.* 72, and *Fasc.* 76. On Rickmansworth Road; *Melv.* 95. Near Ruislip; *Herb. Harr.* A single plant on a farm building on the road from Harefield to Ruislip, 1866.

III. Abundant about Twickenham, Whitton, Hounslow, &c., J. E. Bowman; *Francis,* 44. Not so now.

IV. Hampstead; *Blackst. Spec.* 72. Wall below Jack Straw's Castle; *Irv. MSS.* Hedges bet. Stanmore and Harrow; *Melv.* 95. West side of Hampstead Heath; *Burnett,* 105.

V. About Brentford, J. E. Bowman; *Francis,* 44.

VI. Highgate; on Finchley Church; *Mart. App. P. C.* 66, 70. Bank opp. Hyde field bet. Edmonton and Bush Hill; *Forst. Midd.* [Wall of a melon frame, Warren Lodge, Edmonton.] [Brick wall near top of Muswell Hill, about 1856; *Church.*]

VII. Ken Wood; *Irv. MSS.*

First record: *Gerarde,* 1597? Seems to have been formerly much more frequent than is now the case. *Hermionitis sterilis;* (II.) on the stone walls of Hampton Court in the garden of Master Huggens, keeper of the said palace; *Ger.* 977, is perhaps a young state. A form *fol. multifidis,* lane at Tanner's End, Edmonton; *Forst. Midd.* and *Herb. Mus. Brit.*

[CETERACH, *Willd.*

851. C. officinarum, *Willd.*
Asplenium sive Ceterach off., J. B. (Blackst.).
Cyb. Br. iii. 248. Moore, t. 93.
Walls; very rare. P. June—September.

II. 'Commonly brought from Hampton Court by botanists in the early part of the present century. My specimens were obtained there by Dr. Hooper. Whether planted there or not, I cannot say;' *Varenne.* Tower of the church at Harmondsworth, 1850, fide John Lloyd; *Pamplin.*

VII. On the wall of Brook House, Hackney, Dr. Wilmer; *Blackst. Spec.* 5.

First record: *Wilmer,* 1746. Perhaps planted in Middlesex.]

BLECHNUM, *Linn.*

852. B. boreale, *Sw.* Hard Fern.
Lonchitis aspera (Ger., Blackst.). *Lonchitis altera Diosc.* (Johns.). *Osmunda Spicant* (Curt.).
Cyb. Br. iii. 248. Curt. F. L. f. 2 (drawn from a Middlesex specimen).
Heaths and woods; rare. P. July—September.

I. On Harefield Common, abundantly; *Blackst. Fasc.* 54. Harrow Weald Common!; *Melv.* 96.

III. Hounslow Heath, rare; *Hemsley.*

IV. Hampsteede Heath!; *Ger.* 979, *Johns. Eric.,* &c. Bishop's Wood.

VI. Wood near Highgate; *Herb. Dillenius.*

VII. Environs of Ken Wood; *Curt. F. L.* [Margin of pond in Vale of Health, Hampstead; *Irv. MSS.*]

First record: *Gerarde,* 1597.

PTERIS, *Linn.*

853. P. aquilina, *L.	Common Fern. Brakes. Bracken.* (*Female Fern* of old authors).

Filix fœmina, Ger. (Blackst.).

Cyb. Br. iii. 285.   Moore tt. 98, 99.

Heaths and open uncultivated places, also woods and walls; common. P.   June, July.

I. Harefield!; *Blackst. Fasc.* 28.   Elstree.   Bet. South Mims and Potter's Bar.

II. Towing path, Hampton Court!; *Newb.*   Bushey Park, abundant.

III. Hounslow Heath, Drilling Ground, &c., abundant.

IV. Harrow Weald; *Melv.* 96.   Bet. Finchley and Hendon!; *Newb.* North Heath, Hampstead.

VI. Winchmore Hill Wood.   Enfield.

VII. [Hyde Park, about 1825; *Pamplin.*]   South Heath, Hampstead.

First record: *Merrett,* 1666.   Mr. Moore mentions that he finds on Hampstead Heath his varieties *integerrima* and *crispa* (ii. 242). Young seedlings frequently grow in the chinks of brick walls; they then have a delicate appearance very unlike the usual form, and are barren.   The following refer to this state: *Filix ramosa sive fœmina minor nostras,* on old walls by the Thames side; *Budd. MSS.* [*Filix saxatilis crispa, Park.*; over the gates as you go into King Street, (Westminster); *Merrett,* 39.] [Walls of Savoy, Gray's Inn, Westminster, Royal Gardens, &c.; *Pet. Midd.*] *F. sax. ramosa maritima nostras,* on the wall of Chelsea physic garden; *Dill.* in *R. Syn.* iii. 125. Mr. Dan. Hanbury also informs us that it still usually appears in wet seasons in such places, and that in 1866 it was abundant on the walls of the Tower.

[OSMUNDA, *Linn.*

854. O. regalis, *L.	Flowering Fern. Royal Fern.*

Filix florida sive O. vulg. (Ger., Johns., Park., Ray, Pet.).

Cyb. Br. iii. 290.   Curt. F. L. f. 6.

Bogs on heaths; very rare.   P.   July—September.

IV. In the midst of a bog at the further end of Hampsteede Heath, at the bottom of a hill adjoining to a small cottage; *Ger.* 969, *Johns. Enum., Park. Theat.* 1039.   Of late it is all destroyed; *Johns. Ger.* 1131.   Towards the north side of Hampstead Heath; *Pet. Midd.* On the low part of the Heath, sparingly; *MS. note* (*Alchorne's*) *quoted in Phyt.* iii. 166.

VI. Beyond Highgate ; *Lob. Ill.* 68.

VII. Ken Wood, Mr. Hunter; *Park Hampst.* 30.

First record: *Gerarde,* 1597 ; last, *Hunter,* 1813.]

OPHIOGLOSSUM, *Linn.*

855. O. vulgatum, *L.*   *Adder's-Tongue.*

Ophioglossum (Ger., Dill., Blackst.).

Cyb. Br. iii. 292.   Moore t. 113.

Damp meadows ; rather common. P.   April—June.

I. Near Harefield, abundantly ; *Blackst. Fasc.* 66.   Cowley, in a meadow just beyond the Church ; *Blackst. Spec.* 62.   Ruislip Moor, abundant.

II. Medowes by Colbrooke ; *Ger.* 327.

III. Wike Farm, Sion Lane, Isleworth ; *Francis,* 55.

IV. Hampstead Heath ; *Huds.* i. 382.   Ibid., 1865 ; *Syme.*   Football Field, Harrow ; *Melv.* 96.

V. Osterley Park, near the ladder-stile, Mr. J. Beevis ; *Francis,* 55.   About Brentford ; *Hemsley.*   Greenford ; *Melv.* 96.   .Abundant in several meadows by the Brent near Perivale ; *Lees.*

VII. [Meadow near the preaching Spittle adjoining to London ; the Mantells by London ; *Ger.* 327.]   Near the Mill by Bow, Hackney Marsh, Mr. Newton ; *R. Syn.* iii. 128.   [St. Marylebone, 1818 ; *Herb. G. & R.*]   [Wet meadows bet. Parson's Green and the Thames; *Pamplin.*]   Lower part of the grounds of Branch Hill Lodge, Hampstead ; *Jewitt.*   Slope of the meadow adjoining Millfield Lane, by Ken Wood grounds, 1866.

First record : *Gerarde,* 1597.

MARSILEACEÆ.

PILULARIA, *Linn.*

856. P. globulifera, *L.*   *Pillwort.*

Graminifolia repens palustris vasc. granorum Pip. æmulis (Ray, Blackst.).

Cyb. Br. iii. 299.   Newman, 393.

Boggy places on heaths and commons ; rare.   P.   July, August.

I. Moorish ground one and a half miles before you come to Uxbridge ; *MS. note (Alchorne's) quoted in Phyt.* iii. 166.   On Hillingdon Common ; *Sm. Fl. Brit.* 1143.

III. On Hounslow Heath towards Hampton, Sam. Doody ; *R. Syn.* ii. 344.   Near the powder-mills, Mr. Watson ; *Blackst. Spec.* 28.

IV. Hampstead Heath ; *Huds.* i. 393.

VI. Enfield Chase, half a mile from the town towards Barnet ; *MS. note (Alchorne's) quoted in Phyt.* iii. 166.

First record: *Doody,* 1696.   Not seen for many years.

LYCOPODIACEÆ.

LYCOPODIUM, *Linn.*

857. L. clavatum, *L.*　　　*Club Moss.　Wolf's Claw.*
Muscus clavatus (Ger., Merr., Johns., &c.).
Cyb. Br. iii. 293.　Newman, 353.
Dry heaths ; very rare.　P.　July, August.
III. Hounslow Heath, abundantly ; *Huds.* i. 394.
IV. Hampstead Heath, nigh unto a little cottage ; *Ger.* 1373, *Johns. Eric.,*
　　&c.　Wilson's Folly, Hampstead Heath, 1865 ; *Grugeon* (*v. s.*).
　　Bishop's Wood ; *Budd. MSS.*
First record : *Gerarde,* 1597 ; also first as a British plant.

(L. annotinum, *L.*　Cyb. Br. iii. 293, 520.　III. Hounslow Heath ; *Forst.*
Midd.　No doubt an error.)

858. L. inundatum, *L.*
Muscus terrestris repens, clavis singularibus foliosis erectis (Ray).
Cyb. Br. iii. 294.　Newman, 369.
Damp places on heaths ; very rare.　P.　July, August.
I. Harefield Common, rather plentiful.
III. Hounslow Heath ; *Huds.* ii. 463.
IV. [On Hampstead Heath ; *R. Cat.* i. 215.　1828, *Pamplin.*　By the
　　pond on the Hendon side of the Heath, Golding Bird ; *Lond. Fl.*
　　83.　Ibid., Newman ; *Phyt.* i. 50.]
First record : *Ray,* 1670 ; also first as a British plant.

PHANEROGAMIA :
　　Dicotyledones　.　.　.　.　.　.　.　638
　　Monocotyledones .　.　.　.　.　.　.　196
　CRYPTOGAMIA (*Ferns, &c.*)　.　　.　.　.　25
　　　　　　　　　　　　　　　　　　　　　　859

SUMMARY.

I. The 859 species may be thus classified :—

Native 768
Introduced, and more or less completely naturalised, marked with an
asterisk (*) in the Flora 91
Total 859

There are also about 120 other species mentioned which have been met with as casuals, garden escapes, or accidental introductions ; but these are omitted in all the following summaries and comparisons.

II. No less than 58 species are probably extinct. This indicates that many localities have been destroyed. The same thing is illustrated by the very large proportion of 'rare' species; though this may be lessened by further investigation. The plants of Middlesex, when arranged in accordance with their relative frequency (see p. 2), give this result :—

Very common	129	
Common	126	377
Rather common	122	
Rather rare	132	
Rare	159	424
Very rare	133	
		801
Extinct		58
Total . . .		859

The probably extinct species are these :—

22.	Aquilegia vulgaris.	233.	Lythrum Hyssopifolia.
48.	Sisymbrium Irio.	250.	Herniaria glabra.
49.	Sophia.	251.	hirsuta.
60.	Cochlearia anglica.*	258.	Cotyledon Umbilicus.
84.	Dianthus prolifer.	267.	Cicuta virosa.
85.	Armeria.	281.	Bupleurum rotundifolium.
88.	Cucubalus bacciferus.	295.	Tordylium maximum.
119.	Althea officinalis.	319.	Galium anglicum.
128.	Hypericum Elodes.	325.	Centranthus Calcitrapa.
165.	Trigonella ornithopodioides.	393.	Arnoseris pusilla.
168.	Trifolium ochroleucum.	403.	Lactuca saligna.
171.	scabrum.	411.	Sonchus palustris.
173.	glomeratum.	420.	Xanthium strumarium.
188.	Lathyrus Aphaca.	426.	Campanula Rapunculus.
226.	Mespilus germanica.	440.	Gentiana Pneumonanthe.

* Still exists (see Addenda).

445.	Cuscuta europæa.
449.	Cynoglossum montanum.
469.	Lathræa squamaria.
482.	Scrophularia Ehrharti.
507.	Mentha pubescens.
512.	gentilis.
514.	Pulegium.
543.	Utricularia vulgaris.
544.	minor.
552.	Anagallis tenella.
571.	Atriplex marina.
574.	Rumex maritimus.
613.	Salix pentandra.
630.	Myrica Gale.

638.	Juniperus communis.
645.	Orchis purpurea.
646.	militaris.
647.	ustulata.
650.	pyramidalis.
651.	Gymnadenia conopsea.
654.	Ophrys apifera.
666.	Leucoium æstivum.
722.	Cyperus fuscus.
729.	Scirpus Tabernæmontani.
731.	cæspitosus.
793.	Avena pubescens.
851.	Ceterach officinarum.
854.	Osmunda regalis.

There is a likelihood of some of these being refound; but, on the other hand, there are some 'very rare' species which, in a few years' time, must be added to this list. Such are some Ferns and those semi-maritime plants which still maintain their ground in the Isle of Dogs, e.g. *Petroselinum segetum*, *Aster*, *Samolus*, *Scirpus maritimus*, *Sclerochloa distans*.

III. *Comparison with the 'Cybele Britannica.'*—In vol. iv. of the *Cybele* the number of species for the whole of Great Britain is reckoned as 1,425. Adopting the same specific limits, and fitting the foregoing enumeration to Mr. Watson's list (*Cyb. Br.* iv. pp. 175–233), we have in Middlesex 826 species.

In the list just referred to the plants are distinguished as respectively belonging to certain 'types of distribution' (explained in *Cyb. Br.* iv. pp. 499–519, and better in *Comp.* pp. 23–32); of these the following table shows the numbers for the whole of Great Britain, and for Middlesex:—

Types	Great Britain	Middlesex
British	532	465
English	409	300
Intermediate	37	4
Scottish	81	5
Highland	120	0
Germanic	127	44
Atlantic	70	3
Local or doubtful	49	5
Total . . .	1,425	826

At pp. 234–281 of the same vol. of the *Cybele* is a list of the 1,425 British species, arranged in a series according to their relative frequency, tested by the occurrence in, or absence from, the 38 sub-provinces into which Great Britain has been divided by Mr. Watson.

The list begins with the *commonest* plants, or those which occur in all the sub-provinces. We want in Middlesex the following:—

Of 120 British species, found in all (38) sub-provinces :
Anthyllis Vulneraria.

Of 54, found in 37 sub-provinces :
Carex ampullacea.

Of 44, found in 36 sub-provinces :
Narthecium ossifragum.   Comarum palustre,

Of 41, found in 35 sub-provinces :
Glaux maritima.

Of 46, found in 34 sub-provinces :
Gentiana campestris.   Armeria maritima.
Pinguicula vulgaris.   Lycopodium Selago.
Habenaria viridis.   Triglochin maritimum.

Of 39, found in 33 sub-provinces :
Geum rivale.   Schœnus nigricans.
Eriophorum vaginatum.

Of 36, found in 32 sub-provinces :
Arabis hirsuta.   Triticum junceum.
Spergularia marina.   Atriplex Babingtonii.

Of 51, found in 31 sub-provinces :
Silene maritima.   Ammophila arundinacea.
Rhyncospora alba.

Of 40, found in 30 sub-provinces :
Helianthemum vulgare.   Carex arenaria.
Gnaphalium dioicum.   Salicornia herbacea.
Epipactis latifolia ?   Thalictrum minus.
Honkenya peploides.

Of 38, found in 29 sub-provinces :
Vicia sylvatica.   Glyceria maritima.
Carex Œderi.   Eleocharis multicaulis?
Cakile maritima.   Cerastium tetrandrum.

Of 37, found in 28 sub-provinces :
Cystopteris fragilis.   Schoberia maritima.
Ranunculus Lingua.   Zostera marina.
Cochlearia officinalis.   Ruppia maritima.
Scirpus pauciflorus.   Taraxacum palustre.

Of 32, found in 27 sub-provinces :
Polypodium Phegopteris.   Silene anglica.
Carex curta.   Carex distans.
Geranium sanguineum.

348 SUMMARY.

Of 28, found in 26 sub-provinces:

Galeopsis versicolor.	Sagina maritima.
Vaccinium Oxycoccos.	Salsola Kali.
Lathyrus sylvestris.	Potamogeton heterophyllus.

Of 26, found in 25 sub-provinces:

Empetrum nigrum.	Festuca rubra.
Juncus maritimus.	Carex extensa.
Sedum anglicum.	Anagallis cærulea.
Drosera intermedia.	Potamogeton rufescens.
Asplenium marinum.	

Of 28, found in 24 sub-provinces:

Pyrola minor.	Eriophorum latifolium.

Of 27, found in 23 sub-provinces:

Polypodium Dryopteris.	Eryngium maritimum.
Rubus saxatilis.	Carex teretiuscula.

Of 31, found in 22 sub-provinces:

Viola lutea.	Phleum arenarium.
Glaucium luteum.	Beta maritima.
Lycopodium alpinum.	Cochlearia danica.
Allosorus crispus.	Hippocrepis comosa.
Juncus obtusifolius.	Convolvulus Soldanella.
Lepturus incurvatus.	

Of 26, found in 21 sub-provinces:

Myrrhis odorata.	Pyrethrum maritimum.
Listera cordata.	Artemisia maritima.
Vaccinium Vitis-Idæa.	Centaurea nigrescens.

Of 17, found in 20 sub-provinces:

Trollius europæus.	Statice Limonium.
Hymenophyllum Wilsoni.	Crambe maritima.
Drosera anglica.	Polygonum Raii.
Carduus Eriophorus.	Anchusa sempervirens.

Of the rarer species, found in less than 20 sub-provinces, it would take up too much space to particularise our wants.

IV. *Comparison with Adjacent Counties.**—The following is a list of those species—native and naturalised—recorded in one or more, but not in all, of the group of five counties of which Middlesex forms the centre. They are numbered consecutively; a ? implies doubt; a 0 signifies that the species is a mere casual. Many maritime species are restricted to Essex; the flora of Bucks is as yet imperfectly known.

* A very small portion of Kent may be considered 'adjacent' to the Isle of Dogs, from which it is separated by the Thames. To have included the flora of Kent in the following comparison, however, would have extended it very considerably, as that county possesses many additional species.

	Midd.	Herts	Essex	Surrey	Bucks
Thalictrum saxatile, 1	...	...	1	...	...
minus, 2	...	2	...	...	...
Anemone Pulsatilla, 3	...	3	3	...	...
apennina, 4	0	0	...	4	0
Ranunculus heterophyllus, 5	...	5	5	...	5?
confusus, 6	...	...	6	...	...
Lenormandi, 7	...	...	...	7	...
tripartitus, 8	...	...	...	8	...
Lingua, 9	...	9	...	...	9
hirsutus. 10	10	10	10	10	...
Helleborus fœtidus, 11	...	11	11	11	11
Delphinium Ajacis?, 12	...	12	12	12	0
Papaver hybridum, 13	13	13	13	13	...
somniferum, 14	0	14	0	14	0
Corydalis solida, 15	15	15	15	...	...
claviculata, 16	...	...	16	16	16
Fumaria micrantha, 17	17	17	17	17	...
parviflora. 18	...	18	18	18	...
Vaillantii, 19	...	19	19	19	...
Barbarea præcox, 20	20	20	20	20	...
Arabis hirsuta, 21	...	...	...	21	21
Cardamine impatiens, 22	...	...	...	22	...
Dentaria bulbifera, 23	23	23	...	23	23
Hesperis matronalis, 24	0	0	0	24	...
Sisymbrium Irio, 25	25	...	25	25	...
Sophia, 26	26	26	26	26	...
Erysimum cheiranthoides, 27	27	27	27	27	...
Diplotaxis tenuifolia, 28	28	28	28	28	...
Alyssum calycinum, 29	...	29	29	...	29
Cochlearia officinalis, 30	...	...	30	...	...
danica, 31	...	...	31	...	...
anglica, 32	32	...	32	...	...
Teesdalia nudicaulis, 33	33·	...	33	33	33
Iberis amara, 34	0	34	0	0	34
Lepidium Draba, 35	0	...	35	35	35
Smithii, 36	36	...	...	36	...
ruderale, 37	37	...	37	...	...
latifolium, 38	...	...	38	...	...
Senebiera didyma. 39	39	...	39	39	...
Isatis tinctoria, 40	...	...	40	40	40
Cakile maritima, 41	...	...	41	...	...
Crambe maritima, 42	...	...	42	...	...
Reseda suffruticulosa, 43	43	...	43	...	...
Helianthemum vulgare, 44	...	44	44	44	44
Viola palustris, 45	45	...	45	45	...
Drosera intermedia, 46	...	...	...	46	46
Polygala calcarea, 47	...	...	...	47	...
Frankenia lævis, 48	...	...	48	...	...
Elatine hexandra, 49	...	...	...	49	...
hydropiper, 50	...	...	...	50	...
Dianthus prolifer, 51	51	...	...	...	...

	Midd.	Herts	Essex	Surrey	Bucks
Dianthus deltoides, 52	52	...	52	...	...
Cucubalus bacciferus, 53 . . .	53	...	...	...	...
Silene anglica, 54	0	...	54	54	54
maritima, 55	...	...	55	...	...
Sagina maritima, 56	...	...	56	...	...
ciliata, 57	57	57	57	57	...
subulata, 58	58	...	...	58	58
Honckenya peploides, 59 . . .	...	...	59	...	...
Stellaria nemorum, 60	...	...	...	...	60
Cerastium tetrandrum, 61 . . .	...	...	61	...	...
Lepigonum salinum, 62 . . .	...	...	62	...	...
marinum, 63 . . .	...	...	63	...	...
Althæa officinalis, 64	64	...	64	...	...
Hypericum dubium, 65 . . .	65?	65	65	...	...
montanum, 66 . . .	...	66?	...	66	66
Elodes, 67 . . .	67	...	67	67	67
Geranium Phæum, 68	0	68	...	...	0
sanguineum, 69 . . .	...	...	69	...	0
pyrenaicum, 70 . . .	70	70	70	70	...
rotundifolium, 71 . . .	71	...	71	71	71
Erodium maritimum, 72 . . .	...	...	...	72	...
Linum angustifolium, 73 . . .	...	...	73	73	...
perenne, 74	...	...	74	...	...
Impatiens fulva, 75 . . .	75	...	...	75	75
parviflora, 76 . . .	76	...	...	76	...
Oxalis stricta, 77	77	77	77	77	...
Medicago falcata, 78 . . .	...	78	78	...	...
maculata, 79 . . .	79	...	79	79	...
denticulata, 80 . . .	80	...	80	...	...
Melilotus arvensis, 81	81	81	81	...	...
Trifolium medium, 82	82	82	82	82	...
ochroleucum, 83 . . .	83	83	83	...	...
scabrum, 84	84	...	84	84	...
maritimum, 85 . . .	...	...	85	...	...
glomeratum, 86 . . .	86	...	86	86	...
hybridum, 87 . . .	87	...	87	87	...
Anthyllis Vulneraria, 88 . . \ .	...	88	88	88	88
Astragalus Hypoglottis, 89 . . .	...	89	89	...	...
Glycyphyllus, 90 . . .	...	90	90	90	90
Vicia bithynica, 91	...	...	91	...	...
lathyroides, 92 . . .	92	...	...	92	92
gracilis, 93	?	93	93	93	93
Lathyrus hirsutus, 94 . . .	...	...	94	94	...
tuberosus, 95 . . .	...	...	95	...	...
sylvestris, 96 . . .	...	96	...	96	...
palustris, 97 . . .	...	...	...	97?	...
Hippocrepis comosa, 98 . . .	...	98	98	98	98
Prunus Padus, 99 . . .	99	99	99	99	...
Sanguisorba officinalis, 100 . . .	100	100	...	...	100
Poterium muricatum, 101 . . .	...	...	101	...	101
Agrimonia odorata, 102 . . .	...	...	102	102	...

	Midd.	Herts	Essex	Surrey	Bucks
Comarum palustre, 103	...	103	103	103	103
Geum rivale, 104	...	104	104	...	...
Rosa spinosissima, 105	105	...	105	105	...
Sabini, 106	106	...	...	...	...
tomentosa, 107	107	107	107	107	...
villosa, 108	108	...	...	108	108
systyla, 109	109	109	109	...	...
Pyrus Aria, 110	110	110	...	110	110
Lythrum Hyssopifolia, 111	111	111	...	111	...
Epilobium lanceolatum, 112	...	...	...	112	...
roseum, 113	113	113	113	113	...
obscurum, 114	114	?	114	114	114
tetragonum, 115	115	115	115	115	...
Chrysosplenium alternifolium, 116	...	...	116	116	...
Myriophyllum alterniflorum, 117	117	...	117	117	...
Herniaria glabra, 118	118	...	...	...	...
hirsuta, 119	119	...	...	...	...
Cotyledon Umbilicus, 120	120	...	...	120	...
Ribes nigrum, 121	121	121	121	121	...
Parnassia palustris, 122	122	122	122	...	122
Eryngium maritimum, 123	...	...	123	...	...
Cicuta virosa, 124	124	124	...	...	...
Apium graveolens, 125	125	125	125	125	...
Carum Carui, 126	126	126	126	...	126
Bunium Bulbocastanum, 127	...	127	...	...	...
Bupleurum tenuissimum, 128	128	...	128	128	...
falcatum, 129	...	...	129	129	...
Œnanthe pimpinelloides, 130	...	...	130	...	...
Lachenalii, 131	131	131	131	...	...
silaifolia, 132	132	...	...	132	132
Fœniculum vulgare, 133	133	133	133	133	...
Seseli Libanotis, 134	...	134	...	...	...
Peucedanum officinale, 135	...	...	135	...	...
palustre, 136	...	...	136	...	...
Archangelica officinalis, 137	137	...	...	?	...
Tordylium maximum, 138	138	...	...	...	138
Caucalis daucoides, 139	...	139	139	139	...
Myrrhis odorata, 140	...	...	...	140?	...
Smyrnium Olusatrum, 141	141	141	141	141	...
Coriandrum sativum, 142	...	...	...	142	...
Asperula Cynanchica, 143	...	143	143	143	143
Galium tricorne, 144	...	144	144	144	...
Vaillantii, 145	...	...	145	...	...
anglicum, 146	146	146	146	...	...
erectum, 147	...	...	147	147	...
Centranthus Calcitrapa, 148	148	...	148	...	...
Valerianella carinata, 149	149	...	149	149	...
auricula, 150	...	150	150	150	...
Petasites fragrans, 151	151	...	...	151	...
Aster Tripolium, 152	152	...	152	...	...
Erigeron acris, 153	0	153	153	153	153

	Midd.	Herts	Essex	Surrey	Bucks
Erigeron Canadensis, 154 . . .	154	...	154	154	...
Inula Helenium, 155	155	155	155	155	...
crithmoides, 156 . . .	...	...	156	...	...
Pulicaria vulgaris, 157 . . .	157	157	157	157	...
Diotis maritima, 158 . . .	...	...	158?	...	...
Artemisia Absinthium, 159 . .	159	159	159	159	...
maritima, 160 . . .	...	...	160	...	...
Filago apiculata, 161	...	161	161	161	...
spathulata, 162 . . .	162	?	162	162	162
gallica, 163	?	163	163	...	163
Antennaria dioica, 164 . .	...	164	...	?	...
Doronicum Pardalianchès, 165 .	165	165	0	...	...
plantagineum, 166 . .	...	...	166	...	...
Senecio viscosus, 167 . . .	0	...	167	...	...
campestris, 168 . . .	...	168	168	168	...
Serratula tinctoria, 169 . .	169	169	169	169	...
Centaurea Jacea, 170 . . .	170	...	...	...	...
solstitialis, 171 . .	...	171	171	171	...
Calcitrapa, 172 . .	172	172	172	172	...
Carduus tenuiflorus, 173 . .	173	...	173	...	...
eriophorus, 174 . .	...	174	174	...	...
Arnoseris pusilla, 175 . .	175	...	175	175	...
Hypochæris glabra, 176 . .	...	176	...	176	...
maculata, 177 . .	...	...	177	...	...
Lactuca saligna, 178 . .	178	...	178	...	...
scariola, 179 . .	179	...	179	179	...
Leontodon palustre, 180 . .	...	180	...	...	...
Sonchus palustris, 181 . . .	181	...	181	...	...
Crepis taraxacifolia, 182 . .	...	...	182	182	...
setosa, 183 . . .	...	183	183	183	183
fœtida, 184 . . .	...	...	184	184	184
biennis, 185 . .	0	185	185	185	...
Hieracium murorum, 186 . .	186	186	...	186	186
maculatum, 187 . .	187	...	...	187	...
tridentatum, 188 . .	188	...	188	188	...
umbellatum, 189 . .	189	189	189	189	...
Xanthium strumarium, 190 . .	190	...	...	190	190
Campanula latifolia, 191 . .	191	191	...	...	...
Vaccinium Vitis-Idæa, 192 . .	...	...	192	...	...
Oxycoccos, 193 . .	...	...	193	193	...
Pyrola minor, 194 . . .	...	194	...	194	194
media, 195 . . .	...	?	...	...	195?
Monotropa Hypopitys, 196 . .	...	196	196	196	196
Erythræa pulchella, 197 . .	...	197	...	197	...
Gentiana Pneumonanthe, 198 .	198	...	...	198	...
campestris, 199 . .	...	199	...	...	199
Villarsia nymphæoides, 200 . .	200	0	200	200	200
Convolvulus Soldanella, 201 . .	...	...	201	...	...
Cuscuta Epilinum, 202 . .	...	202	202	...	202
Hassiaca, 203 . .	...	...	203	...	...
Cynoglossum montanum, 204 . .	204	204	204	204	...

	Midd.	Herts	Essex	Surrey	Bucks
Symphytum tuberosum, 205 . . .	...	...	205	0	...
Myosotis sylvatica, 206 . . .	...	?	206	206?	...
Orobanche elatior, 207 . . .	...	207	207	207	...
cærulea, 208 . . .	...	208	...	...	...
Verbascum Thapsiforme, 209 . .	...	...	...	?	...
Lychnitis, 210 . . .	0	210	...	210	...
pulverulentum, 211 .	...	...	...	211	...
virgatum, 212 . . .	0	0	...	...	212
Linaria repens, 213 . . .	...	213	...	...	213
Scrophularia Ehrharti, 214 . .	214	...	...	...	...
vernalis, 215 . .	?	0	215	215	0
Melampyrum cristatum, 216. .	...	216	216	...	216
arvense, 217 . .	...	217	217	...	...
Mimulus luteus, 218 . . .	...	218	...	218	...
Mentha pubescens, 219 . .	219	...	...	...	...
piperita, 220 . . .	220	220	220	220	...
gracilis, 221 . . .	221	...	...	...	...
gentilis, 222 . . .	222	...	...	222	...
acutifolia, 223 . . .	...	...	...	223?	...
Calamintha Nepeta, 224 . .	...	224	224	224	224
Galeopsis ochroleuca, 225 . .	...	...	225	...	...
versicolor, 226 . .	0	226	226	0	...
Stachys ambigua, 227 . . .	227	...	227	227	...
Teucrium Botrys, 228 . . .	...	...	...	228	...
Leonurus Cardiaca, 229 . .	...	...	...	229	...
Ajuga Chamæpitys, 230 . .	...	230	230	230	...
Pinguicula vulgaris, 231 . .	...	231	...	...	231
Utricularia neglecta, 232 . .	...	...	232	...	...
intermedia, 233 . .	...	...	...	...	233?
minor, 234 . . .	234	...	...	234	...
Primula elatior, 235 . . .	...	...	235	...	...
Anagallis cærulea, 236 . .	...	236	236	236	...
Glaux maritima, 237 . . .	...	...	237	...	...
Samolus Valerandi, 238 . .	238	238	238	238	...
Statice Limonium, 239 . .	...	...	239	...	...
Bahusiensis, 240 . .	...	...	240	...	...
occidentalis, 241 . .	...	...	241	...	...
Armeria maritima, 242. . .	...	...	242	...	...
Plantago maritima, 243 . .	...	...	243	...	...
Amaranthus retroflexus, 244. .	244	244	...	244	...
Suæda fruticosa, 245 . . .	...	...	245	...	...
maritima, 246 . .	...	...	246	...	...
Salsola Kali, 247 . . .	...	...	247	...	...
Chenopodium urbicum, 248 . .	248	248	248	248	...
ficifolium, 249. .	249	...	249	249	249
murale, 250 . .	250	250	250	250	...
hybridum, 251. .	251	251	251	251	...
botryoides, 252 . .	...	...	252	...	...
glaucum, 253 . .	253	...	253	253	...
Beta maritima, 254 . . .	...	...	254	...	...
Salicornia herbacea, 255 . .	...	...	255	...	...

A A

	Midd.	Herts	Essex	Surrey	Bucks
Salicornia radicans, 256 . . .	...	...	256	...	...
Atriplex littoralis, 257	...	...	257	...	...
marina, 258	258	...	258	...	...
deltoidea, 259	259	...	259	259	259
Babingtonii, 260 . . .	?	...	260	...	...
arenaria, 261	...	...	261	...	...
Obione portulacoides, 262 . . .	...	...	262	...	...
Rumex maritimus, 263	263	...	263	263	263
palustris, 264	264	264	264	264	...
nemorosus, 265	265	265	265	265	...
Polygonum nodosum, 266 . . .	266	266	266	266	...
mite, 267	267	...	267	267	...
minus, 268	268	268	268	268	...
dumetorum, 269	...	269	...	269	...
Daphne Mezereum, 270 . . .	...	270	270	270	270
Thesium humifusum, 271 . . .	...	271	271	271	271
Asarum europæum, 272 . . .	...	272	...	...	0
Euphorbia platyphylla, 273 . . .	273	273	273	273	...
Cyparissias, 274 . . .	...	...	274	...	...
Paralias, 275 . . .	...	...	275	...	...
Buxus sempervirens, 276 . . .	0	...	...	276	276
Mercurialis annua, 277 . . .	277	277	277	277	...
Callitriche pedunculata, 278 . . .	278	278	278	278	...
Salix pentandra, 279 . . .	279	279	279	279	...
fragilis, 280	280	280	280	280	...
undulata, 281	281	...	281	281	...
triandra, 282	282	282	282	282	...
purpurea, 283	...	...	283	...	...
Lambertiana, &c, 284 . .	284	284	284	284	...
Helix, 285	285	285	285	285	...
stipularis, 286 . . .	286	...	286	...	...
Smithiana, 287 . . .	287	287	287	287	...
rubra, 288	288	...	288	288	...
ambigua, 289 . . .	...	...	289	...	...
Myrica Gale, 290	290	...	...	290	290
Stratiotes aloides, 291 . . .	...	0	291	...	...
Orchis purpurea, 292 . . .	292	...	...	292	292
militaris, 293 . . .	293	293	...	...	293
Simia, 294	...	...	...	294?	...
ustulata, 295 . . .	295	295	295	295	...
latifolia, 296 . . .	?	...	296	296?	...
hircina, 297 . . .	...	...	...	297?	...
Habenaria viridis, 298 . . .	...	298	298	298	298
Aceras anthropophora, 299 . . .	...	299	299	299	299
Ophrys arachnites, 300 . . .	...	...	...	300	...
aranifera, 301 . . .	...	...	301	301	...
Herminium Monorchis, 302 . . .	...	302	302	302	302
Epipactis latifolia, 303 . . .	?	303	303	303	303
Cephalanthera grandiflora, 304	...	304	304	304	304
ensifolia, 305 . . .	...	305?	...	305	...
Crocus vernus, 306 . . .	306	306	306	306	...

	Midd.	Herts	Essex	Surrey	Bucks
Leucoium æstivum, 307	307	...	307	...	307
Galanthus nivalis, 308 .	...	308	0	308	...
Asparagus officinalis, 309	...	...	309	...	...
Maianthemum bifolium, 310 .	310	...	...	...	...
Tulipa sylvestris, 311 .	311	0	311	311	0
Lilium Martagon, 312 .	0	0	312	312	...
Scilla autumnalis, 313 .	313	...	...	313	...
Allium oleraceum, 314 .	314	314	314	...	...
Muscari racemosum ?, 315	0	315	0	...	...
Narthecium ossifragum, 316 .	...	...	...	316	316
Juncus maritimus, 317 .	...	...	317	...	...
glaucus, 318 .	318	318	318	318	...
diffusus, 319 .	...	319	319	319	...
obtusiflorus, 320	...	320	320	320	...
supinus, 321 .	321	321	321	321	...
compressus 322	322	...	322	322	...
Gerardi, 323 .	...	...	323	...	...
Triglochin maritimum, 324 .	...	...	324	...	...
Typha angustifolia, 325	325	325	325	325	...
Sparganium minimum, 326 .	...	326	326	326	...
Wolffia arrhiza, 327 .	327	...	327	327	...
Potamogeton polygonifolius, 328 .	328	328	328	328	...
plantagineus, 329	...	329	...	...	...
rufescens, 330 .	...	330	330	330	...
heterophyllus, 331 .	...	331	331	...	...
prælongus, 332 .	...	...	332	332	...
zosterifolius, 333 .	333	...	333	333	...
acutifolius, 334 .	...	334	...	334	...
obtusifolius, 335 .	335	335	335	335	...
compressus, 336 .	336	...	336	336	...
pectinatus, 337 .	337	337	337	337	...
densus, 338 .	338	338	338	338	...
Ruppia maritima, 339 .	...	...	339	...	...
rostellata, 340 .	...	...	340	...	...
Zostera marina, 341 .	...	...	341	...	...
Cyperus fuscus, 342 .	342	...	...	342	...
Schœnus nigricans, 343 .	...	343	343?	343	...
Rhyncospora alba, 344 .	...	...	...	344	344
Eleocharis multicaulis, 345 .	?	345	345	345	...
Scirpus maritimus, 346 .	846	...	346	346	...
sylvaticus, 347 .	347	347	347	347	...
triqueter, 348 .	348	...	...	348	...
carinatus, 349 .	349	...	...	349	...
Tabernæmontani, 350 .	350	...	350	350	...
cæspitosus, 351 .	351	...	351	351	...
pauciflorus, 352 .	...	352	352	...	352
setaceus, 353 .	353	353	353	353	...
Blysmus compressus, 354 .	354?	354	354	354	...
Eriophorum vaginatum, 355 .	...	...	...	355	...
latifolium, 356 .	...	356	356	356	...
gracile, 357 .	...	...	...	357	...

A A 2

	Midd.	Herts	Essex	Surrey	Bucks
Carex dioica, 358 .	...	358	...	...	...
disticha, 359	359	359	359	359	...
arenaria, 360	...	...	360	...	...
divisa, 361 .	361	...	361	...	...
divulsa, 362	362	362	362	362	...
teretiuscula, 363 .	...	363	363	363	...
paniculata, 364 .	364	364	364	364	...
Bönninghauseniana, 365	...	365	...	365	...
axillaris, 366	366	366	366	366	...
elongata, 367	...	...	367	367	...
curta, 368 .	...	...	...	368	...
stricta, 369 .	...	369	369	...	...
vulgaris, 370	370	370	370	370	...
pallescens, 371	371	371	371	371	...
panicea, 372	372	372	372	372	...
strigosa, 373	373	373	373	373	...
pendula, 374	374	374	374	374	...
præcox, 375	375	375	375	375	...
pilulifera, 376	376	376	376	376	...
flava, 377 .	377	377	377	377	...
Œderi, 378 .	...	378?	378	378	...
extensa, 379	...	...	379	...	...
distans, 380	...	...	380	...	...
binervis, 381	381	381	381	381	...
lævigata, 382	382	382	382	382	...
depauperata, 383	...	...	...	383	...
pseudo-Cyperus, 384 .	384	...	384	384	...
ampullacea, 385 .	...	385	385	385	385
vesicaria, 386	386	386	386	386	...
Digitaria sanguinalis, 387	387	387	...	387	...
humifusa, 388	...	...	...	388	...
Echinochloa Crus-galli, 389 .	389	0	0	389	...
Setaria viridis, 390	390	0	390	390	...
glauca, 391	391	0	...	391	...
Spartina stricta, 392	...	...	392	...	...
Phleum Böhmeri, 393 .	...	393	393	...	...
arenarium, 394	...	...	394	...	...
Alopecurus fulvus, 395 .	395	395	395	395	...
bulbosus, 396	...	...	396	...	...
Leersia oryzoides, 397 .	...	...	...	397	...
Psamma arenaria, 398 .	...	...	398	...	...
Calamagrostis lanceolata, 399	...	399	399	399	...
Epigeios, 400	400	400	400	400	...
Apera Spica-venti, 401 .	401	401	401	401	...
Agrostis setacea, 402	...	...	...	402	...
canina, 403	403	403	403	403	...
Polypogon monspeliensis, 404	...	...	404	...	...
littoralis, 405	...	...	405	...	...
Gastridium lendigerum, 406 .	...	406	406	406	...
Aira caryophyllea, 407 .	407	407	407	407	...
præcox, 408 .	408	408	408	408	...

	Midd.	Herts	Essex	Surrey	Bucks
Avena pratensis, 409	409	409	409	409	...
strigosa, 410	...	0	0	410	...
Poa compressa, 411	411	411	411	411	...
Glyceria plicata, 412	412	412	412	412	...
Sclerochloa maritima, 413	...	...	413	...	...
distans, 414	414	...	414	...	...
Borreri, 415	...	...	415	...	...
procumbens, 416	...	...	416	...	...
loliacea, 417	...	...	417	...	...
Catabrosa aquatica, 418	418	418	418	418	...
Festuca uniglumis, 419	...	...	419	...	...
duriuscula, 420	420	420	420	...	...
rubra, 421	...	...	421	...	...
gigantea, 422	422	422	422	422	...
Bromus diandrus, 423	...	...	...	423	...
Serrafalcus secalinus, 424	424	424	424	424	...
racemosus, 425	425	425	425	425	...
mollis, 426	426	426	426	426	...
Brachypodium pinnatum, 427	...	427	427	427	...
Triticum pungens, 428	...	...	428	...	...
acutum, 429	...	...	429	...	...
junceum, 430	...	...	430	...	...
Elymus arenarius, 431	...	...	431	...	...
Hordeum sylvaticum, 432	...	432	432	...	432
maritimum, 433	...	...	433	...	...
Lepturus incurvatus, 434	...	...	434	...	...
Equisetum hyemale, 435	...	...	...	435	...
Polypodium Robertianum, 436	...	...	...	...	436
Lastrea Thelypteris, 437	...	...	437	437	...
Oreopteris, 438	438	438	438	438	...
Cystopteris fragilis, 439	...	...	...	?	...
Osmunda regalis, 440	440	...	440	440	440
Botrychium Lunaria, 441	...	...	441	441	441
Pilularia globulifera, 442	442	442	442	442	...
Lycopodium Selago, 443	...	...	...	443	443
inundatum, 444	444	...	444	444	444

Of the above species, 244 are absent from Middlesex, whilst occurring in one or more of the surrounding counties: of these, however, only 19 are found in all four adjacent counties, and most of them are chalk plants.

Of the 200 Middlesex species which are wanting in one or more of the surrounding counties, only 11 are absent from all four.

V. *Comparison with M. Thurmann's 'Essai de Phytostatique'* (1849).—
M. Thurmann, from an examination of the Jura and adjacent districts of France and Switzerland, arrived at the conclusion that the distribution of plants is influenced chiefly by the mechanical properties of the subjacent rocks, and but very slightly by their chemical composition.

He divides rocks into two classes: '*dysgéogènes*' and '*eugéogènes*,' possessing opposite physical properties. Dysgeogenous rocks, though very permeable to fluids, ar.' but slightly absorbent, and form a dry shallow soil. The Jura has mostly these characters, which are also found in the Permian, Oölite, and Cretaceous districts of England; in Middlesex they only appear in the very small portions of chalk in the N. and N.W. These rocks are usually calcareous, and rain-water, in percolating through them, becomes hard in consequence of the solvent action of the carbonic dioxide (carbonic acid) which it contains. Eugeogenous rocks, on the contrary, are very absorbent, and but slightly permeable; they readily disintegrate under atmospheric influences, and form a deep open soil with moisture at no great depth from the surface. Tracts in the Vosges and valley of the Saône exhibit these conditions, and they are abundantly shown in England, nearly the whole of Middlesex presenting them in a very characteristic manner. In chemical nature they are either silicious or aluminous, and their soluble constituents having generally disappeared in the process of disintegration, rain-water has little solvent action upon them.

M. Thurmann has classified soils derived from eugeogenous rocks into '*psammiques*' and '*péliques*;' the latter differ from the former in having their particles in a finer state of subdivision, the rocks from which they are produced being softer, and therefore yielding more completely to the disintegrating forces. These differences, however, result in reality from chemical composition, the former including the *silicious* gravels and sands, the latter the *aluminous* pure loams and clays.*

The characteristic plants of dysgeogenous tracts M. Thurmann calls '*xérophiles*;' those of eugeogenous districts being termed '*hygrophiles*.' He gives two contrasting lists, each containing 50 species typical respectively of the floras of the Jura (xerophilous) and of the Vosges (hygrophilous). A comparison of these lists with the flora of Great Britain demonstrates the predominance of the hygrophilous character in the latter, and this character is still more marked in the flora of Middlesex.

Of the list of 50 typical 'xérophiles,' Great Britain possesses 23 (= 46 per cent.), whilst it has as many as 41 (= 82 per cent.) of the 50 'hygrophiles.'

Middlesex has only 8 of the first list (= 16 per cent.):—

Orchis pyramidalis.	Mercurialis perennis.
Calamintha officinalis.	Fagus sylvatica.
Daphne Laureola.	Rosa rubiginosa.
Euphorbia amygdaloides.	Orchis militaris.

whilst it has 33 (= 66 per cent.) of the 'hygrophiles':—

Lathyrus macrorrhizus.	Aira flexuosa.
Sarothamnus scoparius.	Ononis spinosa.
Alnus glutinosa.	Hypericum pulchrum.

---

* Whilst the physical characters of eugeogenous rocks almost exclusively determine the character of the vegetation, this is probably not absolutely the case with dysgeogenous; the suitability of the soil they form to the growth of some at least of the plants peculiar to them, is to be attributed partly to its chemical condition.

Centaurea Calcitrapa.	Hieracium boreale.
Verbascum Blattaria.	Jasione montana.
Filago minima.	Stellaria Holostea.
Alopecurus pratensis.	Trifolium fragiferum.
Rumex acetosella.	Luzula multiflora.
Nardus stricta.	Aira cæspitosa.
Hypericum humifusum.	Triodia decumbens.
Senecio aquaticus.	Montia fontana.
Lotus major.	Pulicaria vulgaris.
Juncus squarrosus.	Senecio sylvaticus.
Galium saxatile.	Lepigonum rubrum.
Betula alba.	Vaccinium Myrtillus.
Quercus sessiliflora.	Digitalis purpurea.
Calluna vulgaris.	

M. Thurmann more fully elucidates his views by giving other lists, containing 149 xerophilous species. Of these 67 (= 44·96 per cent.) are British plants.

In Middlesex only 29 species (= 19·46 per cent.) are found :—

Aquilegia vulgaris.	Vinca minor.
Inula Conyza.	Verbascum Thapsus.
Orchis purpurea.	Spiranthes autumnalis.
O. ustulata.	Crocus vernus.
Gymnadenia conopsea.	Carduus acaulis.
Ophrys apifera.	Herminium monorchis.
O. muscifera.	Ruscus aculeatus.
Cerastium arvense.	Festuca duriuscula.
Hypericum hirsutum.	Melica uniflora.
Rosa spinosissima.	Ceterach officinarum.
Carlina vulgaris.	Scolopendrium vulgare.

And the eight species in the list * on the preceding page.

The 'hygrophiles' are also shown in more detail by three lists in the 'Essai.' The first is of 'hygrophiles' in general, and contains 134 names ; the other two are contrasting—the one contains typical *sand-loving* 'hygrophiles' (82), and the other specially *clay-loving* 'hygrophiles' (39), and these correspond to a great extent to the two classes of eugeogenous soils called 'psammiques' and 'péliques' respectively. In all, the three lists contain 255 names. The British Isles contain of the first list 91, of the second 59, and of the third 27 species—in all 177 (= 69 per cent.).

Middlesex possesses of the plants in the first list 59, in the second 44, and the third 24—in all 127 (= 49·80 per cent.).

It is not necessary to mention those in the first, a list of ordinary 'hygrophiles' having been already given.

---

* Except *Orchis militaris*, which is omitted by Thurmann in this list.

The second list, '*psammophiles*,' contains the following Middlesex species:—

Teesdalia nudicaulis.
Erysimum cheiranthoides.
Ornithopus perpusillus.
Lepigonum rubrum.
Radiola millegrana.
Montia fontana.
Galium saxatile.
Filago minima.
Senecio sylvaticus.
Jasione montana.
Juncus squarrosus.
Echinochloa Crus-galli.
Aira caryophyllea.
Phragmites communis.
Festuca Myurus.
Aira præcox.
Myosurus minimus.
Sisymbrium Sophia.
Lepidium ruderale.
Dianthus prolifer.
Ulex europæus.
Sarothamnus scoparius.

Ononis spinosa.
Melilotus vulgaris.
Herniaria glabra.
H. hirsuta.
Saxifraga granulata.
Scandix Pecten.
Centaurea Calcitrapa.
Arnoseris pusilla.
Thrincia hirta.
Xanthium Strumarium.
Myosotis versicolor.
Verbascum Blattaria.
Digitalis purpurea.
Antirrhinum Orontium.
Rumex acetosella.
Amaranthus retroflexus.
Salix viminalis.
Populus nigra.
Betula alba.
Cyperus fuscus.
Scirpus setaceus.
Osmunda regalis.

In the list of '*pélophiles*' we find the following in Middlesex:—

Stellaria Holostea.
Trifolium fragiferum.
Lysimachia nemorum.
Alisma Plantago.
Lotus major.
Senecio aquaticus.
Alnus glutinosa.
Ranunculus Flammula.
Alsine tenuifolia.
Hypericum humifusum.
H. pulchrum.
Lythrum Hyssopifolia.

Lonicera Periclymenum.
Tussilago Farfara.
Bidens cernua.
Pulicaria vulgaris.
Specularia hybrida.
Veronica scutellata.
Chlora perfoliata.
Centunculus minimus.
Salix aurita.
Luzula multiflora.
Eleocharis acicularis.
Triodia decumbens.

Of the lists of Thurmann's '*psammophiles*' and '*pélophiles*' respectively :

	Ps.	Pél.
Great Britain has, per cent.	71·95	69·23
Middlesex ,,	53·65	61·53

showing that our county is a district especially rich in pelophilous species, as indeed would be expected from its geological features.

To sum up: of M. Thurmann's three categories of Mid-Europe species, there are in Middlesex:—

			per cent.
Xerophilous	.	.	19·20 (29 out of 149)
Psammophilous	.	.	53·65 (44 out of 82)
Pelophilous	.	.	61·53 (24 out of 39)

On examining the lists of Middlesex species which we have given, it will, however, be seen that their relationship to the subjacent rocks here is by no means in every case similar to that which obtains in Central Europe.

Let us first look at the lists of *xerophilous* species. Out of the 29 species there enumerated not more than 9, i. e. the first 2 of the former list and the first 7 of the latter, are with us restricted to the calcareous district; the remaining 20 are by no means confined to the chalk of the county, though a few of them show more or less partiality for it. We have, however, some other species, whose range seems confined to our chalk districts:—

Dentaria bulbifera.	Specularia hybrida.
Reseda lutea.	Gentiana Amarella.
Viola hirta.	Chlora perfoliata?
Onobrychis sativa.	Neottia Nidus-avis.
Bupleurum rotundifolium.	Bromus erectus.
Campanula Trachelium?	Triticum caninum.

These, which all occur in the Jura, are not considered xerophilous in Thurmann's 'Essai.'

In the list of *psammophilous* species also the majority do not show a decided bias for any soil. The first 16 names, however, are well-marked sand and gravel species in Middlesex, and to these may be added as equally characteristic (though not so in Central Europe) the following 9 (making with the others 25):—

Lepidium Smithii.	Trifolium striatum.
Viola canina.	Erigeron canadensis.
Sagina ciliata.	Plantago Coronopus.
Mœnchia erecta.	Carex panicea.
Ulex nanus.	

Analogous remarks may be made with regard to the third category, that of *pelophilous* species. Only the first 7 names in the list are well-marked clay plants in our county, the remainder being more nearly sand-loving species here, and two (*Specularia* and *Chlora*) confined to the chalk. The following species may be added to the 7 to make up 25 characteristic '*pélophiles*' in Middlesex for comparison with the '*psammophiles*' above:—

Hypericum hirsutum.	Genista tinctoria.
Rhamnus Frangula.	Lathyrus macrorrhizus.

Sanguisorba officinalis.
Pyrus torminalis.
Sanicula europæa.
Viburnum opulus.
Senecio erucifolius.
Lactuca virosa.
Primula vulgaris.

Orchis Morio.
Epipactis purpurata.
Fritillaria meleagris.
Luzula sylvatica.
Carex pendula.
C. sylvatica.
Ophioglossum vulgatum.

A SKETCH

OF THE

PROGRESS OF BOTANICAL INVESTIGATION IN MIDDLESEX.

WITH BIOGRAPHICAL NOTICES.

In this chapter those botanists who have published on the Flora of Middlesex are mentioned in chronological order, the nature and extent of their investigations are stated, and a short account of their writings given.

In the case of some of those who died previously to the commencement of the present century, particulars of their lives are added : *Turner, Johnson, Plukenet, Doody, Petiver, Buddle, Blackstone* and *Curtis* are so treated. In this part of the subject we have drawn largely on the collection of letters and other MSS. formerly belonging to Sir H. Sloane, Petiver and others, now in the British Museum. These are quoted as *Sloane MSS.* ; there is an excellent catalogue of them by Ayscough, printed in 1782. Besides this inexhaustible mine of information, we have derived much assistance from the following books among others : *—

> *Philosophical Letters between the late learned Mr. Ray and several of his ingenious Correspondents.* Published by W. Derham. Lond. 1718.†
> *Historical and Biographical Sketches of the Progress of Botany in England.* By Richard Pulteney, M.D., F.R.S. 2 vols. Lond. 1790.
> *Biographical Notices of various Botanists in Rees' Cyclopædia.* [By Sir J. E. Smith.] Lond. 1819-20 (but many of the volumes really published several years before).
> *Literary Illustrations of the 18th Century.* By John Nichols, F.S.A. Vol. i. 1817, and vol. iv. 1822.
> *Correspondence of Linnæus and other Naturalists.* Edited by Sir J. E. Smith. 2 vols. Lond. 1821.
> *Extracts from the Literary and Scientific Correspondence of Richard Richardson, M.D.; F.R.S.* [Edited by Dawson Turner.] Yarmouth, 1835.

---

* As the following sketch does not profess to give a full history of the progress of British Botany, so neither are the biographical notices intended as complete lives. Some new—i.e. unpublished—matter will, however, be found here, which it is hoped may be of use to future biographers.

† In the *Correspondence of John Ray,* published in 1848 by the Ray Society, are some additional letters to Petiver and Sir Hans Sloane, printed from the originals in the Sloane MSS.

Of any observations in Natural History made before the introduction of printing into this country, little indeed has been preserved; if there were in England any persons occupied in original investigation, all record of their labours so far as we can discover, has been lost.

But it indeed appears that very little interest in Science existed in the end of the fifteenth and beginning of the sixteenth century, for though books were printed in England in Edward IV.'s reign (1474, if not earlier), it was not till 1516 that any work on Natural History issued from the press. This was a botanical book in small folio, *The Grete Herball*, illustrated with rough block wood-engravings, and printed at Southwark by Peter Treveris. It appears to be a translation of a French book, the *Grand Herbier*, itself only translated from the celebrated *Ortus Sanitatis*, long the mine from which the early authors dug material for their semi-botanical treatises.

There is little enough of anything like science in this old herbal, yet nothing better was produced in England for many years. Several editions * were indeed published, but without improvements; that of 1526 is the best known. No attempt is made to distinguish between native and foreign species, nor are any localities given.

A taste for botanical studies, however, soon began to appear, for we are told by a contemporary writer that ' there have bene in England & there are now also certain learned men which have as muche knowledge in herbes, yea & more than diverse Italianes & Germanes, whyche have set forthe in prynte herballes & bokes of simples. I mean of Doctor *Clement*, Doctor *Mendy* & Doctor *Owen*, Doctor *Wotton* & Master *Falconer*. Yet none of these set forth any thyng.' This extract is from the ' Prologe' to the first part of *A New Herball*, which was printed in London in 1551; the author was WILLIAM TURNER, who is well known as the ' Father of British Botany.'

This eminent man was born at Morpeth, Northumberland, about the beginning of the sixteenth century, and is supposed to have been the son of a tanner of that place; the exact date of his birth is not known. He was educated at Cambridge at Lord Wentworth's expense, and was elected Fellow of Pembroke College in 1530.† We know from his own statement (Preface to *Herball* of 1568) that he held that fellowship in 1538, in which year he published his first botanical work. It is called *Libellus de Re Herbaria novus*, and was printed in London in the form of a quarto tract of twenty pages. Although he says in the preface that he was at this time very young, the contents show that he had already studied botany practically, for he notices the localities of several plants in his native county of Northumberland. With regard to these, it is interesting to remark that they are the earliest printed records of the kind in England.

Turner about this time became Latimer's disciple, embraced with en-

---

* We have seen copies dated 1526, 1529, 1539 (without figures). One is said to have appeared so late as 1561.

† B.A. 1529–30, junior treasurer of his college 1532, commenced M.A. 1533, senior treasurer 1538.—*Cooper's Athenæ Cantabrigienses*, i. p. 256.

thusiasm the principles of the Reformation, and published on the subject; he left Cambridge about 1540, and travelled about England preaching. In consequence probably of refusal to subscribe to ' the Six Articles ' he was imprisoned and remained so for a considerable time ; on his release, about 1542, he was obliged to leave England. For some years he lived in various parts of Germany and Italy, and seems to have given very much of his time to botany. He did not return to England till the death of Henry VIII., and during his residence abroad is said to have published at Cologne, in 1544, *Historia de Naturis Herbarum Scholiis et Notis vallata.** He also took the degree of M.D. at Ferrara, and became intimate with the great naturalist Gesner, with Lucas Gynus, the ' reder of Dioscorides in Bonony,' and the works if not the persons of the botanists Fuchs, Brunfels, &c.

On his return to England in 1547, he seems to have been high in favour. The new king made him a Prebend † of York and Canon of Windsor,‡ the Duke of Somerset, Lord Protector, appointed him his physician, and the University of Oxford granted him the degree of M.D. on his appointment. In 1548 he published *The Names of Herbes in Greke, Latin, Englishe, Duche and Frenche, &c.* This was printed in London, and the preface is dated from Sion House, then the residence of the Lord Protector, to whom the book is dedicated. In it the author says that he had finished a herbal in Latin about two years before, but was advised not to publish it till he had seen more of the indigenous plants of England, especially those growing in the west country. The book contains 126 pages, and the names are arranged alphabetically. Several localities are given, sixteen of which are in Middlesex, chiefly about Sion ; they are the foundation of our county Flora. As if he had not enough preferment, in 1550 he was made Dean of Wells, though he was not ordained a priest till December 21, 1552 by Ridley.§ He was also a member of the House of Commons.

It was at this time that he published the *first part* of his *Herball*, already mentioned, which is dedicated to his patron. It represents a great advance on all previous attempts, and is evidently the result of original research ; the figures though rough are often characteristic. The *second part* did not appear till 1562 ; and in the interval between them Edward VI. had died, and Mary succeeding, Turner (in 1553) had again to seek refuge abroad, perhaps in Denmark (v. *Johns. Ger.* 94), from his persecutors. He returned in 1558 on the accession of Elizabeth, who restored him all his church preferments, and in 1563 presented him to the rectory of Wedmore, Somersetshire. The *second part*, however, was printed at Cologne, as was also the *third part*, which did not appear till 1566, though the Preface is dated ' At Welles, 1564, the 24 day of June.'

The latter years of his life appear to have been passed at Wells and in

* On this book see list of Turner's works below.
† Prebendarius de Bottevant ; *Tanner's Bibliotheca*, where is much valuable matter relating to Turner.
‡ This is doubtful (v. *Ath. Cantab.*).
§ He had the college title for orders March 20, 1536–37, and was probably made a deacon then (*Ath. Cantab.*).

London, where he had a house 'in the crossed Fryers.' From this he dates
the Preface to the collected three parts of his *Herball*, republished in one
volume early in 1568, and dedicated to Queen Elizabeth, who had always
stood his powerful friend, even when suspended for nonconformity in 1564.
Upwards of 300 species are given as natives of England.

He died on July 7, in the same year, and was buried on the 9th, in the
south aisle of St. Olave's, Hart Street, Crutched Friars. A stone erected
by his widow is let into the corner of the east wall, on which the following
inscription is still easily legible :—

CLARISSIMO . DOCTISSIMO . FORTISSIMOQVE . VIRO .
GVLIELMO . TVRNERO . MEDICO . AC . THEOLOGICO . PERITISSI-
MO . DECANO   WELLENSI . PER . ANNOS . TRIGINTA . IN . VTRAQVE .
SCIENTIA . EXERCITATISSIMVS . ECCLESIÆ . ET . REI . PVBLICÆ .
PROFVIT . ET . CONTRA . VTRIVSQVE . PERNITIOSISSIMOS . HOS-
TES . MAXIME . VERO . ROMANVM . ANTICHRISTVM . FORTISSIMVS .
JESV . CHRSTI . MILES . ACERRIME . DIMICAVIT . AC . TANDEM . COR-
PVS . SENIO . ET . LABORIBVS . CONFECTVM . IN . SPEM . BEATISSIM :
RESVRRECTIONIS . HIC . DEPOSVIT . ANIMAM . IMMORTALEM .
CHARISSIMO . EIVSQVE . SANCTISSIMO . DEO . REDDIDIT . ET . DEVICTIS :
CHRISTI . VIRTVTE . MVNDI . CARNISQVE . VIRIBVS . TRIVMPHAT . IN . ÆTERNVM .

MAGNVS . APOLLINEA . QVONDAM . TVRNERVS . IN . ARTE .
MAGNVS . ET . IN . VERA . RELIGIONE . FVIT .
MORS . T MEN . OBREPENS . MAIOREM . REDDIDIT . ILLVM .
CIVIS . ENIM . CÆLI . REGNA . SVPERNA . TENET .

OBIIT . 7 . DIE . IVLII . AN . DOM . 1568

According to the *Fasti Oxonicnses*, he married the daughter of George
Ander, an alderman of Cambridge. He left two daughters and a son,
Peter, who was also educated at Cambridge, took his M.D. at Heidelberg
in 1581, and incorporated at Oxford 1599 (where his son, also Peter, was
Geometry professor in 1649). At the end of the copy of Turner's *Herball*,
in the library of the Linnæan Society, is a long printed list of errata and
corrigenda by the author's son. At the end he says, 'if it please God to
lend me lyfe & health,' he intended to have it reprinted, 'augmented and
increased,' from which it would appear that this Peter Turner had some
knowledge of botany. He died in 1614, and was buried, says the St. Olave's
register, 'Maie 28 . . . in ye south Ile of ye church, closs by his Father,'
where there is a handsome monument to his memory.

The species observed and recorded by Turner in Middlesex are : *Lepidium
campestre, Hypericum quadrangulum, Spiræa Filipendula, Sanguisorba,
Sherardia, Anthemis nobilis, Onopordum, Tragopogon pratensis, Veronica
serpyllifolia, Mentha Pulegium, Calamintha Clinopodium, Daphne Laureola,
Euphorbia amygdaloides, Mercurialis annua, Spiranthes autumnalis* and
Endymion nutans.

Of Turner as a theological controversialist this is not the place to speak ;
he was the author of many anti-Romish tracts, the titles of some of which
now seem somewhat amusing. It was, no doubt, this facility of writing
which made him so much an object of persecution by the Roman Catholics.

He also wrote several small medical treatises, of no great importance now. Subjoined is an approximately correct list of all his publications, compiled from various sources. Those marked * are in the library of the British Museum.

* *A Cōparison betwene the olde learnynge & the Newe.* Translated out of Latin [of *Urbanus Regius*] in Englysh. Southwarke, 1537. 12mo. [Also in 1538.]

The Abridgement of Unio dissidentium, containing the agreement of the doctors with Scripture; and also of the doctors with themselves. Lond. 1538. 8vo.

* *Libellus De Re herbaria novus*, in quo herbarum aliquot nomina græca, latina et anglia habes, una cum nominibus officinarum in gratiam studiose juventutis nunc primum in lucem æditus. Lond. 1538. 4to.

* *The huntyng and fyndyng out of the Romyshe Foxe* which more than seuen yeares hath bene hyd among the bisshoppes of Englonde, after that the Kynges Hyghnes had commanded hym to be dryven owt of hys Realme. Basyll, 1543. 12mo. Published under the pseudonym of *Willm. Wraghton*, and dedicated to King Henry VIII.

Historia de Naturis Herbarum Scholiis et notis vallata. Coloniæ, 1544. 8vo.†

* *Avium præcipuarum quarum apud Plinium et Aristotelem mentio est brevis et succincta historia* . . . adiectis nominibus Græcis, Germanicis et Britannicis. Coloniæ, 1544. 12mo. Dedicated to Edward, Prince of Wales.

The Rescuynge of the Romishe Fox; otherwise called the examination of the hunter, devised by Steph. Gardiner. Winchester, 1548 or 1545. Under the name of *Will. Wraghton.*

* *The Names of Herbes in Greke, Latin, Englishe, Duche, and Frenche,* wyth the commune names that Herbaries and Apotecaries use. London. Preface dated March 15, 1548. 12mo.

* *The olde Learnyng and the new compared together* whereby it may easely be knowē which of them is better and more agreyng wyth the everlasting word of God. Lond. 1548. 12mo.

* *A New Dialogue wherein is conteyned the examinatiō of the Masse* and of that kind of priesthod whiche is ordeyned to saye masse; and to offer up for remission of Synne the bodye and bloude of Christ againe. [No place or date.] 12mo. 1550 (*Ames*).

✓ * *A New Herball* wherein are conteyned the names of Herbes in Greke, Latin, Englysh, Duch, Frenche, and in the Potecaries and Herbaries Latin, with the properties, degrees, and naturall places of the same, gathered & made by me Wylliam Turner. Lond. 1551, fol. This is the first edition of the 1st part of the *Herball,* and contains 88 folios.

* *A Preservative or Triacle agaynst the Poyson of Pelagius* lately renewed and styrred up agayn by the furious secte of the Annabaptistes. [No place or date.] 12mo. 1551 (*Tanner*). Another edition in 1561 (*Ames*).

Stephani Winton. Episcopi [Gardiner] De vera Obedentia oratio. Translated, 'with additions and a most vindictive preface,' by Dr. Turner, and 'published abroad.' Rouen, 1553. (See *Biograph. Brit.* iii. 2124.)

* *The Huntyng of the Romyshe Vuolfe.* No place or date. [1554?] 12mo. Lond. 1561 (*Ath. Cant.*).

* *A new Booke of Spirituall Physik* for dyuerse diseases of the nobilite and gentlemen of Englande. Rome, 1555. 12mo.

✓ *The seconde parte of Vuilliam Turner's Herball,* wherein are conteyned, &c. Collen, 1562, fol.

Hereunto is ioyned also a *Booke of the bath of Baeth* in England, and of the virtues of the same.‡ Collen, 1562, fol. The preface dated Basyll, March 10, 1557.

A Book of the natures and properties as well of the Baths of England as of other baths in Germany and Italy.‡ With the last (*Tanner*).

† This is only mentioned by Bumaldus (*Bib. Bot.* 18), unless Bale's *De Naturis Herbarum* be it. Neither Seguier, Haller, or Pritzel were able to meet with it.

‡ These were reprinted in *1568, and are usually bound up with the *Herball*. An abridged form in 4to was printed in the *Englishman's Treasure* in *1587, *1626, and *1633.

Of the Nature of all Waters. Ibid. (*Tanner*).
The third part of William Turner's Herball. Collen, 1566, fol. (According to Pulteney, but the usual date seems to be 1568.)

* *A new Boke of the Natures and Properties of all Wines* that are commonlye used here in England, with a confutation of an errour of some men that holde that Rhennish and other small white wines ought not be drunken of them that either have or are in daunger of the stone, the renine, and diverse other diseases. Whereunto is annexed the booke of the natures and vertues of Triacles, newly corrected and set foorth againe. Lond. 1568. 12mo. Dedicated to Sir William Cecill.

* *The first and seconde partes of the Herball of William Turner, Doctor in Phisick, lately oversene, corrected, and enlarged, with the thirde parte, lately gathered, and now set out,* with the names, &c. Collen, 1568, fol. The first part is entirely rewritten, and contains 224 pages. The second part has 171 folios. The title of the third part is *The thirde parte of Vuilliam Turner's Herball,* wherein are conteined the herbes, trees, rootes, and fruhtes, wherof is no mention made of Dioscorides, Galene, Plinye, and other olde authores. It is dedicated to the Company of Surgeons of London, and contains 81 pages.

Tanner gives the following without date :—

Homely against Gluttony and Drunkenness.†
Treatise on Original Sin, written against Rob. Coccheus.

And Wood :—

The Hunting of the Fox and the Wolf, because they did make havock of the sheep of Jesus Christ.

The following are in Bale's *Summarium* :—

De Arte Memorativa.   *De Metallicis.*
De Hierosolymorum Excidio.   *Imagines Stirpium.*
Epigrammata varia.   *Contra Gardineri Technas.*
De Baptismo Parvulorum.   *Contra quendam Arrianum.*
De Lapidibus.

And these in the *Athenæ Cant.* :—

Gisberti Longolii Epitaphium.
Carmen jocosum ad papam pro Joanne Standicio.

This was published after his death :—

Palsgravi Catechismum, cap. xix., *translated* by William Turner. Lond. 1572. 8vo.

Turner was part author of the two following :—

* *Dialogus de Avibus et earum nominibus* Græcis, Latinis, et Germanicis . . . per *Dn. Gybertum Longolium* . . . paulo ante mortem conscriptus. Coloniæ, 1544. 12mo.
* *The Sū of Divinitie* drawen out of the holy scripture, very necessarye, not onlye for Curates and yong studentes in divinitie, but also for al christen men and women what so ever age thei be of. Drawen out of Latine into Englishe by *Roberte Hutten.* Lond. 1548. 12mo.

He also prepared for the press William of Newburgh's *Historia Rerum Anglicanarum,* from a MS. in the library at Wells. This he sent to Antwerp to be printed by W. Sylvius. It appeared there in 1567, with several chapters omitted, and a preface by Sylvius inserted instead of Turner's.

Mr. Hazlitt states (*Handbook of Brit. Lit.* p. 618) that an autograph commonplace-book of Turner's was sold at Puttick and Simpson's, July 7,

† This may be the homily so called in the *Booke of Homilies* of 1562, but the style of that composition is little like our author's.

1861 ; he can, however, afford no further information about it, nor can we trace it in any way. Turner also greatly assisted in the translation of the Bible by comparing it with the Hebrew, Greek, and Latin texts.

In 1546 a proclamation was issued by Henry VIII., forbidding any persons to read the works of certain authors, among whom we find Turner mentioned ; and in 1555 his writings were again prohibited by royal proclamation. It is likely that many copies of his books were destroyed, and hence their rarity.

A few Middlesex additions are found in the *Adversaria Nova*, a joint production of Matthias de Lobel and Peter Pena, printed in London in 1570–71. In Lobel's *Observationes* (Antwerp, 1576) are also a few similar notes.* Lobel spent, at all events, the latter part of his life in England, and is said to have had charge of a physic garden at Hackney. The fragmentary *Illustrationes* of this author was not published till 1655, by Dr. How ; in this volume numerous Cyperaceæ and Grasses are recorded as found about Highgate, many of which it is now impossible to identify. Lobel's connection with Highgate is found in his daughter, who is said to have married a Mr. James Coel of that place. Lobel died in London in 1616.†

The additions to our flora by Lobel are: *Bidens cernua, Gnaphalium sylvaticum, Villarsia, Scutellaria galericulata* and *S. minor, Stachys palustris, Hydrocharis, Juncus bufonius, Luzula sylvatica* and *L. pilosa, Sagittaria, Butomus, Carex hirta, Phalaris arundinacea, Catabrosa*, and *Equisetum sylvaticum.*

The next work to which we are indebted is the well-known ' *Herball*, or general historie of plants gathered by John Gerarde, of London, Master in Chirurgerie,' printed in London in 1597. Gerarde lived in Holborn ' within the suburbs of London,' where he had an extensive garden.‡ He seems to have botanised much about London, especially in the northern outskirts, Hampstead, Islington, &c., and 73 species are first noticed as inhabitants of our county by him, of which the majority are still to be found in his stations. *Diplotaxis tenuifolia, Armoracia amphibia, Rhamnus Frangula, Ornithopus, Pyrus torminalis, Bunium flexuosum, Solidago, Erica cinerea* and *E. Tetralix, Vaccinium Myrtillus, Melampyrum pratense, Lamium Galeobdolon, Plantago Coronopus, Salix repens, Listera ovata, Convallaria majalis, Eriophorum polystachion, Blechnum, Ophioglossum* and *Lycopodium clavatum*, are among his discoveries. He died about 1607.

It is to THOMAS JOHNSON, however, that we are especially indebted for advancing the local botany of our county at this period. Of this indefatig-

---

* The *Adversaria* and *Observationes* are commonly met with bound in one volume as *Historia Stirpium*, Antwerp, 1576, and a second edition in 1605.

† Paul de Lobell, an apothecary who lived in Lyme Street, and was employed to give the medicines to Sir Thos. Overbury in the Tower by which he was poisoned in 1615, was perhaps a son of the botanist. He married the sister of Dr. Mayerne, physician to James I., (See Amos. *The Great Oyer of Poisoning*, pp. 167–169).

‡ An alphabetical catalogue of the plants of this garden was published in 1596 in the form of a quarto tract. The list occupies eighteen pages in double columns, and contains 1,039 names. It is now very scarce. A second edition was printed in 1599 (*Cat. Bibl. Bodl.*).

able botanist and excellent man we have been able to gain little further information than that contained in his own works, and Wood's *Athenæ* and *Fasti Oxonienses*.

The exact date of his birth, which was at Selby, in Yorkshire, we cannot ascertain, but it was probably at the beginning of the seventeenth century; nor do we know where he received his education, which, to judge from his writings, must have been a good one. At the date of his first published work, in 1629, he was an apothecary in London, already of some note, and a prominent member of the Society of Apothecaries. His house of business was on Snow Hill. The title of this book is *Iter Plantarum Investigationis ergo susceptum a decem Sociis in Agrum Cantianum Anno Dom.* 1629, *Julii* 13, and is a pleasantly written account of one of the herborising excursions which for some years it had been the practice of the Company to make at intervals. This is the first printed account of a botanical excursion in England. An appendix of three pages to this little book is of special interest, *Ericetum Hamstedianum seu Plantarum ibi crescentium observatio habita Anno eodem* 1 *Augusti*, which is an account of a similar excursion to Hampstead Heath on August 1st, 1629. The party consisted of the following besides Johnson: Jonas Styles, William Broad, Leonard Buckner, Robert Larking, John Sotherton, John Marriott, Thomas Crosse, and two Edward Browns, of whom one was Broad's servant. They left London early in the morning and proceeded to Kentish Town, then a country place enough, whence they walked to Highgate, where they were caught in a heavy shower, but, nothing daunted, made their way into the wood,* and then on to the heath. They returned to London by way of Hampstead village and Kentish Town. Three lists of the plants observed are given, the names being those of Lobel, Dodoens, and Gerarde, and the very common species being omitted; there is also a short list of some others, which Johnson had seen on the same ground on May 1 of the same year. The whole number of flowering plants observed was 72. The author expresses a hope that this excursion would be a prelude to others in succeeding years; and the hope seems to have been fulfilled, for in 1632 he published (as an appendix to the *Descriptio Itineris Plantarum Investigationis ergo suscepti in Agrum Cantianum*) the *Enumeratio Plantarum in Ericeto Hampstediano locisq. vicinis crescentium.* This consists of seven pages, six of which are occupied by a catalogue of names. It was intended to have included only those species not given in the former lists, but out of the 97 flowering plants enumerated, 28 are catalogued in the previous account. This leaves 69 new ones, which, added to the 72 species in the lists of 1629, make the whole number of Hampstead plants recorded by Johnson to be 141. Many of these had, however, been previously observed there by Gerarde. This catalogue may be considered as the first *Flora* of a small district printed in England.

In the following year, 1633, Johnson published his great work, a new edition of Gerarde's *Herball*. Thirty-six years had passed since the original edition appeared, and so well satisfied were the public with it that, although

* Probably Bishop's and Ken Woods, the latter not then enclosed.

botany had made great progress abroad during the period, as is well shown by the publication of Columna's *Ecphrasis*, Clusius's works, and especially of Bauhin's *Pinax* in 1623, yet nothing but Johnson's little books had been issued in England. *The Herball, &c., by John Gerarde . . . . very much enlarged and amended by Thomas Johnson, Citizen and Apothecarye of London*, may be looked upon almost as a new work, so great are the additions and improvements. More than 800 new plants were added, and 700 figures; and, according to Pulteney, the book contains descriptions of about 2,850 plants and 2,717 figures. But of equal importance with the additions are the corrections which Gerarde's work much needed. Johnson, who is very modest in his pretensions, says in his preface, it is Gerarde's 'matter, not method,' he endeavours to amend; and he has so thoroughly revised and corrected it throughout, that the book almost deserves the title of Gerarde Emaculatus bestowed on it by Ray. A very great convenience in using the book is found in the editor's additions being distinguished from the original text; yet we often find the whole book alluded to, and portions quoted by modern writers, as if entirely Gerarde's. Another edition was printed in 1636, but with no alterations. The additions to the flora of our county are not numerous in this book, and are chiefly in the immediate vicinity of London and Westminster.

In 1634, Johnson published *Mercurius Botanicus, sive Plantarum Gratiâ suscepti Itineris anno 1634 Descriptio*, to which is added *De Thermis Bathonicis Tractatus*; and in 1641, *Mercurii Bot. Pars altera, sive Plantarum Gratia suscepti Itineris in Cambriam seu Walliam Descriptio, &c.* No Middlesex localities are found in these, but in the last many Welsh plants are added to the British flora. He also translated the works of Ambrose Paré, the great French surgeon; they were published in 1643.

During this time we may suppose Johnson to have quietly pursued his vocation, but the stirring times of the civil wars led him to show his zeal for the king by entering the army. He seems to have been as good a soldier as a botanist, for he distinguished himself greatly in the war, and became lieutenant-colonel to Sir Marmaduke Rawdon. In 1642, the University of Oxford made him a Bachelor of Physic, and in the next year he proceeded to M.D. He did not, however, live long to practise his profession as a physician, for on September 14, 1644, during a skirmish with the rebels under Colonel Richard Norton, at the siege of Basing House, he received a shot in the shoulder, 'whereby contracting a feaver, he died a fortnight after.'* He was much regretted, being, we are informed, 'no less eminent in the garrison for his valour and conduct as a soldier, than famous through the kingdom for his excellency as an herbalist and physician.'†

No less than 88 species of plants were recorded for the first time by Johnson in Middlesex; amongst them *Caltha, Alliaria, Drosera rotundifolia, Oxalis acetosella, Medicago maculata, Lathyrus Nissolia, Pyrus Aria, Hydrocotyle, Silaus, Lactuca virosa, Menyanthes, Digitalis, Primula vulgaris* and *P. veris, Callitriche platycarpa, Carpinus, Arum, Scirpus sylvaticus*, and *Nardus.*

The smaller botanical works of Johnson are all very scarce books; in 1847 they were collected and republished by Mr. Pamplin, under the title of *Opuscula omnia botanica Thomæ Johnsoni*, and edited by T. S. Ralph, A.L.S., the volume being dedicated to Edward Forster and William Borrer.

The *Theatrum Botanicum, or Theater of Plants*, by John Parkinson, 'the King's herbarist,' was published in 1640,* during Johnson's lifetime. Parkinson was an older man than Johnson, having been born in 1567 ; and like him, was an apothecary in London. He had a garden in Long Acre (v. *Park. Theat.* 609). At the time of the publication of this book, Parkinson was seventy-three years old, and probably did not live long after; he was certainly dead before 1656. Though the *Theatrum* was in date of publication seven years subsequent to the 'emaculate' edition of Gerarde, Parkinson's observations ought probably to be considered as antecedent to those of Johnson. It is no doubt a more original work than the latter, as it is certainly more extensive. The following are all the additions (13) to the Middlesex flora made by Parkinson : *Brassica polymorpha (Rapa), Dianthus Armeria, Centaurea Calcitrapa, Carduus arvensis, Hieracium Pilosella, H. umbellatum,* and *H. boreale, Cuscuta Epithymum, Stachys sylvatica, Iris fœtidissima, Scilla autumnalis, Scirpus maritimus,* and *Calamagrostis Epigeios.*

The *Phytologia Britannica* of William How (commonly called Dr. How) appeared without the author's name in 1650, and is the first attempt at a Flora of England ; all previous works having been general systems of botany, including all known plants. Few localities are given for Middlesex species ; the following are all the additions to our flora : *Dianthus prolifer, Silybum Marianum, Vinca major,* and *Glyceria fluitans.* How died in 1656, and was buried at St. Margaret's, Westminster.

Dr. Christopher Merrett's *Pinax Rerum Naturalium Britannicarum,* which was published first in 1666,† contains nearly 200 more names than How's work, but there are not a few errors, and many varieties and exotics are reckoned. It must be admitted that there was material existing at the time for a better Flora of Great Britain than this. Merrett was born in 1614, took his degree of M.D. at Oxford in 1642, was one of the earliest members of the Royal Society founded in 1663, and died at his house in Hatton Garden in 1695, his body being buried in St. Andrew's, Holborn. His dried plants are preserved in vols. 14, 19, 29, 30, 33, 34 and 288 of the Sloane Herbarium and there are several MSS. notes by him in a copy of the *Pinax* in the British Museum.

Several (36) plants stand first vouched for in Middlesex by Merrett :

---

* Parkinson had previously, in 1629, published his '*Paradisus terrestris* or garden of pleasant flowers,' one of the earliest horticultural treatises.

† The date 1667 is much more common in copies. Perhaps, as the Rev. W. W. Newbould suggests, the greater part of the original stock was destroyed in the great fire of London. The re-issue was printed without the author's sanction, as appears from the first vol. of the *Phil. Trans.* p. 448 (in the number published April 8, 1867), where Dr. Merrett informs the public ' that within the space of four moneths he shall republish his *Pinax* . . . with many additions, and in his proposed new method ; and that he wholly disclaims the *Second Edition* of that book, as being printed and published without his knowledge.'

Ranunculus hirsutus, Sisymbrium Irio, Lythrum Hyssopifolia, Bupleurum tenuissimum, Vinca minor, Limosella, Veronica montana, Rumex pulcher, Paris, Tamus, Echinochloa, Melica uniflora, and *Equisetum limosum* are among them.

We must not entirely omit the name of one who, though he never published, yet contributed in his time to forward British botany. Thomas Willisel was, says Pulteney, 'an unlettered man;' he seems to have gained his living by travelling about England collecting plants, and in this capacity his services are acknowledged by Merrett, Morison, W. Sherard, and Ray; * the Royal Society also employed him, and he went to America for a similar object. He discovered in Middlesex, *Lactuca saligna, Lamium incisum,* and *Apera.* Willisel died before 1686.

Between the publication of How's *Phytologia* and Dr. Merrett's *Pinax,* another botanical author had appeared. This was John Ray, who in 1660 published his first book, a catalogue of plants growing about Cambridge, with an appendix, in 1663. This little book must have at once established the author as a botanist of repute, and it no doubt contributed greatly to set on foot that spirit of original investigation which led to the rapid and extensive strides in the knowledge of British plants made during the sixty years following.

With the writings of Ray, then, we may say, commenced a new era, which, in 1670, the appearance of the *Catalogus Plantarum Angliæ* fully introduced. This Flora was in every way so superior to Merrett's, that the latter must have been at once superseded.

A second edition, greatly augmented, came out in 1677,† and in 1688, a small tract of twenty-seven pages of additional species, called *Fasciculus Stirpium Britannicarum.* In all these the plants are arranged alphabetically, but schemes of classification had long occupied the mind of their author, and in 1682 the *Methodus nova* was given to the world; this was quickly followed by Ray's great systematic work on universal botany, the *Historia Plantarum,* of which the two first volumes appeared in 1686 and 1688, though the third (a supplement to the two preceding) was not published till 1704.

The first *systematic* Flora of Great Britain came out in 1690, under the title of *Synopsis Methodica Stirpium Britannicarum.* This accurate and invaluable volume, of which a much improved second edition appeared in 1696,‡ became at once the text-book of English botanists, and the model for all subsequent floras.

Our great naturalist died February 17, 1705-6, at seventy-eight years of age. He seems to have been but rarely in London, and very few Middlesex localities are recorded by him. He added, however, to our flora, *Cicuta virosa, Sonchus palustris, Juncus supinus* and *J. squarrosus, Avena pubescens,* and *Lycopodium inundatum.*

---

* There is a letter to Willisel from Ray in *Ray's Letters,* p. 358, written in September 1661.

† In the library of the Brit. Mus. (968, f. 7) is a copy of the 2nd edit. of the *Catalogus,* full of MS. additions and corrections in Ray's neat handwriting.

‡ It was really published at the end of 1695 (v. *Sloane MSS.* 3332, fol. 169).

It is, however, to Ray indirectly that the botany of our county owes very much. The example of his zeal, and the perusal of his writings, stirred up an enthusiasm for botanical pursuits in London, as well as in all parts of England, so that Linnæus termed this 'the Golden Age of Botany,' and a number of ardent workers appeared. Among these coadjutors and correspondents of Ray, Samuel Doody, James Petiver, and Leonard Plukenet appear prominently; and by all of these, but especially by the two first named, the vegetation of the neighbourhood of London was closely studied; and probably at the end of the seventeenth century, no part of England, with the exception of the country round the two old Universities, had been so thoroughly investigated.

LEONARD PLUKENET* was born in 1642, and is supposed by Smith to have been of French extraction.† Pulteney implies that he was educated at Cambridge, but his name does not occur in the matriculation lists of the time. He seems to have taken an M.D. degree possibly abroad, as his name is not given in a list of 386 English medical graduates printed in 1695; nor was he a fellow or member of the College of Physicians in that year.

His letters in the Sloane MSS.‡ throw very little light on his life. The earliest dated were written in 1672 and 1673, and are addressed to a cousin, Captain Anthony Irby, at Boston. At this time he was living in St. Margaret's Lane, near Old Palace Yard, Westminster, and there he seems to have continued all his life. Plukenet mentions his wife in a letter dated '23 June, 1673.' Most of the letters, which are eighteen in number, want both date and address, and are besides, mostly on trivial affairs. The botanical ones are addressed to Ray (a somewhat fulsome piece of eulogy), to Bobart, Professor of Botany at Oxford, to Matthew Dodsworth,§ and to Sir Hans Sloane (probably). There are also letters to Smith, the bookseller in St. Paul's Churchyard; to Col. Byrd, 'Mr. Bannister's patron,' and to a female cousin.

We may suppose that he practised medicine as a regular physician; but it is probable that he was not very successful in his profession. There is no evidence that he was connected with the Apothecaries' Company.

He must have been engaged in botanical studies for some time previously to 1688, for he assisted Ray in the second volume of the *Historia Plantarum*. In the preface to the first edition of the *Synopsis* (1690), Ray speaks of him as a botanist of the very highest order.

About this time, as appears from a letter of Sherard's in the *Richardson Correspondence* (p. 5),‖ he was appointed supervisor of the king's gardens at Hampton Court. Queen Mary, consort of William III., from whom he received much favour, no doubt gave him this appointment.

It was not till Plukenet was nearly fifty that he published. In 1691 ap-

---

* Also written *Plucknett*, *Pluknet*, and *Pluckenett*.

† 'Plus que net,' latinized 'plus quam nitidus.' (*Rees' Cyclopædia*.)

‡ Vols. 4043, 4055.

§ *Sloane MSS.* 4043, fols. 64 and 66, are two interesting letters from Dodsworth to Pulteney, dated 'Cowick,' 1680 and 1681.

‖ So also Petiver, in a letter to Breynius, apparently written in 1693. (*Sloane MSS.* 4055, fol. 165.)

peared the two first parts of the *Phytographia*. The first, dedicated to Henry Compton, Bishop of London, consists of 72 plates; and the second, inscribed to George Bentinck, Earl of Portland, of 48 (73–120). The next year was published the third part, containing 130 more plates (121–250), and dedicated to King William. This contains an index to the whole three. In 1696, 78 more plates were added (251–328), with letterpress relating to the whole. This book has the title of *Almagestum Botanicum*. A *Mantissa*, or supplement, to the *Almagestum* with 22 more plates (329–350) was printed in 1700; it also contains an index to the whole book.

Hitherto, all his works had been printed entirely at his own expense and without a publisher; but his last book, the *Amaltheum*, which appeared in 1705, and by the addition of 104 new plates brought up the whole to 454, was helped into the world by the subscription of fifty-five guineas, contributed by a few celebrated men, including the Earl of Portland.

All these volumes may be considered as parts of one great work on new and rare plants from all parts of the world, containing, according to Pulteney, 2,740 figures, with descriptive letterpress.* Though, of course, chiefly devoted to exotic species, several British plants were first figured in these plates. The figures are generally good, though stiff; a large number of varieties are given. The original specimens of many are in the British Museum

Plukenet died on July 12, 1706, aged 64, and was buried in his parish church, St. Margaret's, Westminster. In the parochial register he is described as 'Dr. Leonard Pluckenett, Queen's Botanist.'

A portrait of our author in his forty-eighth year, prefixed to the *Phytographia*, gives the impression of an amiable man, which his letters confirm. He was a staunch defender of Protestantism before the accession of William III., and seems to have been a sincerely pious man. From the contents of two very humble letters addressed to the Earl of Portland, one of which is a refutation of some malicious insinuations made by 'Mrs. Dorothy,' a servant in his lordship's household, one would gather him to have been in rather poor circumstances; but these, which are without date, were probably written in his earlier years.

His great devotion and extraordinary application to his favourite study are sufficiently evident from his writings. In the *Mantissa* he is somewhat harsh in his criticisms of Petiver; but it must be allowed that his strictures are well-deserved (see biographical notice of Petiver).

The backs of several letters in the Sloane MSS. are covered with attempts to form an anagram out of his own name, in which he was at last successful: thus, 'Leonardus Plucenetius,' ' Ut pene nullus sic ardeo.' This he adopted as his motto, and it appears in the title-page of each of the parts of the *Phytographia*.

Plukenet left two sons—one, Richard, was at Cambridge in 1696 (*Pulteney*); the other, Robert, was educated at Eton.

* The whole were reprinted in 1720, and again in 1769, in four vols. An index of Linnæan names was published in 1779 by Dr. Giseke.

His medical MSS. are to be found in Sloane MSS. 1502, but are chiefly prescriptions. His botanical MSS. came into the hands of Mr. Hudson, by whom they were given to Sir J. E. Smith. Plukenet's vast herbarium of more than 8,000 plants, mostly passed into the hands of Dr. Moore, Bishop of Norwich (one of the subscribers to the *Amaltheum*), who bought them of his executors; from Dr. Moore they were purchased by Sir Hans Sloane, and now form vols. 60, 61, 83–105, 145 and 242 of the Sloane Herbarium in the British Museum.*

Plukenet has recorded very few species for Middlesex, though he evidently knew the plants of the neighbourhood of London well, and has figured many for the first time. His time was so fully occupied with exotic and garden botany as to leave him little leisure for a critical study of indigenous plants. In his earlier life, however, he seems to have done so, for there are in the British Museum copies of Ray's *Catalogus* (both editions), full of his MS. notes; and his criticisms on and additions to the first edition of the *Synopsis* are printed in Ray's *Letters*, pp. 226–235. He first observed *Trifolium subterraneum* in our county; but this is apparently the only addition he made to its flora.

SAMUEL DOODY was born in Staffordshire on May 28, 1656, and was the eldest son, by his second wife, of John Doody,† an apothecary in that county, and afterward in London.‡ Another son, Joseph, a younger brother of Samuel, born in 1658, was, in 1703, settled in Stafford.§

Doody assisted his father in his business, and succeeded him at his death (which, however, did not happen till after 1696); their shop was in the Strand, 'over against Salesbury House.'

It is probable that his business took up most of his time, and prevented him from following out to its full extent his inclination towards botanical pursuits. We have found a small commonplace book with the date 1687 written in it (Sloane MSS. 3361), which contains numerous extracts from books on botany, lists of plants, &c., showing that at that time he had already begun the study.

The first printed record of him is the acknowledgment of his help by Ray in the preface to vol. ii. of the *Historia* (1688); the next in the *Synopsis* of 1690; and in the appendix to the latter book is a list of his observations. These are chiefly, as might be expected, in the neighbourhood of London, and several new mosses are noticed. At this time he seems to have been intimate with Plukenet and Petiver. The Cryptogamia had then been very little studied; Doody applied himself especially to them, and soon became an authority on the subject. In a letter to Dr. Richardson in 1691, W. Sherard says that Doody was ' putting out a small treatise of Musci ;' no

---

* ' Part of Plukenet's Herbarium was in the possession of Philip Carteret Webb, Esq., and was disposed of at the sale of his books.—Mr. Dryander.' (*Pulteney, Addenda to* vol. ii.)

† J. D. often spelt his name ' Doodie.'

‡ *Sloane MSS.* 2353.

§ A letter from him to Petiver in *Sloane MSS.* 4043, fol. 90.

such treatise was ever published, but a MS. in Doody's hand (Sloane MSS. 2315) in the British Museum seems to be a rough draught of it.

In 1693, Doody undertook, at a salary of 100*l.* per annum, the care and expenses of the Apothecaries Society's garden at Chelsea, in the room of Mr. John Watts. It is probable that he continued the management till his death.* He does not seem to have married.

His fame as a cryptogamic botanist steadily increased. In 1695 he was elected a F.R.S., and the following year, on the appearance of the second edition of the *Synopsis*, his diligence was made known to the world by the long list of new plants and localities which he contributed to its pages, and his accuracy vouched for by the deserved praise of Ray. The mosses continued to occupy the most part of his attention. Buddle says in 1698, that Doody knew them 'the best of any man;' and there is no doubt that he was in this aspect, as Pulteney terms him, 'the Dillenius of his time.' But into his labours in this field we cannot follow him. In phanerogamic botany we shall find that he has contributed much that is new.

For the next ten years he continued to botanise about town, his companions being frequently his friends Buddle and Petiver, and entered his remarks and discoveries in a copy of ed. ii. of the *Synopsis*, which still exists. There are several letters in Sloane MSS. 4043, written by him at this time to Petiver and Sloane, but they contain little of interest.

He mentions that he suffered much from gout, and it may have been of this malady, in one of its forms, that in 1706, at fifty years of age, he died. He was buried at Hampstead on December 3,† where his funeral sermon was preached by Buddle. Of this interesting discourse a part, viz. 'all that related to him and Botany,' may be found in Sloane MSS. 2972.

In this sermon Buddle says that Doody 'was in Botany very particular, very singular, none before him ever knew so much;' and again, 'every botanist cannot be a Doody,' which remarks show the very high opinion in which he was held by his contemporaries.

Of his character some idea may be formed from the following extract, also from Buddle's sermon. He was 'very slow of speech, even in his mother tongue, and at first sight you would take him to be of as little sense as eloquence . . . . he generally wanted words to express his wisdom, but when he did or could exert himself, his discourse was always full of argument and sound reasoning, plain and improving. . . . His notions of God and religion were very sublime . . . . his faith was orthodox. . . . His probity and integrity were visible in all his actions . . . . the plainness and simplicity of his soul were very conspicuous.' Buddle adds that poor Doody had 'a vice, which . . . . he seemed industriously to make known, and indeed it was very notorious, whilst his virtues, like his learning, lay hid to all but his particular friends and acquaintances.' From various considerations, it seems not unlikely that the 'vice' in question was an over-fondness for the bottle.

* Field's *Memoirs of the Botanick Garden at Chelsea*, pp. 15–18.
† In the parish register the entry is 'Samuel Dooty—from London.'

He contributed one paper to the Royal Society, a ' Case of Dropsy of the Breast' (*Phil. Trans.* xix. (1697), p. 390); and this seems to have been the extent of his literary productions.

Though scarcely an author in the science, few men have advanced the knowledge of the British flora more than Doody: he is a good example of that class of scientific men who, without themselves printing, are yet of the greatest help to those who do ; such men may exercise as great an influence on the progress of science as the most prolific authors.

We have already mentioned Doody's interleaved copy of the second edition of Ray's *Synopsis*. This valuable volume is in the library of the British Museum (969, f. 21); it seems to have belonged to Doody's father, and has his autograph, ' John Doody,' as well as Petiver's, to whom the book afterwards belonged. It is alluded to by W. Sherard in a letter to Dr. Richardson in 1723, where he says, ' My brother* copied Mr. Doody's observations on the *Synopsis* for his own use: 'twas in the hands of Mr. Petiver, and I suppose now in Sir Hans', where all things centre. You know how exact and diligent a botanist he was, for which reason I have entered† almost all his queries (which are generally of plants about London), that the new set of botanists may go to the places mentioned and examine them.' Most of these notes were accordingly incorporated into the new edition of the *Synopsis*, and their source acknowledged by Dillenius in the preface ; but a considerable number were omitted, on what grounds does not appear. These, so far as they relate to Middlesex, we have supplied from the original MSS. There are several other botanical books in the British Museum of Doody's in which he has written notes, and some books of medical receipts and prescriptions.

His dried plants are to be found in vols. 60, 61, 145, 271, 286 and 288 of the Sloane Herbarium. Some are foreign, probably from Chelsea gardens.

With regard to Middlesex, he was the first to notice the following flowering plants, *Ranunculus parviflorus, Teesdalia, Silene inflata, Hypericum Elodes, Medicago denticulata, Trigonella, Lathyrus Aphaca, Myriophyllum alterniflorum, Pimpinella magna, Archangelica, Tordylium, Arnoseris, Verbascum Blattaria, Utricularia minor, Plantago major, Polygonum nodosum, P. aviculare, Euphorbia platyphylla, E. exigua, Populus alba, Acorus, Eleocharis acicularis, Scirpus carinatus, S. Tabernæmontani, Carex paniculata, C. pilulifera, Dactylis, Festuca gigantea* and *F. elatior.* Of cryptogamic plants he added, *Lastrea Filix-mas* and *Pilularia*, besides very many mosses, lichens, and fungi.

He seems to have paid much attention to Hounslow Heath and its neighbourhood, and has been quoted for some statements of discoveries there and elsewhere, which are very improbable ; e.g. *Buffonia, Carduus heterophyllus, Crepis fœtida, Euphorbia hiberna*: reference to the pages of this Flora will in several cases tend to absolve the accurate Doody from these blunders, and to lay the blame of them on the right shoulders.

---

* James Sherard, M.D., F.R.S., of Eltham, Kent.
† i.e., for the third edition of the *Synopsis*.

JAMES PETIVER, the best known of Ray's correspondents, was the son of Mr. James Petiver, of Hillmorton near Rugby, and Mary his wife, and was born between 1660 and 1670. He tells us that he had his 'juvenile education at Rugby Free School,* in Warwickshire, under ye patronage of a kind grandfather, Mr. Richard Elborowe ; since which' he continues, ' I have often bewailed my not being allowed, after that time, academical learning ' (Sloane MSS. 3339, fol. 10, written in 1713).

In course of time he was apprenticed to Mr. Feltham, apothecary to St. Bartholomew's Hospital, and commenced the study of medicine. The earliest record of him in connection with botany is a small collection of ' plants growing in the fields and gardens about London, gathered about the year 1683 or 1684 ' in vol. clxxiv. of the Sloane Herbarium.

At what time he first set up in practice on his own account does not appear; but in 1692 we find him ' at the White Cross, near Long Lane in Aldersgate Street,' and in the same street, if not the same house, he continued to reside all his life. He was already recognised as a botanist of some repute, and very intimate with Ray, who was indebted to him for several notes in the second vol. of the *Historia*, published in 1688, and in the *Synopsis* of 1690, acknowledged in the prefaces to those books. Pulteney tells us, that in 1692 he made a tour through the Midland counties ; but it is evident that his profession rarely allowed him to go far from town.

His investigations about London, however, were so well known that in Gibson's edition of Camden's *Britannia*, published in 1695, the list of the ' more rare plants growing in Middlesex ' (pp. 335–340) was written by him, all the other county lists being contributed by Ray. That Petiver was hurried in the matter appears from a letter to Mr. Scampton,† written ' July 4, 1695.' ' I had but one day & a halfe to compose ye Catalogue of Middlesex, wch. if I might have had more time it should have been somewt. more perfect.' This was his first appearance as an author, and the list is also interesting as the first for the whole county. It contains 108 plants (including a few cryptogams), and Doody, Plukenet and Sloane are mentioned as contributors.

In October of the same year, Petiver was made a Fellow of the Royal Society, and in November he published the first ' century ' of the *Museum Petiverianum*, a small octavo pamphlet, containing descriptions of 100 specimens of plants, animals and fossils, both British and foreign. Of this work ten centuries were printed, the last in January 1703.

There is a list of new plants, &c., by Petiver, in the appendix to the second edition of the *Synopsis* in 1696. These are chiefly among the mosses and allied plants. That the notes are not more extensive is owing to the fact that during this year he was devoted to the study of insects.

At this time Petiver was in a good practice, and was also apothecary to

---

* He entered the school in 1676 (*Rugby School Register*, p. 1).
† *Sloane MSS.* 3332, fol. 129.

the Charterhouse ; he, nevertheless, found time to correspond with naturalists in all parts of the world, and to form a large museum of specimens in
all departments of science.   In 1697 Petiver's collection of plants numbered
'between 5 & 6 thousand different specimens' (Sloane MSS. 3333, fol. 255).
Neither did he neglect home botany, making frequent expeditions to Hampstead and the neighbourhood, with his friends Doody and Buddle.   In
1699* we find him visiting Ray, now past seventy years old, at Black Notley
in Essex.

The *Gazophylacium Naturæ et Artis*, of which the first part was published
in 1702, was of a similar character to the *Museum* already mentioned, consisting of descriptions of objects of natural history, in octavo. Each 'decade,'
however, was accompanied by ten plates. Two volumes (of five decades)
were published, the last decade appearing in 1709. 'Catalogues,' or
indices, to each volume were subsequently printe l. The work was published by subscription, and each plate is dedicated to a subscriber. The
mingling of birds, snakes, insects, plants, fossils, and even antiquities,
with an occasional impossibility,† gives these plates a most unscientific
appearance.

Petiver contributed to the third volume of Ray's *Historia*, in 1704, lists
of little known Asiatic and African plants.   Of these, Dr. Pulteney
remarks that they will ' remain a lasting testimony of the early and extreme
diligence of this indefatigable collector.'

In 1707, the year after Doody's death, the Society of Apothecaries, of
which Petiver was an active member, being in difficulties with respect to
the Physic Garden at Chelsea, he, together with Messrs. Wyche and Andrews,
were appointed to inspect it.   A plan was ultimately agreed upon by which
a lease was made out between the Society on the one part, and twenty members (amongst whom were Petiver, Rand, and Joseph Miller) on the other
part, who agreed to subscribe 100*l.* or more per annum, to keep the garden
in repair, for seven years from Michaelmas 1707.   The lease was granted
for 5*l.* per annum ; but the plan does not seem to have answered.‡   Petiver
' officiated as demonstrator of plants to the Society as early as 1709, but
how long prior to that period cannot now be ascertained.   He probably
resigned it to, or at least was assisted in that office by, Mr. Rand.'§

In 1707 Petiver commenced a monthly periodical called, *The Monthly
Miscellany ; or, Memoirs for the Curious . . . by several hands.*   As it seems
to be exceedingly scarce,‖ a brief account of it will be of interest.   The
first volume, from January to December 1707, contains nothing botanical
except a review of the first volume of the *Gazophylacium.*   In the June and
July numbers of the second volume (1708) is a paper ¶ by ' J. P.' entitled
' Mr. John Ray, his method of English Plants illustrated; in a letter to

---

* *Sloane MSS.* 4039, fol. 275.
† Such as the ' Papilio' in tab. 10, fig. 6, named by Linnæus *Ecclipis.*
‡ Field's *Memoirs of the Botanick Garden at Chelsea,* p. 21.   § Ibid. p. 25.
‖ The only complete (?) copy we have seen is in the British Museum Library (*P.P.* 5420).
¶ The MS. of this is in *Sloane MSS.* 3337, ff. 20–25.

Mr. S. D.'* Vol. iii. (commencing January, 1709) is imperfect, and perhaps the publication was discontinued before its end. The number for August 1709 was not printed till 1710, with a new publisher.

In the September number Petiver commenced a catalogue, which, if ever completed, has eluded our search. It is called ' *Botanicum Londinense ; or, the London Herbal.* Giving the names, descriptions, and virtues, &c. of such Plants about London, as have been observed in the several monthly herbor- izings made for the use of the young apothecaries, and others, students in the science of Botany, or knowledge of plants.' ' The first division or walk,' containing ' such plants as grow plentifully nearest the city, viz.: within a mile, as bet. London and Islington, &c. or about that distance ' occupies the whole number, except the first two pages. The plants, eighty-eight in number, are arranged after Ray.

The number for October † is imperfect. The first two pages are on British Bees, and then the *Botanicum Londinense* continues thus : ' Having given you in our last an account of such plants as most plentifully grow on the north side of the City . . . . we now proceed to the second walk or division, which contains such plants as are commonly found west of London, viz. between Westminster and Chelsea, &c.' Of this we have seen no complete copy ; many portions cut out from it lie loose in Petiver's her- barium, in which are also a few with the printed date of ' November 1709,' apparently from a subsequent number.

The *Memoirs* seem to have died out gradually ; yet in 1715 the *London Herbal* is advertised ‡ as a separate publication, ' price 2s. 6d.' Was this a reprint? We have never seen it alluded to, nor have we met with a copy. If it exists, it must be of value as an account of the botany of the suburbs at the time.

In the year 1711 Petiver made a journey into Holland, returning in the autumn. At Leyden he had a friend, named Judkin (from whom are letters in Sloane MSS. 4046) with whom he seems to have stayed. A full account of this visit will be found in a letter to Patrick Blair (Sloane MSS. 3338, fol. 28), where he tells us that the chief object of his going over was to purchase for Dr. Sloane the best part of Dr. Hermann's Museum.§

In the autumn of 1712 he made a ' trip to the Bath and Bristow,' and in 1715 he went with James Sherard to Cambridge, Newmarket, &c., of which latter tour there is a description in Sloane MSS. 2330, fol. 914.

A set of figures of British plants was taken in hand by Petiver in 1713. The first volume contains 50 plates on copper, containing more than 600 figures, and in 1715 this was supplemented by 22 more, increasing the

* Samuel Daniell, ' the person who composed a voyage to the Levant,' and collected for Petiver there.
† Alluded to by Petiver in 1712, in a letter to Blair, as ' now very scarce to be had.' (*Sloane MSS.* 3338, fol. 86.)
‡ At end of the *Hortus Medicinalis Peruvianus.*
§ 109 lots were purchased for apparently little more than 14l. See a letter from Petiver to Sloane in *MSS.* 3337, fol. 160, dated ' June 29, 1711 ;' and *Sloane MSS.* 4055, 155, dated ' June 18, Leyden.'

figures to nearly 880. The book is found with either a Latin or an English title : *Herbarii Brittannici clariss. D. Raii Catalogus cum Iconibus ad vivum delineatis* ; or, *Catalogue of Mr. Ray's English Herball, &c.* These figures, though small, are very accurately drawn, but they have the disadvantage of being all of one size, instead of being drawn to a scale. The name accompanies each figure, and there are references to Ray's *Historia* and both editions of the *Synopsis*. Many varieties are figured, especially amongst the Compositæ and Polygonaceæ. and these are useful in determining Ray's plants. The last plant figured is Dodder. The letter L is placed against those species seen within ten miles of London, and H distinguishes those growing about Hampstead.* Petiver also intended to publish an ' iconical supplement,' or figures to Ray's *Historia*, but this extensive scheme was never fulfilled.

The *Graminum, Muscorum, Fungorum, Submarinorum, &c., Britannicorum Concordia*' was printed in 1716, and is a catalogue of British Grasses and Cyperaceæ, Mosses, &c. It contains numerous localities, few of which are original ; much was copied from Buddle's MSS. The only two copies of this tract we have seen contain but twelve pages (375 species), and appear unfinished. Sir J. E. Smith, however, says he had 'never seen more.'

We have now enumerated the chief contributions of Petiver to scientific literature, but besides these there are various small tracts and single sheets published by him, a list of which, so far as we have been able to obtain their names, is given below ; many seem to have been scarcely really published, but printed for distribution among the author's friends. Our list is probably incomplete :—

Plantarum Generæ Catalogus	. 1709
Pterigraphia Americana. Pp. 4, tab. 20	. 1712
Aquat. Animalium Amboinæ Catalogus. Pp. 4, tab. 22 . .	. 1713
Plantarum Etruriæ rariorum Catalogus. Pp. 4 . . .	. 1715
Plantarum Italiæ Marinarum et Graminum Icones, &c. P. 1, tab. 5 .	1715
Hortus Peruvianus Medicinalis, or the South-Sea Herbal. Pp. 3, tab. 7 .	1715
Monspelii Desideratarum Plantarum Catalogus. Pp. 4 . .	. 1716
Proposals for the Continuation of an Iconical Supplement to Mr. John Ray his Universal History of Plants. 1 sheet . .	. 1716
Petiveriana, seu Naturæ Collectanea III. Pp. 4 . .	. 1716–17
Plantarum Ægyptiacarum rariarum Icones. P. 1, tab. 2 .	. 1717
Plantæ Silesiacæ rariores ac desideratæ. 1 sheet .	. 1717
English Butterflies. Pp. 2, tab. 6	. 1717

And the following without date :—

Botanicum Anglicum (labels for the herbarium). Pp. 2.
Hortus Siccus Pharmaceuticus (labels). Pp. 12.
Rudiments of English Botany. P. 1, tab. 4.
James Petiver his Book, being Directions for gathering Plants. 1 sheet.
Brief Directions for the easie making and preserving Collections. 1 sheet.
Plants engraved for Ray's English Herball. 1 sheet.

as well as many plates of birds, fishes, insects, and plants, from different countries.†

* Fifty-seven have H affixed.
† In 1764 all Petiver's books having become very scarce, those that could be obtained were reprinted with the title, ' *Jacobi Petiveri Opera Historiam naturalem spectantia,*' in

Petiver also contributed as many as twenty-one papers to the Royal Society, which appear in vols. 19 to 29 of the *Philosophical Transactions*, from 1697 to 1717.

On the private life and character of this energetic man the Sloane MSS. throw considerable light. He was never married. Of his father we know nothing, but his mother (who had remarried a Mr. Glentworth) survived him. He had a maternal uncle very rich, but somewhat eccentric, Richard Elborowe,* of Rugby, who, besides sending frequent presents to Petiver, granted him an annuity of 40*l*. per annum. It would appear, however, that Petiver in some way managed to offend this gentleman, for the latter made Elborowe Glentworth, James's younger half-brother, sole executor of his will. In answer to Petiver's protests Mr. Elborowe writes (April 8, 1704), ' I have left you a very fair and plentyfull legacy,† besides other provisions made for you in my will, which will keep you from the condition mentioned in your letter, which, if you live to receive, I hope you will make better improvement than hitherto you have done' (Sloane MSS. 4043, fol. 282). This uncle died in 1707 ; his last letter is dated April 5, and his will is dated July 31, and was proved on December 4 of that year. In a letter, not dated, to his brother, Elborowe Glentworth, Petiver bitterly complains that ' on ye 9th of next month, which is now neer at hand, our very good and generous Uncle has been dead three years, and yet I have not, in all this time, received of you one Farthing, either of principall or Interest, of the Legacy he left me' (Sloane MSS. 3331, fol. 608). Nor was it ever paid, for there are many other letters, even as late as 1714 (Sloane MSS. 3330, 937), asking for it, apparently without effect. Of this brother, who inherited much of Mr. Elborowe's property, and must have been well off, we can discover nothing further than that he was also at Rugby School,‡ and that he married, and lived 'at the old barge-house against the Temple ' in 1714. Petiver wrote to him, however, in 1711, from the Hague, and the brothers were at that time evidently on good terms.

Petiver seems to have been rather unfortunate in money matters ; many of his letters are requests for the payment of debts, either of money lent, or in payment of professional attendance. In 1706 he lost ' near 800 pound ' by the ' breaking of Mr. Ayrey' to whom he had lent it (see Sloane MSS. 3335, f. 9).

Many of his relations were in poor circumstances. § One William Pettiver ‖ (as he always spells his name) was in 1703–4 at Merton College, Oxford ; he was one of a large family who also lived at Hillmorton, and

three vols. (two folio, and one octavo) for six guineas. A good many plates not before published were added to this reprint.

* Mr. Elborowe built a school-house at Rugby, and endowed it for ever, for the instruction of thirty poor children, and an almshouse for six poor widows.

† 7,000*l*., to be paid within twelve months after the death of the testator, as appears from a codicil to Elborowe's will.   ‡ Entered in 1692 (*Register*, p. 6).

§ It is supposed that he was connected with Sherard, whom he certainly addresses as ' coz,' ' kinsman,' &c.

‖ Also at Rugby. Entered the school in 1689 (*Register* ; where are also the entries of Richard, John, and Nicholas Petiver in the years 1677, 1682, and 1687, respectively).

was probably Petiver's cousin. He tried to get a fellowship, Mr. Elborowe using his influence with the warden, Dr. Marlin, in his behalf, but without success, and so he was ' forced to leave beloved Oxford for good and all.' [*] He died about 1713, as appears from a letter from his brother Henry, written in that year.[†]

Petiver's sister, Jane, married before 1707 Mr. Thomas Woodcock, and lived in Leicestershire ; their son, Thomas Woodcock, is often mentioned by Petiver, and after the latter's death was in correspondence with Sir Hans Sloane about the purchase of his uncle's collections.[‡] Thomas Woodcock and his wife were both dead in 1727.

Before going to Holland, in 1711, Petiver wrote the following letter to Sir Hans Sloane :—

Hond. Sir,—In case I should dy before my return from Holland, I make you sole possessor of all my collections of naturall things whatsoever, as well duplicates as single samples, and of all my manuscripts relating thereto, excluding my printed books, on consideration of your cancelling my bond, and paying 500 pounds to be disposed of as follows, viz., 200 pounds to my brother Mr. Edward Woodcock, for ye use of his son Edward when he comes to age.—100 pounds to charity, viz., 50 pounds to be put out to interest to buy the schoolboyes of Mr. Elborow's Charity at Rugby, Psalters, Testaments, or other books they may have occasion for, the whole and overpluss to be disposed of as my brother Mr. Edward Woodcocke, and at his decease who he shall appoint, may think fit. The other 60 pounds to be paid after ye expiration of one year, for ye use of ye aged poor of ye parish of St. Botolph, Aldersgate, London, at ye discretion of ye said Mr. Edward Woodcock and Mr. Will. Litler of the aforesaid parish as they shall see fitt.—100 pounds for ye discovery and collecting naturall productions, to be reposited in ye Museum of ye Royall Society ; the President, Dr. Sloane, and 2 others they shall nominate, to be Trustees and Curators for ye improvement of ye same.—Lastly, 100 pounds for ye benefitt and improvement of ye Physick Garden at Chelsea, in stocking it annually with such exotics, and uncommon plants and goods as they shall want, and to incourage and promote ye monthly Botanick Herbarizings and Lectures of Botany. This last donation to be managed by four such skilfull members of ye Company of Apothecaries as shall be approved of by the four persons in trust for that of the Royall Society. Both these last legacies to be maintained out and from ye interest that such of the hundred pounds shall annually produce. This is my will.

JAMES PETIVER.

It is evident that at this time, then (1711), Petiver had nothing to bequeath but his collections. These were no doubt of great value ; Pulteney states that ' Sir Hans Sloane . . . . offered Petiver 4,000 pounds for his museum some time before his death ; ' but the above letter shows that at this time the latter was willing to let Sloane have it for 500*l.*, on condition that a certain bond was cancelled.

Petiver lived seven years after the above was written, and seems to have always spent fully up to his income, if not beyond it. Shortly after this he began to have bad 'health (see his letters, *passim*), and early in 1717 was beyond any active exertion (J. Sherard in *Richardson Corresp.* 127). According to Pulteney (ii. 42) he died at his house in Aldersgate Street, April 20, 1718. ' His body was carried to Cooke Hall,[§] where, agreeably to the custom of the time, it lay in state. The pall was supported by Sir

---

* *Sloane MSS.* 4045, f. 179.   † Ibid. 3340. f. 17.   ‡ Ibid. 4066, fols. 61–63.
§ Probably *Cook's Hall*, then in Aldersgate Street, opposite St. Botolph's Church.

Hans Sloane, Dr. Levit, physician to the Charterhouse, and four other physicians.' He was buried 'in the cancell' of St. Botolph's Church, Aldersgate Street. In the register of the parish the date of burial is given 'April 10, 1718;' if so, the day of his death given in Pulteney cannot be correct.

Petiver's will is dated August 13, 1717. There are small legacies to his mother, 'Mrs. Mary Glentworth,' his nephew, Thomas Woodcock, and his three nieces Woodcock; also 50l. for his funeral expenses, 5l. to the beadle of the Apothecaries Society, 'one guinea and a ring of 20 shillings' price' to the clergyman who preached his funeral sermon,* and 5l. to the charity children of St. Botolph's who should attend his funeral; with a few other trifling bequests. All the remainder of his property whatever was bequeathed to his sister, Jane Woodcock, whom he also made sole executrix.

His collections, MSS., books, &c., came into Sir Hans Sloane's possession; some he purchased 'for a considerable sum.' † All are now in the British Museum; the herbarium, consisting of plants from all parts of the world, forms a very large part of the Sloane Herbarium; the British plants are found in vols. 150–152. The letters and MSS. are intimately mixed up with those to Sloane himself; there are besides fifteen volumes of medical MSS.,‡ and many commonplace pocket-books full of rough notes, which would doubtless repay perusal. The printed books are also full of marginal notes and additions, and occasionally drawings.

Petiver possessed great powers of observation, and a quick perception, without a great amount of botanical judgment or accuracy. As the first who in this country attempted to make science popular, he is deserving of great praise. He felt the want of a more liberal education than he had received; his Latin was (at all events at times) composed for him by Dr. Tancred Robinson.§ To Buddle, also, he was very much indebted, and he seems to have used much of his matter without due acknowledgment. As is often the case in a prolific writer his publications are very unequal in merit, and while one admires his diligence, one cannot but perceive the haste with which some of his compositions were written. In the *Museum* he censures Plukenet with considerable severity for 'rash conjectures' and 'false references.' Sir J. E. Smith, too, tells us that, 'in the collections of Tournefort and Vaillant, at Paris,' he saw 'various tickets' in which Petiver displays 'malignity and coarseness' in his criticisms of Plukenet.

He seems to have possessed administrative powers of value, and was a very active member both of the Apothecaries Company and the Royal Society. He must also have had tact and a power of making friends, for he became the centre of communication for the naturalists of the time. His

* Dr. Brady, according to Pulteney (ii. 42) who says that five guineas were so left, and 'fifty pounds to the charity school of St. Ann's Aldersgate:' of this we find no evidence.
† Letter to Richardson.—*Nichols*, i. p. 276.
‡ See *Ayscough's Catalogue*, ii. 574 and 669.
§ See a letter to him in *Sloane MSS.* 3330.

medical practice was not probably of a high order, for he advertised six quack nostrums—the *Indian Purge, Purging Marmalade, Ambretta, Golden Aqua Mirabilis, Paul Royal,* and *Syrup of Manna or Cordial Purge*: the sale of these, however it may have filled his pockets, can scarcely have added to his reputation.

As regards Middlesex, a very full list of species growing about London can be obtained from his books, in all amounting to about 340; the majority of these would be inhabitants of our county. Petiver stands as the earliest observer of as many as 66 species, certainly natives of Middlesex; amongst which we find *Ranunculus auricomus, Papaver dubium, Dianthus deltoides, Moenchia, Radiola, Sanicula, Adoxa, Asperula odorata, Filago minima, Lactuca muralis, Jasione montana, Littorella, Polygonum lapathifolium, P. minus, Salix triandra, Potamogeton lucens, Zannichellia, Carex ovalis, C. pallescens, C. pseudo-Cyperus, Milium,* and *Asplenium Ruta-muraria.*

We have next to speak of a botanist who, though he never published, has left us an excellent MS. Flora, and a herbarium of large extent illustrating its pages, and thus conferred a great benefit on his successors in botanical science. Of ADAM BUDDLE we have been able to gather but scanty information. Beyond a casual allusion to his collections, his name is not mentioned in Pulteney's sketches; and though he is often spoken of in letters of the period, the amount of definite information respecting him is very small. He was born at Deeping St. James, Lincolnshire (*v. Budd. MSS.* sub Prunus, 6), but we have not been able to discover in what year. He was educated at Catherine Hall, Cambridge, and took his B.A. in 1681, and M.A. in 1685. He seems not to have commenced his botanical studies till some time after, as he is not mentioned in either of Ray's editions of the *Synopsis.* In 1696, however, the date of the second edition, he must have been working at Mosses, for the next year he is mentioned in a letter of Petiver's [*] as then well versed in them. In what year he took holy orders we do not know. In 1698 he was living at Henley in Suffolk, and during that year and the next kept up a correspondence with Doody and Petiver, also sending them his collections of grasses and mosses, at that time the best in the kingdom. So valuable were they that they were transmitted to the French botanist Tournefort, at Paris. Buddle was also intimate with Dale of Braintree, one of the best British botanists of the time; and in 1699 he paid a visit to Ray, which he describes in Sloane MSS. 4039, fol. 275.

He first became especially connected with Middlesex in 1700, when he went to Hampstead, though under what circumstances we cannot discover. Nichols (i. p. 269), indeed, states that at about this period he 'had a living given him by Lord Keeper Wright,' but as the perpetual curate of Hampstead during this period was Samuel Nalton, B.D.,[†] Buddle could not have held that position. It is possible that he may have been curate there temporarily,[‡] for, as already mentioned, he in 1706 preached there a funeral sermon on his friend Doody.   At all events, he continued to live either there or in London till his death.

There are as many as nineteen letters, or parts of letters, written by Buddle (to Petiver chiefly) in the Sloane MSS.; many are mere scraps, and the majority without direction or date.   The chief facts obtained from them are that Buddle was married* and had children,† and that he was a sufferer from gout.   His special botanical friends were Petiver and Doody, with whom he was constantly engaged in exploring the neighbourhood of town, and many of his letters consist of appointments to meet for that purpose.   The last with a date was written on ' 13 May, 1702 ; ' but there is another, in a weak and trembling hand, evidently much later.   In the ' Richardson Correspondence ' are seven letters of Buddle's, mostly ' filled with lists of plants,' the last, ' May 28th, 1709,' has the heading 'Gray's Inn,' which goes to confirm the statement of Nichols, that he was ' reader' there.‡   The editor of the ' Richardson Correspondence ' quotes (p. 87, note) from a letter of Buddle's, ' apparently written about' 1703 : ' I have not . . . this half year . . . minded anything of Botany, nor listened to anything of Naturall Philosophy, *being too intent upon the necessities of life to think of the Nugæ,*' from which, says he, ' I am afraid he was very poor.'

Despite this drawback, Buddle found time before the year 1708 to write an entirely new and complete English Flora.   This manuscript, which exhibits well the accuracy, diligence and knowledge of its author, occupies twelve volumes (bound in three) of the Sloane MSS., 2970–2980.§   The title runs, ' *Hortus Siccus Buddleanus sive Methodus nova Stirpium Britannicarum*, auct. A. Buddle,' and the book is dedicated to the Bishop of Carlisle and others. The index contains the derivation of the generic names, and there is also a collection of the local names of plants.   We find this alluded to by Lhwyd, who writes in 1708, ' The Doctor ‖ tells me that Mr. Buddle hath drawn up a new Synopsis Plant. Brit., but that he doubts whether he can get it printed ; tho' he supposes it a very considerable improvement of Mr. Ray's, who, he says, wanted many things to compleat his.   He adds that he improves the method by the help of Tournefort, Rivinus, &c., and that he often refers to figures and corrects vicious ones.'¶   Buddle himself says of it in 1709, ' As for my being an author, I have prepared a book ready ; but we can't agree about a method.   I have jumbled Mr. Ray's and M. Tournefort's together (they are both dead).   Some think I favour too much M.

---

* His wife's name was Elizabeth.   An undated letter from her (*Sloane MSS.* 4039, fol. 293) to Petiver was written when Buddle was very ill, and ' all hopes of his recovery ' gone.   She long survived her husband, for in 1741 we find her ill and in poverty, writing to Sir Hans Sloane from Suffolk.   She did not die till 1752, and was buried at Henley.

† Two children, a son and daughter, were baptised at Henley in 1696 and 1697.

‡ No list of the ' readers ' (or curates) has been preserved at Gray's Inn.

§ There is also a rough draft, *Sloane MSS.* 2201, and a duplicate of vol. iv. (containing the Grasses), *Sloane MSS.* 2306, which belonged to Petiver, and contains references to the *Concordia Graminum.*

‖ W. Sherard (?).

¶ *Rich. Corresp.* 95.

Tournefort, which is a reflection upon Mr. Ray which I am sure I do not design; neither would I offend any of his living admirers; but I find he that would please everybody must never print.'* And the book never was printed, doubtless for the very reason here hinted at, that the public esti-mation of Ray's system and works was so great as to allow of no improve-ment being made. And the same state of things is still more forcibly shown, when, fifteen years after, Dillenius was compelled to suppress his own name as editor of the third edition of the *Synopsis*, which still bears Ray's name alone on the title.

All hope of the publication of his Flora was destroyed by Buddle's death, the exact date of which we have not been able to ascertain, but which occurred between the years 1714 and 1716,† when he could not have much passed middle age, if he had attained it.

At the time of his decease his collections and MSS. were in Petiver's hands. Buddle bequeathed them to Sir Hans Sloane, who obtained them from Petiver with some difficulty, and the latter subsequently borrowed them again of Sir Hans. It is but doing justice to Buddle's great merits to remark that Petiver had the use of all his materials, and sometimes appro-priated his observations.

Buddle's labours are duly acknowledged by Dillenius in his preface to his edition of the *Synopsis*, which appeared in 1724, and Sir Hans Sloane is thanked for lending the *Hortus Siccus*: it is to be regretted that more use was not made of it. This herbarium now forms vols. cxvi. to cxxv. of the Sloane Herbarium.‡ Its value can be scarcely overrated as a means of determining the plants of the *Synopsis*, and as the MS. flora contains references to it throughout, all Buddle's plants can be determined with ease. It is undoubtedly the most trustworthy and accurately-named herbarium of the period which exists.

Buddle was the first to notice many of the less attractive species about town. Hampstead, Isleworth, and Hounslow were his chief hunting-grounds. His name stands as the first recorder of twenty-five plants in this flora; amongst others, of *Geranium rotundifolium, Ulex nanus, Trifolium medium, T. filiforme, Lotus major, Epilobium parviflorum, Arctium minus, Carduus tenuiflorus, Sonchus oleraceus, Veronica Anagallis, Mentha rotundi-folia, M. piperita, M. hirsuta, M. sativa, M. rubra, M. gentilis, M. arvensis, Chenopodium album, C. ficifolium, Potamogeton pusillus, Eleocharis palustris, Alopecurus agrestis, Agrostis vulgaris, Equisetum palustre*, and *Lastrea dilatata*.

ISAAC RAND, an indefatigable London botanist, like Buddle, paid particular attention to the inconspicuous plants about town. We know nothing more of him than that he was an apothecary and curator of the gardens at Chelsea from 1724 to 1743. Plukenet in 1700 (*Mant.* 112) calls him a botanist of great promise, and Buddle frequently mentions him. Dillenius also acknowledges his help in the preface of his edition to Ray's *Synopsis*.

---

* *Rich. Corresp.* 103.
† He was not living when the *Concordia Graminum* was published in that year.
‡ Vol. liv. also contains ' English Mosses and Grasses' collected and named by Buddle.

He published in 1730 *Index Plantarum Officinalium in Horto Chelseiano*, and in 1739 *Horti Chelsciani Index compendiarius*. We have not been able to find the year of his death. He first recorded in our county, *Matricaria inodora, Mentha pubescens, Plantago media, Chenopodium glaucum, Rumex maritimus, R. palustris* and *Carex paludosa*.

After Petiver's death in 1718 little seems to have been done in British botany for some years. It is true that in 1724 appeared the third edition of Ray's *Synopsis*, a book which advanced the science to some extent; but the improvements on the previous edition were made by the men whose labours we have been recording, and by their contemporaries. Dillenius, who edited the book, did not come to England from his native country, Germany, till 1721. He added but little to the stock of knowledge of British plants, being at first occupied at James Sherard's house at Eltham, on the *Hortus Elthamensis*, published in 1732; and then at Oxford, from 1728, with his duties as Professor of Botany, the *Historia Muscorum*, 1741, and the great 'Pinax' of the plants of the world, which never appeared. Dillenius died in 1747. He first noticed and distinguished in Middlesex *Sagina apetala, Cerastium semidecandrum, Geranium pusillum, Œnanthe fluviatilis, Linaria Cymbalaria*, and *Festuca sciuroides*.

The other botanists connected with Middlesex whose assistance is acknowledged in the preface to the *Synopsis* by Dillenius are James Newton[*] and William Stonestreet, M.A., who were dead before the publication of the work; and—besides Rand and William Sherard—Charles Du Bois of Mitcham, Thomas Manningham, D.D.; Matthew Dodsworth, William Vernon, Thomas Dandridge, and John Martyn, who were all living in 1724. Particulars of all these would occupy too much space; the pages of this Flora bear witness to their activity and zeal, and the Sloane Herbarium contains much of the results of their explorations.

During this period, at which botany generally must be considered to have been at a somewhat low ebb in England, a little book was published on local botany of some interest as among the earliest of its class, and especially important to us as the first separate work devoted to Middlesex plants. The author was JOHN BLACKSTONE, of whom we would gladly give particulars, but can find little to record.

His two letters to Dr. Richardson, printed in the *Correspondence* (pp.

* Dillenius was indebted to John Martyn, who transcribed them 'from an obscure manuscript,' for the observations of Newton (see Preface to *Mart. Tourn.*). *James Newton* was probably in the medical profession, as he wrote a paper in the *Phil. Trans.* (xx. pp. 263–4) on the effects of *Papaver corniculatum luteum* eaten in mistake for Eryngo. He observed many plants about London, and entered localities in the margin of his copy of Ray's *Catalogus*. We have made use of a professed transcript of some of these in an unknown hand in the possession of the Rev. W. W. Newbould. He was also, according to Dryander, the author of '*Enchiridion universale plantarum*, or an universal and complete History of Plants, with their *Icons* in a *Manual*,' of which only 42 pages and 15 plates were printed, comprehending 'Liber 1. *De Arboribus Pomiferis*;' the date of this seems about 1680. James Newton first noticed in Middlesex *Erythræa Centaurium, Pedicularis palustris, Setaria verticillata, Avena fatua,* and *Polystichum aculeatum.* (It is necessary to mention that Pritzel confounds him with a later *James Newton,* who in 1752 published *A Complete Herbal.*)

351–355), are indeed the first notices of him which we have met with; they bear the dates Dec. 11 and Dec. 18, 1736. At this time he lived in the Strand, and was probably, as Dawson Turner suggests, 'in the course of his apprenticeship' to an apothecary. He says that he was much attached to botany, and had prepared a small work for the press, which he desired to publish by subscription, the price not likely to exceed five shillings. This was accordingly printed in the autumn of the next year, 1737, under the title *Fasciculus Plantarum circa Harefield sponte nascentium*, and dedicated to Sir Hans Sloane, from whom the author had received much kindness and encouragement. Harefield is a parish in the extreme north-west of Middlesex, and Blackstone's Flora includes little beyond its limits. He enumerates 524 plants, in alphabetical order, using the names of Gerarde and C. Bauhin's Pinax; special localities are given for the rarer species. The bulk of the book is in Latin, but there is an English appendix. A few varieties and cryptogams are mentioned.

Blackstone's connection with Harefield is partly explained in the latter of the letters mentioned above, in which he says, ' I have for these last three years been employ'd in making a collection of the native English plants; and having an opportunity of going to see my friends pretty often, I made it my business to see as many of the adjacent places as my time would permit, and to collect such plants as offer'd themselves in the course of my walks, without ever intending to publish anything on this subject. But, being detain'd last summer by a long illness, near four months, on the spot, I found so many rare plants that I thought it worth while to make a catalogue of them.* . . . The plants there mention'd were gather'd almost solely by myself . . . the catalogue is not general, being only intended as an essay for a more particular search thereabout.' His ' friends ' were the Ashbys, of Breakspears—at least, it was at Mr. Francis Ashby's house that he stayed during the spring and autumn of the next year, 1737 † (*Sloane MSS.* 4038).

In August 1737, Blackstone was living at the house of Alex. Benet, Esq., Maiden Ash, near Chipping Ongar, Essex, in which neighbourhood he noticed many plants.

At this time his health was bad, and he took a gloomy view of his prospects in life; in December, however, his health was much better, and he was ' determined to get into business some way.' After this we lose sight of him till 1740, in which year he was established as an apothecary in Fleet Street, ' at the Griffin near Salisbury Court ; ' ‡ and though ' trade is dull,

* By the kindness of Mr. Pamplin we have been favoured with the loan of an interleaved copy of Johnson's *Mercurius Botanicus* which belonged to Blackstone. It contains many notes in his hand of the localities of plants near Harefield, bearing date 1734, 1735, and 1736. All, or nearly all, are printed in the *Fasciculus*.

† Mr. Ashby is mentioned at p. 29 of the *Fasciculus* as having observed the Fritillary growing at Ruislip ' above forty years.' The parish church contains numerous monuments of this ancient family, which became extinct in the male line in 1800.

‡ In the British Museum is a folio tract of two pages, *The Modest Reply of J. Blackstone, Apothecary, to the abusive Reflections cast on him in a late Anonymous Paper*. No date.

and money very hard to get at,' he already contemplated a publication of the localities collected together since 1737.* These were printed in 1746, in the *Specimen Botanicum, quo plantarum plurium rariorum Angliæ indigenarum loci natales illustrantur,* which consists of original localities in various parts of England of 366 more or less rare species in alphabetical order. Many of the localities in the *Fasciculus* are repeated here, but there are besides numerous additions to the Middlesex Flora. There are two plates of Fungi in the volume. Not a few of the localities in our county were contributed by correspondents—John Wilmer, William Watson,† John Hill, Mr. Hurlock, and Mr. Nichols. This, says Pulteney, 'I consider as the last book published in England on the indigenous botany before the system of Linnæus had gained the ascendency over that of Mr. Ray.'

Of Blackstone's relatives we know nothing; his mother was living in 1737. He died in 1753. His dried plants are now in the British Museum, and the copy of Ray already noticed is in some sort a guide to them, Blackstone having marked with a cross all those species of which he possessed specimens.

In the Harefield Catalogue a very large number of plants, chiefly the commoner species, were first recorded as natives of our county; and, in all, Blackstone was the first to put on record no less than 232 species as Middlesex inhabitants. Amongst these are the chalk plants, including a number of orchids, many of which have since become extinct. Our own explorations at Harefield have corroborated so many of Blackstone's observations as to make us trust him for the rest; and though he cannot be considered as a botanist of a high order, he seems to have been a good and careful observer in the field. His name is connected with the discovery of *Dentaria bulbifera, Lathræa,* and *Fritillaria,* all at Harefield.

We have now arrived at the period of the writings of Linnæus, destined to work so great a revolution in science. The first sketch of the sexual system of classification was published at Leyden in 1735, in the *Systema Naturæ,* and this was followed by the *Genera Plantarum,* in 1737. The year before (1736), Linnæus himself had paid a visit to this country.‡ The *Species Plantarum* appeared first in 1753 at Stockholm, and soon after this the views of its illustrious author began to gain favour in England.

The first § British flora arranged on the new system was published in 1760, the author being the celebrated and versatile Dr. (commonly called Sir) John Hill. The *Flora Britannica* is a slovenly performance, in which the matter of the Dillenian edition of the *Synopsis* is re-arranged under

* In the library of the British Museum is a copy of the second edition of Ray's *Synopsis* (969, f. 18) with Blackstone's autograph, and the date '1736,' in which he entered these localities as he obtained them.

† Afterwards Sir William Watson, F.R.S.; for an account of whom see *Pulteney's Sketches,* ii. pp. 295-340.

‡ In his copy of the Dillenian edition of Ray's *Synopsis* (now in the library of the Linnæan Society) Linnæus has written, 'emi Londin. 1736.'

§ Grufberg, one of Linnæus' pupils at Upsal, had already in 1754 arranged the plants of the *Synopsis* by their Linnæan names. The list is printed in the *Amœnitates Academicæ,* iv. pp. 88-111.

Linnæan classes, orders, and genera. The latter are frequently misapplied, and the trivial names of Linnæus are not given. In this book are several new localities, and three additional species for Middlesex. Mr. Clements contributed many of them. Hill* died in 1775. A volume of 100 specimens of British plants preserved by him ' in a manner which prevents the common misfortune of their being destroyed by insects,' is in the British Museum.

This inadequate work was quickly superseded by the *Flora Anglica* of William Hudson, F.R.S., an apothecary living in Panton Street (afterwards in Jermyn Street), Haymarket, and sub-librarian of the British Museum from 1757 to 1758.† This appeared in 1762, and, as Sir J. E. Smith says, it 'marks the establishment of Linnæan principles of Botany in England.' From this time the artificial scheme reigned almost supreme in this country for more than sixty years, fostered by Smith and his followers, and owing to the facilities of study which its simplicity conferred, to which the binominal system of nomenclature still more contributed, numerous workers appeared, and the dead period was in turn succeeded by a time of activity.

Hudson's Flora contains many Middlesex localities,‡ especially about Highgate and Hornsey ; the majority stand on his own authority, but there are some contributed by Drs. Watson and Wilmer, already noticed as correspondents of Blackstone, and some by Stanesby Alchorne § and Peter Collinson. ‖ A second edition of this was published with considerable additions in 1778. Hudson died in 1793, and left his herbarium to the Apothecaries Society. The plants which he first noticed in this county are *Viola palustris, Geranium pyrenaicum, Herniaria glabra, Sedum dasyphyllum, Datura, Mentha viridis, Potamogeton compressus, P. pectinatus, Poa pratensis*, and *P. compressa*.

Between the two editions of the *Flora Anglica* several books, bearing more or less on our subject, were published. Of these, the first in point of time is the *Plantæ Cantabrigienses* of Thomas Martyn, M.A., printed in 1763, in the appendix to which are 'Lists of the more rare plants growing in many parts of England and Wales.' These are arranged in counties, and Middlesex occupies pp. 64–74. Very few of them are original, being taken chiefly from Blackstone and the Dillenian *Synopsis* ; a serious omission is found in the want of all quotations of the authorities. T. Martyn was born at Chelsea in 1735, and was perpetual curate of Edgware in our county ; he succeeded his father, John Martyn, as Professor

---

* For an account of the extraordinary life of this versatile but unscrupulous genius see the *Biographia Dramatica*, vol. i. p. 341.

† For some particulars of Hudson see vol. xviii. of *Rees' Cyclopædia*.

‡ His own copy, in the library of the British Museum (969, f. 25), has many MS. notes.

§ Of the Mint ; died 1800. His dried plants are preserved in the British Museum.

‖ Died 1768. Mr. Pamplin has kindly lent us a copy of Blackstone's *Fasciculus*, containing MS. notes by Collinson, who lived at Mill Hill, on Middlesex plants. It afterwards belonged to his son, *Michael Collinson*, who has added many notes of interest on the orchids formerly found at Harefield.

of Botany at Cambridge in 1761, and died in 1825, at the advanced age of eighty-nine. He performed a very useful work in editing the ninth edition of Miller's *Gardener's Dictionary*, which is a treasury of information on botanical matters.*

Sir T. G. Cullum, Bart., commenced an English Flora on the Linnæan system in 1774. It was discontinued, after 104 pages were printed, on the appearance of the second edition of Hudson's book. In this, as well as in James Jenkinson's *Generic and Specific Description of British Plants,* 1775 (which excludes all trees and the grasses), and in Stephen Robson's *British Flora,* 1777, we have found a few notes on Middlesex plants.

WILLIAM CURTIS,† to whom London botanists are especially indebted, was the eldest son of a Quaker, a tanner by trade, and was born at Alton, Hampshire, in 1746 ; there were three other sons and two daughters. His education was carried on at the small village of Hollybourn, and must have been of a very elementary character. As a boy he was fond of natural history, and made a great friend of an ostler named John Legg (or Lagg), who was well acquainted with indigenous plants.

At the age of fourteen, his grandfather, who was a thriving chemist and druggist at Alton, took him, intending to bring him up to his own trade. After six years, Curtis was sent to London, in 1766, and was at first with Mr. George Vaux of Pudding Lane, but soon left him and became assistant to Thomas Talwin, of 51 Gracechurch Street. All these well-meaning persons endeavoured to turn Curtis from his botanical pursuits, which they considered to be ruining his prospects. The latter partly succeeded, and Curtis attended the lectures of a medical school.

Medicine and practical pharmacy were, however, uncongenial subjects, and though followed from a sense of duty, Curtis's inclinations were ever towards botanical studies. On the death of Mr. Talwin, Curtis succeeded to the whole business, but soon gave up part, and ultimately the whole, of it to Mr. Wavell.‡

On December 15, 1772, Curtis was elected Demonstrator of Plants and Præfectus Horti to the Apothecaries Society, succeeding Mr. Stanesby Alchorne. He proposed to the society to deliver lectures on Botany at Apothecaries Hall, but the plan was not approved of ; on August 27, 1777, he resigned his appointment.

After the relinquishment of his practice, Curtis set up a small garden in Grange Road, Bermondsey, where he cultivated many plants ; though this

---

* See Gorham's *Memoirs of John and Thomas Martyn*, 1830, for further particulars.

† We have obtained our information about Curtis from the following sources :— *Gentleman's Magazine for* 1799, p. 628 ; *Rees' Cyclopædia* ; Sketch of the Life and Writings of the late Mr. Wm. Curtis, by Dr. Thornton (in third vol. of *Curtis's Lectures*), 1805 ; Memoirs of the Life and Writings of the late Mr. Wm. Curtis (prefixed to the *General Indexes to the first series of the Botanical Magazine*), by Samuel Curtis. 1828.

‡ This gentleman had considerable taste for botany, and was accustomed to accompany Curtis in his herbalising excursions ; he married a wealthy connection, and left London for Edinburgh, where he took a medical diploma, and afterwards set up as a physician at Barnstaple.

was a private garden, and was not scientifically arranged, he gave lectures there on Botany, instead of at Apothecaries Hall.

In 1774, some results of his explorations were printed in a *Catalogue of Plants growing Wild in the Environs of London.* This was published anonymously: it contains 632 Phanerogamic and 211 Cryptogamic species (of which 21 are Ferns, &c.). It is of little use, no localities, not even the county. being given, nor any indication of whether or not Curtis had himself observed all of them.

Three years after this, in 1777, he commenced his great work, original in design as in matter, the *Flora Londinensis.* Before entering on this undertaking, he issued an address to the public on the subject, and although not much encouraged, he persevered. This excellent book, of which Sir J. E. Smith remarks that it 'ranks next to Ray's *Synopsis* in original merit and authority upon English plants,' was published in numbers, each of which contained six plates with descriptions in folio, the price 2s. 6d. uncoloured, and 5s. coloured.* Twelve numbers formed a fasciculus, of which six (= 72 numbers) were issued, each with an index. The book is generally bound in two volumes, and the whole number of plates is 435.

It was the labour of his life to make this as perfect as possible, and he devoted to it the profits of his lectures, the subscriptions to his garden, and the money left him by his father. It progressed very slowly, Dr. Thornton says, 'like a funeral,' and was never profitable to the author. Once it came to a full stop; but Dr. Lettsom, finding Curtis in debt, very kindly gave him 500l. unsolicited, and 'without any proper security.' Curtis was persuaded to try an octavo edition, of which a very few numbers appeared.

Three hundred copies of each number were printed; the first volume is dedicated to the Earl of Bute, a great patron of Botany, and the second to Dr. Lettsom, who so materially helped the publication. No pains were spared in its production; the figures are life-size, full-coloured, and almost always very accurate. The artist employed was chiefly Mr. Sydenham Taste Edwards, but many of the drawings of the earlier numbers were by Mr. Kilburn, Mr. Milton, Mr. Sansom, and Mr. Sowerby; the whole of the colouring was done by Mr. William Graves, who was Curtis's constant assistant till the latter's death. The descriptions show a critical acquaintance with plants, and many species were first well defined and separated from their congeners in these pages. Curtis at first intended to include only the plants found within ten miles of London, but he was led, from a feeling of rivalry with *English Botany*, of which the first number appeared in 1790, to figure in the later numbers some from distant parts of England. It is to be regretted that while so many London plants remained to be figured, the consistency of the book should have been destroyed in this manner.†

* ' A few copies highly-finished, 7s. 6d.'—*Smith.*

† After Curtis's death the *Flora Londinensis* was sold (in 1818?) to Mr. Geo. Graves, and in 1815 a continuation was commenced by Professor (afterwards Sir) William J. Hooker, then of Glasgow. This was published in numbers, and the first volume completed in 1821; in 1828 a second volume was finished. As the whole of the original work was at the same time re-issued in three volumes, the two new volumes are generally called vols. iv. and v. In these the illustration of London botany is by no means made a prominent object.

At the request of his friend Dr. Lettsom, in July and August 1782, he made an excursion of six weeks to Settle, Yorkshire, and the same year published a *Catalogue of the Wild Plants* observed there.

About 1778 Curtis set up and opened to the public, on payment of an annual subscription, a botanical garden, and in this place he received much encouragement. From *A Catalogue of the British, Medicinal, Culinary, and Agricultural Plants cultivated in the London Botanic Garden*, printed in 1783, we learn that the position of the garden was 'very near the Magdalen-Hospital, St. George's Fields, in the road from the said hospital to Westminster-Bridge Turnpike through Lambeth Marsh village,' and that the subscription was one guinea a year, or two guineas with the privilege of receiving roots and seeds. In this little book is a complete list of British plants, arranged according to the month of flowering; all those growing in the garden are numbered consecutively, and amount to 1,008; the species additional to Ray and Hudson are printed in italics, and are 45 in number. But what renders the list very useful is, that all the plants found in the environs of London are distinguished by the letter L, from which we find that Curtis had noticed more than 700 species near town. A plan of the garden is also given, a catalogue of the books in the library, and a list of subscribers, amongst whom we find Sir Joseph Banks, the Hon. Daines Barrington, the Earl of Bute, and Drs. Lettsom, Sims, and Watson.

In 1787 Curtis commenced the popular *Botanical Magazine*, by which he is so well known. This was the most lucrative transaction of his life, and counterbalanced the loss occasioned by the *Flora Londinensis*. He himself said that 'one brought him pudding and the other praise.' It was published in monthly numbers at a shilling each, and it is said that as many as 2,000 copies of each were sold during Curtis's lifetime.

On the foundation of the Linnæan Society in 1788, Curtis was among the original members.

The spread of buildings, the new roads, smoke, and bad smells of the Lambeth garden, added to the greatly increased rent demanded by the landlord, compelled Curtis to leave the locality. In 1790 he set up a new and much larger garden at Queen's Elm, Brompton. A plan of this is given in Dr. Thornton's *Sketch* (p. 28); it was three and a half acres in extent, and there were seven acres adjoining for experiments in agriculture; a catalogue of the plants was published every year. After the establishment of this garden, 'he became,' says Thornton, 'sunk in spirits and indolent, I would rather say disappointed;' the *Flora Londinensis* was his favourite occupation, but the *Botanical Magazine* 'he stuck to as a drudgery.' In 1797 he took as his partner Mr. Wm. Salisbury, who, after Curtis's death, improved the garden and continued its management.[*]

After having laboured for about a year under an affection of the chest, he died on July 7, 1799, aged fifty-three years. His body was buried in Battersea churchyard, opposite the west entrance to the church; the stone has

---

[*] Mr. Salisbury subsequently removed the garden to Sloane Street, where it gradually degenerated, and in 1828 formed part of a nursery-ground.

a modest epitaph, but the four lines of rhyme at its conclusion are now (February 1868) quite illegible.

We have not met with the date of Curtis's marriage. He left a widow with one daughter* who in 1801 married a relative, Samuel Curtis, then a florist at Walworth. By his will Curtis left the *Flora Londinensis* and *Botanical Magazine* to his wife and daughter, and the control and management of his works to Dr. John Sims, one of his executors. The *Flora Lond.*, as we have mentioned, was sold to Mr. Graves, but Dr. Sims carried on the *Bot. Mag.* till 1826, when Samuel Curtis became the sole proprietor, and Dr. W. J. Hooker the editor. Samuel Curtis died so recently as 1860, at La Chaise, Jersey, at the age of eighty-one (see *Bot. Mag.* vol. lxxxvi.).

There is a portrait of Curtis prefixed to his *Lectures on Botany*, and also to his *Memoirs*.

Besides the important works mentioned, Curtis was the author of the following:—

Instructions for collecting and preserving Insects	1771
Fundamenta Entomologiæ of Linnæus (translation)	1772
Linnæus's System of Botany (so far as relates to his classes and orders)	1777
Proposals for opening the London Botanic Garden by Subscription	1778
A Short History of the Brown-tail Moth	1782
Assistant Plates to the Materia Medica	1786
An Enumeration of the British Grasses	1787
Companion to the Botanical Magazine	1788
Observations on Curculio Lapathi and Silpha grisen (Linn. Trans. 1. 86)	1788
Practical Observations on British Grasses (a second edition of the Enumeration)	1790
A Third Edition	1798
Proposals for a course of Herborising Excursions. (1 sheet)	1792 ?
Directions for cultivating Crambe maritima or Sea Kale	1799

After his death, his *Lectures on Botany as delivered in the Botanic Garden at Lambeth*, in three volumes, were published in 1805 by his son-in-law, Samuel Curtis; and Sir Joseph Banks edited two more editions of the *British Grasses*, in 1805 and 1812. In the *Linn. Trans.* (vol. vi. 15) is a paper, *Observations on Aphides by the late Mr. W. Curtis*, 'digested' by Sir J. E. Smith.

He contemplated a general Natural History of Great Britain, and made excursions for the purpose of collecting material; in these he was always attended by Mr. Sydenham Edwards, who sketched birds, plants, &c., on the spot, to which Curtis afterwards appended minute descriptions. Many hundred original drawings of Mr. Edwards were bequeathed to Samuel Curtis.

The favourite haunts of Curtis near London were Charlton, in Kent, and Battersea Fields, in Surrey, then very different to their present condition, and very productive of rare plants. In Middlesex, the western suburbs of town, and Hounslow and Hampstead Heaths, were explored by him, as well as the Isle of Dogs, and other parts. His observations in these localities

* She died July 2, 1827, leaving six sons and seven daughters with her husband.

are scattered through the *Flora Londinensis.* Besides many fresh stations for those previously known, Curtis first noticed the following plants in Middlesex : *Fumaria capreolata, Nasturtium palustre, Thlaspi arvense, Sagina subulata, Spergula, Trifolium fragiferum, Epilobium tetragonum, Leucoium æstivum, Sclerochloa distans, S. rigida, Festuca pratensis, Bromus asper.*

In 1789, Mr. Richard Gough, the antiquarian, edited a new edition of Camden's *Britannia.* In this the county lists of plants are considerably more extensive than those in Gibson's edition. In the preface it is stated, '. . . some young friends have exerted their utmost diligence in collecting the plants peculiar to each county from books and the researches of themselves and other botanists, who have multiplied since Ray in the same proportion as the science has improved.' The list of ' rare plants found in Middlesex ' occupies pp. 32–39 of vol. ii. Its foundation is Petiver's list in the former edition (fitted with Linnæan names), with additions from Dillenius's Ray's *Synopsis,* Blackstone, Hudson, and Curtis. There are also several original, chiefly in the eastern portion of the county ; in all, 402 species are enumerated. We are informed by Mr. Pamplin that the compiler of this was the late Edward Forster, the well-known Essex botanist. At the time this catalogue was compiled he was twenty-four years of age.*

The next book we have to notice is called *Indigenous Botany, or Habitations of English Plants ; containing the result of several Botanical Excursions, chiefly in Kent, Middlesex, and the adjacent counties, in* 1790–92, by Colin Milne, LL.D. and Alexander Gordon, 1793. Only one volume was ever published ; it includes the Linnæan classes to the end of Pentandria. This work is very original in design, and the remarks often judicious ; it appears generally trustworthy, though there are some instances of apparently false diagnosis (e. g. *Drosera intermedia, Phyteuma orbiculare, Spartina stricta,* and *Buffonia*). Many original stations in Middlesex are recorded, and a few new species added : *Alchemilla vulgaris, Apium, Œnanthe Phellandrium, Borago, Potamogeton obtusifolius ?, Alopecurus geniculatus, Agrostis alba.*

James Dickson, F.L.S., a nurseryman in Covent Garden, published in 1793 the first fasciculus of his *Hortus Siccus Britannicus,* consisting of twenty-five dried plants, named and mounted. Nineteen numbers were published, the last in 1802. There are a few Middlesex localities given, and three new plants, *Herniaria hirsuta, Gentiana Pneumonanthe,* and *Aira Caryophyllea.* He died in 1822 (see a short memoir in *Trans. Hort. Soc. Lond.* vol. v.).

The *Botanist's Guide,* by Dawson Turner and Lewis Weston Dillwyn, F.R.SS., was published in two volumes in 1805. The Middlesex list is contained in vol. ii. pp. 399–414, and is not very extensive, 144 species only being noticed. The names are those of Smith's *Flora Britannica,* and the original authority for each locality is always stated. Of the new ones a few were supplied by the authors, but more by Mr. Joseph Woods, who recorded nine species new to the county. Edward Forster's stations are mostly the

* Forster died in 1849, æt. 84. His herbarium, containing also plants of Mr. Borrer's, Mr. Lambert's, and Mr. W. Wilson's collecting, is now in the British Museum. A short biography is to be found in Gibson's *Flora of Essex,* Appendix vi.

same as those in Gough's *Camden*. Other contributors were Sir J. Banks, Rev. Dr. Goodenough of Ealing, Mr. Dickson, and Messrs. Borrer, Lambert, Teesdale, Wood, and Woodward.

A little-known tract of thirty-six pages, printed in 1813, has also the title *The Botanist's Guide*. It is 'a catalogue of scarce plants found in the neighbourhood of London.' There is no author's name, but it is dedicated to J. E. Smith, M.D., ' by the Editor.' In Sir J. E. Smith's own copy, now in the library of the Linnæan Society, he has written, 'from his honoured friend Jos. Cockfield, dedication copy,' from which it seems that Mr. Cockfield * was the author. There is very little original in this pamphlet; the Middlesex list occupies pp. 9–17, and is mainly from Turner and Dillwyn. A second list (pp. 29–32) contains Blackstone's Harefield plants, but by an unfortunate misprint, ' Hounslow ' instead of Harefield, the usefulness of this portion is much impaired.

By the kindness of Mr. E. G. Varenne, of Kelvedon, Essex, we have been furnished with the labels from the herbarium of two medical practitioners, who between 1815 and 1823 botanised much about London. Their names were William Frederick Goodger and Richard Rozea, the former ' resident apothecary to the parochial infirmary of St. Marylebone from 1811 to 1832 ; ' the latter a surgeon of Marylebone, who died about forty years ago. The herbarium was presented to Mr. Varenne six or seven years back ; ' as might be imagined, the specimens were injured by long neglect, but, nevertheless, were quite good enough for recognition and examination. Had the localities been attached to every specimen, a very complete list of the common plants of Middlesex might have been formed' (Varenne in litt. May 1867). They have put on record twelve species not before noticed in our county.

The *Topography and Natural History of Hampstead*, by John James Park, was printed by subscription in 1813, but published with additions in 1818. Pp. 22–42 are occupied by an account of the vegetable productions, which consists of Johnson's *Iter* of 1629, given entire, with a few of his plants determined, some quotations from Gerarde, and ' a catalogue of the rarer plants now found on the heath and in the meadows, woods, and ponds in its vicinity . . . compiled from the personal observations of Mr. Bliss and Mr. Hunter, Lord Mansfield's steward at Ken Wood House.' This list is further said to have been revised by Mr. Wheeler, ' Professor of Botany to the Society of Apothecaries,' but is a very faulty affair ; 61 species are enumerated, of which 19 are mosses, &c.; of the remaining 42, not a few mentioned as growing in Ken Wood, if ever there, must have been planted : such are *Staphylea*, *Pyrus domestica*, *Empetrum*, *Scilla verna*, *Polygonatum verticillatum*, *P. officinale*, and *Stipa pennata*. *Maianthemum*, however, is noticed there for the first time (as *Convallaria bifolia*).

At p. 22 of Faulkner's *Historical and Topographical Account of Fulham*, 1813, is a list of a few plants found in 1811 'on the banks of the Thames and elsewhere ' in this parish. It contains nothing of importance.

The first British Flora arranged on the modern views of natural affinities

* This gentleman lived at Upton, and was a friend of Forster.

was Samuel F. Gray's *Natural Arrangement of British Plants*, published in 1821. This valuable book has never attracted the attention it deserves; it was strongly opposed and decried by the votaries of the Linnæan system, and set aside by the *English Flora* of Sir J. E. Smith, arranged on the Linnæan system, which appeared only three years afterwards, in 1824. Gray's book is certainly superior to Dr. Lindley's *Synopsis of the British Flora, arranged according to the Natural Orders*, printed in 1829, in the preface to which it is deliberately ignored.

Mr. Alexander Irvine, now of Chelsea, prepared a MS. catalogue of plants noticed in the years from 1825 to 1834, within two miles of Hampstead Heath. We have been kindly furnished by the author with this list, which contains upwards of 600 plants, some of which, however, are certainly not now to be met with. The same botanist published in 1838 *The London Flora*, which, notwithstanding its title, includes the whole S.E. of England, and even extends into the midland counties. Many original Middlesex localities are given in this book, generally trustworthy. Mr. Pamplin, Dr. Golding Bird, and Mr. Ralph contributed some of them. The *Handbook of British Plants*, by the same author, published in 1858, has been found useful. Mr. Irvine was among the first to pay special attention to introduced exotics and naturalised species; many of those found in our county were first noticed by him. He also first recorded in Middlesex the occurrence of *Barbarea præcox, Erysimum cheiranthoides, Lepidium ruderale, Sagina ciliata, Anthemis arvensis, Myosotis cæspitosa, Chenopodium hybridum, Lemna trisulca, L. gibba, Carex binervis, Aira flexuosa*, and other commoner species.

In the *Medical Botany* of Dr. Stephenson and Mr. Churchill, published in 1828–31, there is in vol. iii., under tab. 105, a list of plants growing on and near Hampstead Heath: 113 names are given; they are chiefly from Martyn, Curtis, Blackstone, the *Botanist's Guide*, and other previously published books.

The first volume of Mr. Hewett C. Watson's *New Botanist's Guide* was published in 1835. The list for ' Middlesex and London ' is in pp. 97–103. In addition to Mr. Watson's own observations, many of Mr. Pamplin's are here given, and the MS. notes of Mr. J. Winch in his copy* of the *Flora Britannica* are entered. The second volume, published in 1837, contains at pp. 586–589 a supplementary list of Middlesex localities. Several of these were contributed by G. E. Dennes and R. Castles. Of the invaluable *Cybele Britannica* by Mr. Watson we need not speak; it is quoted systematically throughout this Flora. Mr. Watson was the first to observe *Diplotaxis muralis, Cerastium arvense, Œnanthe Lachenalii, Filago spathulata*, and *Allium oleraceum*, in Middlesex.

The *Flora Metropolitana*, by Daniel Cooper, was printed in 1836, between the issue of the two volumes of the *New Botanist's Guide*. It is ' the result of numerous excursions, made in 1833, 1834, and 1835,' within thirty miles of

---

* This book is in the library of the Linnæan Society. We have entered the MS. notes from the original source. They refer to years bet. 1800 and 1830.

London, and is dedicated to J. Forbes Young, Esq., M.D.* The author regrets (in the preface) that he had given little attention to our county, and the lists for Middlesex (pp. 98–115) are compilations; a star is affixed to the plants Cooper had himself seen, and from this indication it appears that the only parts of the county he had visited, were Greenford, Fulham, Highgate, and Hampstead. Authorities, however, are generally quoted, and we also find the names of the following contributors: James Carter, W. Chatterley, W. Cullen, Miss Gawler, J. L. Gowler, W. Lloyd, Mrs. Lowe, T. W. Mann, and Dr. James Mitchell. A *Supplement* ' containing many additional localities procured last summer,' appeared in 1837. The Middlesex additions are few (pp. 11 and 12); they were all contributed by E. H. Button, G. E. Dennes, and G. Francis. Though at the time perhaps a useful publication, the *Flora Metropolitana* must be considered as one of the least trustworthy books of its class. The errors contained in it are numerous, but seem to be due less to faulty observation on the part of the author, than to want of judgment in entering the information received from others or extracted from books.

In 1860 the Rev. W. M. Hind, now incumbent of Pinner, published in the *Harrow Gazette*, for January 16 and February 3, a list of plants growing about the town. These papers were reprinted in the *Phytologist* for 1860 (*N. S.* vol. iv. pp. 107–119), under the title, ' The Flora of Harrow and its vicinity;' the list contains 385 species and varieties. In vol. v. (1861) of the same periodical (pp. 198–204), Mr. Hind added 162 more plants, and corrected a few errors. In all, he contributed fifteen unrecorded species to our Flora, *Carex axillaris* being the most interesting.

Mr. J. C. Melvill's *Flora of Harrow* came out in 1864. This interesting little book, which contains also lists of the Birds and Lepidoptera of the district, was ' entirely drawn up by Harrow boys.' The Flora is founded on Mr. Hind's lists, but by the further researches of Mr. Melvill the number of species is increased to 568, and 49 varieties are mentioned. The Rev. F. W. Farrar, of Harrow School, wrote the preface, and continues to foster a taste for natural science in the school. In connection with the *Harrow Flora* we have consulted the prize herbarium collected by its author, and now preserved in the school library at Harrow: this has in several cases served to clear up difficulties.

Nothing further, that we know of, has been published, beyond isolated notes in the journals. Those botanists who have contributed information in other ways are enumerated in another place, and their observations, with our own, which extend over the past nine years, form with the published records all that is known of the local botany of Middlesex.

---

* Dr. Young's herbarium of British plants is now at Kew. It contains many specimens collected in Middlesex between 1833 and 1844 by Dr. Young, Mr. Cooper, Henry Kingsley, Dr. McCreight, J. Taylor, and others.

APPENDIX.

LISTS OF CRYPTOGAMIA.

I. MUSCI.

This list is chiefly composed of notes communicated by Dr. Braithwaite and Rev. W. M. Hind. It is arranged after Wilson's *Bryologia Britannica*, but is a mere outline.

SPHAGNUM *cymbifolium*, Dill. Harrow Weald; *Hind.* Heaths; *Braithwaite.*

S. acutifolium, Ehrh. Heaths; *Braithwaite.*

S. cuspidatum, Dill. Harrow Weald; *Hind.* Heaths; *Braithwaite.*

S. contortum, Schultz. *Var. β.* subsecundum. Heaths; *Braithwaite.*

PHASCUM *serratum*, Schreb. First discovered in England by Mr. Dickson on the north side of Muswell Hill; *E. B.* 460.

P. muticum, Schreb. Common on moist banks about London, *Curt F. L. f.* 4. Hampstead Heath; *Braithwaite.*

P. cuspidatum, Schreb. Fallow-fields; *Braithwaite. Var. β.* piliferum. Near London, Mr. Dickson; *E. B.* 1888.

P. nitidum, Hedw. Kensington Gardens (the specimen figured); *E. B.* 1036. North side of Muswell Hill, Mr. Dickson; *Burnett* 105.

P. subulatum, L. Hampstead Heath; *Syme & Braithwaite.*

WEISSIA *contraversa*, Hedw. Banks near Highgate; *Braithwaite.* Hampstead; *Syme.*

W. cirrhata, Hedw. Trees on east heath, Hampstead; *Syme.*

DICRANUM *crispum*, Hedw. Woods near Southgate; *With.* ii. 97.

D. varium, Hedw. Harrow Weald; *Hind.* Clay banks; *Braithwaite.*

D. cerviculatum, Hedw. On Hampstead Heath, Dickson; *With.* iii. 813.

D. heteromallum, Hedw. Harrow Weald; *Hind.* Hampstead Heath; *Braithwaite.* Bishop's Wood, abundant.

D. scoparium, Hedw. Pinner; *Hind.* Hampstead Heath in fruit!; *Curt. F. L. f.* 4.

D. majus, Turner. Pinner; *Hind.*

LEUCOBRYUM *glaucum*, Hampe. Harefield (in fruit), Herb. Ward; *Braithwaite.* Harrow Weald; *Hind.*

CERATODON *purpureus*, Bridel. Harrow Weald; *Hind.* Hampstead; *Syme.* Heaths, everywhere; *Braithwaite.*

CAMPYLOPUS *torfaceus*, Br. & Sch. Hampstead Heath; *Braithwaite.*

POTTIA *minutula*, Br. & Sch.   Fallow fields ; *Braithwaite.*

P. truncata, Br. & Sch.   Harrow.   Hampstead Heath ; *Syme.*   Near Bishop's Wood.

DIDYMODON *rubellus*, Br. & Sch.   Harrow ; *Hind.*

TORTULA *unguiculata*, Hedw.   Harrow Weald ; *Hind.*   Near Highgate ; *Braithwaite.*

T. fallax, Hedw.   Ruislip ; *Hind.*

T. Hornschuchiana, Schultz.   Laleham ; *Braithwaite.*

T. muralis, Timm.   Pinner ; Harrow! ; *Hind.*   Hampstead ; *Syme.*   Everywhere ; *Braithwaite.*

T. subulata, Bridel.   Pinner ; *Hind.*   Hampstead ; *Syme.*

T. ruralis, Hedw.   In fruit on roof of shed opp. 'Spaniards,' Hampstead ; *Syme.*   Common ; *Braithwaite.*

CINCLIDOTUS *fontinaloides*, P. Beauv.   Thames about London, frequent ; *Huds.* i. 398.   On stones among the mud at low water in the bed of the river opposite Lambeth Palace ; *Mackay.*

ENCALYPTA *vulgaris*, Hedw.   Harefield! ; *Newb.*

GRIMMIA *pulvinata*, Smith.   Common ; *Braithwaite.*   Harrow Weald ; *Hind.*

RACOMITRIUM *lanuginosum*, Bridel.   Hampstead Heath ; *Burnett* 105.

R. canescens, Bridel.   Hampstead ; Enfield Chase, Dillenius ; *With.* ii. 95.

ORTHOTRICHUM *affine*, Schrad.   Pinner ; *Hind.*

O. diaphanum, Schrad.   On trees ; *Braithwaite.*

O. crispum, Hedw.   Pinner ; *Hind.*

TETRAPHIS *pellucida*, Hedw.   Staines ; *Braithwaite* ; abundant by side of carriage-drive, Ken Wood ; *Mackay.*

ATRICHUM *undulatum*, P. Beauv.   Harrow ; *Hind.*   Hampstead Heath ; *Braithwaite.*   Bishop's Wood.

POGONATUM *nanum*, Bridel.   Not uncommon about London ; *Curt. F.L.* f. 2.   Staines ; *Braithwaite.*

P. aloides, Bridel.   Pinner ; Harrow Weald ; *Hind.*   Hampstead Heath ; *Braithwaite.*

POLYTRICHUM *commune*, L.   Pinner ; *Hind.*   Hampstead Heath.   *Var. γ.* minus.   Harrow Weald Common ; *Herb. Harr.*

P. juniperinum, Hedw.   Hampstead Heath (the specimen figured) ; *E. B.* 1200 & *Braithwaite.*

P. piliferum, Schreb.   Hampstead Heath ; *Syme & Braithwaite.*

AULACOMNION *palustre*, Schwaegr.   Hampstead Heath ; *Huds.* i. 403.   With pseudopodia, on West Heath ; *Syme.*

A. androgynum, Schwaegr.   Hampstead Heath ; *Huds.* i. 403.   Near Hampstead ; *Braithwaite.*

LEPTOBRYUM *pyriforme*, Schimp.   In 1790, and often since, on the inside of the walls of greenhouse frames in Lee & Kennedy's garden, Hammersmith ; *E. B.* 389.   On a bank at Clapton, E. Forster ; *B. G.* 413.

BRYUM *nutans*, Schreb.   Pinner ; *Hind.*   Hampstead Heath ; *Braithwaite.*

B. capillare, Hedw.   Pinner ; *Hind.*

B. cæspiticium, L.   Pinner, *Hind.*   Harrow.

B. atropurpureum, Web. & Mohr.   East Heath, Hampstead ; *Syme.*

B. argenteum, L.   Pinner ; *Hind.*   Hampstead ; *Syme.*   Common in and round London.

MNIUM *rostratum*, Schwaegr. Harrow Weald; *Hind.*

M. hornum, L. Pinner; *Hind.* In fruit about Hampstead; *Curt. F. L.* f. 1. Hampstead; *Syme.*

M. undulatum, Hedw. Pinner; Harrow; *Hind.* Renter's Lane, Hendon; *Davies.*

FUNARIA *hygrometrica*, Hedw. Pinner; Harrow; *Hind.* Hampstead; *Syme.* Edmonton. Winchmore Hill Wood.

PHYSCOMITRIUM *pyriforme*, Br. & Sch. Near Highgate; *Braithwaite.*

BARTRAMIA *fontana*, Bridel. Bog on North Heath, Hampstead; *Davies.* Bogs; *Braithwaite.*

B. pomiformis, Hedw. On Hampstead Heath, Dillenius; *Huds.* i. 404. Heaths; *Braithwaite.*

FISSIDENS *exilis*, Hedw. Boggy ground in Enfield Chase, Dickson; *Herb. Mus. Brit. & Wilson*, 302.

F. bryoides, Hedw. Hampstead; *Syme.* Harrow.

F. adiantoides, Hedw. Shady pit in Kensington Gardens, towards Bayswater Gate; *E. B.* 426.

F. taxifolius, Hedw. With the last in Kensington Gardens, *E. B.* 426. Harrow; *Hind.* Near Highgate; *Braithwaite.*

LEUCODON *sciuroides*, Schwaegr. Harrow; *Hind.*

ANTITRICHIA *curtipendula*, Bridel. On stumps in Enfield Forest; near Southgate; *With.* ii. 134.

PTEROGONIUM *gracile*, Sw. On beeches in Enfield Chase, abundantly; *Huds.* i. 430. Hampstead; *S. & C.* 105.

ISOTHECIUM *myurum*, Dill. On trees; *Braithwaite.*

I. myosuroides, Dill. Walls at Hampstead, Dill.; *With.* ii. 138. Harrow; *Hind.*

I. alopecurum, Dill. Harrow; *Hind.*

LESKEA *polycarpa*, Ehrh. On trunks of Willows, at Chiswick, Mr. Teesdale; *E. B.* 1922. Harrow Weald; *Hind.*

L. sericea, Dill. Harrow: *Hind.*

HYPNUM *albicans*, Dill. Hampstead Heath; *Braithwaite.*

H. glareosum, Bruch. Harrow; *Hind.*

H. lutescens, Dill. On walls; *Braithwaite.*

H. plumosum, Swartz. On trunks of trees in Enfield Chase; walls about London; *Huds.* i. 423.

H. populeum, Hedw. On trees; *Braithwaite.*

H. velutinum, Dill. Hampstead; *Braithwaite.*

H. rutabulum, Dill. Harrow!; *Hind.* Hampstead; *Braithwaite.*

H. prælongum, Dill. Common; *Braithwaite.* Harrow.

H. striatum, Hedw. Pinner; *Hind.*

H. ruscifolium, Dill. Harrow; *Hind.* Staines; *Braithwaite.*

H. confertum, Dicks. Banks; *Braithwaite.*

H. serpens, Dill. Harrow; *Hind.*

H. riparium, Dill. Pinner; *Hind.*

H. stellatum, Dill. Bogs; *Braithwaite.*

H. chrysophyllum, Bridel. Hampstead; *J. Bagnall*, who showed specimens at Birmingham Nat. Hist. Soc. 1868.

H. palustre, Dill. Hampstead Heath; *Burnett*, 105.

H. stramineum, Dicks. West side of Hampstead, Mr. Dickson; *B. G.* 2405. Fruits sparingly in the bog on West Heath; *Syme.*

H. cordifolium, Swartz. Harrow; Pinner; *Hind.*

H. cuspidatum, Dill. Hampstead Heath; *S. & C.* 105, and *Syme*. Meadow between Finchley and Bishop's Wood; *Davies.* Harrow Weald; *Hind.*

H. Schreberi, Dill. Pinner; *Hind.*

H. purum, Dill. Harrow Weald; *Hind.*

H. tamariscinum, Hedw. Common about London; *Curt. F. L.* f. 4. Pinner; *Hind.* Bishop's Wood.

H. splendens, Dill. Woods; *Braithwaite.*

H. triquetrum, Dill. Harrow; *Hind.*

H. loreum, Dill. Harrow; *Hind.*

H. squarrosum, Dill. Harrow; *Hind.*

H. fluitans, Dill. Ditches near Hackney; *With.* ii. 120. North Heath, Hampstead; *Davies & Braithwaite.* Harrow; *Hind.*

H. uncinatum, Hall. Harrow Weald; *Hind.*

H. molluscum, Dill. Banks; *Braithwaite.*

H. cupressiforme. Dill. Harrow; *Hind.* Common; *Braithwaite. Var. β. compressum*, wood near Tottenham, Dill.; *Huds.* i. 423.

H. undulatum, Dill. Bishop's Wood, Buddle; *Huds.* i. 420. Pinner; *Hind.*

H. sylvaticum, Dill. Harrow; *Hind.*

H. denticulatum, Dill. Hampstead; *Syme.* Harrow Weald; *Hind.*

HOMALIA *trichomanoides*, Dill. Hillingdon; *Braithwaite.*

NECKERA *complanata*, Br. Eur. Hampstead Wood, Dill; *Huds.* i. 419. Harrow; *Hind.*

N. crispa, Dill. Hampstead Heath; *S. & C.* 105.

CRYPHÆA *heteromalla*, Dill. On Enfield Chase, Dill.; *Huds.* i. 396.

FONTINALIS *antipyretica*, L. Harrow; *Hind.* Duke's River, Twickenham.

F. squarrosa, L. On old planks in the Thames, Martyn; *B. G.* 413.

II. HEPATICÆ.

JUNGERMANNIACEÆ.

SCAPANIA *undulata*, Nees. Hampstead Heath; *Budd. MSS.* Harrow Weald Common; *Hind.*

PLAGIOCHILA *asplenioides*, N. & M. Pinner; *Hind.*

JUNGERMANNIA *albicans*, L. Woods at Hampstead; *Budd. MSS.*

J. crenulata, Sm. Harrow Weald Common; *Hind.*

? *J. sphærocarpa*, Hook. About Highgate rivulet in Old-Fall Wood; *With.* iii. 872.

LOPHOCOLEA *bidentata*, Nees. Pinner Wood; *Hind.* North Heath, Hampstead; *Davies.*

L. heterophylla, Nees. (= *J. cuspidata*, E.B.) Pinner Wood; *Hind.* Hornsey Wood, 1795, in full fructification; *E. B.* 281. Hampstead; *Burnett* 105.

CALYPOGEIA *Trichomanes*, Cord. Hampstead, *Huds.* i. 401. Pinner Wood; *Hind.*

LEPIDOZIA *reptans*, Nees. Bishop's Wood; *Budd. MSS.* Hampstead, Martyn; *B. G.* 413.

Trichocolea *tomentella*, Nees. Wood between Highgate and Hornsey; *Budd. MSS.*

Ptilidium *ciliare*, Nees. By water running through Old-Fall Wood, between Highgate and Muswell Hill, Mr. Dandridge; *R. Syn.* iii. 111.

Martinellia *complanata*, Gray. (Radula, *Dum.*) Pinner Wood; *Hind.*

Frullania *dilatata*, Nees. Harrow; *Hind.*
? *F. Tamarisci*, Nees. Hampstead; *Burnett* 105.

Fossombronia *pusilla*, Nees. Bishop's Wood, Martyn; *B. G.* 413.

Pellia *calycina*, Nees. Harrow; *Hind.*
P. epiphylla, Nees. Hampstead; *Burnett* 105. Harrow Weald Common; *Hind.*

Blasia *pusilla*, L. On Hounslow Heath; *Huds.* ii. 519.

Riccardia *multifida*, Gray. (Aneura, *Dum.*) Cane Wood, Doody; *Huds.* i. 437.

Metzgeria *furcata*, Nees. In a deep dry ditch between 'Mother Huff's' and Cane Wood, Hampstead; *Budd. MSS.*

Anthoceros *punctatus*, L. On Hounslow Heath; *Huds.* i. 519. About London; *E. B.* 1537.

MARCHANTIACEÆ.

Marchantia *polymorpha*, L. Hampstead Heath. Bishop's Wood.

RICCIACEÆ.

Riccia *fluitans*, L. Pinner; Harrow; *Hind.* Pond in the meadows between Bishop's Wood and Finchley, 1866, abundant; *Davies.* Pond on Hadley Common! ; *Warren.*

III. CHARACEÆ.

Chara *flexilis*, L. Ponds at Hendon; *Quekett.* Near Hornsey; *Huds.* i. 399.
C. syncarpa, Thuill. Ruislip Reservoir, 1861; *Hind.*
C. translucens, Pers. Stanmore Heath, 1827–30; *Varenne.*
C. vulgaris, L. Woodridings, Pinner! ; Roxeth; Apperton; *Hind.* Isle of Dogs; ponds at Notting Hill; *Quekett.* Kilburn! ; *Warren.*
C. hispida, L. Finchley Common, not unfrequent, J. Woods; *B. G.* 399.
C. fragilis, Desv. Pinner Hill; Eastcott; Ruislip Reservoir; *Hind.*

IV. LICHENES.

BY REV. JAMES M. CROMBIE, M.A., F.L.S.

The county of Middlesex is by no means rich in Lichens. In fact, it is the very poorest of any county in Great Britain. The dense and smoky atmosphere surrounding the metropolis is very unfavourable to the growth of this order of cryptogams. Hence it is that the old trees in the parks, which might be expected to yield a fair number of species, present but very scanty traces indeed of any lichen vegetation. Nor even in the remoter suburbs do we find any great improvement, though one or two very common species here begin to appear on the coping of walls about villas. It is only when we reach the higher grounds in the county to the north and west, at some distance from London, that we find, in suitable habitats, several of the more frequent and universally distributed species. Amongst these may be mentioned the following, though doubtless some others may on further examination be detected:—*Cladonia pyxidata*, Linn., not uncommon on heaths; *C. fimbriata*, Linn., infrequent, and its vars. *tubæformis*, Flk., *radiata*, Schreb., as at Enfield and Pinner; *C. furcata*, Hoffm., and its var. *racemosa*, Hoffm., not very uncommon in woods; *C. coccifera*, Ach., common, as on Harrow Weald; *Cladina sylvatica*, Linn., and *C. rangiferina*, Hoffm., on heaths and in woods, but barren; *Bæomyces rufus*, De C., rare on heaths, and scarcely fertile; *Usnea barbata*, Frs. var. *hirta*, Linn., on old trees in woods, barren, and not well developed; *Evernia prunastri*, Ach., on trees in woods, sparingly and barren; *Ramalina calicaris*, Ach., and its vars. *fastigiata*, Pers., *farinacea*, Linn., not uncommon on trees and hedges, but seldom in good condition; *Peltigera canina*, on mossy banks, and on the ground in woods, not uncommon; *P. polydactyla*, Hoffm., in moist places of shady woods, rare, as at Hampstead and Pinner; *Parmelia caperata*, Dill., rare, on trees and old pales in the higher tracts, and always barren; *P. saxatilis*, Linn., on the trunks of old trees in woods, not frequent nor fertile; *P. olivacea*, on trunks and branches of trees, rare, as at Pinner; *P. perlata*, Linn., on the trunks of aged trees, here and there in the north of the county; *Physcia stellaris*, var. *hispida*, Dill., on the ground on heaths, but by no means well developed; *Physcia cæsia*, on old walls and roofs of houses, probably not infrequent; *P. parietina*, Linn., on roofs of houses, pretty common, and its forms *viridis*, Schreb., and *citrinella*, Frs., on old pales, even in the suburbs of London; *Placodium murorvm*, on old brick walls and grave-stones, but scarcely fertile; *Lecanora subfusca*, Linn., on trees and walls in the suburbs and country, common, and its var. *atrynea*, Ach., frequent on old pales; *L. exigua*, on aged trees and old roofs, probably not rare; *L. atra*, Huds., on brick walls, at least in the northern districts; *L. varia*, Ehrh., common on old trees and pales, as also its var. *symmicta*, Ach., but rarely fertile; *Psora ostreata*, Hoffm., on old pales, not unfrequent in the northern tracts, as at Finchley, and abundant near Totteridge, but never fertile; *Lecidea canescens*, Dcks., common on the trunks of old trees, but always barren; *L. denigrata*, Frs., on old pales near water, rare, as on the Finchley Road; *L. sahuletorum*, Flk., on decaying mosses on shaded brick walls, probably not rare; *S. pelidna*, Ach., on calcareous and arenaceous stones, as bridges and churchyards, not unfrequent in the rural districts; *L. uliginosa*, Ach., on clayey soil on heaths frequent, and abundant at Hampstead; *L. parasema*, Ach., on the trunks of trees and on pales, frequent in the north and west of the county; *L. myriocarpa*, De C., on old posts and pales on the borders of fields, apparently not uncommon; *Opegrapha varia*, Pers., on the bark of various trees, pretty frequent; *O. atra*, Pers., on the smooth bark of young trees, not rare on the outskirts of woods; *Graphis scripta*, Linn., on the trunks of trees in woods, common and variable; *Arthonia astroidea*, Ach., on the smooth bark of trees, apparently rare, as at Pinner; *A. melaspermella*, Nyl., on wood (pales) near London, rare; *Stigmatidium crassum*, on the bark of old hornbeams, perhaps not unfrequent in the north of the county; *Calicium curtum*, Turn. & Borr., on old pales and decaying trees, not very uncommon, as at Ealing and near Totteridge; *C. trichiale*, Ach., on old posts, probably not rare, as about Hendon; *Pertusaria communis*, D. C., on the trunks of large trees, chiefly beech, in shady woods, common; *P.*

multipunctata, Turn., with the preceding, and usually in a sorediate condition; *Verrucaria viridula*, Schrad., on the mortar of old walls, not infrequent; *V. muralis*, Ach., with the preceding, and on calcareous stones in the cemeteries; *V. mutabilis*, Borr., on stones and flint pebbles in open fields on a gravelly soil, common; *V. epidermidis*, Ach., on the smooth bark of trees in shady woods, and its var. *atomaria*, Ach., pretty frequent. From this list, which, however, is not to be regarded as a complete one, it will at once be perceived how poor a field the county of Middlesex presents to the lichenologist. Nor would this be less true, even if some few species of *Lecideæ* and *Verrucariæ*, which will no doubt repay further research, were added. Very different is it when we enter Essex, where, in Epping Forest at least, a goodly harvest may be gathered. The only rarity as yet found in Middlesex is *Arthonia melaspermella*, Nyl., a new species recently gathered by Mr. Currey, and recorded by Dr. Nylander in *Flora*, 1865, p. 605.

V. HYMENOMYCETOUS FUNGI.

BY WORTHINGTON G. SMITH, F.L.S.

The sequence of the species is founded on the later views of Fries as expressed in his *Monographia Hymenomycetum Sueciæ*: nearly all the species here enumerated, several being very recent additions to the British Flora, have been gathered by the author within ten miles north of London. This district, though in many places woody, is virtually destitute of fir groves, therefore the plants peculiar to these prolific localities are few and unimportant. When a species is known to be either wholesome or poisonous, its nature is stated, the author's own experience being supplemented by some notes communicated by Mr. J. A. Clark, of Street, Somerset. A few species and localities have been added by Mr. M. C. Cooke and the Rev. W. M. Hind, and a few others extracted from Sowerby's *British Fungi* (1797–1803).

ORDER I. *AGARICINI.*

AGARICUS, *Linn.*

SECTION I. LEUCOSPORI. *Spores white.*

Subgenus 1. AMANITA.

A. (*Am.*) *vaginatus*, Bull. Woods north of London. Common. Esculent.
A. (*Am.*) *Ceciliæ*, B. & Br. Open places in Bishop's Wood; rare.
A. (*Am.*) *vernus*, Bull. In the spring and early summer. Bishop's Wood; rare. Poisonous.
A. (*Am.*) *Phalloides*, Fr. Common in woods north of London. Poisonous.
A. (*Am.*) *Mappa*, Batsch. In woods north of London. Uncommon.
A. (*Am.*) *muscarius*, L. Under birches, Ken Wood; uncommon. Bohun Lodge, Barnet; common. Poisonous.
A. (*Am.*) *excelsus*, Fr. Open places. Highgate Wood. Uncommon.
A. (*Am.*) *pantherinus*, De C. With *A. rubescens*, P. not uncommon, not poisonous.
A. (*Am.*) *rubescens*, P. Common in woods north of London. Esculent.
A. (*Am.*) *asper*, P. Woods north of London. Not common.

Subgenus 2. LEPIOTA.

A. (*Lep.*) *procerus*, Scop. Pastures and roadsides. Not uncommon. Hornsey Wood, Hampstead Heath, &c.; *Curtis.* Esculent.

A. (*Lep.*) *rachodes*, Vitt. In hedges and under trees. Stoke Newington. Esculent, but not equal to *A. procerus*, Scop.

A. (*Lep.*) *excoriatus*, Schæff. Pastures at Hampstead; rare. Sometimes gathered for *A. procerus*, Scop. Not poisonous.

A. (*Lep.*) *gracilentus*, Kromb. Pastures at Hampstead; rare. Sometimes mistaken for *A. procerus*, Scop.

A. (*Lep.*) *acutesquamosus*, Weinm. Old pasture land. Highgate.

A. (*Lep.*) *clypeolarius*, Bull. On a dungheap, Hornsey; rare. Sweet-scented.

A. (*Lep.*) *cristatus*, Fr. Common in meadows, gardens, &c. Odour fœtid. Abundant in the garden of the Royal Horticultural Society.

A. (*Lep.*) *cepæstipes*, Sow. In greenhouses. Common.

A. (*Lep.*) *granulosus*, Batsch. In fields. Not common.

Subgenus 3. ARMILLARIA.

A. (*Ar.*) *ramentaceus*, Bull. Once only. 'The Elms,' Hampstead, Sept. 20, 1862; *M. C. C.*

A. (*Ar.*) *melleus*, Vahl. Common everywhere in stumps; the taste of this species, when raw, is bitter and pungent, and causes constriction of the throat; it is said, however, to be very good when cooked. Mr. J. A. Clarke says it is 'excellent fried, or stewed in milk.'

A. (*Ar.*) *mullus*, Sow. Gathered by Sowerby in Kensington Gardens, Jan. 1796. Sow. 184. This remarkable species appears to be very different from any form of *A. melleus*, Vahl. It is recognised as an *Armillaria* by Fries in his *Epicrisis*, but referred by Berkeley to *Tricholoma*.

A. (*Ar.*) *mucidus*, Fr. Under old beeches, on beech-mast, Ken Wood, Hampstead. This species, though generally found on beech *trunks*, occasionally roots amongst the dead leaves and *débris* of beech trees.

Subgenus 4. TRICHOLOMA.

A. (*Tr.*) *sejunctus*, Sow. Rare, mixed with *A. luridus.* Road sides and woody places.

A. (*Tr.*) *rutilans*, Schæff. Common. Fir stumps. Millfield Lane, &c. Not poisonous. Tried by Mr. J. A. Clarke; taste not agreeable.

A. (*Tr.*) *luridus.* Schæff. Not common. Woody places.

A. (*Tr.*) *terreus*, Schæff. Rare. Bishop's Wood. Tasteless; *J. A. C.*

A. (*Tr.*) *cartilagineus*, Bull. A large abnormal specimen of this species came up under the pavement before a barber's shop, in the Goswell Road, late in 1864. It raised a large paving stone, measuring 4 ft. 1 in. by 2 ft. 1 in., and weighing 2 hundredweight, completely out of its setting. (See *Journal of Botany*, vol. iii. p. 28.)

A. (*Tr.*) *carneus*, Bull. Rare. Fields, Stoke Newington.

A. (*Tr.*) *gambosus*, Fr. Once in a pasture, Holly Lodge, Highgate; *M. C. C.* Esculent.

A. (*Tr.*) *personatus*, Fr. Not common. Fields and roadsides. Esculent.

A. (*Tr.*) *nudus*, Bull. Extremely common and variable. In fields, by roadsides, on dungheaps, rotten trunks, &c.

A. (*Tr.*) *grammopodius*, Bull. Rare. Fields.

A. (Tr.) humilis, Fr. Rare. Shady places.

A. (Tr.) subpulverulentus, P. Fields north of London. Not common. Esculent; *J. A. C.*

Subgenus 5. CLITOCYBE.

A. (Cl.) nebularis, Batsch. Rare ; Ken Wood. Esculent.

A. (Cl.) odorus, Bull. Rare ; woods north of London. Esculent ; *J. A. C.*

A. (Cl.) candicans, Fr. Rare ; woods north of London.

A. (Cl.) dealbatus, P. Not common. Ken Wood, &c. Esculent.

A. (Cl.) elixus, Sow. Common. Sometimes in great abundance in gardens, &c. Seven Sisters Road and Stoke Newington. Found in Kensington Gardens by Sowerby.

A. (Cl.) fumosus, P. and *A. fumosus,* var. *polius.* On cinder-heaps, north of London, and on the ashes left by gipsies in the woods. Abundant in Old Stoke Newington churchyard, Christmas, 1868.

A. (Cl.) infundibuliformis, Schæff. Common. Woods, north of London.

A. (Cl.) geotrupus, Bull. Rare ; under trees in pastures. Millfield Lane. Woods about London; *Sowerby.* Esculent.

A. (Cl.) inversus, Scop. Millfield Lane, Jan. 8, 1863 ; *M. C. C.*

A. (Cl.) flaccidus, Sow. Rare ; Ken Wood. Found in Sowerby's time in same place.

A. (Cl.) cyathiformis, Fr. Common ; woods north of London.

A. (Cl.) brumalis, Fr. Very common; woods and shady places.

A. (Cl.) fragrans, Sow. Rare ; Bishop's Wood. Esculent; *J. A. C.*

A. (Cl.) laccatus, Scop. Very common under trees.

Subgenus 6. COLLYBIA.

A. (Col.) radicatus, Rehl. Very common about stumps.

A. (Col.) platyphyllus, Fr. Rare ; on a stump, Bishop's Wood.

A. (Col.) fusipes, Bull. Very common on stumps. Esculent.

A. (Col.) maculatus, A. & S. Not common. Bishop's Wood; generally about stumps. Lord Mansfield's Woods; *Sowerby.* This is the *A. carnosus* of Curtis's *Flora Londinensis.* Lord Mansfield's small pinewood; *Curtis.* Alperton ; *Rev. W. Hind.* Poisonous, *J. A. C.*

A. (Col.) butyraceus, Bull. Very common; shady places.

A. (Col.) velutipes, Curt. Abundant on trees, stumps and rails ; everywhere in and about London.

A. (Col.) conigenus, P. On fir cones ; pastures, Millfield Lane.

A. (Col.) tuberosus, Bull. Common in Bishop's Wood on dead *Russula nigricans,* Fr.

A. (Col.) dryophilus, Bull. Very common on banks and in shady places.

A. (Col.) tenacellus, P. On fir cones and dead fir leaves ; pastures. Millfield Lane.

Subgenus 7. OMPHALIA.

A. (Omp.) muralis, Sow. Common on old walls; north of London.

A. (Omp.) rufulus, B. & Br. Common in Bishop's Wood.

A. (Omp.) Fibula, Bull. Very common. Woods north of London.

Subgenus 8. MYCENA.

A. (My.) purus, P. Common in woods ; north of London.

A. (My.) lacteus, P. Rare ; fir leaves ; Millfield Lane.

A. (My.) galericulatus, Scop. Very common ; stumps.

A. (My.) polygrammus, Bull. Common ; stumps.

A. (*My.*) *alcalinus*, Fr. Very common in the woody places, parks, and gardens throughout London.

A. (*My.*) *tenuis*, Bolt. Very common in woods and grassy places.

A. (*My.*) *filopes*, Bull. Bishop's Wood.

A. (*My.*) *acicula*, Schæff. Very common about stumps.

A. (*My.*) *sanguinolentus*, A. and S. Not uncommon. Bishop's Wood.

A. (*My.*) *galopus*, Schrad. Ken Wood, Sept. 1862 ; *M. C. C.*

A. (*My.*) *epipterygius*, Scop. Very common in woods ; north of London.

A. (*My.*) *vulgaris*, P. Ken Wood.

A. (*My.*) *roridus*, Fr. On dead brambles. Bishop's Wood.

A. (*My.*) *tenerrimus*, B. Fir leaves. Millfield Lane.

A. (*My.*) *corticola*, Schum. Common on and about stumps.

Subgenus 9. PLEUROTUS.

A. (*Pl.*) *dryinus*, P. Common on elms and oaks. Stoke Newington.

A. (*Pl.*) *ulmarius*, Bull. Very common ; on elms, in the parks, and by roadsides. Esculent.

A. (*Pl.*) *subpalmatus*, Fr. Abundant on squared elms. Hornsey.

A. (*Pl.*) *euosmus*, Berk. Extremely common about London on old elms, early summer, June 1869, &c. The pale lilac spores are characteristic, but its peculiar and powerful odour and tough coriaceous stem point rather to *Lentinus* than *Agaricus*. Once in Millfield Lane on dead holly ; *M. C. C.* Esculent ; *J. A. C.*

A. (*Pl.*) *ostreatus*, Jacq. Common on elms. Stoke Newington, and all the London parks.

A. (*Pl.*) *salignus*, Hoffm. On willows ; side of New River, Stoke Newington.

A. (*Pl.*) *acerosus*, Fr. Not uncommon on wet banks, and in ruts made by cart-wheels in Bishop's Wood.

A. (*Pl.*) *hypnophilus*, P. Bishop's Wood, Sept. 1862 ; *M. C. C.*

SECTION II. HYPORHODII. *Spores salmon-colour.*

Subgenus 10. VOLVARIA.

A. (*Vo.*) *volvaceus*, Bull. Often in stoves at Holly Lodge, Highgate ; *M. C. C.*

Subgenus 11. PLUTEUS.

A. (*Pl.*) *cervinus*, Schæff. Not uncommon on stumps and fallen timber.

A. (*Pl.*) *nanus*, P. On a fallen tree, Bishop's Wood.

A. (*Pl.*) *phlebophorus*, Ditm. Ken Wood, Sept. & Oct. 1862 ; *M. C. C.*

Subgenus 12. ENTOLOMA.

A. (*En.*) *sinuatus*, Fr. This is probably the *A. fertilis*, P. of Berkeley's *Outlines.* Not uncommon in Bishop's Wood. Poisonous.

A. (*En.*) *Bloxami*, B. & Br. Exposed pastures ; north of London.

A. (*En.*) *clypeatus*, L. In pastures. Hampstead. Rare.

A. (*En.*) *nidorosus*, Fr. Not uncommon in woods north of London.

Subgenus 13. CLITOPILUS.

A. (*Cl.*) *prunulus*, Scop. Common ; woods north of London. Esculent.

Subgenus 14. LEPTONIA.

A. (*Lep.*) *chalybæus*, P. Fields; Highgate and Hampstead ; rare.

Subgenus 15. NOLANEA.

A. (*Nol.*) *pascuus*, P. Woods and shady places, north of London ; common.

SECTION III. DERMINI. *Spores brown.*

Subgenus 16. PHOLIOTA.

A. (*Ph.*) *durus*, Bolt. Common in rich pastures.
A. (*Ph.*) *præcox*, P. Not uncommon in rich pastures and roadsides.
A. (*Ph.*) *radicosus*, Bull. About stumps ; rare. Hornsey.
A. (*Ph.*) *pudicus*, Bull. About elm stumps ; not common. Esculent, but tough ; *J. A. C.*
A. (*Ph.*) *leocromus*, Cooke. On stumps; elder, &c. Highgate, 1862 ; *M. C. C.*
A. (*Ph.*) *capistratus*, Cooke. On old stumps, elms, &c. Highgate ; rare ; *M. C. C.* 1862. This species was found by my friend J. Aubrey Clarke, Esq., in some abundance under elms, in company with the usual form of *A. pudicus*, Bull, to which it is closely allied. The spores of both are the same in size and colour. Esculent, but tough ; *J. A. C.*
A. (*Ph.*) *squarrosus*, Müll. About stumps ; very common.
A. (*Ph.*) *adiposus*, Fr. ; not uncommon. On a poplar, Albion Road, Stoke Newington.
A. (*Ph.*) *spectabilis*, Fr. Common, woods north of London. Harrow Road ; *Rev. W. M. Hind* (*A. aureus* of Berkeley's *Outlines*).
A. (*Ph.*) *mutabilis*, Schæff. About stumps in hedges ; not uncommon.

Subgenus 17. HEBELOMA.

A. (*He.*) *fastibilis*, Fr. Common in woods north of London.

(*Inocybe.*)

A. (*In.*) *scaber*, Müll. Common in the paths of woods north of London.
A. (*In.*) *rimosus*, Bull. In woods north of London ; common.
A. (*In.*) *auricomus*, Batsch. In woods north of London on fir leaves.
A. (*In.*) *lucifugus*, Fr. Woods north of London ; uncommon.
A. (*In.*) *geophyllus*, Sow. Woods north of London ; common. Garden, Harrow ; *Rev. W. M. Hind.*

Subgenus 18. FLAMMULA.

A. (*Fl.*) *carbonarius*, Fr. On ashes left by gipsies. Bishop's Wood; not common.
A. (*Fl.*) *filiceus*, Cooke. On dead fern stems. Conservatory, Holly Lodge, Highgate, Aug. 1862 ; *M. C. C.*

Subgenus 19. NAUCORIA.

A. (*Nau.*) *melinoides*, Fr. Very common in fields.
A. (*Nau.*) *semiorbicularis*, Bull. Very common in fields.
A. (*Nau.*) *furfuraceus*, P. Pinner Wood ; *Rev. W. M. Hind.*
A. (*Nau.*) *erinaceus*, Fr. Fallen stems. Bishop's Wood.

Subgenus 20. GALERA.

A. (*Ga.*) *tener,* Schæff. On dung, and in rich pastures, woods, and gardens; very common.

Subgenus 21. CREPIDOTUS.

A. (*Cr.*) *mollis,* Schæff. On stumps and dead branches; very common.
A. (*Cr.*) *variabilis,* P. On dead sticks, grass, &c.; very common.

SECTION IV. PRATELLÆ. *Spores brownish-purple.*

Subgenus 22. PSALLIOTA.

A. (*Ps.*) *cretaceus,* Fr. In pastures north of London; rare.
A. (*Ps.*) *arvensis,* Schæff. In fields north of London; very common. Esculent.
A. (*Ps.*) *campestris,* L. *The Mushroom.* Rich pastures north of London; common in all its varieties. Kensington Gardens; *Sowerby.* Esculent.
A. (*Ps.*) *echinatus,* Roth. In gardens and greenhouses; not common.

Subgenus 23. STROPHARIA.

A. (*St.*) *æruginosus,* Curt. Very common in woods and on stumps. Lord Mansfield's little wood near the 'Spaniards;' *Curtis.* Poisonous.
A. (*St.*) *albocyaneus,* Desm. In pastures; not common.
A. (*St.*) *semiglobatus,* Batsch. Extremely common on horse-dung; comes up in the streets; on rotten straw and rubbish; top of the cornice, London Bridge. Pasture by Hornsey Wood, towards Islington; *Curtis,* under *A. glutinosus.* Probably poisonous.

Subgenus 24. HYPHOLOMA.

A. (*Hyp.*) *sublateritius,* Fr. On old stumps, and in hedges; common. Poisonous.
A. (*Hyp.*) *capnoides,* Fr. About old fir stumps; Ken Wood; rare.
A. (*Hyp.*) *fascicularis,* Huds. About stumps and rails; extremely common and poisonous.
A. (*Hyp.*) *lacrymabundus,* Bull. In woods and fields, north London; very common.
A. (*Hyp.*) *velutinus,* P. About stumps and rails; extremely common.
A. (*Hyp.*) *appendiculatus,* Bull. About stumps, and in pastures; common.
A. (*Hyp.*) *hydrophilus,* Bull. In hedges and pastures, and about stumps; not common.
A. (*Hyp.*) *lanaripes,* Cooke. In a conservatory, Highgate; *M.C.C.*

Subgenus 25. PSILOCYBE.

A. (*Ps.*) *coprophilus,* Bull. On horse dung, pastures. Millfield Lane; rare.
A. (*Ps.*) *semilanceolatus,* Fr. Common in all the parks, and in the woods north of London.
A. (*Ps.*) *spadiceus,* Schæff. In woods, and about stumps and fallen trees north of London; abundant in a gravel pit, Stoke Newington.
A. (*Ps.*) *Fœnisecii,* P. In the parks and fields; common.

Subgenus 26. PSATHYRA.

A. (*Ps.*) *corrugis,* P. In gardens, &c.; not uncommon in London.
A. (*Ps.*) *gossypinus,* Fr. Once only at Highgate, 1861; *M. C. C.*

Section V. Coprinarii. *Spores black.*

Subgenus 27. Panæolus.

A. (Pan.) separatus, L. Dung in fields. Common.
A. (Pan.) fimiputris, Bull. On dung; parks, fields, and gardens. Very common.

Subgenus 28. Psathyrella.

A. (Ps.) gracilis, Fr. In gardens. Mildmay Park.
A. (Ps.) atomatus, Fr. In hedges and gardens. Stoke Newington, &c.
A. (Ps.) disseminatus, P. By roadsides and about rotten and fallen trees. Common.

COPRINUS, *Fr.*

C. comatus, Fr. Common in pastures. Dust yards at Highbury, Kingsland, &c. Esculent.
C. atramantarius, Fr. Very common in fields and roadsides; comes up in gardens and cellars throughout London; common in dustbins. Esculent.
C. fimetarius, Fr. Common on dungheaps, &c.
C. niveus, Fr. Common on dungheaps, in gardens, &c.
C. micaceus, Fr. Very common on stumps.
C. deliquescens, Fr. Very common on stumps.
C. radiatus, Fr. Very common in pastures and gardens.
C. ephemerus, Fr. Very common on dung, &c.
C. plicatilis, Fr. Very common in pastures, by roadsides, in ditches, &c.

BOLBITIUS, *Fr.*

B. titubans, Fr. Not common. Pastures and ditch sides, Highgate. Found by Sowerby in Kensington Gardens.

CORTINARIUS, *Fr.*

Subgenus 1. Phlegmacium.

C. (Ph.) glaucopus, Fr. Ken Wood; not common.
C. (Ph.) callochrous, Fr. Bishop's Wood; not common.
C. (Ph.) purpurascens, Fr. Bishop's Wood; not common.

Subgenus 2. Myxacium.

C. (My.) elatior, Fr. Bishop's Wood; common.

Subgenus 3. Inoloma.

C. (In.) violaceus, Fr. Under grove of trees, Millfield Lane; Ken Wood, not common; common about Barnet. Esculent.
C. (In.) sublanatus, Fr. Hampstead Wood, *Sowerby.*

Subgenus 4. Dermocybe.

C. (Der.) tabularis, Fr. Bishop's Wood; rare.
C. (Der.) anomalus, Fr. Bishop's Wood; rare.

C. (Der.) sanguineus, Fr. Bishop's Wood; rare. Pinner Wood; *Rev. W. M. Hind.*

C. (Der.) cinnamomeus, Fr. Pinner Wood; *Rev. W. M. Hind.*

Subgenus 5. TELAMONIA.

C. (Tel.) torosus, Fr. Bishop's Wood; rare.

C. (Tel.) hinnuleus, Fr. Ken Wood; rare.

C. (Tel.) ileopodius, Fr. Grove of trees, Millfield Lane; not common.

Subgenus 6. HYGROCYBE.

C. (Hyg.) castaneus, Fr. Highgate Wood; rare.

C. (Hyg.) leucopus, Fr. Bishop's Wood; not common.

PAXILLUS, *Fr.*

P. involutus, Fr. Extremely common. Woods north of London, and in gardens. Not poisonous.

HYGROPHORUS, *Fr.*

H. eburneus, Fr. Wood sides. Common.

H. aromaticus, B. Harrow Weald; *Rev. W. M. Hind.*

H. pratensis, Fr. Pastures. Uncommon. Esculent.

H. virgineus, Fr. Pastures. Very common. Kensington Gardens; *Sowerby.* Esculent.

H. niveus, Fr. Pastures. Very common. Esculent.

H. lætus, Fr. Pastures. Common.

H. ceraceus, Fr. Pastures and roadsides. Common. Harrow Cricket-ground, *Rev. W. M. Hind.*

H. coccineus, Fr. Pastures. Common.

H. miniatus, Fr. Bog on Hampstead Heath.

H. conicus, Fr. Pastures. Common. Poisonous.

H. psittacinus, Fr. Pastures. Common. Esculent; *F. C. Penrose, Archt.*

LACTARIUS, *Fr.*

L. torminosus, Fr. Pastures by Ken Wood; rare. Harrow Weald; *Rev. W. M. Hind.* Poisonous.

L. blennius, Fr. Woods and shady places north of London. A very dangerous species.

L. uvidus, Fr. Harrow Weald; *Rev. W. M. Hind.*

L. pyrogalus, Fr. Woods and fields. Not uncommon.

L. acris, Fr. Bishop's Wood; rare; very poisonous.

L. chrysorrhœus, Fr. Bishop's Wood, &c.; very common. Milk changes from white to brilliant yellow.

L. piperatus, Fr. Woods north of London. Very common. Poisonous.

L. vellereus, Fr. Woods north of London. Very common. Poisonous.

L. deliciosus, Fr. One specimen under a grove of fir and other trees, Millfield Lane, Highgate. Esculent. Milk changes from orange to green. Esculent. Requires to be soaked in boiling water before it is broiled.

L. quietus. Fr. Woods north of London. Common. Harmless, *J. A. C.*

L. theiogalus, Bull. Bishop's Wood; common. Poisonous.

L. rufus, Fr. Woods north of London. Not common. One of the most acrid of fungi.

L. fuliginosus, Fr. Bishop's Wood, &c.; common. Flesh changes from white to burnt sienna on being broken.
L. serifluus, Fr. Woods, &c. Extremely common.
L. subdulcis, Fr. Woods, &c. Very common.

RUSSULA, *Fr.*

R. nigricans, Fr. Woods north of London. Common. Kensington Gardens; *Sowerby.*
R. adusta, Fr. Woods north of London. Not uncommon.
R. furcata, Fr. Woods and woody places north of London. Common.
R. rosacea, Fr. Woods north of London. Common. Poisonous; *J. A. C.*
R. virescens, Fr. Woods north of London; rather rare; but extremely common just out of Middlesex, in Essex. Esculent.
R. lepida, Fr. Ken Wood; not common. Esculent.
R. vesca, Fr. Woods north of London. Not common.
R. heterophylla, Fr. Woods north of London. Very common. Esculent.
R. fœtens, Fr. Woods and shady places north of London. Very common. Fragrant when young, fœtid when old. Poisonous.
R. emetica, Fr. Bishop's Wood; rare. Pinner, *Rev. W. M. Hind.* Poisonous.
R. ochrolcuca, Fr. Woods and shady places north of London. Very common.
R. fragilis, Fr. Woods and shady places north of London. Very common. Poisonous.
R. decolorans, Fr. Bishop's Wood; not common.
R. nitida, Fr. Open places in woods. Not common.
R. alutacca, Fr. Woods north of London. Very common. Esculent.
R. lutea, Fr. Woods north of London. Rare.

CANTHARELLUS, *Fr.*

C. cibarius, Fr. (with the white variety). Woods north of London. Not uncommon. Esculent.
C. aurantiacus, Fr. (with the white variety). Woods north of London. Cricket Ground, Harrow; *Rev. W. M. Hind.* Esculent; rather tough, *J. A. C.*
C. tubæformis, Fr. Two or three specimens on a bank in Bishop's Wood, 1863.

NYCTALIS, *Fr.*

N. asterophora, Fr. Woods north of London. Common. Always parasitic on *Russula nigricans*, Fr.
N. parasitica, Fr. Woods north of London. Common. Always parasitic on *Russula fœtens*, Fr.

MARASMIUS, *Fr.*

M. urens, Fr. Woods, fields, and parks throughout London. Not uncommon. Poisonous.
M. peronatus, Fr. Woods north of London. Common in Hampstead. *Sowerby.*
M. porreus, Fr. Ken Wood; not common.

M. oreades, Fr. Woods, roadsides, and parks throughout London. Extremely common. *The Champignon*, Esculent.

M. erythropus, Fr. Ken Wood, Sept. 1863 ; *M. C. C.*

M. ramealis, Fr. Often in Bishop's Wood ; *M. C. C.*

M. rotula, Fr. Shady bank, Millfield Lane ; I have not observed it elsewhere.

M. graminum, B. & Br. On leaves of grass, pasture, Millfield Lane.

M. androsaceus, Fr. Pinner Wood ; *Rev. W. M. Hind.*

M. epiphyllus, Fr. Woods north of London. Very common on dead leaves.

LENTINUS, *Fr.*

L. Dunalii, Fr. Abundant on an ash tree on Stamford Hill.

L. lepideus, Fr. Once on a stump between Tottenham and Kingsland.

L. vulpinus, Fr. On old elms on the road from Seven Sisters Road to Tottenham. Found at Islington in a hollow elm ; *Sowerby.*

PANUS, *Fr.*

P. torulosus, Fr. Stumps. Woods north of London. Not common.

P. stypticus, Fr. Stumps. Woods north of London. Extremely common.

SCHIZOPHYLLUM, *Fr.*

S. commune, Fr. On timber between Shoreditch and Hackney, Mr. B. M. Forster ; *Sowerby.*

LENZITES, *Fr.*

L. betulina, Fr. Woods north of London. Very common on stumps and rails.

L. sepiaria, Fr. On unsquared deals, Thames Dock ; *Sowerby.*

ORDER II. *POLYPOREI.*

BOLETUS, *Fr.*

B. luteus, L. Grove of trees, Millfield Lane ; common.

B. elegans, Schum. Grove of trees, Millfield Lane ; common.

B. flavus, With. Grove of trees, Millfield Lane ; common.

B. sanguineus, With. Borders of Ken Wood, Highgate ; not uncommon.

B. chrysenteron, Fr. Woods and shady places north of London. Common. Probably esculent.

B. subtomentosus, L. Woods north of London. Common. Probably esculent.

B. olivaceus, Schæff. Ken Wood, Oct. 1863 ; *M. C. C.*

B. edulis, Bull. Woods north of London. Very common. Esculent.

B. fragrans. Vitt. Bishop's Wood. Under oaks ; rare. Esculent.

B. æstivalis, Fr. Bishop's Wood. Early summer ; not common. Esculent.

B. luridus, Schæff. Woods and shady places north of London. Common. Poisonous.

B. versipellis, Fr. Woods north of London. Common.

B. scaber, Fr. Woods north of London. Common. Esculent.

B. felleus. Fr. Bishop's Wood ; rare. Poisonous.

B. castaneus, Bull. Borders of wood, Millfield Lane ; not common.

POLYPORUS, *Fr.*

P. squamosus, Fr.   On ash and elm trees ; exceedingly common throughout London.   Esculent when very young ; *J. A. C.*

P. varius, Fr.   On willows.   Stoke Newington ; uncommon.

P. lucidus, Fr.   Barnet, about stumps ; uncommon.   Extremely common in the districts east of Middlesex, generally on or about Hornbeam and Beeches.   Elm tree near Hyde Park ; *Curtis*.   Pinner ; *Rev. W. M. Hind.*

P. intybaceus, Fr.   One large specimen on an oak, Ken Wood, 1862.   I saw this in company with my friend Mr. M. C. Cooke.   Esculent.

P. sulfureus, Fr.   On trees throughout the London district ; common.

P. giganteus, Fr.   Stumps.   Kensington Gardens ; *Sowerby*.

P. epileucus, Fr.   On an elm, near Seven Sisters Road.   This species, first noticed by me as a British plant, has not been observed elsewhere in this country.

P. rutilans, Fr.   Bishop's Wood, common ; a rare species elsewhere.

P. fumosus, Fr.   Hampstead, 1862 ; *M. C. C.*

P. adustus, Fr.   Hampstead ; *M. C. C.*

P. hispidus, Fr.   Elm trees, Foot-ball Field, Harrow ; *Rev. W. M. Hind.*

P. spumeus, Fr.   Kensington Gardens ; *Sowerby*.   Harrow Park ; *Rev. W. M. Hind.*

P. dryadeus, Fr.   On oak stumps.   Bishop's Wood, &c.   Common.

P. betulinus, Fr.   Beech-tree Mount, Roxeth ; *Rev. W. M. Hind.*

P. fomentarius, Fr.   Common.   On a horse-chestnut, Stamford Hill.

P. igniarius, Fr.   Common on stems of trees in woods and gardens.

P. ribis, Fr.   Hampstead ; *M. C. C.*

P. ulmarius, Fr.   St. James's Park ; *Sowerby*.

P. annosus, Fr.   Highgate ; not uncommon ; *M. C. C.*

P. radiatus, Fr.   Bishop's Wood ; uncommon.

P. velutinus, Fr.   Pinner ; *Rev. W. M. Hind.*

P. ferruginosus, Fr.   Holly Lodge, Highgate ; *M. C. C.*   Harrow ; *Rev. W. M. Hind.*

P. sanguinolentus, Fr.   Infesting the naked clay, old disused sawpit, Mildmay Park, 1867.   New to Britain, and unnoticed elsewhere, till Mr. Broome found it on a wet bank in Epping Forest.   See *Journal of Botany*, March, 1869.

P. molluscus, Fr.   Old sleepers, North London Railway ; not uncommon. Pinner ; *Rev. W. M. Hind.*

P. terrestris, P.   On the naked ground, Kentish Town and Highgate.

TRAMETES, *Fr.*

T. suaveolens, Fr.   On a willow, Stoke Newington ; rare.

T. odora, Fr.   On a willow, Stoke Newington.

T. gibbosa, Fr.   On gates and rails.   In a sewer under Temple Bar ; *Dr. Trimen*.   Root of an old poplar, Lambeth ; *Sowerby*.   Common throughout London ; rare elsewhere.

DÆDALEA, *Fr.*

D. quercina, P.   On oak stumps and rails ; very common.

D. unicolor, Fr.   Highgate, Hampstead ; *M. C. C.*

E E

MERULIUS, *Fr.*

M. tremellosus, Schrad.   Willows, Willow Walk, Chelsea ; *Sowerby.*
M. lacrymans, Fr.   *Dry Rot.*   Very common in and about London on decaying trees, and in the roofs, floors, and cellars of private houses and
public buildings.

FISTULINA, *Bull.*

F. hepatica, Fr.   Bishop's Wood, &c.   Usually upon oaks.   Upon an elm,
Millfield Lane.   I have observed this species also on the ash and
beech.   Esculent.

Ord. 3.   *HYDNEI.*

HYDNUM, *L.*

H. repandum, L.   Woods north of London.   Common.   Hornsey Wood ;
Sowerby.   Esculent.
H. auriscalpium, L.   Fir cones, near Old Stoke Newington churchyard.
Pine Wood, opposite Lord Mansfield's House ; *Curtis.*
H. coralloides, Scop.   On an old ash-tree, Bohun Lodge, Barnet, where it
has appeared for several years in succession.   Esculent.
H. ochraceum, P.   On dead branches, Bishop's Wood ; common.
H. membranaceum, Bull.   On dead wood and sticks.   Common in the parks
and woods, and on wooden garden edging.

RADULUM, *Fr.*

R. quercinum, Fr.   Kensington Gardens.   Sowerby's *Hydnum Barba-Jovis*
t. 328, is referred by Berkeley to *R. quercinum.*   Fries (*Epic.* p. 528),
however, refers it to *Odontia.*

Ord. 4.   *AURICULARINI.*

CRATERELLUS, *Fr.*

C. lutescens, Fr.   Bishop's Wood ; rare.
C. cornucopioides, Fr.   Bishop's Wood ; very common.

THELEPHORA, *Fr.*

T. laciniata, P.   Woods north of London.   Common.
T. sebacea, Fr.   Bishop's Wood, running over grass ; common.

STEREUM, *Fr.*

S. purpureum, Fr.   Throughout the London district.   Extremely common.
Generally on poplars.
S. hirsutum, Fr.   Throughout the London district.   Extremely common.
S. rugosum, Fr.   On stumps ; very common.

HYMENOCHÆTE, *Lév.*

H. rubiginosa, Lév.   On stumps and rails.   Extremely common.
H. corrugata, B.   Woods, &c., north of London.   Common.

AURICULARIA, *Fr.*

A. mesenterica, Bull.   On stumps and stems of trees.   Not uncommon in the woods and parks.

CORTICIUM, *Fr.*

C. læve, Fr.   On sticks in woods north of London.   Extremely common.
C. cæruleum, Fr.   On dead branches, &c., in woods north of London. Common.
C. quercinum. Fr.   On oak branches.   Bishop's Wood, &c.
C. cinereum, Fr.   Common ; *M. C. C.*
C. Sambuci, P.   Millfield Lane ; *M. C. C.*

Ord. 5. *CLAVARIEI.*

CLAVARIA, *L.*

C. coralloides, L.   Woods, &c., north of London.   Common.   Harrow Park ; *Rev. W. M. Hind.*   Esculent.
C. cinerea, Bull.   Holly Lodge ; *M. C. C.*
C. cristata, Holmsk.   Woods, &c. north of London.   Common.   Esculent.
C. rugosa, Bull.   Woods and shady places north of London.   Common. Esculent.
C. fusiformis, Sow.   Hampstead Heath and Hornsey Wood ; *Sowerby.*
C. inæqualis, Müll.   Bishop's Wood, amongst dead leaves.
C. vermiculata, Scop.   In pastures and road sides.   Common.
C. fragilis, Holmsk.   Pastures, Highgate, &c.   Wet field, New River, between Stoke Newington and Hornsey ; *Sowerby.*
C. juncea, Fr.   Bishop's Wood.
C. acuta, Sow.   Highgate, 1863 ; *M. C. C.*

CALOCERA, *Fr.*

C. cornea, Fr.   Hampstead and Kensington Gardens; *Sowerby.*   Harrow Park, on timber ; *Rev. W. M. Hind.*

TYPHULA, *Fr.*

T. phacorrhiza, Fr.   Hampstead Heath and Hornsey Wood ; *Sowerby.*

PISTILLARIA, *Fr.*

P. micans, Fr.   On a dead thistle, Highbury.
P. quisquilaris, Fr.   Bishop's Wood ; *M. C. C.*

Ord. 6. *TREMELLINI.*

TREMELLA, *Fr.*

T. frondosa, Fr.   Ken Wood, 1863 ; *M. C. C.*
T. foliacea, P.   Ken Wood, 1863 ; *M. C. C.*
T. mesenterica, Retz.   Extremely common in the woods and parks.
T. albida, Huds.   Bishop's Wood.
T. sarcoides, Sm.   Barnet, &c.   Common.

EXIDIA, *Fr.*

E. recisa, Fr.   Colney Hatch, 1864 ; *M. C. C.*
E. glandulosa, Fr.   Bishop's Wood, &c.   On oak branches.   Very common.

DACRYMYCES, *Nees.*

D. stillatus, Nees.   Very common on sticks and rails throughout London.

ADDITIONS AND CORRECTIONS.

Page 4, line 14, *for* 40 *read* 46.
    Budd. MSS. Some refer probably to years subsequent to 1708, the date given.
    6, after line 26, insert *Herb. Hill* (v. p. 392).
    8, after line 39, insert *Merr. MSS.* (v. p. 372).
    line 36, *for* 1670 *read* 1672.
    13, Anemone apennina, *L.* There is a specimen in the herbarium of the late O. Jewitt labelled 'Bishop's Wood, April 1863.' A transposition of labels seems likely.
    14, **5.** ? Ranunculus Drouetii, *Schultz.* The Rev. W. W. Newbould considers the Middlesex plant to be more likely a state of *R. heterophyllus,* Bab. Man., without floating leaves ; to this species Mr. Warren also refers a plant with floating leaves gathered at Cricklewood.
    18, **13.** R. auricomus, *L.* IV. *add,* Cricklewood! ; *Warren.*
    19, **18.** R. arvensis, *L.* should have the * indicating naturalisation. II. *add* Shepperton ; *Newb.*
    21, **22.** Aquilegia vulgaris, *L.* Rev. W. W. Newbould thinks he has seen this at Harefield.
    23, **26.** Papaver Argemone, *L. add,* IV. Near the Brent Reservoir ! ; *Warren.*
    30, Cardamine impatiens, *L.* The whole account should run on as a continuous paragraph. VII. *add,* In the ditches about Primrose Hill ; *Merr. MSS.*
    33, **48.** Sisymbrium Irio, *L.* A translation of the part of Morison's 'Præludia' relating to the growth of this in 1667 is given in *Science Gossip,* 1865, p. 149.
    49. S. Sophia, *L. add,* IV. Two plants near Cricklewood, 1869, imported with grain ; *Warren.*
    34, **50.** S. Thalianum, *Gaud.* IV. *add,* Near Brent Reservoir ! ; *Warren.* A ' *Var. β* ' is recorded from ' walls about Harefield ' ; *Forst. Midd.*
    35, line 12, *for* ' *napus* ' *read* ' *Napus.*'

Page 36, **54** and **55.** The synonyms from Johns. and Blackst. are to be doubtfully referred here.

    56. Sinapis alba, *L.* I. *add*, Harefield! ; *Warren.*

39, **59.** Draba verna, *L.* II. *add*, Shepperton ; *Newb.*

    60. Cochlearia anglica, *L.* Still exists in the Isle of Dogs, just opposite Greenwich Hospital, April 1869 ; *Warren.*

40, **62.** Armoracia amphibia, *Koch.* VII. *add*, By the Paddington Canal, Kensal Green ; *Warren.*

41, Camelina sativa, *Crantz*, should stand before *Thlaspi.*

    Iberis amara, *L.* The commonly cultivated *I. umbellata* may have been the plant found.

    65. Lepidium Draba, *L.* VII. By the Finchley Road Station of Midland Ry., 1869.

44, **73.** Reseda lutea, *L. add*, II. Shepperton, sparingly ; *Newb.* VII. Waste ground, Finchley Road, introduced.

    75. R. luteola, *L.* II. *add*, Bushey Park ; Jesse's *Gleanings in Nat. Hist.* p. 138.

47, **81.** V. tricolor, *L.* IV. *add*, Near the Brent Reservoir! ; *Warren.*

49, **86.** Dianthus deltoides, *L. add* to synonyms. *D. glaucus* (Forst. Midd.). II. *add*, On a mound in Hampton Court Park ; Jesse's *Gleanings*, p. 143.

51, **90.** Silene noctiflora, *L.* VII. *add*, Waste ground, Finchley Road, 1869.

53, **95.** Sagina procumbens, *L.* VII. *add* Grosvenor Square, 1869 ! ; *Warren.*

54, **99.** Sagina nodosa, *E. Meyer.* I. *add* Harefield Mcor, near Denham ; *Herb. Hill.*

55, **103.** Stellaria media, *With.* The var. *S. neglecta*, Weihe, has been noticed at Chiswick and in the Isle of Dogs.

60, **115.** Spergula arvensis, *L.* β. VII. *add*, Field this side Cane Wood ; *Herb. Hill.*

    First record, *for* ' Curtis, about 1790,' *read* ' Hill, about 1750.'

63, **122.** Hypericum quadrangulum, *L.* VII. *add*, Kilburn! ; *Warren.*

68, **135.** Geranium columbinum, *L.* I. *add*, Harefield! ; *Warren.*

    line 23, *for* ' Dove's Foot. Crane's-bill,' *read* ' Dove's-foot Crane's-bill.'

69, **140.** Erodium cicutarium, *Sm.* II. *add*, Towing-path bet. Kingston Bridge and Hampton Court ! ; *Warren.*

72, **146.** Impatiens parviflora, *DC.* VII. *add*, Fields bet. Hampstead and the Swiss Cottage, on rubbish, abundant, 1869.

    147, *for* acetosella *read* Acetosella.

78, **162.** Melilotus officinalis, *Willd.* VII. *add*, Finchley Road.

    164. M. vulgaris, *Willd.* VII. *add*, Finchley Road. Back of Adelaide Road.

79, **169.** Trifolium arvense, *L.* IV. *add*, Near Brent Reservoir! ; *Warren.*

Page 80, **170.** T. striatum, *L.* II. *add*, Shepperton ; *Newb.* VII. *add*,
On an unused bridge over the North-Western Ry., ¼ mile west
of Edgware Road Station, abundant, 1869 ! ; *Warren.*

81, line 19, *for* ' *Curtis*, about 1780 ' *read* ' *Blackstone*, 1737.'

176. T. repens, *L.* The same dark-flowered plant was seen in the
Park in 1869, and therefore does not seem due to the high
temperature of the previous year ; *Newb.*

84, **183.** Vicia tetrasperma, *Moench.* VII. *add*, Near Edgware Road
Station ; *Warren.*

92, **204.** Alchemilla arvensis, *Scop.* IV. *add*, Near Brent Reservoir! ;
Warren.

94, **209.** Potentilla Fragariastrum, *Ehrh.* I. *add*, Pinner.
Comarum palustre, *L.* Specimens from Hampstead Heath in
Herb. Hill.

210. Fragaria vesca, *L.* For some account of Merrett's plant
see Masters' *Vegetable Teratology*, pp. 276-7 (note), where it is
called the ' Plymouth Strawberry.'

95, **211,** *for* idæus *read* Idæus.
line 24, *after* heaths *add* very common.

96, 4. Rubus discolor, *W. & N.* VII. *add*, Notting Hill ; *Babington's
British Rubi*, 106.

6. R. leucostachys, *Sm. add*, I. Harrow Weald Common, Hind ;
Bab. l.c. 122. VI. Trent Park, Hind ; *Bab. l.c.* 122.

97, 8. R. villicaulis, *W. & N. add*, I. Pinner Wood, Hind ; *Bab. l.c.* 146.

98, 15. R. rudis, *Weihe, add*, VI. Trent Park, Hind ; *Bab. l.c.* 193.

17. *Var.* R. pallidus, *Weihe, add*, VI. Trent Park, Hind ; *Bab.
l.c.* 210.

99, 20. R. Guntheri, *Weihe, add*, VI. Trent Wood, Hind ; *Bab. l.c.* 238.

24. *Var.* β. conjungens *add*, VII. Notting Hill, Hind ; *Bab. l.c.* 273.
line 7 from bottom, *for* ' *Hort* ' *read* ' *Host.* '

104, **225.** 1. Cratægus oxyacanthoides, *Thuill.* I. *add*, Pinner, abundant. IV. *add*, Near Brent Reservoir! ; *Warren.*

2. C. monogyna, *Jacq.* On the growth of the old thorns in Bushey
Park, see Jesse's *Gleanings in Nat. Hist.* p. 152.

106, **228.** Pyrus malus, *L.* VII. *add*, Kilburn, 1869 ; *Warren.*

107, **231.** P. torminalis, *Ehrh.* IV. *add*, By Brent Reservoir ! ;
Cricklewood! ; *Warren.*

109, line 10, *for* ' *Blackstone*, 1737 ' *read* ' *Merrett*, about 1670.' (By
Hamstead Heath ; *Merr. MSS.*)

110, **239.** Epilobium roseum, *Schreb.* VII. *add*, Near Edgware Road
Station of North London Ry.! ; *Warren.*

240. E. tetragonum, *L.* (Bab.). VII. *add*, Kilburn ! ; *Warren.*

112, **245.** Myriophyllum spicatum, *L.* IV. *add*, Bet. Finchley and
Mill Hill ; *Newb.* V. *add*, In the river (= Brent) betwixt
Harrow of ye Hill and London, running to Brentford ; *Merr. MSS.*
First record : *for* ' *Petiver*, 1695 ' *read* ' *Merrett*, about 1670.'

Page 113, **247.** Hippuris vulgaris, *L. add,* II. By the Thames near Shepperton!; *Newb.*

115, **252.** Scleranthus annuus, *L.* IV. *add,* Abundant near Brent Reservoir!; *Warren.*

121, line 20, *for* Herrow *read* Harrow.

124, **274.** Carum Carui, *L. add,* VII. By the Paddington Canal near Bloomfield Road, 1869!; *Warren.*

131, **295.** Tordylium maximum, *L.* should have the [indicating extinction. Add to account, ' *Tordylium,* betwixt St. James and Chelsey'; *Merr. MSS.* If this was *T. maximum* it will make the first record, *Merrett,* about 1670.

133, **299.** Torilis nodosa, *Gaertn.* VII. *add,* Waste ground, Finchley Road, introduced.

300. Scandix Pecten-Veneris, *L.* VII. *add,* Near Brent Reservoir!; *Warren.*

136, **306.** Adoxa moschatellina, *L.* VII. *add,* In the chestnut walks nigh Hampstead; *Merr. MSS.* First record, *for* ' *Petiver,* 1695' *read* ' *Merrett,* about 1670.'

137, **308.** Cornus sanguinea, *L.* VII. *add,* Near Kensal Green!; *Warren.*

Foot note, for another instance see *Notes and Queries,* 1869, p. 589.

139, **315.** Sherardia arvensis, *L.* II. *add,* Towing-path bet. Kingston Bridge and Hampton Court!; *Warren.* VI. *add,* By old Finchley Church; *J. B. George.*

141, **320.** Galium Mollugo, *L.* IV. *add,* Near the Brent Reservoir!; *Warren.*

321. G. verum, *L.* IV. *add,* Bet. Finchley and Mill Hill; *Newb.*

146, **333.** Knautia arvensis, *Coult.* II. *add,* Shepperton; *Newb.*

151, **345.** Inula Conyza, *DC. add to synonyms,* ' *Conyza major montana germanica* (Park.)'; *to localities,* VII. Towards Hampstead; *Park. Theat.* 127; *add,* First record: *Parkinson,* 1640.

153, Before **352** *insert* A. tinctoria, *L.* VII. On waste ground near Finchley Road Ry. Station, 1869, very abundant.

155, **356.** Matricaria inodora, *L.* IV. *add,* Near Brent Reservoir!; *Warren.*

159, **366.** VII. *add,* Specimens from Cane Wood in *Herb. Hill.*

163, **375.** 3. Arctium minus, *Schk. Remove* ' *Bardana minor, Ger.* (Pet.), and Near the bridge at Newington; *Pet. Midd.*' to **420.** Xanthium Strumarium, *L.*

165, **379.** Centaurea Cyanus, *L.* VII. *add,* Rubbish, Finchley Road, 1869.

381. C. Calcitrapa, *L.* VII. *add,* I. of Dogs, 1844. E. Palmer; *Herb. S. P. Woodward.*

before Onopordum *insert* C. diffusa, *Lam.* VII. On waste ground, Finchley Road, 1869. A native of Russia and Turkey.

Page 171, **398.** Apargia autumnalis, *Willd.* IV. *add,* Near the Brent Reservoir! ; *Warren.*

172, line 8, after railway-banks *insert* rare.

173, **404.** Lactuca virosa, *L.* IV. *add,* Cricklewood! ; *Warren.*

175, **410.** Sonchus arvensis, *L.* II. *add,* Shepperton ; *Newb.*

177, line 14 from bottom, *for* 1829 *read* 1629.

188, **444.** Convolvulus sepium, *L.* VII. *add,* Kilburn! ; *Warren.*

191, before '*Echium*' *insert* S. tuberosum, *L.* In Middlesex, a few miles north of London, James W. White ; *Science Gossip,* 1869, p. 138.

   455. Lithospermum arvense, *L.* IV. *add,* Near Cricklewood, imported with grain, 1869 ; *Warren.*

209, **503.** Mentha rotundifolia, *L.* I. *add,* Specimens from near Harefield are in *Herb. Hill.*

211, **509.** M. sativa, *L.* VII. *add,* By the New River side two miles beyond Islington ; *Herb. Hill.*

212, **510.** M. rubra, *Sm.* I. *add,* River-side a mile below Denham, Bucks ; *Herb. Hill.*

219, **533.** β. Galeopsis bifida, *Bönn.* VII. *add,* Finchley Road.

233, Chenopodium opulifolium, *Schrad.* VII. *add,* Waste ground Finchley Road. Back of Adelaide Road.

250, **602.** Mercurialis annua, *L.* VII. *add,* Side of New River, half a mile beyond Islington.

263, **634.** Quercus Robur, *L.* On an oak, in Hampton Court Park, perhaps the oldest in England, see Jesse's *Gleanings in Nat. Hist* pp. 153-4.

264, line 7, *for* pendunculata *read* pedunculata.

271, Herminium Monorchis, *R. Br.* should be enclosed within ( ) ; the latter of which has dropped out.

288, **697.** Butomus umbellatus, *L.* I. *add,* Brook bet. Hillingdon and Uxbridge ; *Herb. Hill.*

292, **707.** Lemna gibba, *L.* VII. *add,* Shepherds Well Fields, Hampstead.

304, **741.** Carex divulsa, *Good.* I. *add,* Near Harefield, 1869! ; *Warren.*

318, **781.** Apera Spica-venti, *Beauv.* VII. *add,* Finchley Road, on waste ground.

329, **817.** Festuca gigantea, *Vill.* VI. *add,* Edmonton.

331, **821.** Bromus asper, *L. add,* VI. Edmonton.

   823. Serrafalcus secalinus, *Bab.* VII. *add,* Waste ground, Finchley Road, 1869.

332, S. arvensis, *Godr. add,* VII. Waste ground, Finchley Road, 1869, abundant.

337, **840.** Lastrea Oreopteris, *Presl.* IV. *add,* Bishop's Wood, 1858 ; *Herb. S. P. Woodward.*

341, **851.** Ceterach officinarum, *Willd.* II. *add,* Specimens from Harmondsworth Church collected in 1843 are in *Herb. S. P. Woodward.*

INDEX

TO GENERA IN THE FLORA.

(Synonyms are in *Italics*.)

———◆———

Acer, 65
Achillea, 153
Aconitum, 21
Acorus, 290
Actinocarpus, 286
Adonis, 13
Adoxa, 135
Ægopodium, 124
Æthusa, 128
Agrimonia, 91
Agrostis, 318
Aira, 320
Ajuga, 222
Alchemilla, 91
Alisma, 286
Alliaria, 34
Allium, 281
Alnus, 262
Alopecurus, 315
Alsine, 54
Althæa, 61
Alyssum, 38
Amaranthus, 230
Ammi, 126
Anacharis, 267
Anagallis, 227
Anchusa, 190
Anemone, 13
Angelica, 129
Angelica, 130
Anthemis, 153
Anthoxanthum, 314
Anthriscus, 133
Anthyllis, 83
Antirrhinum, 198
Apargia, 170
Apera, 317
Apium, 121
Aquilegia, 21
Archangelica, 130
Arabis, 30
Arabis, 34
Arctium, 162
Arenaria, 54
Aristolochia, 247
Armoracia, 39
Arnoseris, 169
Artemisia, 156
Arum, 291
Arundo, 317
Arrhenatherum, 322

Asperula, 139
Asplenium, 339
Aster, 148
Athyrium, 339
Atriplex, 236
Atropa, 194
Avena, 321

Ballota, 221
Barbarea, 29
Bellis, 150
Berberis, 21
Beta, 236
Betula, 262
Bidens, 152
Blechnum, 341
Blysmus, 302
Borago, 189
Brachypodium, 332
Brassica, 35
Brassica, 36, 37
Briza, 326
Bromus, 330
Bromus, 329, 331, 332
Bryonia, 113
Buffonia, 52
Bunium, 123
Bupleurum, 126
Butomus, 288
Buxus, 247

Calamagrostis, 317
Calamintha, 214
Callitriche, 251
Calluna, 181
Caltha, 20
Campanula, 179
Campanula, 181
Capsella, 43
Cardamine, 30
Cardamine, 32
Carduus, 166
Carex, 303
Carlina, 162
Carpinus, 264
Carum, 124
Castanea, 263
Catabrosa, 327
Caucalis, 132, 133

Centaurea, 164
Centranthus, 143
Centunculus, 227
Cerastium, 57
Cerastium, 58
Ceratophyllum, 250
Ceterach, 341
Chærophyllum, 134
Chærophyllum, 133
Cheiranthus, 27
Chelidonium, 25
Chenopodium, 231
Chlora, 185
Chrysanthemum, 156
Chrysanthemum, 155, 157
Chrysosplenium, 119
Cichorium, 169
Cicuta, 120
Circæa, 111
Claytonia, 114
Clematis, 12
Cochlearia, 39
Colchicum, 281
Comarum, 94
Conium, 134
Convallaria, 276
Convolvulus, 187
Coriandrum, 135
Cornus, 136
Coronilla, 88
Coronopus, 43
Corydalis, 25
Corylus, 264
Cotyledon, 117
Cratægus, 104
Crepis, 176
Crocus, 274
Cucubalus, 49
Cuscuta, 188
Cynoglossum, 189
Cynosurus, 327
Cyperus, 298

Dactylis, 328
Daphne, 247
Datura, 195
Daucus, 132
Delphinium, 21
Dentaria, 32

Dianthus, 48
Digitalis, 198
Digitaria, 311
Diplotaxis, 37
Dipsacus, 145
Doronicum, 159
Draba, 38
Drosera, 47

Echinochloa, 312
Echium, 191
Eleocharis, 298
Elodea, 267
Elsholzia, 213
Empetrum, 247
Endymion, 281
Epilobium, 108
Epipactis, 273
Equisetum, 335
Erica, 181
Erigeron, 149
Eriophorum, 302
Erodium, 69
Erysimum, 34
Erythræa, 185
Euonymus, 73
Eupatorium, 147
Euphorbia, 247
Euphrasia, 204

Fagopyrum, 246
Fagus, 263
Falcatula, 78
Festuca, 328
Filago, 158
Fœniculum, 129
Fragaria, 94
Fraxinus, 183
Fritillaria, 278
Fumaria, 26

Galeopsis, 219
Galinsoga, 151
Galium, 140
Gentiana, 186
Genista, 74
Geranium, 66
Geum, 101
Glaucium, 24
Glyceria, 325
Gnaphalium, 159
Gymnadenia, 270

Habenaria, 270
Hedera, 136
Helleborus, 20
Helminthia, 172
Helosciadium, 121
Heracleum, 131
Herminium, 271
Herniaria, 114
Hesperis, 32
Hibiscus, 62
Hieracium, 176
Hippuris, 112
Holcus, 319
Hordeum, 333

Hottonia, 224
Humulus, 254
Hyacinthus, 281
Hydrocharis, 267
Hydrocotyle, 120
Hyoscyamus, 195
Hypericum, 62
Hypochœris, 169

Iberis, 41
Ilex, 183
Impatiens, 71
Inula, 150
Inula, 151
Iris, 273

Jasione, 179
Juncus, 282
Juniperus, 265

Knautia, 146
Köhleria, 322
Königa, 38

Lactuca, 173
Lamium, 217
Lapsana, 168
Lastrea, 337
Lathræa, 196
Lathyrus, 86
Lemna, 291
Lepidium, 41
Lepigonum, 59
Leontodon, 174
Leontodon, 170
Leucoium, 276
Ligustrum, 183
Lilium, 279
Limnanthemum, 186
Limosella, 202
Linaria, 199
Linum, 70
Listera, 272
Lithospermum, 191
Littorella, 230
Lolium, 334
Lonicera, 138
Lotus, 83
Lunaria, 38
Luzula, 284
Lycium, 195
Lychnis, 51
Lycopodium, 344
Lycopsis, 190
Lycopus. 213
Lysimachia, 225
Lythrum, 107

Maianthemum, 277
Malachium, 57
Malva, 60
Marrubium, 222
Matricaria, 155
Meconopsis, 24
Medicago, 76
Melampyrum, 203
Melica, 322

Melilotus, 77
Melissa, 215
Mentha, 209
Menyanthes, 187
Mercurialis, 249
Mespilus, 105
Milium, 316
Moenchia, 58
Montia, 113
Molinia, 323
Muscari, 281
Myosotis, 192
Myosurus, 13
Myrica, 262
Myriophyllum, 111

Narcissus, 275
Nardosmia, 148
Nardus, 316
Nasturtium, 27
Nasturtium, 40
Neottia, 272
Nepeta, 217
Nicandra, 195
Nicotiana, 195
Nuphar, 22
Nymphæa, 22

Œnanthe, 126
Œnothera, 111
Onobrychis, 88
Ononis, 75
Onopordum, 164
Ophioglossum, 343
Ophrys, 271
Orchis, 268
Origanum, 214
Ornithogalum, 279
Ornithopus, 87
Orobanche, 196
Ortegia, 115
Osmunda, 342
Oxalis, 72

Panicum, 313
Panicum, 312
Papaver, 23
Parietaria, 252
Paris, 266
Parnassia, 119
Pastinaca, 130
Pedicularis, 203
Peplis, 108
Petasites, 147
Petroselinum, 121
Phalaris, 313
Phleum, 314
Phragmites, 317
Phyteuma, 179
Picris, 172
Pilularia, 343
Pimpinella, 124
Plantago, 228
Poa, 323
Polemonium, 187
Polygala, 48
Polygonatum, 277

Polygonum, 242
Polypodium, 336
Polypogon, 319
Polystichum, 338
Populus, 261
Potamogeton, 293
Potentilla, 92
Potentilla, 94
Poterium, 91
Primula, 224
Prunella, 216
Prunus, 88
Pteris, 342
Pulicaria, 151
Pulmonaria, 191
Pyrus, 105

Quercus, 263

Radiola, 70
Ranunculus, 14
Raphanus, 44
Reseda, 44
Rhamnus, 73
Rhinanthus, 204
Ribes, 117
Rosa, 101
Rubus, 95
Rumex, 238
Ruscus, 278

Sagina, 52
Sagittaria, 287
Salix, 255
Salvia, 213
Sambucus, 137
Samolus, 227
Sanguisorba, 90
Sanicula, 120
Saponaria, 49
Sarothamnus, 75
Saxifraga, 118
Scabiosa, 146
Scandix, 133
Scilla, 279
Scilla, 281

Scirpus, 299
Scleranthus, 114
Sclerochloa, 326
Scolopendrium, 340
Scrophularia, 201
Scutellaria, 215
Sedum, 115
Sempervirum, 117
Senebiera, 43
Senecio, 160
Serrafalcus, 331
Serratula, 163
Setaria, 312
Sherardia, 139
Sibthorpia, 205
Silaus, 129
Silene, 50
Silene, 51
Silybum, 168
Sinapis, 36
Sinapis, 37
Sison, 123
Sisymbrium, 32
Sisymbrium, 34
Sium, 125
Smyrnium, 135
Solanum, 194
Solidago, 150
Sonchus, 175
Sparganium, 290
Spartina, 313
Specularia, 181
Spergula, 59
Spergularia, 59
Spiræa, 90
Spiranthes, 271
Stachys, 220
Staphylea, 73
Stellaria, 55
Stipa, 316
Symphytum, 190

Tamus, 266
Tanacetum, 157
Taraxacum, 174
Taxus, 265
Teesdalia, 41

Teucrium, 222
Thalictrum, 12
Thlaspi, 40
Thrincia, 170
Thymus, 214
Tilia, 62
Tillæa, 115
Tordylium, 131
Torilis, 132
Tragopogon, 171
Trifolium, 79
Triglochin, 288
Trigonella, 78
Triodia, 322
Trisetum, 321
Triticum, 333
Tulipa, 278
Turritis, 30
Tussilago, 148
Typha, 289

Ulex, 74
Ulmus, 254
Urtica, 253
Utricularia, 223

Vaccinium, 182
Valeriana, 143
Valerianella, 144
Verbascum, 197
Verbena, 223
Veronica, 205
Viburnum, 138
Vicia, 84
Villarsia, 186
Vinca, 184
Viola, 45
Viscum, 137

Wolffia, 293

Xanthium, 179

Zannichellia, 297

www.ingramcontent.com/pod-product-compliance
Lightning Source LLC
Chambersburg PA
CBHW020904210326
41598CB00018B/1763